Undergraduate Texts in Physics

Undergraduate Texts in Physics (UTP) publishes authoritative texts covering topics encountered in a physics undergraduate syllabus. Each title in the series is suitable as an adopted text for undergraduate courses, typically containing practice problems, worked examples, chapter summaries, and suggestions for further reading. UTP titles should provide an exceptionally clear and concise treatment of a subject at undergraduate level, usually based on a successful lecture course. Core and elective subjects are considered for inclusion in UTP.

UTP books will be ideal candidates for course adoption, providing lecturers with a firm basis for development of lecture series, and students with an essential reference for their studies and beyond.

More information about this series at http://www.springer.com/series/15593

Øyvind Grøn

Introduction to Einstein's Theory of Relativity

From Newton's Attractive Gravity to the Repulsive Gravity of Vacuum Energy

Second Edition

 Springer

Øyvind Grøn
OsloMet—Oslo Metropolitan University
Oslo, Norway

ISSN 2510-411X   ISSN 2510-4128   (electronic)
Undergraduate Texts in Physics
ISBN 978-3-030-43861-6   ISBN 978-3-030-43862-3   (eBook)
https://doi.org/10.1007/978-3-030-43862-3

This Springer imprint is published by the registered company Springer Nature Switzerland AG
The registered company address is: Gewerbestrasse 11, 6330 Cham, Switzerland

Preface to the Second Edition

These notes are a transcript of lectures delivered by Øyvind Grøn during the spring of 1997 at the University of Oslo. The manuscript has been revised in 2019. The present version of this document is an extended and corrected version of a set of Lecture Notes which were written down by S. Bard, Andreas O. Jaunsen, Frode Hansen and Ragnvald J. Irgens using LATEX2ε. Sven E. Hjelmeland has made many useful suggestions which have improved the text.

The manuscript has been revised in 2019. In this version, solutions to the exercises have been included. Most of these have been provided by Håkon Enger. I thank all my good helpers for enthusiastic work which was decisive for the realization of the book.

I hope that these notes are useful to students of general relativity and look forward to their comments accepting all feedback with thanks. The comments may be sent to the author by e-mail to oyvind.gron.no@gmail.com.

Oslo, Norway   Øyvind Grøn

Preface to the First Edition

These notes are a transcript of lectures delivered by Øyvind Grøn during the spring of 1997 at the University of Oslo.

The present version of this document is an extended and corrected version of a set of Lecture Notes which were typesetted by S. Bard, Andreas O. Jaunsen, Frode Hansen and Ragnvald J. Irgens using LATEX2ϵ. Svend E. Hjelmeland has made many useful suggestions which have improved the text. I would also like to thank Jon Magne Leinaas and Sigbjørn Hervik for contributing with problems and Gorm Krogh Johnsen for help with finishing the manuscript. I also want to thank Prof. Finn Ravndal for inspiring lectures on general relativity.

While we hope that these typeset notes are of benefit particularly to students of general relativity and look forward to their comments, we welcome all interested readers and accept all feedback with thanks.

All comments may be sent to the author by e-mail.

E-mail: Oyvind.Gron@iu.hio.no   Øyvind Grøn

Contents

List of Figures

List of Definitions

List of Examples

List of Exercises

Chapter 1
Newton's Theory of Gravitation

Abstract In this chapter we first deduce Newton's law of gravitation in its local form as a preparation for comparing Newton's and Einstein's theories, including a discussion of tidal forces. Then we give a presentation of the main conceptual foundation of the general theory of relativity, emphasizing the principle of equivalence and the principle of relativity.

In Newton's theory there is an absolute space and time. They are independent of the content in the universe. Newton wrote: "Absolute space, in its own nature, without regard to anything external, remains always similar and immovable." And further: "Absolute, true and mathematical time, of itself, and from its own nature flows equably without regard to anything external." Thus, every object has an absolute state of motion in absolute space. Hence an object must be either in a state of absolute rest or moving at some absolute speed.

Galileo, however, argued for a relativity of rectilinear motion with constant velocity as least with respect to mechanical phenomena. This principle is obeyed by Newton's theory of gravity.

In Newton's theory an inertial frame is defined as a reference frame moving along a straight line with constant velocity.

The fundamental laws of Newton's theory of gravitation are Newton's three laws plus the law of gravitation (see below). With reference to an inertial frame *Newton's three laws* take the form:

1. If a body is not acted upon by forces, or if the sum of the forces acting upon a body is zero, the body is either at rest of moves along a straight line with constant velocity.
2. The sum of the forces acting upon a body is equal to its (inertial) mass times its acceleration,

$$\sum \vec{F} = m_I \vec{a}. \tag{1.1}$$

3. If a body A acts upon a body B with a force, then B acts back on A with an equally large and oppositely directed force.

© Springer Nature Switzerland AG 2020

Ø. Grøn, *Introduction to Einstein's Theory of Relativity*,
Undergraduate Texts in Physics, https://doi.org/10.1007/978-3-030-43862-3_1

1

In a non-inertial frame with acceleration $\vec{a}_f$ one will experience an "artificial acceleration of gravity" $\vec{g} = -\vec{a}_f$, and Newton's 2. law takes the modified form

$$\sum \vec{F} = m_I(\vec{a} + \vec{a}_f) = m_I(\vec{a} - \vec{g}). \tag{1.2}$$

If no forces act on a body, it is said to be *freely falling*. A freely falling body in a non-inertial frame will have an acceleration $\vec{a} = \vec{g}$.

1.1 The Force Law of Gravitation

Consider two particles with masses M and m, respectively. They are at a distance r from each other and act on each other by a gravitational force F. The situation is shown in Fig. 1.1.

According to Newton's theory the gravitational force between the particles is given by

$$\vec{F} = -mG\frac{M}{r^2}\vec{e}_r, \; \vec{e} = \frac{\vec{r}}{r}, r = |\vec{r}|. \tag{1.3}$$

Let V be the potential energy of m (see Fig. 1.1). Then

$$\vec{F} = -\nabla V(\vec{r}), \; F_i = -\frac{\partial V}{\partial x^i}. \tag{1.4}$$

For a spherical mass distribution $V(\vec{r}) = -mG(M/r)$ with zero potential infinitely far from the centre of M. Newton's law of gravitation is valid for small velocities, i.e. velocities much smaller than the velocity of light and weak fields. Weak fields are fields in which the gravitational potential energy of a test particle is very small compared to its rest mass energy. (Note that here one is interested only in the absolute values of the above quantities and not their sign).

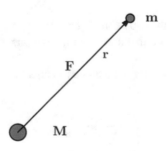

Fig. 1.1 Newton's law of gravitation. Newton's law of gravitation states that the force between two spherical bodies is attractive, acts along the line joining the centres of the bodies, is proportional to the product of the masses and inversely proportional to the distance between the centres of the masses

$$mG\frac{M}{r} \ll mc^2 \Rightarrow r \gg \frac{GM}{c^2}. \tag{1.5}$$

The *Schwarzschild radius* for an object with mass M is $R_S = 2GM/c^2$. Hence, *far outside the Schwarzschild radius the gravitational field is weak*. To get a feeling for the magnitudes, you may insert the mass of the earth. Then you find that the Schwarzschild mass of the Earth is 9 mm. Comparing with the radius of the Earth, which is $R_E \approx 6400$ km, we may conclude that the gravitational field is weak on the surface of the Earth. Similarly the Schwarzschild radius of the Sun is 3 km and the Earth is about 150 million km from the Sun. Hence the gravitational field of the Sun is very weak in most parts of the solar system. This explains, in part, the success of Newtonian gravity for describing the motion of bodies in the gravitational field of the Earth and the Sun.

Example 1.1 (*Two particles falling towards each other*) Two point particles with masses m_1 and m_2 are instantaneously at rest at a distance r_0 from each other in empty space, with no other forces present than the gravitational force between them.

How long time will they fall before they collide?

Newton's 2. law is valid with reference to an inertial frame. Hence we start by introducing a coordinate system fixed with respect to the mass centre of the particles. In this system particles 1 and 2 have coordinates r_1 and r_2, respectively. The equations of motion of the two particles are

$$\ddot{r}_1 = G\frac{m_2}{(r_2 - r_1)^2}, \quad \ddot{r}_2 = -G\frac{m_1}{(r_2 - r_1)^2}. \tag{1.6}$$

where a dot denotes differentiation with respect to time. Subtracting the equations and introducing the distance $r = r_2 - r_2$ between the particles as a new coordinate, we get the differential equation

$$r^2\ddot{r} + G(m_1 + m_2) = 0. \tag{1.7}$$

Writing this as

$$\dot{r}\ddot{r} = -\frac{G(m_1 + m_2)}{r^2}\dot{r} \tag{1.8}$$

and integrating with the boundary condition that $\dot{r} = 0$ for $r = r_0$, we get the energy conservation equation

$$\dot{r}^2 = 2G(m_1 + m_2)\left(\frac{1}{r} - \frac{1}{r_0}\right). \tag{1.9}$$

Hence, the falling time is given by the integral

$$t = \sqrt{\frac{r_0}{2G(m_1 + m_2)}} \int_0^{r_0} \sqrt{\frac{r}{r_0 - r}} \, dr. \tag{1.10}$$

Performing the integration gives

$$t = \frac{\pi}{2} \frac{r_0^{3/2}}{\sqrt{2G(m_1 + m_2)}}. \tag{1.11}$$

Assume that the particles have a mass equal to 1 kg and starts 1 m from each other. Then the falling time is 26.5 h. This illustrates that gravity is indeed a very weak force.

1.2 Newton's Law of Gravitation in Local Form

Consider a gravitational field due to a mass distribution. Let P be a point in the field (see Fig. 1.2) with position vector $\vec{r} = x^i \vec{e}_i$, and let the gravitating mass element be at $\vec{r}' = x^{i'} \vec{e}_{i'}$. *Newton's law of gravitation* for a continuous distribution of mass is

$$\vec{F} = -Gm \int \rho(\vec{r}') \frac{\vec{r} - \vec{r}'}{|\vec{r} - \vec{r}'|^3} d^3 r' = -\nabla V(\vec{r}). \tag{1.12}$$

Note that the ∇ operator acts on the unprimed coordinates, only.
Let us consider Eq. (1.12) term by term

Fig. 1.2 Deduction of Newton's law of gravitation in local form. The dice is a mass element, and P is the field point

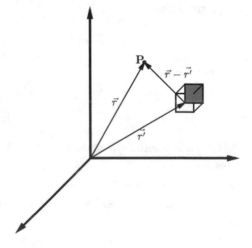

$$\nabla \frac{1}{|\vec{r} - \vec{r}'|} = \vec{e}_i \frac{\partial}{\partial x_i} \frac{1}{\left[\left(x^j - x'^j \right) \left(x_j - x'_j \right) \right]^{1/2}} = \vec{e}_i \frac{\partial}{\partial x_i} \left[\left(x^j - x'^j \right) \left(x_j - x'_j \right) \right]^{-1/2}$$

$$= \vec{e}_i \left(-\frac{1}{2} \right) \left[\left(x^j - x'^j \right) \left(x_j - x'_j \right) \right]^{-3/2} 2 \left(x^k - x'^k \right) \frac{\partial x_k}{\partial x_i}$$

$$= -\vec{e}_i \frac{\left(x^k - x'^k \right) \delta^i_k}{\left[\left(x^j - x'^j \right) \left(x_j - x'_j \right) \right]^{3/2}} = -\frac{\left(x^i - x'^i \right) \vec{e}_i}{\left[\left(x^j - x'^j \right) \left(x_j - x'_j \right) \right]^{3/2}}$$

$$= -\frac{\vec{r} - \vec{r}'}{|\vec{r} - \vec{r}'|^3} \tag{1.13}$$

Equations (1.12) and (1.13) imply

$$V(\vec{r}) = -Gm \int \frac{\rho(\vec{r}')}{|\vec{r} - \vec{r}'|} d^3 r'. \tag{1.14}$$

Hence, the gravitational potential at the point P is

$$\phi(\vec{r}) \equiv \frac{V(\vec{r})}{m} = V(\vec{r}) = -G \int \frac{\rho(\vec{r}')}{|\vec{r} - \vec{r}'|} d^3 r'. \tag{1.15}$$

It follows that

$$\nabla \phi(\vec{r}) = -G \int \rho(\vec{r}') \frac{\vec{r} - \vec{r}'}{|\vec{r} - \vec{r}'|^3} d^3 r', \tag{1.16}$$

and

$$\nabla^2 \phi(\vec{r}) = -G \int \rho(\vec{r}') \nabla \cdot \frac{\vec{r} - \vec{r}'}{|\vec{r} - \vec{r}'|^3} d^3 r'. \tag{1.17}$$

We now calculate the divergence in the integrand

$$\nabla \cdot \frac{\vec{r} - \vec{r}'}{|\vec{r} - \vec{r}'|^3} = \frac{\nabla \cdot \vec{r}}{|\vec{r} - \vec{r}'|^3} + \left(\vec{r} - \vec{r}' \right) \cdot \nabla \frac{1}{|\vec{r} - \vec{r}'|^3} = \frac{3}{|\vec{r} - \vec{r}'|^3} - \left(\vec{r} - \vec{r}' \right)$$

$$\cdot \frac{3 \left(\vec{r} - \vec{r}' \right)}{|\vec{r} - \vec{r}'|^5} = \frac{3}{|\vec{r} - \vec{r}'|^3} - \frac{3}{|\vec{r} - \vec{r}'|^3} = 0, \vec{r} \neq \vec{r}'. \tag{1.18}$$

The condition $\vec{r} \neq \vec{r}'$ means that the field point is outside the mass distribution. Hence, the Newtonian potential at a point in a gravitational field outside a mass distribution satisfies the *Laplace equation*

$$\nabla^2 \phi = 0. \tag{1.19}$$

We shall now generalize this to the case where the field point may be inside a mass distribution. It will then be useful to utilize *the Dirac delta function*. This function has the following properties.

$$\delta(\vec{r} - \vec{r}') = 0, \quad \text{for} \ \ \vec{r} \neq \vec{r}', \tag{1.20}$$

$$\int \delta(\vec{r} - \vec{r}')d^3r' = \begin{cases} 1 & \text{if } \vec{r} = \vec{r}' \text{ is contained in the integration region.} \\ 0 & \text{if } \vec{r} = \vec{r}' \text{ is not contained in the integration region.} \end{cases} \tag{1.21}$$

$$\int f(\vec{r}')\delta(\vec{r} - \vec{r}')d^3r' = f(\vec{r}). \tag{1.22}$$

A calculation of the integral (1.17) which is valid also in the case where the field point is inside the mass distribution is obtained by utilizing Gauss integral theorem,

$$\int_V \nabla \cdot \vec{A}d^3r' = \oint_S \vec{A} \cdot d\vec{S}, \tag{1.23}$$

where S is the boundary surface of the volume V. We shall also need the definition of a *solid angle*,

$$d\Omega \equiv \frac{ds'_\perp}{|\vec{r} - \vec{r}'|^2}, \tag{1.24}$$

where $ds'_\perp$ is the projection of the area normal to the line of sight. It is represented by absolute value of the component of $d\vec{s}'$ along the line of sight, where $d\vec{s}'$ is the normal vector of the surface element of the mass distribution subtending the solid angle $d\Omega$ at the field point P (Fig. 1.3).

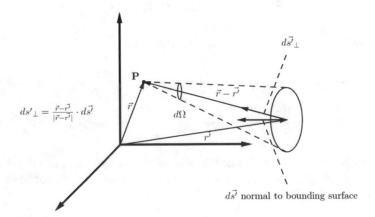

Fig. 1.3 Solid angle. Solid angle $d\Omega$ is defined such that the surface of a sphere subtends the angle 4π at the centre

Applying the Gauss integral theorem we have

$$\int_V \nabla \cdot \frac{\vec{r} - \vec{r}'}{|\vec{r} - \vec{r}'|^3} d^3 r' = \oint_S \frac{\vec{r} - \vec{r}'}{|\vec{r} - \vec{r}'|^3} \cdot d\vec{s}' = \oint_S \frac{ds'_\perp}{|\vec{r} - \vec{r}'|^2} = \oint_S d\Omega \qquad (1.25)$$

Hence

$$\int_V \nabla \cdot \frac{\vec{r} - \vec{r}'}{|\vec{r} - \vec{r}'|^3} d^3 r' = \begin{cases} 4\pi & \text{if } P \text{ is inside the mass distribution} \\ 0 & \text{if } P \text{ is outside the mass distribution} \end{cases}. \qquad (1.26)$$

This may be written in terms of the Dirac delta function as

$$\int_V \nabla \cdot \frac{\vec{r} - \vec{r}'}{|\vec{r} - \vec{r}'|^3} d^3 r' = 4\pi \delta(\vec{r} - \vec{r}'). \qquad (1.27)$$

We now have

$$\nabla^2 \phi(\vec{r}) = -G \int \rho(\vec{r}') \nabla \cdot \frac{\vec{r} - \vec{r}'}{|\vec{r} - \vec{r}'|^3} d^3 r' = G \int \rho(\vec{r}') 4\pi \delta(\vec{r} - \vec{r}') d^3 r'$$

$$= 4\pi G \rho(\vec{r}), \qquad (1.28)$$

showing that the Newtonian gravitational potential obeys the Poisson equation. This means that Newton's gravitational theory can be expressed in the following way:

- Mass generates a gravitational potential according to

$$\nabla^2 \phi = 4\pi G \rho. \qquad (1.29)$$

- The gravitational potential generates acceleration of gravity $\vec{g}$ according to

$$\vec{g} = -\nabla \phi. \qquad (1.30)$$

1.3 Newtonian Incompressible Star

We shall apply Eqs. (1.29) and (1.30) to calculate the gravitational field of a Newtonian incompressible star. Let the gravitational potential be $\phi(r)$. In the spherically symmetric case Eq. (1.29) then takes the form

$$\frac{1}{r^2}\frac{d}{dr}\left(r^2\frac{d\phi}{dr}\right) = 4\pi G\rho. \tag{1.31}$$

Assuming that $\rho = $ constant and integrating gives

$$r^2\frac{d\phi}{dr} = \frac{4\pi}{3}G\rho r^3 + K_1 = M(r) + K_1. \tag{1.32}$$

where $M(r)$ is the mass inside a sphere with radius r. According to Eq. (1.30) the gravitational acceleration is

$$\vec{g} = -\nabla\phi = -\frac{d\phi}{dr}\vec{e}_r, \tag{1.33}$$

or

$$g = \frac{M(r)}{r^2} + \frac{K_1}{r^2} = \frac{4\pi}{3}G\rho r + \frac{K_1}{r^2}. \tag{1.34}$$

Finite g in $r = 0$ gives $K_1 = 0$

$$g = \frac{4\pi}{3}G\rho r, \quad \frac{d\phi}{dr} = \frac{4\pi}{3}G\rho r. \tag{1.35}$$

Assume that the mass distribution has a radius R. A new integration then leads to

$$\phi = \frac{2\pi}{3}G\rho r^2 + K_2. \tag{1.56}$$

Demanding continuous potential at $r = R$ gives.

$$\frac{2\pi}{3}G\rho R^2 + K_2 = \frac{M(R)}{R} = -\frac{4\pi}{3}G\rho R^2. \tag{1.37}$$

Hence

$$K_2 = -2\pi G\rho R^2. \tag{1.38}$$

Thus the potential inside the mass distribution is

$$\phi = \frac{2\pi}{3}G\rho(r^2 - 3R^2). \tag{1.39}$$

The star is in hydrostatic equilibrium that is the pressure forces are in equilibrium with the gravitational forces.

Consider a mass element $dm = \rho dV = \rho dAdr$, in the shell drawn in Fig. 1.4.

The pressure force on the mass element is $dF = dAdp$, and the gravitational force is

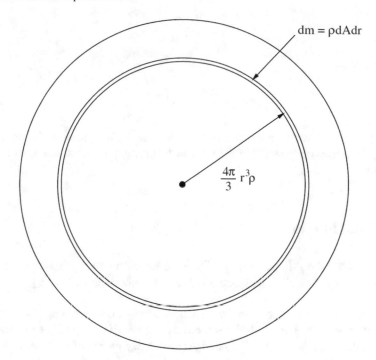

$$dm = \rho dA dr$$

$$\frac{4\pi}{3} r^3 \rho$$

Fig. 1.4 Mass shell in a star. A shell with thickness dr is affected by both gravitational and pressure forces

$$dG = g dm = \frac{Gm(r)}{r^2} dm, \qquad (1.40)$$

where $m(r)$ is the mass inside the shell. With constant density $m(r) = (4\pi/3)\rho r^3$. Hence

$$dG = g dm = \frac{4\pi}{3} G\rho^2 r dA dr. \qquad (1.41)$$

Equilibrium, d$F = -$dG, demands that

$$dp = -\frac{4\pi}{3} G\rho^2 r dr. \qquad (1.42)$$

Integrating gives

$$p = K_3 - \frac{2\pi G}{3}\rho^2 r^2. \qquad (1.43)$$

Vanishing pressure at the surface of the mass distribution, $p(R) = 0$, gives the value of the constant of integration

$$K_3 = \frac{2\pi G}{3} \rho^2 R^2 \qquad (1.44)$$

which leads to

$$p(r) = \frac{2\pi G}{3} \rho^2 (R^2 - r^2). \qquad (1.45)$$

No matter how massive the star is, it is possible for the pressure forces to keep the equilibrium with gravity. In Newtonian theory, gravitational collapse is not a necessity.

1.4 Tidal Forces

Tidal forces are the difference of gravitational force on two neighbouring particles in a gravitational field. The tidal force is due to the inhomogeneity of a gravitational field.

Figure 1.5 shows two point masses, each with mass m, with a separation vector $\vec{\varsigma}$ and position vectors $\vec{r}$ and $\vec{r} + \vec{\varsigma}$, respectively, where $|\vec{\varsigma}| << |\vec{r}|$. The gravitational forces on the mass points are $\vec{F}(\vec{r})$ and $\vec{F}(\vec{r} + \vec{\varsigma})$. By means of a Taylor expansion to the lowest order in $|\vec{\varsigma}|$ we get for the i-component of the tidal force

$$f_i = F_i(\vec{r} + \vec{\varsigma}) - F_i(\vec{r}) = \varsigma^j \left(\frac{\partial F_i}{\partial x^j} \right)_{\vec{r}}. \qquad (1.46)$$

The corresponding vector equation is

$$\vec{f} = (\varsigma \cdot \nabla)_{\vec{r}} \vec{F}. \qquad (1.47)$$

Using that

Fig. 1.5 Tidal forces. The separation vector $\vec{\varsigma}$ between two mass points 1 and 2 acted upon by gravitational forces $\vec{F}_1$ and $\vec{F}_2$

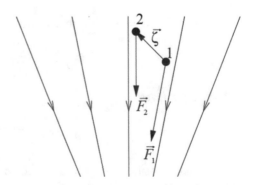

$$\vec{F} = -m\nabla\phi, \tag{1.48}$$

The tidal force may be expressed in terms of the gravitational potential according to

$$\vec{f} = -m(\vec{\varsigma} \cdot \nabla)_{\vec{r}}\nabla\phi. \tag{1.49}$$

It follows that in a local Cartesian coordinate system the i-component of the relative acceleration of the particles is

$$\frac{d^2\varsigma_i}{dt^2} = -\left(\frac{\partial^2\phi}{\partial x^i \partial x^j}\right)_{\vec{r}}\varsigma^j. \tag{1.50}$$

Example 1.2 (*Tidal force on a system consisting of two particles*) We shall first consider two test particles with a vertical separation vector in the gravitational field of a particle with mass M. Let us introduce a small Cartesian coordinate system at a distance R from the mass (Fig. 1.6). The particles are separated from each other by a distance $z \ll R$.

According to Eq. (1.3) a test particle with mass m at a point $(0, \ 0, \ z)$ is acted upon by a gravitational force

$$F_z(z) = -m\frac{GM}{(R+z)^2}, \tag{1.51}$$

While an identical particle at the origin is acted upon by a force

$$F_z(0) = -m\frac{GM}{R^2} . \tag{1.52}$$

Fig. 1.6 Horizontal tidal force. A small Cartesian coordinate system at a distance R from a particle with mass M

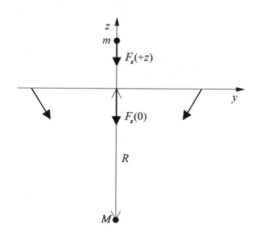

Secondly, we shall first consider two particles in the same height in an inhomogeneous gravitational field.

If this little system is falling freely towards M, an observer at the origin will say that the particle at $(0, \ 0, \ z)$ is acted upon by a force

$$f_z = F_z(z) - F(0) \approx \frac{2GmM}{R^3} z \qquad (1.53)$$

directed away from the origin along the positive z-axis. This is the tidal force.

In the same way particles at the same height in the gravitational field, at positions $(x, \ 0, \ 0)$ and $(0, \ y, \ 0)$ are attracted towards the origin by tidal forces

$$f_x = -\frac{GmM}{R^3} x, \ f_y = -\frac{GmM}{R^3} y. \qquad (1.54)$$

Note that the tidal force increases faster in strength with decreasing distance from the mass which produces the gravitational field than the gravitational force.

Equations (1.53) and (1.54) have among others the following consequence: if an elastic circular ring is falling freely in the gravitational field of the Earth, it will be stretched in the vertical direction and compressed in the horizontal direction (Fig. 1.7).

In general tidal forces cause changes of shape.

Example 1.3 (*The tidal field on the Earth due to the Moon*) The Earth–Moon system is illustrated in Fig. 1.8. (Actually the distance between the Earth and the Moon is much greater compared to the magnitude of the Earth.) The tidal force due to the Moon on the surface of the Earth is the difference between the gravitational force at A and C in the gravitational field of the Moon.

From the extended Pythagorean law we have, with reference to Fig. 1.8

$$r_1^2 = R^2 - 2rR \cos \psi + r^2. \qquad (1.55)$$

Fig. 1.7 An elastic ring originally circular, falling freely in the gravitational field of the Earth

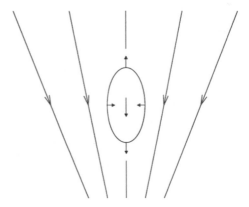

Fig. 1.8 The Earth-Moon
system

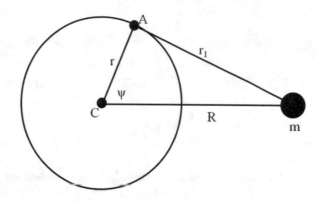

A series expansion of

$$\frac{1}{r} = \frac{1}{R}\left[1 - 2\frac{r}{R}\cos\psi + \left(\frac{r}{R}\right)^2\right]^{-1/2} \tag{1.56}$$

to second order in r/R gives

$$\frac{1}{r} = \frac{1}{R}\left[1 + \frac{r}{R}\cos\psi + \frac{1}{2}\left(\frac{r}{R}\right)^2(3\cos^2\psi - 1)\right]. \tag{1.57}$$

Hence, the potential at a point A on the surface of the Earth in the gravitational field of the Moon is

$$V_m = -\frac{Gm}{r} = -\frac{Gm}{R}\left[1 + \frac{r}{R}\cos\psi + \frac{1}{2}\left(\frac{r}{R}\right)^2(3\cos^2\psi - 1)\right]. \tag{1.58}$$

The first term, $V_C = -Gm/R$, is the potential in the gravitational field of the Moon at the centre, C, of the Earth. The second term is $V_A = -(Gm/R^2)_r\cos\psi = -g_{Moon}x_A$, where x_A is component of the separation between A and C along the gravitational field. This is the difference of the potential at C and A in the gravitational field of the Moon if one neglects the inhomogeneity in the Moon's field at the Earth. Hence the sum of the first two terms is then the potential at A in the gravitational field of the Moon. This means that the third term,

$$V_t = -\frac{Gm}{2R^3}r^2(3\cos^2\psi - 1), \tag{1.59}$$

is the difference between the potential at A in the Moon's gravitational field if the field is considered homogeneous with the value at the centre of the Earth and the actual potential at A. This difference is due to the inhomogeneity of the gravitational field of the Moon at the Earth, i.e. it is due to the tidal gravitational field. It is therefore called *the tidal potential* at A.

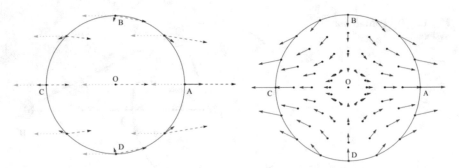

Fig. 1.9 Tidal acceleration field. *The Tidal acceleration field* (red) at the surface of the Earth due to the Moon is the acceleration of gravity at the surface (black) of the Earth minus the acceleration of gravity at the centre (green) of the Earth in the Moon's gravitational field

The height difference, Δh, between flood and ebb due to the Moon's tidal field is given by

$$g\Delta h = V_t(0) - V_t(\pi/2), \tag{1.60}$$

where $V_t = V_t(\psi)$, and g is the acceleration of gravity at the surface of the Earth. This gives

$$\Delta h = \frac{3}{2}\frac{Gm}{g}\frac{r^2}{R^3}. \tag{1.61}$$

For a numerical result we need the following values:

$$M_{\text{Moon}} = 7.35 \cdot 10^{22}\,\text{kg}, \ g = 9.81\,\text{m/s}^2, \ R = 3.85 \cdot 10^5\,\text{km}, \ r_{\text{Earth}} = 6378\,\text{km}$$

Inserting this into Eq. (1.46) gives the height differences on the ocean of the Earth due to the Moon, neglecting the effects of ocean currents and coast lines, $\Delta h = 53\,\text{cm}$. The tidal field is shown in Fig. 1.9.

1.5 The Principle of Equivalence

Galilei investigated experimentally the motion of freely falling bodies. He found that they moved in the same way, regardless of what sort of material they consisted of and what mass they had. In Newton's theory of gravitation mass appears in two different ways; as gravitational mass, m_G, in the law of gravitation, analogously to charge in Coulomb's law, and as inertial mass, m_I, in Newton's second law.

Newton's 2. law applied to a freely falling body with gravitational mass m_G and inertial mass m_I in the field of gravity from a spherical body with mass M then takes the form

$$m_I \frac{d^2 \vec{r}}{dt^2} = -G m_G \frac{M}{r^3} \vec{r}. \tag{1.62}$$

The results of Galilei's measurement indicated that the acceleration is independent of the constitution of the bodies, and hence, the gravitational and inertial mass must be the same for all bodies,

$$m_G = m_I. \tag{1.63}$$

Measurements performed by the Hungarian baron Eötvös around the year 1900 indicated that this equality holds with an accuracy better than 10^{-8}.

A parameter which is often used to specify the accuracy of tests of the equality of gravitational and inertial mass is

$$\Delta \equiv \frac{m_G}{m_I} - 1. \tag{1.64}$$

A very accurate test was published on 18 January 2018 [1]. A team of physicists reported about tests based on 7 years with observational data from the MESSENGER space observatory. They deduced from the MESSENGER data that $\Delta = (-4.1 \pm 4.6) \cdot 10^{-15}$. This is the most accurate test of the equality of gravitational and inertial mass to date.

Einstein assumed the exact validity of Eq. (1.63). He considered this as a consequence of a fundamental principle, *the principle of equivalence*, namely that the physical effects of a gravitational field due to an acceleration (including rotation) of the reference frame are equivalent to the physical effects of a gravitational field due to a mass distribution.

A consequence of this principle is the possibility of removing locally the acceleration of gravity by entering a laboratory in free fall. In order to clarify this, Einstein considered a homogeneous gravitational field in which the acceleration of gravity, $\vec{g}$, is independent of the position. Using Eq. (1.2) in a freely falling non-rotating reference frame in this field, with a given by

$$m_I a = \frac{G m_G M}{R^2}, \tag{1.65}$$

a free particle moves according to

$$m_I \frac{d^2 \vec{r}}{dt^2} = m_G g - m_I g = 0, \tag{1.66}$$

where we have used Eq. (1.63). This means that an observer in such a freely falling reference frame will experience that a particle which is not acted upon by non-gravitational forces will move along a straight line with constant velocity.

According to Newton's theory the particle is acted upon by a gravitational force. In Newton's theory a free particle is a particle which is acted upon by a gravitational force only. Furthermore, a reference frame which falls freely in a gravitational field is accelerated according to Newton's theory.

It is not inertial.

All of this is different according to the general theory of relativity. According to Einstein's theory gravitation is not reckoned as a force, and a free particle is not acted upon by any forces. Furthermore, *the general definition of an inertial frame, valid both in Newton's and Einstein's theory, is that an inertial frame is a frame where Newton's 1. law is valid.* We have seen above that a free particle moves along a straight line with constant velocity in a freely falling reference frame. According to Einstein's theory it is not acted upon by any force. Hence Newton's 1. law is valid in the freely falling frame. This means that *according to the general theory of relativity an inertial frame falls freely.* Also, *there is no acceleration of gravity in an inertial frame.* All of these are consequences of the principle of equivalence according to Einstein's theory.

The principle of equivalence has also been formulated in an "opposite way." An observer at rest in a homogeneous gravitational field and an observer in an accelerated reference frame far from any mass distribution will obtain identical results when they perform similar experiments. The physical effects of a gravitational field caused by the motion of the reference frame are equivalent to the physical effects of a gravitational field caused by a mass distribution.

One often hears that there is a connection between gravity and spacetime curvature according to Einstein's theory. The concept spacetime curvature will be thoroughly introduced later, but a few words may be in order already here, so that possible misunderstanding can be avoided at this initial point.

The experience of acceleration of gravity has nothing to do with spacetime curvature. It depends upon the motion of the observer's reference frame. Acceleration of gravity is experienced when the reference frame of the observer is not inertial. It is independent both of spacetime curvature and whether one is close to a mass distribution. When we experience acceleration of gravity at the surface of the Earth, it is because being at rest on this surface means not being in an inertial reference frame. We accelerate upwards relative to an inertial frame when we are at rest on the surface of the Earth. Therefore we experience a downwards acceleration of gravity.

The Newtonian force which *is* related to spacetime curvature is the tidal force as described mathematically in Eq. (1.50). The relativistic generalization of this equation is the equation of geodesic deviation (see Chap. 6) which contains the components of the spacetime curvature.

Tidal forces represent the inhomogeneity of the Newtonian gravitational field. In order to observe this inhomogeneity by physical measurements, one needs to perform an experiment that requires some extension in space and time.

The principle of equivalence as formulated above has only a local validity. The word local here means that the extension in space and time is so small that tidal effects cannot be measured. Hence the principle of equivalence is valid only in the limit that the gravitational field can be considered homogeneous. In a geometrical language the principle of equivalence is valid only as far as spacetime curvature cannot be measured.

1.6 The General Principle of Relativity

The principle of equivalence led Einstein to a generalization of the special principle of relativity. In his general theory of relativity Einstein formulated a general principle of relativity, which says that not only velocities are relative, but accelerations, too.

Let us consider two formulations of the special principle of relativity

S1. All laws of nature are the same (may be formulated in the same way) in all inertial frames.

S2. Every inertial observer can consider himself to be at rest.

These two formulations may be interpreted as different formulations of a single principle. But the generalization of S1 and S2 to the general case, which encompasses accelerated motion and non-inertial frames, leads to two different principles G1 and G2.

G1. The laws of nature are the same in all reference frames.

G2. Every observer can consider himself to be at rest.

In the literature both G1 and G2 are mentioned as the general principle of relativity. But G2 is a stronger principle (i.e. stronger restriction on natural phenomena) than G1. Generally the course of events of a physical process in a certain reference frame depends upon the laws of physics, the boundary conditions, the motion of the reference frame and the geometry of spacetime. The two latter properties are described by means of a metrical tensor. By formulating the physical laws in a metric-independent way, one obtains that G1 is valid for all types of physical phenomena. Even if the laws of nature are the same in all reference frames, the course of events of a physical process will, as mentioned above, depend upon the motion of the reference frame. As to the spreading of light, for example, the law is that light follows null-geodesic curves (see Chap. 4). This law implies that the path of a light particle is curved in non-inertial reference frames and straight in inertial frames.

The question whether G2 is true in the general theory of relativity has been thoroughly discussed recently, and the answer is not clear yet [2].

1.7 The Covariance Principle

The principle of relativity is a physical principle. It is concerned with physical phenomena. This principle motivates the introduction of a formal principle, called *the covariance principle*: the equations of a physical theory shall have the same form in every coordinate system. This principle is not concerned directly with physical phenomena. The principle may be fulfilled for every theory by writing the equations in a form invariant, i.e. covariant way. This may be done by using tensor (vector) quantities, only, in the mathematical formulation of the theory.

The covariance principle and the equivalence principle may be used to obtain a description of what happens in the presence of gravitation. We then start with the physical laws as formulated in the special theory of relativity. Then the laws are written in a covariant form, by writing them as tensor equations. They are then valid in an arbitrary, accelerated system. But the inertial field (fictive force) in the accelerated frame is equivalent to a gravitational field. So, starting within a description referred to an inertial frame, we have obtained a description valid in the presence of a gravitational field.

The tensor equations have in general a coordinate-independent form. Yet, such form-invariant, or covariant, equations need not fulfil the principle of relativity. This is due to the following circumstances. A physical principle, for example the principle of relativity, is concerned with observable relationships. Therefore, when one is going to deduce the observable consequences of an equation, one has to establish relations between the tensor components of the equation and observable physical quantities. Such relations have to be defined; they are not determined by the covariance principle.

From the tensor equations, that are covariant, and the defined relations between the tensor components and the observable physical quantities, one can deduce equations between physical quantities. The special principle of relativity, for example, demands that the laws which these equations express must be the same in every inertial frame

The relationships between physical quantities and tensors (vectors) are theory dependent. The relative velocity between two bodies, for example, is a vector within Newtonian kinematics. However, in the relativistic kinematics of four-dimensional spacetime, an ordinary velocity, which has only three components, is not a vector. Vectors in spacetime, so-called 4-vectors, have four components. Equations between physical quantities are not covariant in general. For example, Maxwell's equations in three-vector form are not invariant under a Galilei transformation. However, if these equations are rewritten in tensor form, then neither a Galilei transformation nor any other transformation will change the form of the equations.

If all equations of a theory are tensor equations, the theory is said to be given a manifestly covariant form. A theory which is written in a manifestly covariant form will automatically fulfil the covariance principle, but it need not fulfil the principle of relativity.

1.8 Mach's Principle

Einstein gave up Newton's idea of an absolute space. According to Einstein all motion is relative. This may sound simple, but it leads to some highly non-trivial and fundamental questions. Imagine that there are only two particles connected by a spring in the universe. What will happen if the two particles rotate about each other? Will the spring be stretched due to centrifugal forces? Newton would have confirmed that this is indeed what will happen. However, when there is no longer any absolute space that the particles can rotate relatively to, the answer is not so obvious. If we, as observers, rotate around the particles, and they are at rest, we would not observe any stretching of the spring. But this situation is kinematically equivalent to the one with rotating particles and observers at rest, which leads to stretching.

Such problems led Mach to the view that all motion is relative. The motion of a particle in an empty universe is not defined. All motion is motion relatively to something else, i.e. relatively to other masses. According to Mach this implies that inertial forces must be due to a particle's acceleration relatively to the great masses of the universe. If there were no such cosmic masses, there would not exist inertial forces, like the centrifugal force. In our example with two particles connected by a string, there would not be any stretching of the spring if there were no cosmic masses that the particles could rotate relatively to.

Another example may be illustrated by means of a turnabout. If we stay on this, while it rotates, we feel that the centrifugal forces lead us outwards. At the same time we observe that the heavenly bodies rotate. According to Mach identical centrifugal forces should appear if the turnabout is static and the heavenly bodies rotate.

Einstein was strongly influenced by Mach's arguments, which probably had some influence, at least with regards to motivation, on Einstein's construction of his general theory of relativity. Yet, it is clear that general relativity does not fulfil all requirements set by Mach's principle. For example there exist general relativistic, rotating cosmological models, where free particles will tend to rotate relative to the cosmic masses of the model.

However, some Machian effects have been shown to follow from the equations of the general theory of relativity. For example, inside a rotating massive shell the inertial frames, i.e. the free particles, are dragged on and tend to rotate in the same direction as the shell. This was discovered by Lense and Thirring in 1918 and is therefore called the *Lense–Thirring effect*. More recent investigations of this effect have, among others, led to the following result [3]: "a massive shell with radius equal to its Schwarzschild radius has often been used as an idealized model of our universe. Our result shows that in such models local inertial frames near the centre cannot rotate relatively to the mass of the universe. In this way our result gives an explanation in accordance with Mach's principle, of the fact that the fixed stars are at rest on the heaven as observed from an inertial reference frame."

1.9 Exercises

1.1 *A tidal force pendulum*

A tidal force pendulum consists of two points with the same mass that are connected by a stiff, massless rod as shown in Fig. 1.10.

The length of the rod is 2ℓ. The pendulum oscillates with respect to the centre of the rod, which is fixed at a constant distance from the centre of the Earth. The pendulum oscillates in a vertical plane. The Earth is considered as a spherically

Fig. 1.10 A tidal force
pendulum

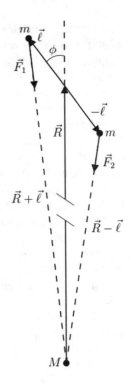

symmetric mass distribution. The gravitational forces acting on the two mass
points are $\vec{F}_1$ and $\vec{F}_2$ as shown in the figure.

Find an expression for the period of the oscillation of the rod. What is the
period of the tidal force pendulum at the surface of the Earth? What happens in
the limit of a homogeneous gravitational field?

1.2 Newtonian potential for a spherically symmetric body

(a) Calculate the Newtonian potential $\phi(r)$ outside and inside a spherical shell
     of matter with mass M and radius R.
(b) Let M and R be the mass and radius of the Earth. Assume that the Earth
     has constant mass density. Find the potential outside and inside the earth.
(c) Calculate the acceleration of gravity at the surface of the Earth.

1.3 Frictionless motion in a tunnel through the Earth

(a) We neglect the rotation of the Earth and air resistance and friction. Assume
     that a hollow straight tube has been drilled through the centre of the Earth.
     A small ball is dropped into the tube from the surface of the Earth and will
     perform an oscillating motion in the tube.

     Find the position of the ball as a function of time. What is the period of the
     motion?

(b) We now consider a straight tube from one point on the surface of the Earth to another, not passing through the centre of the Earth. A ball is dropped into the tube from a point on the surface of the Earth.

Find the period of the motion. Does it depend upon the direction of the tube?

(c) How long time does a satellite moving in a circular path around the Earth at the surface of the Earth use to move around the Earth?

1.4 The Earth–Moon system

(a) Assume that the Earth and the Moon are spherical objects isolated from the rest of the solar system. Write down the equations of motion for the Earth–Moon system. Show that there is a solution where the Earth and the Moon are moving in circular orbits around their common centre of mass. What are the radii of the orbits in terms of the masses of the Earth and the Moon and the period of the motion?

(b) Find the Newtonian potential along the line connecting the two bodies. Show the result graphically and find the point on the line where the gravitational forces from the bodies cancel each other.

(c) Calculate the difference of the gravitational force from the Moon upon a 1 kg particle on the points on the Earth that are closest to the Moon and farthest away from the Moon.

1.5 The Roche limit

(a) A spherical Moon with mass m and radius R is orbiting a planet with mass M. Show that if the Moon is closer to the centre of the planet than

$$r = \left(\frac{2M}{m}\right)^{1/3} R,$$

then a stone lying freely on the surface of the Moon will be elevated due to tidal forces.

(b) The comet Shoemaker–Levy 9, that in July 1994 collided with Jupiter, was ripped apart already in 1992 after having passed near Jupiter. The comet had a mass of $m = 2.0 \cdot 10^{12}$ kg, and the closest passage in 1992 was at a distance of $s = 96\,000$ km from the centre of Jupiter. The mass of Jupiter is $M = 1.9 \cdot 10^{27}$ kg.

Use this information to estimate the size of the nucleus of the comet.

References

1. Genova, A., et al.: Solar system expansion and strong equivalence principle as seen by the NASA MESSENGER mission. Nature Commun. **1**, 10–11 (2018)
2. Grøn, Ø., Vøyenli, K.: On the foundations of the principle of relativity. Found. Phys. **29**, 1695–1793 (1999)
3. Brill, D.R., Cohen, J.M.: Rotating masses and their effect on inertial frames. Phys. Rev. **143**, 1011–1015 (1966)

Chapter 2
The Special Theory of Relativity

Abstract This chapter gives a concise and yet rather complete introduction to the special theory of relativity. Minkowski diagrams are used to illustrate several concepts such as the relativity of simultaneity. Special relativity is a theory of flat space-time admitting accelerated and rotating reference frames. In this chapter we also show how magnetism appears as a 2 order effect in v/c of electricity due to the Lorentz transformation.

In this chapter we shall give a short introduction to the fundamental principles of the special theory of relativity and deduce some of the consequences of the theory.

The special theory of relativity was presented by Albert Einstein in 1905. It was founded on two postulates:

1. The laws of physics are the same in all Galilean frames.
2. The velocity of light in empty space is the same in all Galilean frames and independent of the motion of the light source.

Einstein pointed out that these postulates are in conflict with Galilean kinematics, in particular with the Galilean law for the addition of velocities. According to Galilean kinematics two observers moving relative to each other cannot measure the same velocity for a certain light signal. Einstein solved this problem by a thorough discussion of how two distant clocks should be synchronized.

2.1 Coordinate Systems and Minkowski Diagrams

The most simple physical phenomenon that we can describe is called an event. This is an incident that takes place at a certain point in space and at a certain point in time. A typical example is the flash from a flashbulb.

A complete description of an event is obtained by giving the position of the event in space and time. Assume that our observations are made with reference to a reference frame. We introduce a coordinate system into our reference frame. Usually it is advantageous to employ a Cartesian coordinate system. This may be thought of as a cubic lattice constructed by measuring rods. If one lattice point is chosen as

© Springer Nature Switzerland AG 2020
Ø. Grøn, *Introduction to Einstein's Theory of Relativity*,
Undergraduate Texts in Physics, https://doi.org/10.1007/978-3-030-43862-3_2

origin, with all coordinates equal to zero, then any other lattice point has three spatial coordinates equal to the distances of that point along the coordinate axes that pass through the origin. The spatial coordinates of an event are the three coordinates of the lattice point at which the event happens. It is somewhat more difficult to determine the point of time of an event. If an observer is sitting at the origin with a clock, then the point of time when he catches sight of an event is not the point of time when the event happened. This is because the light takes time to pass from the position of the event to the observer at the origin.

Since observers at different positions have to make different such corrections, it would be simpler to have (imaginary) observers at each point of the reference frame such that the point of time of an arbitrary event can be measured locally. But then a new problem appears. One has to synchronize the clocks, so that they show the same time and go at the same rate. This may be performed by letting the observer at the origin send out light signals so that all the other clocks can be adjusted (with correction for light travel time) to show the same time as the clock at the origin. These clocks show the *coordinate time* of the coordinate system, and they are called *coordinate clocks*.

By means of the lattice of measuring rods and coordinate clocks, it is now easy to determine four coordinates (ct, x, y, z) for every event. (We have multiplied the time coordinate t by the velocity of light c in order that all four coordinates shall have the same dimension.) This coordinatization makes it possible to describe an event as a point P in a so-called *Minkowski diagram*. In this diagram we plot ct along the vertical axis and one of the spatial coordinates along the horizontal axis.

In order to observe particles in motion, we may imagine that each particle is equipped with a flashlight and that they flash at a constant frequency. The flashes from a particle represent a succession of events. If they are plotted into a Minkowski diagram, we get a series of points that describe a curve in the continuous limit. Such a curve is called a *world line* of the particle. The world line of a free particle is a straight line, as shown to the left of the time axis in Fig. 2.1.

A particle acted upon by a net force has a curved world line as the velocity of the particle changes with time. Since the velocity of every material particle is less than the velocity of light, the tangent of a world line in a Minkowski diagram will always make an angle less than 45° with the time axis. A flash of light gives rise to a light front moving onwards with the velocity of light. If this is plotted in a Minkowski diagram, the result is a light cone. In Fig. 2.1 we have drawn a light cone for a flash at the origin. It is obvious that we could have drawn light cones at all points in the diagram. An important result is that *the world line of any particle at a point is inside the light cone of a flash from that point*. This is an immediate consequence of the special principle of relativity and is also valid locally in the presence of a gravitational field.

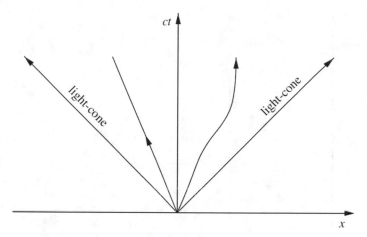

Fig. 2.1 World lines. World lines of particles moving slower than light are inside the light cone

2.2 Synchronization of Clocks

There are several equivalent methods that can be used to synchronize clocks. We shall here consider the radar method. Then a mirror is placed on the x-axis and emits a light signal from the origin at time t_A. This signal is reflected by the mirror at t_B and received again by the observer at the origin at time t_C. According to the second postulate of the special theory of relativity, the light moves with the same velocity in both directions, giving

$$t_B = \frac{1}{2}(t_A + t_C). \qquad (2.1)$$

When this relationship holds we say that the clocks at the origin and at the mirror are *Einstein synchronized*. Such synchronization is presupposed in the special theory of relativity. The situation corresponding to synchronization by the radar method is shown in Fig. 2.2.

The radar method can also be used to measure distances. The distance L from the origin to the mirror is given by

$$L = \frac{c}{2}(t_C - t_A). \qquad (2.2)$$

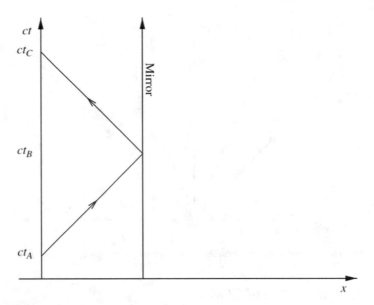

Fig. 2.2 Clock synchronization by the radar method

2.3 The Doppler Effect

Consider three observers (1, 2 and 3) in an inertial frame. Observers 1 and 3 are at rest, while 2 moves with constant velocity along the x-axis. The situation is shown in Fig. 2.3.

Each observer is equipped with a clock. If observer 1 emits light pulses with a constant period τ_1, then observer 2 receives them with a longer period τ_2 according to his or her clock. The fact that these two periods are different is a well-known phenomenon, called the *Doppler effect*. The same effect is observed with sound; the frequency of a receding vehicle is lower than that of an approaching one.

We are now going to deduce a relativistic expression for the Doppler effect. Firstly, we see from Fig. 2.3 that the two periods τ_1 and τ_2 are proportional to each other,

$$\tau_2 = K\tau_1. \tag{2.3}$$

The constant $K(v)$ is called Bondi's K-factor. Since observer 3 is at rest, the period τ_3 is equal to τ_1 so that

$$\tau_3 = \frac{1}{K}\tau_2. \tag{2.4}$$

These two equations imply that if 2 moves away from 1, so that $\tau_2 > \tau_1$, then $\tau_3 < \tau_2$. This is because 2 moves towards 3.

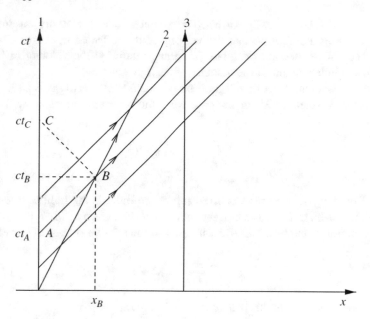

Fig. 2.3 The Doppler effect

The K-factor is most simply determined by placing observer 1 at the origin, while letting the clocks show $t_1 = t_2 = 0$ at the moment when 2 passes the origin. This is done in Fig. 2.3. The light pulse emitted at the point of time t_A is received by 2 when his clock shows $\tau_2 = K t_A$. If 2 is equipped with a mirror, the reflected light pulse is received by 1 at a point of time $t_C = K\tau_2 = K^2 t_A$. According to Eq. (2.1) the reflection event then happens at a point of time

$$t_B = \frac{1}{2}(t_C + t_A) = \frac{1}{2}(K^2 + 1)t_A. \tag{2.5}$$

The mirror has then arrived at a distance x_B from the origin, given by Eq. (2.2),

$$x_B = \frac{c}{2}(t_C - t_A) = \frac{c}{2}(K^2 - 1)t_A. \tag{2.6}$$

Thus, the velocity of observer 2 is

$$v = \frac{x_B}{t_B} = \frac{K^1 - 1}{K^1 + 1}c. \tag{2.7}$$

Solving this equation with respect to K gives

$$K = \left(\frac{c + v}{c - v}\right)^{1/2}. \tag{2.8}$$

This result is relativistically correct. The special theory of relativity was included through the tacit assumption that the velocity of the reflected light is c. This is a consequence of the second postulate of special relativity; the velocity of light is isotropic and independent of the velocity of the light source.

Since the wavelength λ of the light is proportional to the period τ, Eq. (2.3) gives the observed wavelength λ' for the case when the observer moves away from the source,

$$\lambda' = K\lambda = \left(\frac{c+v}{c-v}\right)^{1/2}\lambda. \tag{2.9}$$

This Doppler effect represents a redshift of the light. If the light source moves towards the observer, there is a corresponding blueshift given by K^{-1}.

It is common to express this effect in terms of the relative change of wavelength,

$$z = \frac{\lambda'-\lambda}{\lambda} = K - 1 \tag{2.10}$$

which is positive for redshift. If $v \ll c$, Eq. (2.9) gives

$$\frac{\lambda'}{\lambda} = K \approx 1 + \frac{v}{c} \tag{2.11}$$

To this order the redshift is

$$z = v/c. \tag{2.12}$$

This expression of the Doppler shift is well known in non-relativistic physics.

2.4 Relativistic Time Dilation

Every periodic motion can be used as a clock. A particularly simple clock is called the *photon clock*. This is shown in Fig. 2.4.

The clock consists of two parallel mirrors that reflect a light pulse back and forth. If the period of the clock is defined as the time interval between each time the light pulse hits the lower mirror, then $\Delta t' = 2L_0/c$.

Assume that the clock is at rest in an inertial reference frame Σ' where it is placed along the y-axis, as shown in Fig. 2.4. If this system moves along the ct-axis with a velocity v relative to another inertial reference frame Σ, the light pulse of the clock will follow a zigzag path as shown in Fig. 2.5.

The light signal follows a different path in Σ than in Σ'. The period Δt of the clock as observed in Σ is different from the period $\Delta t'$ which is observed in the rest frame. The period Δt is easily found in Fig. 2.5. Since the pulse takes the time

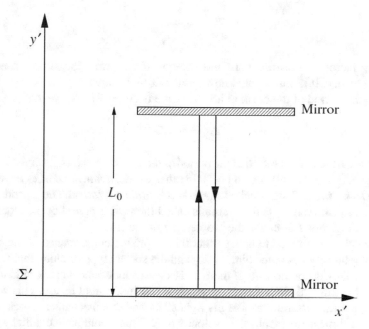

Fig. 2.4 Photon clock at rest

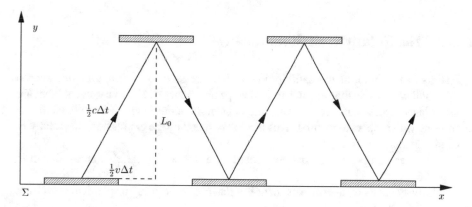

Fig. 2.5 Moving photon clock

$(1/2)\Delta t$ from the lower to the upper mirror and since the light velocity is always the same, we find

$$\left(\frac{1}{2}c\Delta t\right)^2 = L_0^2 + \left(\frac{1}{2}v\Delta t\right)^2. \tag{2.13}$$

i.e.

$$\Delta t = \gamma \frac{2L_0}{c}, \quad \gamma = \frac{1}{\sqrt{1 - v^2/c^2}}. \tag{2.14}$$

The γ factor is a useful shorthand notation for a term which is often used in relativity theory. It is commonly known as the *Lorentz factor*.

Since the period of the clock in its rest frame is $\Delta t' = 2L_0/c$, we get

$$\Delta t = \gamma \Delta t'. \tag{2.15}$$

Thus, we have to conclude that the period of the clock when it is observed to move (Δt) is greater that its rest period ($\Delta t'$). In other words: *a moving clock goes slower than a clock at rest*. This is called *the relativistic time dilation*. The period $\Delta t'$ of the clock as observed in its rest frame is called the proper period of the clock. The corresponding time t' is called the proper time of the clock.

One might be tempted to believe that this surprising consequence of the special theory of relativity has something to do with the special type of clock that we have employed. This is not the case. If there had existed a mechanical clock in Σ that did not show the time dilation, then an observer at rest in Σ might measure his velocity by observing the different rates of his light clock and this mechanical clock. In this way he could measure the absolute velocity of Σ. This would be in conflict with the special principle of relativity.

2.5   The Relativity of Simultaneity

Events that happen at the same point of time are said to be *simultaneous events*. We shall now show that according to the special theory of relativity, events that are simultaneous in one reference frame are not simultaneous in another reference frame moving with respect to the first. This is what is meant by the expression "the relativity of simultaneity".

Consider again two mirrors connected by a line along the x'-axis, as shown in Fig. 2.6.

Halfway between the mirrors there is a flash-lamp emitting a spherical wavefront at a point of time t_C.

The points at which the light front reaches the left-hand and the right-hand mirrors are denoted by A and B, respectively. In the reference frame Σ' of Fig. 2.6 the events A and B are simultaneous.

If we describe the same course of events from another reference frame Σ, where the mirror moves with constant velocity v in the positive x-direction, we find the Minkowski diagram shown in Fig. 2.7. Note that the light follows world lines making an angle of $45°$ with the axes. This is the case in every inertial frame.

In Σ the light pulse reaches the left mirror, which moves towards the light, before it reaches the right mirror, which moves in the same direction as the light. In this reference frame the events when the light pulses hit the mirrors are not simultaneous.

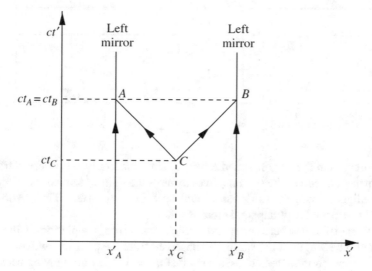

Fig. 2.6 Simultaneous events A and B

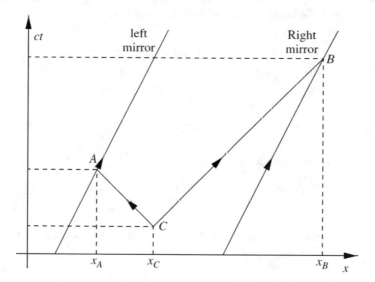

Fig. 2.7 The simultaneous events of Fig. 2.6 in another frame

As an example illustrating the relativity of simultaneity, Einstein imagined that the events A, B and C happen in a train which moves past the platform with a velocity v. The event C represents the flash of a lamp at the mid-point of a wagon. A and B are the events when the light is received at the back end and at the front end of the wagon, respectively. This situation is shown in Fig. 2.8.

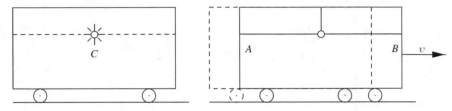

Fig. 2.8 Light flash in a moving train

As observed in the wagon, A and B happen simultaneously. As observed from the platform the rear end of the wagon moves towards the light which moves backwards, while the light moving forwards has to catch up with the front end. Thus, as observed from the platform A will happen before B.

The time difference between A and B as observed from the platform will now be calculated. The length of the wagon, as observed from the platform, will be denoted by L. The time coordinate is chosen such that $t_C = 0$. The light moving backwards hits the rear wall at a point of time t_A. During the time t_A the wall has moved a distance vt_A forwards, and the light has moved a distance ct_A backwards. Since the distance between the wall and the emitter is $L/2$, we get

$$\frac{L}{2} = vt_A + ct_A. \tag{2.16}$$

Thus

$$t_A = \frac{L}{2(c+v)}. \tag{2.17}$$

In the same manner one finds

$$t_B = \frac{L}{2(c-v)}. \tag{2.18}$$

It follows that the time difference between A and B as observed from the platform is

$$\Delta t = t_B - t_A = \gamma^2 \frac{vL}{c^2}. \tag{2.19}$$

As observed from the wagon A and B are simultaneous. As observed from the platform the rear event A happens at a time interval Δt before the event B. This is the *relativity of simultaneity*.

2.6 The Lorentz Contraction

During the first part of the nineteenth century the so-called luminiferous ether was introduced into physics to account for the propagation and properties of light. After J. C. Maxwell showed that light is electromagnetic waves and the ether was still needed as a medium in which electromagnetic waves propagated.

It was shown that Maxwell's equations do not obey the principle of relativity, when coordinates are changed using the Galilean transformations. If it is assumed that the Galilean transformations are correct, then Maxwell's equations can only be valid in one coordinate system. This coordinate system was the one in which the ether was at rest. Hence, Maxwell's equations in combination with the Galilean transformations implied the concept of "absolute rest". This made the measurement of the velocity of the Earth relative to the ether of great importance.

An experiment sufficiently accurate to measure this velocity to order v^2/c^2 was carried out by Michelson and Morley in 1887. A simple illustration of the experiment is shown in Fig. 2.9.

Our earlier photon clock is supplied by a mirror at a distance L along the x-axis from the emitter. The apparatus moves in the x-direction with a velocity v. In the rest frame Σ' of the apparatus, the distance between A and B is equal to the distance between A and C. This distance is denoted by L_0 and is called the *rest length* between A and B.

Light is emitted from A. Since the velocity of light is isotropic and the distances to B and C are equal in Σ', the light reflected from B and that reflected from C have the same travelling time. This was the result of the Michelson–Morley experiment, and it seems that we need no special effects such as the Lorentz contraction to explain the experiment.

However, before 1905 people believed in the physical reality of absolute velocity. The Earth was considered to move through the "ether" with a velocity that changed

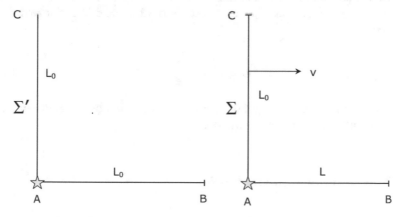

Fig. 2.9 Length contraction

with the seasons. The experiment should therefore be described under the assumption that the apparatus is moving.

Let us therefore describe an experiment from our reference frame Σ, which may be thought of as at rest in the "ether". Then according to Eq. (2.14) the travel time of the light being reflected at C is

$$\Delta t_C = \gamma \frac{2L_0}{c}. \tag{2.20}$$

For the light moving from A to B we may use Eq. (2.18) and for the light from B to A Eq. (2.17). This gives

$$\Delta t_B = \frac{L}{c-v} + \frac{L}{c+v} = \gamma^2 \frac{2L}{c}. \tag{2.21}$$

If length is independent of velocity, then $L = L_0$. In this case the travelling times of the light signals will be different. The travelling time difference is

$$\Delta t_B - \Delta t_C = \gamma(\gamma - 1)\frac{2L_0}{c}. \tag{2.22}$$

To the lowest order in v/c we have $\gamma \approx 1 + (1/2)(v/c)^2$, so that

$$\Delta t_B - \Delta t_C \approx \frac{1}{2}\left(\frac{v}{c}\right)^2. \tag{2.23}$$

which depends upon the velocity of the apparatus.

According to the ideas involving an absolute velocity of the Earth through the ether, if one lets the light reflected at B interfere with the light reflected at C (at the position A), then the interference pattern should vary with the season. This was not observed. On the contrary, observations showed that $\Delta t_B = \Delta t_C$.

Assuming that length varies with velocity, Eqs. (2.20) and (2.21), together with this observation, gives

$$L = \sqrt{1 - \frac{v^2}{c^2}} L_0. \tag{2.24}$$

Hence, $L < L_0$, i.e. the length of a rod is less when it moves than when it is at rest. This is called the *Lorentz contraction*.

2.7 The Lorentz Transformation

An event P has coordinates $(t', x', 0, 0)$ in a Cartesian coordinate system associated with a reference frame Σ'. Thus the distance from the origin of Σ' to P measured with a measuring rod at rest in Σ' is x'. If the distance between the origin of Σ' and the position at the x-axis where P took place is measured with measuring rods at rest in a reference frame moving with velocity v in the x-direction relative to Σ', one finds the length $\gamma^{-1}x$ due to the Lorentz contraction. Assuming that the origin of Σ and Σ' coincided at the point of time $t = 0$, the origin of Σ' has an x-coordinate vt at a point of time t. The event P thus has an x-coordinate

$$x = vt + \gamma^{-1}x' \tag{2.25}$$

or

$$x' = \gamma(x - vt). \tag{2.26}$$

The x-coordinate may be expressed in terms of x' and y' by letting $v \to -v$,

$$x = \gamma(x' + vt'). \tag{2.27}$$

The y- and z-coordinates are associated with axes directed perpendicular to the direction of motion. Therefore, they are the same in the two-coordinate systems

$$y = y', \quad z(z'). \tag{2.28}$$

Substituting x' from Eq. (2.26) into Eq. (2.27) reveals the connection between the time coordinates of the two-coordinate systems,

$$t' = \gamma\left(t - \frac{vx}{c^2}\right) \tag{2.29}$$

and

$$t = \gamma\left(t' + \frac{vx'}{c^2}\right). \tag{2.30}$$

The latter term in this equation is nothing but the deviation from simultaneity in Σ for two events that are simultaneous in Σ'.

The relations (2.27)–(2.30) between the coordinates of Σ and Σ' represent a special case of the *Lorentz transformations*. The above relations are special since the two-coordinate systems have the same spatial orientation, and the x- and x'-axes are aligned along the relative velocity vector of the associated frames. Such transformations are called *boosts*.

For non-relativistic velocities, $v \ll c$, the Lorentz transformations (2.27)–(2.30) pass over into the corresponding Galilei transformations.

The Lorentz transformation gives a connection between the relativity of simultaneity and the Lorentz contraction. The *length* of a body is defined as the difference between the coordinates of its end points, *as measured by simultaneity in the rest frame of the observer.*

Consider the wagon of Sect. 2.5. Its rest length is $L_0 = x'_B - x'_A$. The difference between the coordinates of the wagon's end points, $x_A - x_B$ as measured in Σ, is given implicitly by the Lorentz transformation

$$x'_B - x'_A = \gamma[x_B - x_A - v(t_B - t_A)].\tag{2.31}$$

According to the above definition the length L of the moving wagon is given by $L = x_B - x_A$ with $t_B = t_A$.

From Eq. (2.3) we then get

$$L_0 = \gamma L.\tag{2.32}$$

which is equivalent to Eq. (2.24).

The Lorentz transformation will now be used to deduce the relativistic formulae for velocity addition. Consider a particle moving with velocity u along the x'-axis of Σ'. If the particle was at the origin at $t' = 0$, its position at t' is $x' = u't'$. Using this relation together with Eqs. (2.27) and (2.28) we find the velocity of the particle as observed in Σ

$$u = \frac{x}{t} = \frac{u' + v}{1 + u'v/c^2}.\tag{2.33}$$

A remarkable property of this expression is that by adding velocities less than c one cannot obtain a velocity greater than c. For example, if a particle moves with a velocity c in Σ' then its velocity in Σ is also c regardless of the velocity of Σ relative to Σ'.

Equation (2.33) may be written in a geometrical form by introducing the so-called *rapidity η'* defined by

$$\tanh \eta' = \frac{u'}{c}\tag{2.34}$$

for a particle with velocity u'. Similarly the rapidity, $\bar{\eta}$, of Σ' relative to Σ is given by

$$\tanh \bar{\eta} = \frac{v}{c}.\tag{2.35}$$

Since

$$\tanh(\eta' + \bar{\eta}) = \frac{\tanh \eta' + \tanh \theta}{1 + \tanh \eta' \tanh \theta},\tag{2.36}$$

the relativistic velocity addition formula, Eq. (2.33), may be written as

$$\eta = \eta' + \bar{\eta}. \tag{2.37}$$

Since rapidities are additive, their introduction simplifies some calculations and they have often been used as variables in elementary particle physics.

With these new hyperbolic variables we can write the Lorentz transformation in a particularly simple way. Using Eq. (2.35) in Eqs. (2.27) and (2.30) we find

$$x = x' \cosh \bar{\eta} + ct' \sinh \bar{\eta}, \quad ct = x' \sinh \bar{\eta} + ct' \cosh \bar{\eta}. \tag{2.38}$$

2.8 Lorentz Invariant Interval

Let two events be given. The coordinates of the events, as referred to two different reference frames Σ and Σ', are connected by a Lorentz transformation. The coordinate differences are therefore connected by

$$\Delta t = \gamma\left(\Delta t' + \frac{v}{c^2}\Delta x'\right), \quad \Delta x = \gamma(\Delta x' + v\Delta t'), \quad \Delta y = \Delta y', \quad \Delta z = \Delta z'. \tag{2.39}$$

This leads to

$$-(c\Delta t)^2 + (\Delta x)^2 + (\Delta y)^2 + (\Delta z)^2 = -(c\Delta t')^2 + (\Delta x')^2 + (\Delta y')^2 + (\Delta z')^2, \tag{2.40}$$

showing that the quantity

$$(\Delta s)^2 = -(c\Delta t)^2 + (\Delta x)^2 + (\Delta y)^2 + (\Delta z)^2 \tag{2.41}$$

is invariant under a Lorentz transformation. The quantity Δs is called a *spacetime interval*, or only an *interval*. Due to the minus sign in Eq. (2.40), the square of the interval between two events may be positive, zero or negative. These three types of intervals are termed *space-like* if $(\Delta s)^2 > 0$, xero or *light-like* if $(\Delta s)^2 = 0$ and *time-like* if $(\Delta s)^2 < 0$.

The reasons for these names are the following. Given two events with a space-like interval (A and B in Fig. 2.10), there exists a Lorentz transformation to a new reference frame where A and B happen simultaneously. In this frame the distance between the events is purely spatial. Two events with a light-like interval (C and D in Fig. 2.10) can be connected by a light signal, i.e. one can send a photon from C to D. The events E and F have a time-like interval between them and can be observed from a reference frame in which they have the same spatial position, but occur at different points of time.

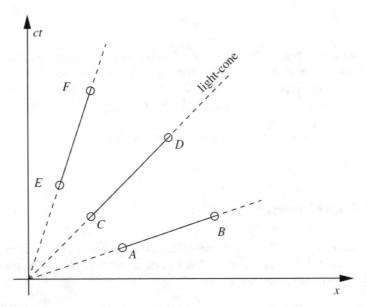

Fig. 2.10 Space-like, light-like and time-like intervals. Interval between A and B is space-like, between C and D light-like and between E and F time-like

Since all material particles move with a velocity less than that of light, the points on the world line of a particle are separated by time-like intervals. The curve is then said to be time-like. All time-like curves through a point pass inside the light cone from that point.

For a particle with velocity $u = \Delta x / \Delta t$ Eq. (2.40) gives

$$(\Delta s)^2 = -\left(1 - \frac{u^2}{c^2}\right)(c\Delta t)^2. \tag{2.42}$$

In the rest frame Σ' of the particle $\Delta x' = 0$, giving

$$(\Delta s)^2 = -(c\Delta t)^2. \tag{2.43}$$

The time t' in the rest frame of the particle is the same as the time measured on a clock carried by the particle. It is called the *proper time* of the particle and denoted by τ. From Eqs. (2.42) and (2.43) it follows that

$$\Delta\tau = \sqrt{1 - \frac{u^2}{c^2}}\,\Delta t = \gamma^{-1}\Delta t, \tag{2.44}$$

which is an expression of the relativistic time dilation.

Equation (2.43) is important. It gives the physical interpretation of a time-like interval between two events. The interval is a measure of the proper time interval

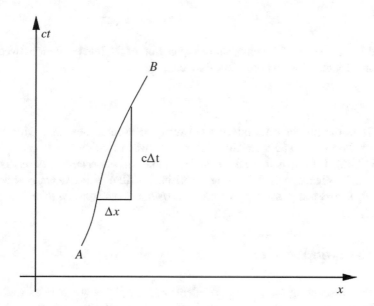

Fig. 2.11 World line of an accelerating particle

between the events. This time is measured on a clock that moves such that it is present at both events. In the limit $u \to c$ (the limit of a light signal), $\Delta\tau = 0$. This shows that $(\Delta s)^2 = 0$ for a light-like interval.

Consider a particle with a variable velocity, $u(t)$, as indicated in Fig. 2.11. In this situation we can specify the velocity at an arbitrary point of the world line. Equation (2.44) can be used with this velocity, in an infinitesimal interval around this point,

$$d\tau - \sqrt{1 - \frac{u^2(t)}{c^2}} dt. \qquad (2.45)$$

This equation means that the acceleration has no local effect upon the proper time of the clock. Here the word "local" means as measured by an observer at the position of the clock. Such clocks are called *standard clocks*.

If a particle moves from A to B in Fig. 2.11, the proper time as measured on a standard clock following the particle is found by integrating Eq. (2.45)

$$\tau_B - \tau_A = \int_{t_A}^{t_B} \sqrt{1 - \frac{u^2(t)}{c^2}} dt. \qquad (2.46)$$

The relativistic time dilation has been verified with great accuracy by observations of unstable elementary particles with short lifetimes [1].

An infinitesimal spacetime interval

$$ds^2 = -c^2dt^2 + dx^2 + dy^2 + dz^2. \tag{2.47}$$

is called a *line element*. The physical interpretation of the line element between two infinitesimally close events on a time-like curve is

$$ds^2 = -c^2d\tau^2, \tag{2.48}$$

where $d\tau$ is the proper time interval between the events, measured with a clock following the curve. The spacetime interval between two events is given by the integral (2.46). It follows that *the proper time interval between two events is path-dependent*. This leads to the following surprising result: *A time-like interval between two events is greatest along the straightest possible curve between them*.

2.9   The Twin Paradox

Rather than discussing the lifetime of elementary particles, we may as well apply Eq. (2.46) to a person. Let her name be Eva. Assume that Eva is rapidly accelerating from rest at the point of time $t = 0$ at origin to a velocity v along the x-axis of a (ct, x) coordinate system in an inertial reference frame Σ (Fig. 2.12).

At a point of time t_P she has come to a position x_P. She then rapidly decelerates until reaching a velocity v in the negative x-direction. At a point of time t_Q, as measured on clocks at rest in Σ, she has returned to her starting location. If we

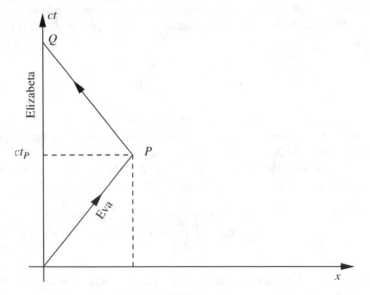

Fig. 2.12 Twin paradox world lines. World lines of the twin sisters Eva and Elisabeth

neglect the brief periods of acceleration, Eva's travelling time as measured on a clock which she carries with her is

$$t_{\text{Eva}} = \left(1 - \frac{v^2}{c^2}\right)^{1/2} t_Q. \tag{2.49}$$

Now assume that Eva has a twin sister named Elizabeth who remains at rest at the origin of Σ.

Elizabeth has become older by $\tau_{\text{Elisabeth}} = t_Q$ during Eva's travel, so that

$$\tau_{\text{Eva}} = \left(1 - \frac{v^2}{c^2}\right)^{1/2} t_{\text{Elisabeth}}. \tag{2.50}$$

For example, if Eva travelled to Alpha Centauri (the Sun's nearest neighbour at four light years) with a velocity $v = 0.8c$, she would be gone for 10 years as measured by Elizabeth. Therefore Elizabeth has aged 10 years during Eva's travel. According to Eq. (2.50), Eva has only aged 6 years. According to Elizabeth, Eva has aged less than herself during her travels.

The principle of relativity, however, tells that Eva can consider herself as at rest and Elizabeth as the traveller. According to Eva it is Elizabeth who has only aged by 6 years, while Eva has aged by 10 years during the time they are apart.

What happens? How can the twin sisters arrive at the same prediction as to how much each of them age during the travel? In order to arrive at a clear answer to these questions, we shall have to use a result from the general theory of relativity. The twin paradox will be taken up again in Chap. 5.

2.10 Hyperbolic Motion

With reference to an inertial reference frame it is easy to describe relativistic accelerated motion. The special theory of relativity is in no way limited to describe motion with constant velocity.

Let a particle move with a variable velocity $u(t) = dx/dt$ along the x-axis in Σ. The frame Σ' moves with velocity v in the same direction relative to Σ. In this frame the particle velocity is $u'(t') = dx'/dt'$. At every moment the velocities u and u' are connected by the relativistic formula for velocity addition, Eq. (2.33). Thus, according to Eq. (2.30), a velocity change du' in Σ' and the corresponding velocity change du in Σ are related by

$$dt = \frac{dt' + (v/c^2)dx'}{\sqrt{1 - v^2/c^2}} = \frac{1 + u'v/c^2}{\sqrt{1 - v^2/c^2}} dt'. \tag{2.51}$$

Combining these expressions we obtain the relationship between the acceleration of the particle as measured in Σ and Σ'

$$a = \frac{du}{dt} = \frac{\left(1 - v^2/c^2\right)^{3/2}}{\left(1 + u'v/c^2\right)^3} a'.$$ (2.52)

Let Σ' be the rest frame of the particle at a point of time t. Then $u' = 0$ at this moment, giving

$$a = \left(1 - \frac{u^2}{c^2}\right)^{3/2} a'.$$ (2.53)

Here a' is the acceleration of the particle as measured in its instantaneous rest frame. It is called *the rest acceleration* of the particle. Equation (2.53) can be integrated if we know how the rest acceleration of the particle varies with time.

We shall now focus on the case where the particle has uniformly accelerated motion and moves along a straight path in space. The rest acceleration of the particle is constant, say $a' = g$. Integration of Eq. (2.53) with $u(0) = 0$ then gives

$$u = \left(1 + \frac{g^2}{c^2}t^2\right)^{-1/2} gt.$$ (2.54)

Integrating once more gives

$$x = \frac{c^2}{g}\left(1 + \frac{g^2}{c^2}t^2\right)^{1/2} + x_0 - \frac{c^2}{g},$$ (2.55)

where the integration constant x_0 is equal to the position at $t = 0$.

Equation (2.55) can be given the form

$$\left(x - x_0 + \frac{c^2}{g}\right)^2 - c^2 t^2 = \frac{c^4}{g^2}.$$ (2.56)

This is the equation of a hyperbola in the Minkowski diagram (Fig. 2.13).

Since the world line of a particle with uniformly accelerated, rectilinear motion has the shape of a hyperbola, this type of motion is called *hyperbolic motion*.

Using the proper time τ of the particle as a parameter, we may obtain a simple parametric representation of its world line. Substituting Eq. (2.54) into Eq. (2.45) we get

$$d\tau = \frac{dt}{\sqrt{1 + (c^2/g)t^2}}.$$ (2.57)

Fig. 2.13 World line of
particle with constant rest
acceleration

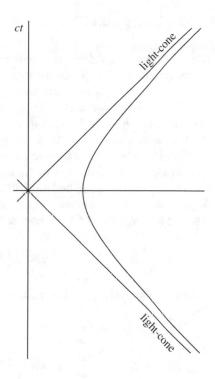

Integration with $\tau(0) = 0$ gives

$$\tau = \frac{c}{g}\text{arc sinh}\left(\frac{gt}{c}\right). \tag{2.58}$$

or

$$t = \frac{c}{g}\sinh\left(\frac{gt}{c}\right). \tag{2.59}$$

Inserting this expression into Eq. (2.55) we get

$$x = \frac{c^2}{g}\cosh\left(\frac{gt}{c}\right) + x_0 - \frac{c^2}{g}. \tag{2.60}$$

These expressions shall be used later when describing uniformly accelerated
reference frames.

Note that *hyperbolic motion* results when the particle moves with *constant rest
acceleration*. Such motion is usually called *uniformly accelerated motion*. Motion
with constant acceleration as measured in the "laboratory frame" Σ gives rise to the
usual parabolic motion.

2.11 Energy and Mass

The existence of an electromagnetic radiation pressure was well known before Einstein formulated the special theory of relativity. In black body radiation with mass density ρ there is an isotropic pressure $p = (1/3)\rho c^2$. If the radiation moves in a certain direction (laser), then the pressure in this direction is $p = \rho c^2$. Einstein gave several deductions of the famous equation connecting the inertial mass of a body with its energy content. A deduction he presented in 1906 is as follows.

Consider a box with a light source at one end. A light pulse with radiation energy E is emitted to the other end where it is absorbed (see Fig. 2.14).

The box has a mass M and a length L. Due to the radiation pressure of the shooting light pulse the box receives a recoil. The pulse is emitted during a time interval Δt. During this time the radiation pressure is

$$p = \rho c^2 = \frac{E}{V} = \frac{E}{Ac\Delta t}, \tag{2.61}$$

where V is the volume of the radiation pulse and A the area of a cross section of the box.

The recoil velocity of the box is

$$\Delta v = -a\Delta t = -\frac{F}{M}\Delta t = -\frac{pA}{M}\Delta t = -\left(\frac{E}{Ac\Delta t}\right)\left(\frac{A\Delta t}{M}\right) = -\frac{E}{Mc} \tag{2.62}$$

The pulse takes the time L/c to move to the other side of the box. During this time the Box moves a distance

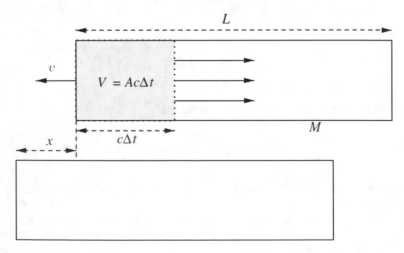

Fig. 2.14 Light pulse in a box

$$\Delta x = \Delta v \frac{L}{c} = -\frac{EL}{Mc^2}. \tag{2.63}$$

Then the box is stopped by the radiation pressure caused by the light pulse hitting the wall at the other end of the box.

Let m be the mass of the radiation. Before Einstein one would put $m = 0$. Einstein, however, reasoned as follows. Since the box and its contents represent an isolated system, the mass centre has not moved. The mass centre of the box with mass M has moved a distance Δx to the left, and the radiation with mass m has moved a distance L to the right. Thus

$$mL + M\Delta x = 0 \tag{2.64}$$

which gives

$$m = -\frac{M}{L}\Delta x = -\left(\frac{M}{L}\right)\left(-\frac{EL}{Mc^2}\right) = \frac{E}{c^2} \tag{2.65}$$

or

$$E = mc^2. \tag{2.66}$$

Here we have shown that radiation energy has a mass given by Eq. (2.65). Einstein derived Eq. (2.66) using several different methods showing that it is valid in general for all types of systems.

The energy content of even small bodies is enormous. For example, by transforming 1 g of matter to heat, one may heat 300,000 metric tons of water from room temperature to the boiling point. (The energy corresponding to a mass m is enough to change the temperature by ΔT of an object of mass M and specific heat capacity $c_v : mc^2 = Mc_v\Delta T$).

2.12 Relativistic Increase of Mass

In the special theory of relativity, force is defined as rate of change of momentum. We consider a body that gets a change of energy dE due to the work performed on it by a force F. According to Eq. (2.66) and the definition of work (force times distance) the body gets a change of mass dm, given by

$$c^2 dm = dE = F ds = F v dt = v d(mv) = mv dv + v^2 dm, \tag{2.67}$$

which gives

$$\int_{m_0}^{m} \frac{dm}{m} = \int_{0}^{v} \frac{v dv}{c^2 - v^2},$$

(2.68)

where m_0 is the rest mass of the body—i.e. its mass as measured by an observer co-moving with the body—and m its mass when its velocity is equal to v. Integration gives

$$m = \frac{m_0}{\sqrt{1 - v^2/c^2}} = \gamma m_0.$$

(2.69)

In the case of small velocities compared to the velocity of light we may use the approximation

$$\sqrt{1 - \frac{v^2}{c^2}} = 1 + \frac{1}{2}\frac{v^2}{c^2}.$$

(2.70)

With this approximation Eqs. (2.66) and (2.69) give

$$E \approx m_0 c^2 + \frac{1}{2}m_0 v^2.$$

(2.71)

This equation shows that the total energy of a body encompasses its rest energy $m_0 c^2$ and its kinetic energy. In the non-relativistic limit the kinetic energy is $(1/2)m_0 v^2$. The relativistic expression for the kinetic energy is

$$E_K = E - m_0 c^2 = (\gamma - 1)m_0 c^2.$$

(2.72)

Note that $E_K \to \infty$ when $v \to c$.

According to Eq. (2.33), it is not possible to obtain a velocity greater than that of light by adding velocities. Equation (2.72) gives a dynamical reason that material particles cannot be accelerated up to and above the velocity of light.

Using Eq. (2.69) the energy and momentum of a particle with rest mass m_0 and velocity v are

$$E = \gamma m_0 c^2, \quad p = \gamma m_0 v$$

(2.73)

Using the identity

$$\gamma^2 = \gamma^2 \frac{v^2}{c^2} + 1,$$

(2.74)

It follows from the expressions (2.73) that

$$E^2 = p^2 c^2 + m_0^2 c^4,$$

(2.75)

or

$$m_0^2 c^4 = E^2 - p^2 c^2. \tag{2.76}$$

Since the rest mass m_0 of a particle is a velocity-independent quantity and hence *Lorentz invariant*, it follows that the quantity $E^2 - p^2 c^2$ of a particle is a Lorentz invariant having the same value in all inertial reference frames even if the energy and momentum of the particle depend upon the frame.

2.13   Lorentz Transformation of Velocity, Momentum, Energy and Force

We shall write down the Lorentz transformation of velocity, momentum, energy and force. (Detailed derivations are given by W. G. V. Rosser in An Introduction to the Theory of Relativity, Butterworths, 1964). Let Σ and Σ' be two inertial frames with a relative velocity v in the x-direction and with co-moving Cartesian coordinates. A particle is moving with velocity components u_x, u_y, u_z in Σ and u'_x, u'_y, u'_z in Σ'. The transformation formulae for the velocity component are

$$u'_x = \frac{u_x - v}{1 - vu_x/c^2}, \quad u'_y = \frac{u_y\sqrt{1 - v^2/c^2}}{1 - vu_x/c^2}, \quad u'_z = \frac{u_z\sqrt{1 - v^2/c^2}}{1 - vu_x/c^2}. \tag{2.77}$$

It follows from these formulae that

$$u'^2 = u'^2_x + u'^2_y + u'^2_z = \frac{(u_x - v)^2 + (u^2 - u_x^2)(1 - v^2/c^2)}{(1 - vu_x/c^2)^2}. \tag{2.78}$$

As shown by Rosser this leads to

$$\frac{1}{\sqrt{1 - u'^2/c^2}} = \frac{1 - vu_x/c^2}{\sqrt{1 - v^2/c^2}\sqrt{1 - u^2/c^2}}. \tag{2.79}$$

In the frames Σ and Σ' the momentum of a particle with rest mass m_0 has components

$$p_i = \frac{m_0 u_i}{\sqrt{1 - u^2/c^2}}, \quad p'_i = \frac{m_0 u'_i}{\sqrt{1 - u'^2/c^2}}, \tag{2.80}$$

and energies

$$E = \frac{m_0 c^2}{\sqrt{1 - u^2/c^2}}, \quad E' = \frac{m_0 c^2}{\sqrt{1 - u'^2/c^2}}. \tag{2.81}$$

Using the first of Eqs. (2.77) and (2.79) gives

$$p'_x = \gamma(mu_x - mv), \tag{2.82}$$

where γ and m is given in Eq. (2.79). Since $mu_x = p_x$ and $m = E/c^2$ we get

$$p'_x = \gamma(p_x - Ev/c^2). \tag{2.83}$$

Using the second and third of Eq. (2.77) we get

$$p'_y = p_y, \quad p'_z = p_z. \tag{2.84}$$

Inserting Eq. (2.79) in the second of Eq. (2.81) we obtain

$$E' = \gamma(E - vp_x). \tag{2.85}$$

At this stage we define force as a 3-vector

$$\mathbf{f} = \frac{d\mathbf{p}}{dt}, \tag{2.86}$$

or

$$f_x = \frac{dp_x}{dt} \text{ etc.} \tag{2.87}$$

where p is the momentum as given in Eq. (2.73). Similarly we have

$$f'_x = \frac{dp'_x}{dt'} \text{ etc.} \tag{2.88}$$

Using the transformation Eq. (2.83) we then have

$$f'_x = \gamma \frac{d}{dt'}\left(p_x - \frac{vE}{c^2}\right) = \gamma \frac{dt}{dt'}\left(\frac{dp_x}{dt} - \frac{v}{c^2}\frac{dE}{dt}\right). \tag{2.89}$$

From the Lorentz transformation $t' = \gamma(t - vx/c^2)$ we have

$$\frac{dt}{dt'} = \frac{1}{\frac{dt'}{dt}} = \frac{1}{\frac{d}{dt}\gamma\left(t - \frac{vx}{c^2}\right)} = \frac{1}{\lambda\left(1 - \frac{vu_x}{c^2}\right)}. \tag{2.90}$$

Furthermore we insert $dp_x/dt = f_x$ in Eq. (2.89) and

$$\frac{dE}{dt} = \mathbf{f} \cdot \mathbf{u} = f_x u_x + f_y u_y + f_z u_z. \tag{2.91}$$

Inserting this in Eq. (2.89) and rearranging gives

$$f'_x = f_x - \frac{vu_y}{c^2 - vu_x} f_y - \frac{vu_z}{c^2 - vu_x} f_z. \tag{2.92}$$

Since $p'_y = p_y$ we get

$$f'_y = \frac{dp'_y}{dt'} = \frac{dp_y}{dt'} = \frac{dt}{dt'} \frac{dp_y}{dt} = \frac{dt}{dt'} f_y. \tag{2.93}$$

Using Eq. (2.90) we obtain the transformations equations for the y- and z-components of the force

$$f'_y = \frac{f_y}{\gamma(1 - vu_x/c^2)}, \quad f'_z = \frac{f_z}{\gamma(1 - vu_x/c^2)}. \tag{2.94}$$

The inverse transformation is

$$f_x = f'_x + \frac{vu'_y}{c^2 + vu'_x} f'_y + \frac{vu'_z}{c^2 + vu'_x} f'_z, \quad f_y = \frac{f'_y}{\gamma(1 + vu'_x/c^2)},$$

$$f_z = \frac{f'_z}{\gamma(1 + vu'_x/c^2)}. \tag{2.95}$$

If Σ' is the rest system of the particle, so that $u'_x = u'_y = u'_z = 0$, the transformation reduces to

$$f_x = f'_x, \quad f_y = \sqrt{1 - v^2/c^2} f'_y, \quad f_z = \sqrt{1 - v^2/c^2} f'_z. \tag{2.96}$$

Example 2.13.1 (*The Lever paradox*) A right-angled lever with arms of equal length L' is at rest at the origin of a Cartesian coordinate system co-moving with an inertial reference frame Σ'. It is oriented with the arms along the x'- and y'-axes. At the end of the x'-arm there acts a force f' in the y-direction, and at the end of the y'-arm there acts an equal force in the x'-direction. Hence the torques acting in the opposite directions are equal, so the lever is in rotational equilibrium and will not start rotating.

An inertial frame Σ moves in the negative x-direction with velocity v relative to Σ'. In this frame the length of the x-arm is Lorentz contracted to $L = \sqrt{1 - v^2/c^2} L'$, but the arm in the y-direction is unchanged. Furthermore, according to the transformation Eq. (2.95) the force in the x-direction is unchanged, but the force in the y-direction is diminished to $f_y = \sqrt{1 - v^2/c^2} f'_y$. Hence in this frame there is a net torque $f'L' - \sqrt{1 - v^2/c^2} f'\sqrt{1 - v^2/c^2} L' = (v^2/c^2) f'L'$, so one is inclined to conclude that the lever will start rotating in the clockwise direction.

This is in conflict with the analysis in the rest frame of the lever, and this conflict is what is called the lever paradox.

What will happen? Shall we trust the analysis in the rest frame of the lever or in the laboratory frame where the lever moves?

The answer is related to the relativity of simultaneity and what we mean by a physical object. If an extended physical object exists at a certain moment, it is made up of a set of simultaneous events. Due to the relativity of simultaneity these events are not simultaneous in a frame moving relatively to the first one. Hence the lever in the laboratory frame and the rest frame of the lever consists of different sets of events.

The set of simultaneous events in the rest frame of an object determines what will happen to the object. We therefore have the rule: *In order to determine what happens to an extended body acted upon by forces one has to perform the calculations in the rest frame of the body.*

This is not in conflict with the principle of relativity. The rule does not introduce any absolute velocity. If one transforms the description in the rest frame of the body to the laboratory frame, there will appear unusual terms depending upon the relative velocity of the object and the laboratory frame. This does not mean that the laws of nature are different in different inertial reference frames. But it means that the laws contain this relative velocity, due to the relativity of simultaneity, because the objects that are described by simultaneity in different reference frames do not consist of the same sets of events.

2.14   Tachyons

Particles cannot pass the velocity barrier represented by the velocity of light. However, the special theory of relativity permits the existence of particles that have *always* moved with a velocity $v > c$. Such particles are called *tachyons*.

Tachyons have special properties that have been used in the experimental searches for them. There is currently no observational evidence for the physical existence of tachyons.

There are also certain theoretical difficulties with the existence of tachyons. The special theory of relativity applied to tachyons leads to the following paradox. Using a tachyon telephone a person, A emits a tachyon to B at a point of time t_1. B moves away from A. The tachyon is reflected by B and reaches A before it was emitted; see Fig. 2.15. If the tachyon could carry information it might bring an order to destroy the tachyon emitter when it arrives back at A.

To avoid similar problems in regards to the energy exchange between tachyons and ordinary matter, a reinterpretation principle is introduced for tachyons. For certain observers a tachyon will move backwards in time, i.e. the observer finds that the tachyon is received before it was emitted. Special relativity tells us that such a tachyon is always observed to have negative energy.

According to the reinterpretation principle, the observer will interpret his observations to mean that a tachyon with positive energy moves forward in time. In this way,

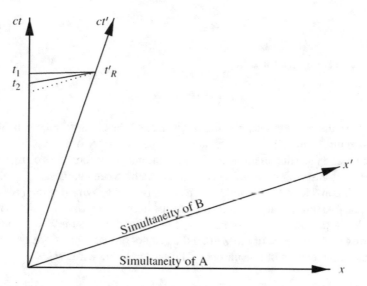

Fig. 2.15 Tachyon paradox. A emits a tachyon at a point of time t_1. It is reflected at B and arrives back at A at a point of time t_2. Note that the arrival event at A is later than the reflection event as measured by B

one finds that the energy exchange between tachyons and ordinary matter proceeds in accordance with the principle of causality.

However, the reinterpretation principle cannot be used to remove the problems associated with the exchange of information between tachyons and ordinary matter. The tachyon telephone paradox cannot be resolved by means of the reinterpretation principle. The conclusion is that if tachyons exist, they cannot be carriers of information in our slowly moving world.

2.15   Magnetism as a Relativistic Second-Order Effect

Electricity and magnetism are described completely by Maxwell's equations of the electromagnetic field,

$$\nabla \cdot \mathbf{E} = \frac{1}{\varepsilon_0} \rho_q, \tag{2.97}$$

$$\nabla \cdot \mathbf{B} = 0, \tag{2.98}$$

$$\nabla \times \mathbf{E} = -\frac{\partial \mathbf{B}}{\partial t}, \tag{2.99}$$

$$\nabla \times \mathbf{B} = \mu_0 \mathbf{j} + \frac{1}{c^2} \frac{\partial \mathbf{E}}{\partial t}, \tag{2.100}$$

together with Lorentz's force law

$$F = q(\mathbf{E} + \mathbf{v} \times \mathbf{B}). \tag{2.101}$$

However, the relation between the magnetic and the electric force was not fully understood until Einstein had constructed the special theory of relativity. Only then could one clearly see the relationship between the magnetic force on a charge moving near a current-carrying wire and the electric force between charges.

We shall consider a simple model of a current-carrying wire in which we assume that the positive ions are at rest while the conducting electrons move with the velocity v. The charge per unit length for each type of charged particle is $\hat{\lambda} = Sne$ where S is the cross-sectional area of the wire, n the number of particles of one type per unit length and e the charge of one particle. The current in the wire is

$$J = Snev = \hat{\lambda}v. \tag{2.102}$$

The wire is at rest in an inertial frame $\widehat{\Sigma}$. As observed in $\widehat{\Sigma}$ it is electrically neutral. Let a charge q move with a velocity u along the wire in the opposite direction of the electrons. The rest frame of q is Σ. The wire will now be described from Σ (see Figs. 2.16 and 2.17).

Note that the charge per unit length of the particles as measured in their own rest frames, Σ_0, is

$$\lambda_{0-} = \left(1 - \frac{v^2}{c^2}\right)^{1/2} \hat{\lambda}, \quad \lambda_{0+} = \hat{\lambda} \tag{2.103}$$

Since the distance between the electrons is Lorentz contracted in $\widehat{\Sigma}$ compared to their distance in Σ_0.

The velocities of the particles as measured in Σ are

Fig. 2.16 Current carrying wire seen from its own rest frame

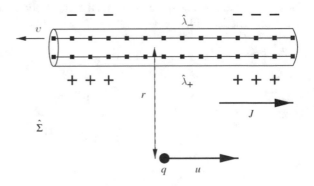

Fig. 2.17 Current carrying wire seen from the frame of a moving charge

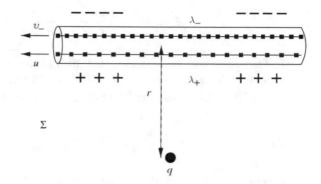

$$v_- = -\frac{v + u}{1 + uv/c^2}, \quad v_+ = -u. \tag{2.104}$$

The charge per unit length of the negative particles as measured in Σ is

$$\lambda_- = \left(1 - \frac{v_-^2}{c^2}\right)^{-1/2} \lambda_{0-}. \tag{2.105}$$

Substitution from Eqs. (2.103) and (2.104) gives

$$\lambda_- = \gamma\left(1 + \frac{uv}{c^2}\right)\hat{\lambda}, \tag{2.106}$$

where $\gamma = \left(1 - u^2/c^2\right)^{-1/2}$. In a similar manner the charge per unit length of the positively charged particles as measured in Σ is

$$\lambda_+ = \gamma\hat{\lambda}. \tag{2.107}$$

Thus as observed in the rest frame of q the wire has a net charge per unit length

$$\lambda = \lambda_- - \lambda_+ = \gamma\frac{uv}{c^2}\hat{\lambda}. \tag{2.108}$$

As a result of the different Lorentz contractions of the positive and negative ions when we transform from their respective rest frames to Σ, a current-carrying wire which is electrically neutral in the laboratory frame is observed to be electrically charged in the rest frame of the charge q.

As observed in this frame there is a radial electrical field with field strength

$$E = \frac{\lambda}{2\pi\varepsilon_0 r}. \tag{2.109}$$

Then a force F acts on q this is given by

$$F = qE = \frac{q\lambda}{2\pi\varepsilon_0 r} = \frac{\hat{\lambda}v}{2\pi\varepsilon_0 c^2 r}\gamma qu. \tag{2.110}$$

If a force acts upon q as observed in $\widehat{\Sigma}$ then a force also acts on q as observed in Σ. According to the relativistic transformation of a force component normal to the direction of the relative velocity between $\widehat{\Sigma}$ and Σ, this force is

$$\widehat{F} = \gamma^{-1} F = \frac{\hat{\lambda}v}{2\pi\varepsilon_0 c^2 r}qu. \tag{2.111}$$

Inserting $J = \hat{\lambda}v$ from Eq. (2.102) and using $c^2 = (\varepsilon_0\mu_0)^{-1}$, where μ_0 is the permeability of vacuum, we obtain

$$\widehat{F} = \frac{\mu_0 J}{2\pi r}qu. \tag{2.112}$$

This is exactly the expression obtained if we calculate the magnetic flux density $\widehat{B}$ around the current-carrying wire using Ampere's circuit law

$$\widehat{B} = \mu_0 \frac{J}{2\pi r} \tag{2.113}$$

and use the force law Eq. (2.101) for a charge moving in a magnetic field

$$\widehat{F} = qu\widehat{B}. \tag{2.114}$$

We have seen here how a magnetic force appears as a result of an electrostatic force and the special theory of relativity. The considerations above have also demonstrated that a force which is identified as electrostatic in one frame of reference is observed as a magnetic force in another frame. In other words, the electric and the magnetic force are really the same. What an observer names it depends upon his state of motion.

Exercises

2.1. *Robb's Lorentz invariant spacetime interval formula* (A. A. Robb, 1936)

Show that the spacetime interval between the emission event at the point of time t_A and the reflection event at t_B in Fig. 2.2 can be expressed as $\Delta s = c\sqrt{t_A t_C}$, where t_C is the point of time when the reflected light signal arrives back at the emitter.

2.2. *The twin paradox*

On New Year's day 2004, an astronaut (A) leaves Earth on an interstellar journey. He is travelling in a spacecraft at the speed of $v = (4/5)c$ heading towards Alpha

Centauri. This star is at a distance of 4 ly (ly = light years) measured from the reference frame of the Earth. As A reaches the star, he immediately turns around and heads home. He reaches the Earth New Year's day 2016 (in Earth's time frame). The astronaut has a brother (B), who remains on Earth during the entire journey. The brothers have agreed to send each other a greeting every New Year day with the aid of radio telescope.

(a) Show that A only sends 6 greetings (including the last day of travel), while B sends 10.
(b) Draw a Minkowski diagram where A's journey is depicted with respect to the Earth's reference frame. Include all the greetings that B is sending. Show with the aid of the diagram that while A is outbound, he only receives one greeting, while on his way home he receives nine.
(c) When does B receive signals from A?
(d) Show how the results from (b) and (c) can be deduced from the Doppler effect.

2.3. *Faster than the speed of light?*

The quasar 3C273 emits a jet of matter that moves with the speed v_0 towards Earth making an angle ϕ to the line of sight (see Fig. 2.18).

(a) Assume that two signals are sent towards the Earth simultaneously, one from A and one from B. How much earlier will the signal from B reach the Earth compared to that from A?
(b) Find an expression of the transverse distance that the emitted part has moved when it reaches B. How much time (relative to the Earth) has this part been travelling?
(c) Calculate the velocity v_0 of the light source in terms of v and θ, and find the value of v_0 if $v = 10c$ and $\theta = 10°$. How large must v_0 be in order that the observed transverse velocity shall be larger than c?

2.4. *Time dilation and Lorentz contraction*

Fig. 2.18 Light cone due to Cherenkov radiation

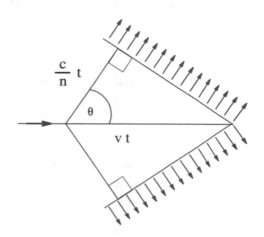

(a) At what speed does a clock move if its rate of time is 0.6 times the rate when it is at rest?

(b) A rod moves in the x-direction. An observer following the rod measures that it makes an angle $\pi/4$ with the x-axis. What is the speed of the rod if an observer at rest on the x-axis finds that it makes an angle $\pi/3$ with the x-axis, due to the Lorentz contraction of its length component in the x-direction?

2.5. Atmospheric mesons reaching the surface of the Earth

Atmospheric muons are formed when molecules in the Earth's atmosphere are hit by particles in the cosmic rays about $L_0 = 10$ km above the surface. The half-life of a muon as measured by an observer co-moving with the muon is $t_0 = 1.56 \times 10^{-6}$ s. The average velocity of the muons are $v = 0.98\,c$.

(a) According to a non-relativistic calculation, how many of ten million muons formed at 10 km height reach the Earth's surface?

(b) Taking the relativistic time dilation into account, how many will then reach the surface of the Earth?

(c) How is this explained by an observer following the muons?

2.6. Relativistic Doppler shift

The relativistic formula for the Doppler effect of an object moving along the direction of sight with a velocity v away from the observer is

$$\lambda_r = \sqrt{\frac{1 + v/c}{1 - v/c}}\lambda_e,$$

where λ_e is the frequency measured by an observer co-moving with the object, and λ_r is the frequency measured for the light received by an observer at rest. If the object moves towards the observer the signs are interchanged.

For an object moving away from an observer there is an increase in the measured wavelength—a redshift. The *redshift* of an object is defined as

$$z = \frac{\lambda_r - \lambda_e}{\lambda_e}.$$

Hence the redshift due to the Doppler effect is

$$z = \sqrt{\frac{1 + v/c}{1 - v/c}} - 1.$$

Positive value of z means redshift and negative value blueshift.

The measured value of z for the centre of our neighbour galaxy, the Andromeda Galaxy, is $z = -0.0004$.

Determine the velocity of the Andromeda Galaxy along the line of sight from this measurement. Which way does the Andromeda Galaxy move relative to the Milky Way?

2.7. *The velocity of light in a moving medium*

Light moves more slowly in a material medium than in empty space. The index of refraction, n, of the medium is defined as the ratio of the velocity of light in the medium when it is at rest, u_0, and the velocity of light in empty space. Then

$$n = \frac{u_0}{c}.$$

We now consider a medium moving with a velocity v in the same direction as the light. The speed of light in this medium relative to the laboratory frame is related to the speed of light in a frame co-moving with the water by the relativistic velocity addition law,

$$u = \frac{u_0 + v}{1 + \frac{u_0 v}{c^2}}.$$

Find the velocity of light in a moving medium in terms of its index of refraction and velocity.

2.8. *Cherenkov radiation*

When a particle moves through a medium with a velocity v greater than the velocity of light in the medium, it emits a cone of radiation with a half-angle θ given by $\cos \theta = c/nv$.

From Wikipedia: https://en.wikipedia.org/wiki/Cherenkov_radiation#/media/File:Cherenkov.svg.

(a) What is the threshold kinetic energy (in MeV) of an electron moving through water in order that it shall emit Cherenkov radiation? The index of refraction of water is $n = 1.3$. The rest energy of an electron is $m_e c^2 = 0.5.11$ MeV.
(b) What is the limiting half-angle of the cone for high-speed electrons moving through water?

2.9. *Relativistic form of Newton's 2 law*

In order to simplify the calculation we shall consider motion along the x-direction only. Let the particle have rest mass m_0 and velocity v. Its momentum is $p = \gamma m_0 v$, where $\gamma = \left(1 - v^2/c^2\right)^{-1/2}$.

The relativistic form of Newton's 2 law is $F = dp/dt$, where F is the force acting on the particle, and t is the coordinate time.

Calculate the form of this law as expressed in terms of v and dv/dt.

2.10. *Lorentz transformation of electric and magnetic fields*

It follows from the Lorentz transformations (2.27), (2.28) and (2.30) that the partial derivatives transform as

$$\frac{\partial}{\partial t} = \gamma\left(\frac{\partial}{\partial t'} + v\frac{\partial}{\partial x'}\right), \quad \frac{\partial}{\partial x} = \gamma\left(\frac{\partial}{\partial x'} + \frac{v}{c^2}\frac{\partial}{\partial t'}\right), \quad \frac{\partial}{\partial y} = \frac{\partial}{\partial y'}, \quad \frac{\partial}{\partial z} = \frac{\partial}{\partial z'}.$$

Deduce the transformation equations for electric and magnetic fields by using the transformation equations for the partial derivatives together with the requirement that Maxwell's equations shall be Lorentz invariant.

Reference

1. Frisch, D.H., Smith, J.H.: Measurement of the relativistic time dilation using μ-mesons. Am. J. Phys. **31**, 342–355 (1963)

Chapter 3
Vectors, Tensors and Forms

Abstract In this chapter we develop the main mathematical concepts used in this book. First vectors, not only as quantities with length and direction, but as differential operators. Then tensors of arbitrary rank are introduced. As a preparation for using Cartan's formalism we introduce forms, i.e. antisymmetric covariant tensors. This antisymmetric tensor formalism is most effective when we introduce an orthonormal basis field. Using an orthonormal basis co-moving with an observer, i.e. where the time-like vector is equal to the 4-velocity of the observer, simplifies the physical interpretation of the calculations.

3.1 Vectors

An expression of the form $a^\mu \vec{e}_\mu$, where a^μ, $\mu = 1, 2, ..., n$ are real numbers, is known as a *linear combination* of the vectors $\vec{e}_\mu$.

The vectors $\vec{e}_1, ..., \vec{e}_n$ are said to be *linearly independent* if there does not exist real numbers $a^\mu \neq 0$ such that $a^\mu \vec{e}_\mu = 0$ (Fig. 3.1).

This has a geometrical interpretation: A set of vectors are linearly independent if it is not possible to construct a closed polygon of the vectors (even by adjusting their lengths).

A set of vectors $\vec{e}_1, ..., \vec{e}_n$ are said to be *maximally linearly independent* if $\vec{e}_1, ..., \vec{e}_n, \vec{v}$ are linearly dependent for all vectors $\vec{v} \neq \vec{e}_\mu$, i.e. if there exist real numbers a^μ such that $\vec{v} + a^\mu \vec{e}_\mu = 0$. Hence the vector $\vec{v}$ may be written as a linear combination of the vectors $\vec{e}_\mu$,

$$\vec{v} = v^\mu \vec{e}_\mu. \tag{3.1}$$

We define the *dimension* of a vector-space as the number of vectors in a maximally linearly independent set of vectors of the space. The vectors $\vec{e}_\mu$ in such a set are known as the *basis vectors* of the space. The numbers v^μ are called the *components of the vector $\vec{v}$ in this basis.*

© Springer Nature Switzerland AG 2020
Ø. Grøn, *Introduction to Einstein's Theory of Relativity*,
Undergraduate Texts in Physics, https://doi.org/10.1007/978-3-030-43862-3_3

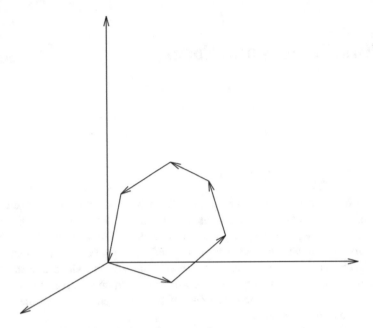

Fig. 3.1 Closed polygon (linearly dependent vectors)

3.1.1 Four-Vectors

Four-vectors (also denoted 4-vectors) are vectors which exist in (4-dimensional) spacetime. A 4-vector equation represents four independent component equations.

As a first example of a 4-vector we shall consider the 4-velocity, i.e. the velocity of a particle through spacetime.

Definition 3.1.1 (*Four-velocity*) The *four-velocity* of a particle is a 4-vector with components equal to the derivatives of the coordinates with respect to the proper time τ of the particle,

$$\vec{u} = u^{\mu}\vec{e}_{\mu} = \frac{\mathrm{d}x^{\mu}}{\mathrm{d}\tau}\vec{e}_{\mu}. \tag{3.2}$$

One often writes the 4-velocity in the following way

$$\vec{u} = \left(u^{0}, u^{1}, u^{2}, u^{3}\right), \tag{3.3}$$

where u^{0} is the time component of the 4-velocity, and $u^{i}, i = 1, 2, 3$ are the spatial components.

The proper time of the particle is the time measured by a standard clock carried by the particle. From Eq. (2.45) we have

$$\frac{dt}{d\tau} = \frac{1}{\sqrt{1 - v^2/c^2}} = \gamma, \tag{3.4}$$

where v is the ordinary velocity through 3-dimensional space. Note that v is a relative velocity which can be transformed away by referring to the rest frame of the particle. On the other hand the 4-velocity, which is a vector in spacetime, cannot be transformed away. In the rest frame of the particle it still has a non-vanishing time component.

With reference to a Cartesian coordinate system with coordinates $\left(x^0 = ct,\ x^1 = x,\ x^2 = y,\ x^3 = z\right)$ the 4-velocity has the component form

$$\vec{u} = c\frac{dt}{d\tau}\vec{e}_t + \frac{dx}{d\tau}\vec{e}_x + \frac{dy}{d\tau}\vec{e}_y + \frac{dz}{d\tau}\vec{e}_z. \tag{3.5}$$

This may be written

$$\vec{u} = \frac{dt}{d\tau}\left(c\vec{e}_t + \frac{dx}{dt}\vec{e}_x + \frac{dy}{dt}\vec{e}_y + \frac{dz}{dt}\vec{e}_z\right). \tag{3.6}$$

The ordinary velocity through 3-space is

$$\vec{v} = \frac{dx}{dt}\vec{e}_x + \frac{dy}{dt}\vec{e}_y + \frac{dz}{dt}\vec{e}_z. \tag{3.7}$$

Using Eqs. (3.4) and (3.7) the 4-velocity can be written

$$\vec{u} = \gamma(c\vec{e}_t + \vec{v}), \tag{3.8}$$

or

$$\vec{u} = \gamma\left(c, v_x, v_y, v_z\right). \tag{3.9}$$

In a co-moving reference frame where the particle is at rest, the 3-velocity vanishes and $\gamma = 1$. In this frame the four-velocity is

$$\vec{u} = c\vec{e}_t. \tag{3.10}$$

In its rest frame the particle moves only in the time direction.

Definition 3.1.2 (*Four-momentum*) The *four-momentum* of a particle is equal to the rest mass of the particle times its four-velocity,

$$\vec{P} = m_0\vec{u}, \tag{3.11}$$

where m_0 is the rest mass of the particle. The components of the 4-momentum are

$$\vec{P} = (E/c, \ \vec{p}), \quad \vec{p} = \gamma m_0 \vec{v} = m\vec{v}, \tag{3.12}$$

where E is the relativistic energy of the particle.

Definition 3.1.3 (*Minkowski force*) The *Minkowski force* acting on a particle is equal to the derivative of the 4-momentum with respect to the proper time of the particle,

$$\vec{F} = d\vec{P}/d\tau. \tag{3.13}$$

Its components are

$$\vec{F} = \gamma \left(\frac{1}{c} \vec{f} \cdot \vec{v}, \ \vec{f} \right), \tag{3.14}$$

where

$$\vec{f} = \frac{d\vec{p}}{dt} \tag{3.15}$$

is the ordinary force.

Definition 3.1.4 (*Four-acceleration*) The *four-acceleration* of a particle is equal to the derivative of its four-velocity with respect to its proper time,

$$\vec{A} = \frac{d\vec{U}}{d\tau}. \tag{3.16}$$

   In the general theory of relativity gravitation is not considered a force. A particle in free fall is in Newtonian gravitational theory said to be influenced by a gravitational force. According to the general theory of relativity the particle is not influenced by any force. Such a particle has no 4-acceleration. *The equation $\vec{A} \neq 0$ implies that the particle is not in free fall.* It is then influenced by non-gravitational forces. It is important to distinguish between the ordinary 3-acceleration, which represents the acceleration of a particle through 3-space relative to an observer, and the 4-acceleration which represents deviation from free fall. In the context of the general theory of relativity the words "a non-accelerating particle" usually mean "a particle which is in free fall".

3.1.2 Tangent Vector Fields and Coordinate Vectors

In a curved space position vectors with finite length do not exist (see Fig. 3.2). Different points in a curved space have different tangent planes. Finite vectors do only exist in these tangent planes (see Fig. 3.3). However, infinitesimal position vectors $d\vec{r}$ do exist.

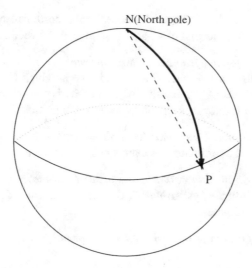

Fig. 3.2 No finite position vector in curved space. Vectors can only exist in tangent planes. The vectors in the tangent plane of N do not contain the vector NP (dashed line)

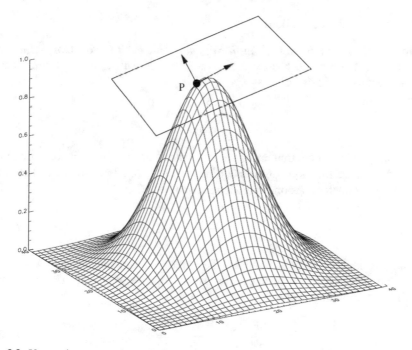

Fig. 3.3 Vectors in tangent planes. In curved space vectors can only exist in tangent planes. The figure shows the tangent plane of a point P in a curved surface

We shall now define the concepts reference frame, coordinate system, co-moving coordinate system, orthonormal basis and coordinate basis vector.

Definition 3.1.5 (*Reference frame*) A *reference frame* is a set of particles with specified motion. An *inertial reference frame* is a frame in which Newton's first law is valid. In the general theory of relativity, where gravity is not a force, this means that an inertial reference frame is a non-rotating set of free particles.

Definition 3.1.6 (*Coordinate system*) A *coordinate system* is a continuum of 4-tuples giving a unique set of coordinates for events in spacetime.

Definition 3.1.7 (*Co-moving coordinate system*) A *co-moving coordinate system* in a frame is a coordinate system where the particles in the reference frame have constant spatial coordinates.

Definition 3.1.8 (*Orthonormal basis*) An *orthonormal basis* $\{\vec{e}_{\hat{\mu}}\}$ in spacetime is defined by

$$\vec{e}_{\hat{t}} \cdot \vec{e}_{\hat{t}} = -1(c = 1), \quad \vec{e}_{\hat{i}} \cdot \vec{e}_{\hat{j}} = \delta_{\hat{i}\,\hat{j}}, \tag{3.17}$$

where $\hat{i}$ and $\hat{j}$ are space indices.

Definition 3.1.9 (*Preliminary definition of coordinate basis vector*) Our preliminary definition is: A coordinate basis vector is the partial derivative of the position vector with respect to a coordinate,

$$\vec{e}_{\mu} = \frac{\partial \vec{r}}{\partial x^{\mu}}. \tag{3.18}$$

A *vector field* is a continuum of vectors, where the components are continuous and differentiable functions of the coordinates. Let $\vec{u}$ be a tangent vector field to a curve with parameter λ (coordinate along the curve). Then

$$\vec{u} = \frac{d\vec{r}}{d\lambda}. \tag{3.19}$$

The position vector of a point on the curve is a function of the coordinates which are again functions of the curve parameter, $\vec{r} = \vec{r}[x^{\mu}(\lambda)]$. The chain rule for differentiation then yields

$$\vec{u} = \frac{\partial \vec{r}}{\partial x^{\mu}} \frac{dx^{\mu}}{d\lambda} = \frac{dx^{\mu}}{d\lambda} \vec{e}_{\mu} = u^{\mu} \vec{e}_{\mu}. \tag{3.20}$$

Thus, the components of the tangent vector field along a curve parameterized by λ are given by

$$u^{\mu} = \frac{dx^{\mu}}{d\lambda}. \tag{3.21}$$

In the theory of relativity the invariant parameter of the world line of a particle is usually chosen to be the proper time of the particle,

$$u^\mu = \frac{dx^\mu}{d\tau}. \tag{3.22}$$

This means that the tangent vector field of the world line of a particle is made up of the four-velocity of the particle.

3.1.3 Coordinate Transformations

Let there be given two coordinate systems, $\{x^\mu\}$ and $\{x^{\mu'}\}$ covering a region of spacetime with basis vectors

$$\vec{e}_\mu = \frac{\partial \vec{r}}{\partial x^\mu}, \quad \vec{e}_{\mu'} = \frac{\partial \vec{r}}{\partial x^{\mu'}}. \tag{3.23}$$

Suppose there exists a coordinate transformation such that the primed coordinates a functions of the unprimed, $x^{\mu'}(x^\mu)$, and an inverse transformation such that the unprimed are functions of the primed, $x^\mu(x^{\mu'})$. Applying the chain rule of differentiation we then obtain

$$\vec{e}_{\mu'} = \frac{\partial \vec{r}}{\partial x^{\mu'}} = \frac{\partial \vec{r}}{\partial x^\mu} \frac{\partial x^\mu}{\partial x^{\mu'}} = \vec{e}_\mu \frac{\partial x^\mu}{\partial x^{\mu'}}. \tag{3.24}$$

This is the transformation equation for the basis vectors. The quantities $\partial x^\mu/\partial x^{\mu'}$ are the elements of the transformation matrix. Indices that are not summation indices are called free indices. We have the following rule: In all terms on each side of an equation the free indices should appear in the same way (as an upper index or a lower index).

Applying this rule we can now find the inverse transformation

$$\vec{e}_\mu = \vec{e}_{\mu'} \frac{\partial x^{\eta'}}{\partial x^\mu}. \tag{3.25}$$

A vector itself is invariant. Only its components and the basis vectors transform. Hence,

$$\vec{v} = v^\mu \vec{e}_\mu = v^{\mu'} \vec{e}_{\mu'} = v^{\mu'} \vec{e}_\mu \frac{\partial x^\mu}{\partial x^{\mu'}}. \tag{3.26}$$

Thus, the components of a vector transform as follows

$$v^{\mu} = v^{\mu'}\frac{\partial x^{\mu}}{\partial x^{\mu'}}, \quad v^{\mu'} = v^{\mu}\frac{\partial x^{\mu'}}{\partial x^{\mu}}. \tag{3.27}$$

The *directional derivative* along a curve parameterized by λ is

$$\frac{d}{d\lambda} = \frac{\partial}{\partial x^{\mu}}\frac{dx^{\mu}}{d\lambda} = v^{\mu}\frac{\partial}{\partial x^{\mu}}, \tag{3.28}$$

where $v^{\mu} = dx^{\mu}/d\lambda$ are the components of the tangent vector of the curve. The directional derivative along an x^{ν}-axis is found by inserting $\lambda = x^{\nu}$. This gives

$$\frac{\partial}{\partial x^{\mu}}\frac{\partial x^{\mu}}{\partial x^{\nu}} = \delta^{\mu}_{\nu}\frac{\partial}{\partial x^{\mu}} = \frac{\partial}{\partial x^{\nu}}. \tag{3.29}$$

Hence, the directional derivative along a coordinate axis is equal to the partial derivative with respect to the coordinate of the axis. From the chain rule of differentiation we get in the primed coordinate system

$$\frac{\partial}{\partial x^{\mu'}} = \frac{\partial x^{\mu}}{\partial x^{\mu'}}\frac{\partial}{\partial x^{\mu}}. \tag{3.30}$$

Comparing with Eq. (3.18) shows that the partial derivatives transform in the same way as the basis vectors.

A weakness of the preliminary definition (3.18) of basis vectors is that it involves the position vectors that are not defined in a curved space, only in the tangent plane of the space. We would therefore like to have a general definition of basis vectors not involving the position vector. The transformation (3.30) of the partial derivatives motivates to the following definition.

Definition 3.1.10 (*General definition of coordinate basis vectors*) We define the *coordinate basis vectors* as partial derivatives,

$$\vec{e}_{\mu} = \frac{\partial}{\partial x^{\mu}}. \tag{3.31}$$

This is the general definition of coordinate basis vectors. It applies in curved as well as in flat spaces.

Since an arbitrary vector can be written as a linear combination of basis vectors, we now have a new way of thinking about vectors. From now on *we can think of a vector as a differential operator*, not only a quantity with magnitude and direction, i.e. an arrow.

Example 3.1.1 (*Transformation between Cartesian- and plane polar coordinates*) The basis vectors of a Cartesian coordinate system and those of plane polar coordinates are shown in Fig. 3.4. We see from the figure that the coordinates are related by the transformation

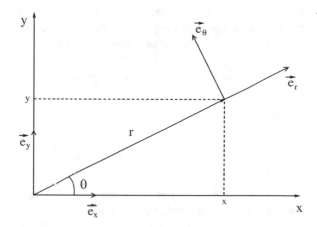

Fig. 3.4 Basis vectors in Cartesian- and plane polar coordinates

$$x = r\cos\theta, \quad y = r\sin\theta \tag{3.32}$$

The transformation of the basis vectors is found from the chain rule for differentiation

$$\vec{e}_r = \frac{\partial}{\partial r} = \frac{\partial x}{\partial r}\frac{\partial}{\partial x} + \frac{\partial y}{\partial r}\frac{\partial}{\partial y} = \cos\theta\vec{e}_x + \sin\theta\vec{e}_y,$$

$$\vec{e}_\theta = \frac{\partial}{\partial\theta} = \frac{\partial x}{\partial\theta}\frac{\partial}{\partial x} + \frac{\partial y}{\partial\theta}\frac{\partial}{\partial y} = -r\sin\theta\vec{e}_x + r\cos\theta\vec{e}_y. \tag{3.33}$$

The basis vectors in the Cartesian coordinate system are unit vectors. The magnitudes of the basis vectors in the plane polar coordinate system are

$$|\vec{e}_r| = \sqrt{\vec{e}_r \cdot \vec{e}_r} = \sqrt{\cos^2\theta + \sin^2\theta} = 1,$$

$$|\vec{e}_\theta| = \sqrt{\vec{e}_\theta \cdot \vec{e}_\theta} = \sqrt{r^2\cos^2\theta + r^2\sin^2\theta} = r. \tag{3.34}$$

This shows that *the coordinate basis vectors need not be unit vectors.*

Definition 3.1.11 (*Orthonormal basis*) An *orthonormal basis* is a vector basis consisting of unit vectors that are normal to each other.

We shall denote orthonormal basis by writing a hat over the indices. Note that an orthonormal basis will not in general be a coordinate basis, i.e. the basis vectors of an orthonormal basis are not just partial derivatives.

The orthonormal basis of the plane polar coordinate system is

$$\vec{e}_{\hat{r}} = \vec{e}_r, \quad \vec{e}_{\hat{\theta}} = \frac{1}{r}\vec{e}_\theta. \tag{3.35}$$

Example 3.1.2 (*Relativistic Doppler effect*) Consider a photon with energy E and momentum p in the rest frame of the emitter and the observer. The energy of the photon as observed in a frame moving with velocity v away from the observer is found from the transformation Eq. (2.85). Inserting the energy $E = h\nu$ and momentum $p = h\nu/c$ in the rest frame of the emitter gives

$$\nu' = \gamma\left(\nu - \frac{v}{c}\nu\right), \tag{3.36}$$

Giving

$$\nu' = \sqrt{\frac{c - v}{c + v}}\,\nu. \tag{3.37}$$

This change of observed frequency depending upon the observed velocity of the emitter is often called the *Doppler shift*.

3.1.4   Structure Coefficients

Let us first define a mathematical quantity which is essential in connection with the structure coefficients.

Definition 3.1.12 (*Commutator of vectors*) The *commutator* of two vectors is

$$[\vec{u}, \vec{v}] = \vec{u}\vec{v} - \vec{v}\vec{u}, \tag{3.38}$$

where $\vec{u}\vec{v}$ is defined as

$$\vec{u}\vec{v} = u^{\mu}\vec{e}_{\mu}(v^{\nu}\vec{e}_{\nu}). \tag{3.39}$$

Hence, $\vec{u}\vec{v}$ means that $\vec{u}$ acts upon $\vec{v}$ as a differential operator. The commutator of two vectors is itself a vector.

Definition 3.1.13 (*Structure coefficients*) The *structure coefficients* $c^{\rho}_{\mu\nu}$ are the components of the commutators of the basis vectors,

$$[\vec{e}_{\mu}, \vec{e}_{\nu}] = c^{\rho}_{\mu\nu}\vec{e}_{\rho}. \tag{3.40}$$

It follows that the structure coefficients are antisymmetric in their lower indices,

$$c^{\rho}_{\nu\mu} = -c^{\rho}_{\mu\nu}. \tag{3.41}$$

Calculating the structure coefficients in a coordinate basis we get

$$\left[\frac{\partial}{\partial x^{\mu}}, \frac{\partial}{\partial x^{\nu}}\right] = \frac{\partial^2}{\partial x^{\mu}\partial x^{\nu}} - \frac{\partial^2}{\partial x^{\nu}\partial x^{\mu}} = 0. \tag{3.42}$$

Hence *the structure coefficients vanish in coordinate basis.*

Example 3.1.3 (*Structure coefficients in plane polar coordinates*) We shall calculate the structure coefficients of an orthonormal basis in a system of plane polar coordinates. Using Eq. (3.35) we have

$$
\left[\vec{e}_{\hat{r}}, \vec{e}_{\hat{\theta}}\right] = \left[\frac{\partial}{\partial r}, \frac{1}{r}\frac{\partial}{\partial \theta}\right] = \frac{\partial}{\partial r}\left(\frac{1}{r}\frac{\partial}{\partial \theta}\right) - \frac{1}{r}\frac{\partial}{\partial \theta}
$$

$$
= -\frac{1}{r^2}\frac{\partial}{\partial \theta} + \frac{1}{r}\frac{\partial^2}{\partial r \partial \theta} - \frac{1}{r}\frac{\partial^2}{\partial \theta \partial r} = -\frac{1}{r^2}\vec{e}_\theta = -\frac{1}{r}\vec{e}_{\hat{\theta}}. \tag{3.43}
$$

Since

$$
\left[\vec{e}_{\hat{r}}, \vec{e}_{\hat{\theta}}\right] = c^{\hat{\rho}}_{\hat{r}\hat{\theta}}\vec{e}_{\hat{\rho}}, \tag{3.44}
$$

and using the antisymmetry (3.41), we get the only non-vanishing structure coefficients in the orthonormal basis field of a plane polar coordinate system

$$
c^{\hat{\theta}}_{\hat{r}\hat{\theta}} = -c^{\hat{\theta}}_{\hat{\theta}\hat{r}} = -\frac{1}{r}. \tag{3.45}
$$

3.2 Tensors

Definition 3.2.1 (*One-form basis*) A *one-form basis* $\{\underline{\omega}^1, \ldots, \underline{\omega}^n\}$ is defined by letting the basis forms act upon the basis vectors in an operation called *contraction*, according to the rule

$$
\underline{\omega}^\mu(\vec{e}_\nu) - \delta^\mu_\nu. \tag{3.46}
$$

An arbitrary one-form can be expressed as a linear combination of the basis forms

$$
\underline{\alpha} = \alpha_\mu \underline{\omega}^\mu, \tag{3.47}
$$

where α_μ are the components of $\underline{\alpha}$ is the given basis. Using Eqs. (3.46) and (3.47) we find

$$
\underline{\alpha}(\vec{e}_\nu) = \alpha_\mu \underline{\omega}^\mu(\vec{e}_\nu) = \alpha_\mu \delta^\mu_\nu = \alpha_\nu. \tag{3.48}
$$

This means that the contraction of a one form with a basis vector gives the component of the one-form corresponding to the basis vector. The contraction of a one-form $\underline{\alpha}$ with a vector $\vec{v}$ gives

$$\underline{\alpha}(\vec{v}) = \underline{\alpha}\big(v^{\mu}\vec{e}_{\mu}\big) = v^{\mu}\underline{\alpha}\big(\vec{e}_{\mu}\big) = v^{\mu}\alpha_{\mu}. \tag{3.49}$$

This shows that the contraction of a one-form with a vector corresponds to the scalar product of two vectors.

We shall now need the concept multilinear function which has the following meaning. A *multilinear function* is a function which is linear in all its arguments. Then we are ready to define tensors.

Definition 3.2.2 (*Tensors*) A *tensor* is a multilinear function which maps one-forms and vectors into real numbers.

We have three different types of tensors:

- A *covariant tensor* maps vectors
- A *contravariant tensor* maps forms
- A *mixed tensor* maps tensors and forms.

A tensor of *rank* $\begin{pmatrix} N \\ N' \end{pmatrix}$ maps N one-forms and N' vectors into real numbers. It is, however, usual to say that the tensor has a rank $N + N'$. A tensor of rank zero is called a scalar quantity and may be given in terms of real numbers. A vector is a contravariant tensor of rank one, and a one-form is a covariant tensor of rank one.

Definition 3.2.3 (*Tensor product*) Let T and S be two covariant tensors of rank m and n. The *tensor product* of T and S is

$$T \otimes S(\vec{u}_1, \ldots, \vec{u}_m, \vec{v}_1, \ldots, \vec{v}_n) = T(\vec{u}_1, \ldots, \vec{u}_m)S(\vec{v}_1, \ldots, \vec{v}_n). \tag{3.50}$$

$T \otimes S$ is a tensor of rank $m + n$. Let $R = T \otimes S$. Then we have

$$R = R_{\mu_1 \ldots \mu_q}\underline{\omega}^{\mu_1} \otimes \underline{\omega}^{\mu_2} \otimes \cdots \otimes \underline{\omega}^{\mu_q}. \tag{3.51}$$

Notice that $S \otimes T \neq T \otimes S$. The components of a tensor are found by contracting the tensor with the basis elements, i.e. basis vectors, in the case of a covariant tensor,

$$R_{\mu_1 \ldots \mu_q} = R\big(\vec{e}_{\mu_1}, \ldots, \vec{e}_{\mu_q}\big). \tag{3.52}$$

The indices of the components of a covariant tensor are written as lower indices as in Eq. (4.52), and the indices of the components of a contravariant tensor as upper indices.

Example 3.2.1 (*Tensor product of two vectors*) Let $\vec{u}$ and $\vec{v}$ be two vectors and $\underline{\alpha}$ and $\underline{\beta}$ two one-forms

$$\vec{u} = u^{\mu}\vec{e}_{\mu}, \quad \vec{v} = v^{\mu}\vec{e}_{\mu}, \quad \underline{\alpha} = \alpha_{\mu}\underline{\omega}^{\mu}, \quad \underline{\beta} = \beta_{\mu}\underline{\omega}^{\mu}. \tag{3.53}$$

From these we can construct a contravariant tensor of rank 2 by $R = \vec{u} \otimes \vec{v}$ with components

$$R^{\mu_1 \mu_2} = R(\underline{\omega}^{\mu_1}, \underline{\omega}^{\mu_2}) = \vec{u} \otimes \vec{v}(\underline{\omega}^{\mu_1}, \underline{\omega}^{\mu_2}) = \vec{u}(\underline{\omega}^{\mu_1})\vec{v}(\underline{\omega}^{\mu_2}) = u^{\mu_1} v^{\mu_2}. \quad (3.54)$$

3.2.1 Transformation of Tensor Components

We shall not limit the discussion to coordinate transformations but will consider transformations between any type of bases, $\{\vec{e}_\mu\} \rightarrow \{\vec{e}_{\mu'}\}$. The elements of the transformation matrices are denoted by $M^\mu_{\mu'}$. Then the transformations of the basis vectors are written

$$\vec{e}_{\mu'} = \vec{e}_\mu M^\mu_{\mu'}, \quad \vec{e}_\mu = \vec{e}_{\mu'} M^{\mu'}_\mu. \quad (3.55)$$

It follows that

$$M^\mu_{\mu'} M^{\mu'}_\nu = \delta^\mu_\nu. \quad (3.56)$$

For a coordinate transformation the elements of the transformation matrix are

$$M^{\mu'}_\mu = \frac{\partial x^{\mu'}}{\partial x^\mu}. \quad (3.57)$$

3.2.2 Transformation of Basis One-Forms

The basis 1-forms transform inversely relatively to the basis vectors

$$\underline{\omega}^{\mu'} = M^{\mu'}_\mu \underline{\omega}^\mu, \quad \underline{\omega}^\mu = M^\mu_{\mu'} \underline{\omega}^{\mu'}. \quad (3.58)$$

The components of a tensor of higher rank transform such that every contravariant index (upper) transforms as a basis 1-form and every covariant index (lower) as a basis vector. Also, all elements of the transformation matrix are multiplied with one another.

Example 3.2.2 (*A mixed tensor of rank 3*) Tensor components transform homogeneously, which means that the transformed tensor components are linear combinations of the original ones. The components of a mixed tensor of rank 3, for example, transform as follows.

$$T^{\alpha'}_{\mu' \nu'} = M^{\alpha'}_\alpha M^\mu_{\mu'} M^\nu_{\nu'} T^\alpha_{\mu\nu}. \quad (3.59)$$

Tensor transformation of components means that tensors have a basis-*independent* existence. That is, if a tensor has non-vanishing components in a given basis then it has non-vanishing components in all bases. This means that tensor equations have a basis-independent form. *Tensor equations are invariant.* A basis transformation might result in the vanishing of one or more tensor components. *Equations written in component form will differ from one basis to another.* An equation expressed in terms of tensor components can be transformed from one basis to another using the tensor component transformation rules. An equation that is expressed only in terms of tensor components is said to be a *covariant equation*. An equation expressed in terms of the tensors themselves has the same form in every coordinate system and is said to be an *invariant equation*.

3.2.3   The Metric Tensor

Definition 3.2.4 (*The scalar product*) The *scalar product* of two vectors $\vec{u}$ and $\vec{v}$ is denoted by $g(\vec{u}, \vec{v})$ and is defined as a symmetric linear mapping which for each pair of vectors gives a scalar, i.e. a number. The symmetry means that $g(\vec{v}, \vec{u}) = g(\vec{u}, \vec{v})$.

Definition 3.2.5 (*The metric tensor*) The *metric tensor*, $\mathbf{g} = g_{\mu\nu}\underline{\omega}^\mu \otimes \underline{\omega}^\nu$, is a covariant symmetric tensor of rank 2 with components made up of the scalar products of the basis vectors,

$$g_{\mu\nu} = g\left(\vec{e}_\mu, \vec{e}_\nu\right) = \vec{e}_\mu \cdot \vec{e}_\nu. \tag{3.60}$$

The values of the *scalar products* $g\left(\vec{e}_\mu, \vec{e}_\nu\right)$ are given by specifying the scalar products of each pair of basis vectors in a basis. The symmetry of the metric tensor means that its components obey

$$g_{\nu\mu} = g_{\mu\nu}. \tag{3.61}$$

The components of the metric tensor are often called *the metric* and are written as a 2 × 2 matrix. Those with equal indices are on the diagonal of this matrix. If they are the only non-vanishing indices, the metric is said to be diagonal. It follows from Eq. (3.60) that *the metric is diagonal if the basis vectors are orthogonal to each other.*

The scalar product of two arbitrary vectors $\vec{u}$ and $\vec{v}$ is

$$\vec{u} \cdot \vec{v} = g(\vec{u}, \vec{v}) = g(u^\mu \vec{e}, v^\nu \vec{e}_\nu) = u^\mu v^\nu g\left(\vec{e}_\mu, \vec{e}_\nu\right) = u^\mu v^\nu g_{\mu\nu}. \tag{3.62}$$

The usual notation is

$$\vec{u} \cdot \vec{v} = g_{\mu\nu}u^\mu v^\nu. \tag{3.63}$$

The absolute value of a vector is

$$|\vec{v}| = \sqrt{g(\vec{v}, \vec{v})} = \sqrt{\left|g_{\mu\nu} v^{\mu} v^{\nu}\right|}. \tag{3.64}$$

The scalar product of two vectors is an *invariant*, meaning that is has the same value in every coordinate system.

Example 3.2.3 (*Cartesian coordinates in a plane*) In the present case the scalar products of the basis vectors are

$$\vec{e}_x \cdot \vec{e}_x = \vec{e}_y \cdot \vec{e}_y = 1, \quad \vec{e}_x \cdot \vec{e}_y = \vec{e}_y \cdot \vec{e}_x = 0. \tag{3.65}$$

Hence

$$g_{xx} = g_{yy} = 1, \quad g_{xy} = g_{yx} = 0. \tag{3.66}$$

This is often written in matrix form

$$g_{\mu\nu} = \begin{pmatrix} 1 & 0 \\ 0 & 1 \end{pmatrix}. \tag{3.67}$$

Example 3.2.4 (*Plane polar coordinates*) For this coordinate system the scalar product of the basis vectors is

$$\vec{e}_r \cdot \vec{e}_r = 1, \quad \vec{e}_\theta \cdot \vec{e}_\theta = r^2, \quad \vec{e}_r \cdot \vec{e}_\theta = 0. \tag{3.68}$$

Hence, in plane polar coordinates the metric tensor has the components

$$g_{\mu\nu} = \begin{pmatrix} 1 & 0 \\ 0 & r^2 \end{pmatrix}. \tag{3.69}$$

Example 3.2.5 (*Non-orthogonal basis vectors*) We consider a skew angled coordinate system where the basis vectors make an angle θ with each other as shown in Fig. 3.5.

For this coordinate system the scalar products of the basis vectors are

Fig. 3.5 Basis vectors in a skew angled coordinate system. Basis vectors $\vec{e}_1$ and $\vec{e}_2$

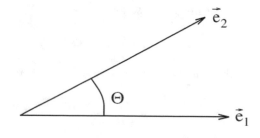

$$\vec{e}_1 \cdot \vec{e}_1 = \vec{e}_2 \cdot \vec{e}_2 = 1, \quad \vec{e}_1 \cdot \vec{e}_2 = \vec{e}_2 \cdot \vec{e}_1 = \cos\theta. \tag{3.70}$$

Hence, the metric tensor has the components

$$g_{\mu\nu} = \begin{pmatrix} 1 & \cos\theta \\ \cos\theta & 1 \end{pmatrix}. \tag{3.71}$$

Definition 3.2.6 (*Contravariant components of the metric tensor*) The contravariant components of the metric tensor, $g^{\mu\nu}$, are defined by

$$g^{\mu\alpha} g_{\alpha\nu} = \delta^{\mu}_{\nu}. \tag{3.72}$$

The *contravariant components* of the metric tensor make up the inverse matrix relative to the matrix made up of the covariant components. It follows from Eq. (3.72) that *the mixed metric tensor can be thought of as a unit tensor of rank 2*,

$$\mathbf{I} = \delta^{\mu}_{\nu} \underline{\omega}^{\nu} \otimes \vec{e}_{\mu} = \underline{\omega}^{\mu} \otimes \vec{e}_{\mu}. \tag{3.73}$$

Note that $\mathbf{I}(\vec{u}) = (\underline{\omega}^{\mu} \otimes \vec{e}_{\mu})(\vec{u}) = \underline{\omega}^{\mu}(\vec{u})\vec{e}_{\mu} = u^{\mu}\vec{e}_{\mu} = \vec{u}$, i.e. the unit tensor applied to a vector gives out the vector.

It is possible to define a mapping between tensors of different type, i.e. covariant or contravariant, using the metric tensor. An example with tensors of rank one is shown in Fig. 3.6.

We can for example map a vector on a one-form

$$v_{\mu} = g(\vec{v}, \vec{e}_{\mu}) = g(v^{\alpha}\vec{e}_{\alpha}, \vec{e}_{\mu}) = v^{\alpha} g(\vec{e}_{\alpha}, \vec{e}_{\mu}) = v^{\alpha} g_{\alpha\mu} = g_{\alpha\mu} v^{\alpha}. \tag{3.74}$$

This is known as *lowering an index. Raising of an index* is made according to

$$v^{\mu} = g^{\mu\alpha} v_{\alpha}. \tag{3.75}$$

We shall now define distance in spacetime along a curve. Let the curve be parametrized by λ (proper time for a time-like curve) with a tangent vector field $\vec{v}$. The squared distance ds^2 between the points along the curve is defined as

$$ds^2 = g(\vec{v}, \vec{v})d\lambda^2. \tag{3.76}$$

Using Eq. (3.56) we get

$$ds^2 = g_{\mu\nu} v^{\mu} v^{\nu} d\lambda^2. \tag{3.77}$$

The tangent vectors have the components $v^{\mu} = dx^{\mu}/d\lambda$, which gives

$$ds^2 = g_{\mu\nu} dx^{\mu} dx^{\nu}. \tag{3.78}$$

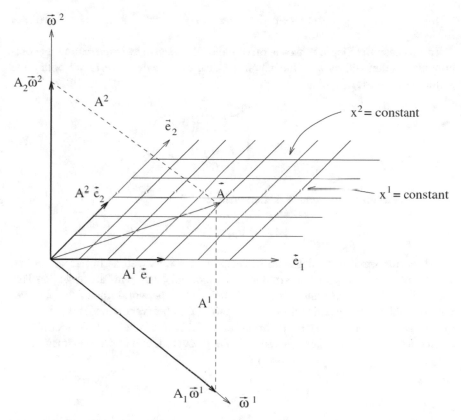

Fig. 3.6 Covariant and contravariant components of a vector

The expression (3.78) is known as the *line-element*.

Example 3.2.6 (*Line element in Cartesian coordinates*) Inserting the metric components (3.66) into Eq. (3.78) we obtain the line element of a plane in a Cartesian coordinate system

$$ds^2 = dx^2 + dy^2. \tag{3.79}$$

Example 3.2.7 (*Line element in plane polar coordinates*) With the metric components (3.61) we get

$$ds^2 = dr^2 + r^2 d\theta^2. \tag{3.80}$$

The line element in (flat) four-dimensional Minkowski spacetime with Cartesian spatial coordinates is required to be Lorentz invariant. It may be shown that the line-element then takes the form

$$ds^2 = -c^2 dt^2 + dx^2 + dy^2 + dz^2. \tag{3.81}$$

In this case the metric tensor is often called the *Minkowski metric*, and its components are denoted by $\eta_{\mu\nu}$. Introducing a time coordinate $x_0 = ct$ we obtain the Minkowski metric,

$$\eta_{\mu\nu} = \begin{pmatrix} -1 & 0 & 0 & 0 \\ 0 & 1 & 0 & 0 \\ 0 & 0 & 1 & 0 \\ 0 & 0 & 0 & 1 \end{pmatrix}. \tag{3.82}$$

The *Minkowski line-element* is then written

$$ds^2 = \eta_{\mu\nu} dx^\mu dx^\nu. \tag{3.83}$$

An orthonormal basis field can be used in curved as well as in flat spacetime. Often one uses a so-called co-moving orthonormal basis with an observer. This means that we have an orthonormal basis field along the world line of the observer with time-like vector equal to the four-velocity of the observer divided by the velocity of light, $\vec{e}_{\hat{t}} = \vec{u}/c$. In every orthonormal basis field there is Minkowski metric with components (3.82) whether spacetime is flat or curved, and the line-element takes the form

$$ds^2 = \eta_{\hat{\mu}\hat{\nu}} dx^{\hat{\mu}} dx^{\hat{\nu}}. \tag{3.84}$$

Example 3.2.8 (*The four-velocity identity*) Let us calculate the scalar product of the four-velocity by itself, using Eqs. (3.9) and (3.82),

$$\vec{u} \cdot \vec{u} = \eta_{\hat{\mu}\hat{\nu}} u^{\hat{\mu}} u^{\hat{\nu}} = -\gamma^2(c^2 - v^2) = -c^2. \tag{3.85}$$

This is called *the four-velocity identity*. In an arbitrary basis it takes the form

$$g_{\mu\nu} u^\mu u^\nu = -c^2. \tag{3.86}$$

3.3   The Causal Structure of Spacetime

The causal structure of spacetime can be illustrated by considering the light cone (Fig. 3.7).

Let us recapitulate some important points. The world lines of material particles or an observer, moving slower than light, are inside the light cone. Such curves are called *time-like*. The invariant parameter of a time-like curve is usually chosen to be

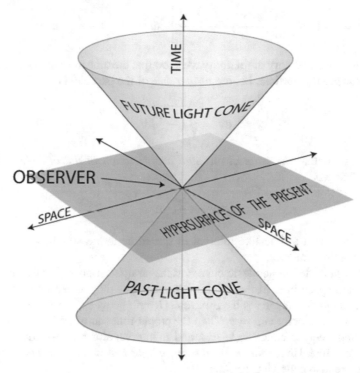

Fig. 3.7 Causal structure of spacetime From: https://en.wikipedia.org/wiki/Light_cone#/media/File:World_line.svg

the proper time τ of an observer following the curve. Then a tangent vector of the curve is the 4-velocity of the observer.

A point in spacetime represents an event. The distance in spacetime between two infinitesimally nearby points in spacetime is called an *interval*. A time-like interval is the interval between two points on a time-like curve. It has $ds^2_{\text{time}} < 0$.

We shall now give a general physical interpretation of the line element for time-like intervals. For this purpose it is sufficient to consider the Minkowski line-element which can be written

$$ds^2 = -c^2 \left\{ 1 - \frac{1}{c^2} \left[\left(\frac{dx}{dt} \right)^2 + \left(\frac{dy}{dt} \right)^2 + \left(\frac{dz}{dt} \right)^2 \right] \right\} dt^2. \tag{3.87}$$

Consider a particle moving with a coordinate velocity

$$\vec{v} = v^x \vec{e}_x + v^y \vec{e}_y + v^z \vec{e}_z = \frac{dx}{dt} \vec{e}_x + \frac{dy}{dt} \vec{e}_y + \frac{dz}{dt} \vec{e}_z. \tag{3.88}$$

Then

$$ds^2 = -\left[1 - \frac{(v^x)^2 + (v^y)^2 + (v^z)^2}{c^2}\right]c^2 dt^2 = -\left(1 - \frac{v^2}{c^2}\right)c^2 dt^2. \qquad (3.89)$$

From the special theory of relativity we know that the time measured by a standard clock following the particle, i.e. the *proper time* of the particle, is

$$d\tau = \sqrt{1 - \frac{v^2}{c^2}}\, dt. \qquad (3.90)$$

Hence we obtain *the general physical interpretation of the line-element for a time-like interval*

$$ds^2 = -c^2 d\tau^2. \qquad (3.91)$$

This means that a time-like interval in spacetime is measured by a clock, it is essentially a time interval.

We shall later define geodesic curves as the straightest possible curves between two events in spacetime. In flat Minkowski spacetime they are straight. It will also be shown that geodesic curves have extremal length between two events.

For light the velocity is $v = c$. Then the proper time vanishes. The world line of light moving freely is called *light-like*, and an interval along the world line of light is called light-like. Hence *the interval along a light-like curve vanishes*, $ds^2_{\text{light}} = 0$. It is therefore also called a zero-interval.

A *spacelike* curve represents the world line of a particle moving faster than light. The interval between two events on such a curve is called a space-like interval and has $ds^2_{\text{space}} > 0$.

3.4 Forms

Definition 3.4.1 (*Antisymmetric tensor*) An *antisymmetric tensor* is a tensor whose sign changes under an arbitrary exchange of two arguments,

$$A(\dots, \vec{u}, \dots, \vec{v}, \dots) = A(\dots, \vec{v}, \dots, \vec{u}, \dots). \qquad (3.92)$$

The components of an antisymmetric tensor changes sign under exchange of two indices,

$$A \dots_\mu \dots_\nu \dots \dots \dots = A \dots_\nu \dots_\mu \dots \qquad (3.93)$$

Definition 3.4.2 (*p-form*) A *p-form* is defined to be an antisymmetric, covariant tensor of rank p.

Definition 3.4.3 (*The wedge product*) The *wedge product* is an antisymmetric tensor product is defined by

$$\underline{\omega}^{[\mu_1} \otimes \cdots \otimes \underline{\omega}^{\mu_p]} \wedge \underline{\omega}^{[\nu_1} \otimes \cdots \otimes \underline{\omega}^{\nu_p]} = \frac{(p+q)!}{p!q!} \underline{\omega}^{[\mu_1} \otimes \cdots \otimes \underline{\omega}^{\nu_q]}, \quad (3.94)$$

where [ ] antisymmetric combination, which is defined by

$$\underline{\omega}^{[\mu_1} \otimes \cdots \otimes \underline{\omega}^{\mu_p]} = \frac{1}{p!} \left(\begin{array}{l} \text{the sum of terms with all possible permutations} \\ \text{of indices with plus for even and minus for an odd} \\ \text{number of permutations} \end{array} \right).$$

$$(3.95)$$

Example 3.4.1 (*Antisymmetric combinations*) The antisymmetric combination of a tensor product of two basis form is

$$\underline{\omega}^{[\mu_1} \otimes \underline{\omega}^{\mu_2]} = \frac{1}{2} \left(\underline{\omega}^{\mu_1} \otimes \underline{\omega}^{\mu_2} - \underline{\omega}^{\mu_2} \otimes \underline{\omega}^{\mu_1} \right). \quad (3.96)$$

The antisymmetric combination of a tensor product of three basis form is

$$\underline{\omega}^{[\mu_1} \otimes \underline{\omega}^{\mu_2} \otimes \underline{\omega}^{\mu_3]} = \frac{1}{6} (\underline{\omega}^{\mu_1} \otimes \underline{\omega}^{\mu_2} \otimes \underline{\omega}^{\mu_3} + \underline{\omega}^{\mu_3} \otimes \underline{\omega}^{\mu_1} \otimes \underline{\omega}^{\mu_2} + \underline{\omega}^{\mu_2} \otimes \underline{\omega}^{\mu_3} \otimes \underline{\omega}^{\mu_1}$$
$$- \underline{\omega}^{\mu_2} \otimes \underline{\omega}^{\mu_1} \otimes \underline{\omega}^{\mu_3} + \underline{\omega}^{\mu_3} \otimes \underline{\omega}^{\mu_2} \otimes \underline{\omega}^{\mu_1} + \underline{\omega}^{\mu_1} \otimes \underline{\omega}^{\mu_3} \otimes \underline{\omega}^{\mu_2}).$$
$$(3.97)$$

Example 3.4.2 (*A 2-form in a 3-space*) In a 3-dimensional space a 2-form may be written in component form as

$$\underline{\alpha} = \alpha_{12}\underline{\omega}^1 \otimes \underline{\omega}^2 + \alpha_{21}\underline{\omega}^2 \otimes \underline{\omega}^1 + \alpha_{13}\underline{\omega}^1 \otimes \underline{\omega}^3 + \alpha_{31}\underline{\omega}^3 \otimes \underline{\omega}^1$$
$$\alpha_{23}\underline{\omega}^2 \otimes \underline{\omega}^3. \quad (3.98)$$

The antisymmetry of $\underline{\alpha}$ means that

$$\alpha_{21} = -\alpha_{12}, \quad \alpha_{31} = -\alpha_{13}, \quad \alpha_{32} = -\alpha_{23}. \quad (3.99)$$

Hence

$$\underline{\alpha} = \alpha_{12} \left(\underline{\omega}^1 \otimes \underline{\omega}^2 - \underline{\omega}^2 \otimes \underline{\omega}^1 \right) + \alpha_{13} \left(\underline{\omega}^1 \otimes \underline{\omega}^3 - \underline{\omega}^3 \otimes \underline{\omega}^1 \right)$$
$$+ \alpha_{23} \left(\underline{\omega}^2 \otimes \underline{\omega}^3 - \underline{\omega}^3 \otimes \underline{\omega}^2 \right), \quad (3.100)$$

Using Einstein's summation convention this may be written as

$$\underline{\alpha} = \alpha_{|\mu\nu|} 2\underline{\omega}^{[\mu} \otimes \underline{\omega}^{\nu]}, \quad (3.101)$$

where $|\mu\nu|$ means summation with $\mu < \nu$, and the bracket denotes antisymmetriza-
tion. We now use the definition (3.94) of the wedge product $\wedge$ with $p = q = 1$. This
gives

$$\underline{\alpha} = \alpha_{|\mu\nu|}\underline{\omega}^{\mu} \wedge \underline{\omega}^{\nu}, \tag{3.102}$$

which is often written as

$$\underline{\alpha} = \frac{1}{2}\alpha_{\mu\nu}\underline{\omega}^{\mu} \wedge \underline{\omega}^{\nu}. \tag{3.103}$$

A tensor of rank 2 can always be split up into a symmetric and an antisymmetric
part,

$$T_{\mu\nu} = \frac{1}{2}\left(T_{\mu\nu} - T_{\nu\mu}\right) + \frac{1}{2}\left(T_{\mu\nu} + T_{\nu\mu}\right) = A_{\mu\nu} + S_{\mu\nu}, \tag{3.104}$$

where

$$A_{\mu\nu} = \frac{1}{2}\left(T_{\mu\nu} - T_{\nu\mu}\right), \quad S_{\mu\nu} = \frac{1}{2}\left(T_{\mu\nu} + T_{\nu\mu}\right). \tag{3.105}$$

We thus have

$$\begin{aligned} S_{\mu\nu}A^{\mu\nu} &= \frac{1}{4}\left(T_{\mu\nu} + T_{\nu\mu}\right)\left(T^{\mu\nu} - T^{\nu\mu}\right) \\ &= \frac{1}{4}\left(T_{\mu\nu}T^{\mu\nu} - T_{\mu\nu}T^{\nu\mu} + T_{\nu\mu}T^{\mu\nu} - T_{\mu\nu}T^{\nu\mu}\right) = 0. \end{aligned} \tag{3.106}$$

This shows that summation over the indices of a product of a symmetric and an
antisymmetric quantity vanishes. In a summation $T_{\mu\nu}A^{\mu\nu}$, where $T_{\mu\nu}$ has no symmetry, and $A_{\mu\nu}$ is antisymmetric, only the antisymmetric part of $T_{\mu\nu}$ contributes. Hence,
in the expression (3.103) only the antisymmetric combinations $\alpha_{[\mu\nu]}$ contribute to
the summation. These antisymmetric combinations are the *form-components*.

Forms are antisymmetric covariant tensors. Because of this antisymmetry a form
with two equal component indices vanishes, $\alpha_{[\mu...\mu...]} = 0$. This implies that on an n-
dimensional space all p-forms with $p > n$ vanish. Hence on a 2-dimensional surface
there exist only up to 2-forms, and in four-dimensional spacetime there exist only
up to 4-forms.

3.4.1   The Volume Form

The antisymmetric Levi-Civita symbol is defined by

$$\varepsilon_{\mu_1\ldots\mu_n} = \begin{cases} 1 & \text{if } \mu_1\ldots\mu_n \text{ is an even permutation of } 1\ldots n \\ -1 & \text{if } \mu_1\ldots\mu_n \text{ is an odd permutation of } 1\ldots n \\ 0 & \text{otherwise} \end{cases} \quad (3.107)$$

It follows that $\varepsilon_{\mu_1\ldots\mu_n}$ if two indices are equal.

The determinant of an $n \times n$-matrix A with elements $A^{\mu\nu}$ may be written

$$A = \det(A) = \varepsilon_{\mu_1\ldots\mu_n} A^{1\mu_1} A^{2\mu_2} \ldots A^{n\mu_n}. \quad (3.108)$$

For example for $n = 2$ this equation gives

$$A = \varepsilon_{\mu_1\eta_2} A^{1\mu_1} A^{2\mu_2} = \varepsilon_{12} A^{11} A^{22} + \varepsilon_{21} A^{12} A^{21} = A^{11} A^{22} - A^{12} A^{21}. \quad (3.109)$$

We shall now consider an n-dimensional space with a metric tensor having components $g_{\mu\nu}$. Let $\{\underline{\omega}^{\hat{\mu}}\}$ be an orthonormal form basis. The volume form $\underline{V}$ is defined by

$$\underline{V} = \underline{\omega}^{\hat{1}} \wedge \cdots \wedge \underline{\omega}^{\hat{n}}. \quad (3.110)$$

Let $M^{\mu}_{\hat{\mu}}$ be the elements of the transformation matrix to an arbitrary basis $\underline{\omega}^{\mu} = M^{\mu}_{\hat{\mu}} \underline{\omega}^{\hat{\mu}}$. Then

$$\underline{V} = M^{\hat{1}}_{\mu_1} \ldots M^{\hat{n}}_{\mu_n} \underline{\omega}^{\mu_1} \wedge \cdots \wedge \underline{\omega}^{\mu_n} = M^{\hat{1}}_{\mu_1} \ldots M^{\hat{n}}_{\mu_n} \varepsilon^{\mu_1\ldots\mu_n} \underline{\omega}^{1} \wedge \cdots \wedge \underline{\omega}^{n}$$
$$= M\underline{\omega}^{1} \wedge \cdots \wedge \underline{\omega}^{n} \quad (3.111)$$

where M is the determinant of the transformation matrix.

It follows from the general Eq. (3.59) for the components of a tensor that the transformation from the components of the metric tensor in an orthonormal basis to the components in an arbitrary basis is

$$g_{\mu\nu} = M^{\hat{\mu}}_{\mu} M^{\hat{\nu}}_{\nu} g_{\hat{\mu}\hat{\nu}}. \quad (3.112)$$

Since the determinant of a matrix is equal to the determinant of the transposed matrix (rows and columns interchanged), it follows from Eq. (3.113) that the corresponding transformation of the determinant made up of the components of the metric tensor is

$$g = M^2 \hat{g}. \quad (3.113)$$

where $\hat{g}$ is the determinant made by the components of the metric tensor in an orthonormal basis. In usual 3-space $\hat{g} = 1$, and in 4-dimensional spacetime $\hat{g} = -1$. Inserting the positive square root of (3.113) into Eq. (3.111) gives

$$\underline{V} = \sqrt{|g|}\underline{\omega}^1 \wedge \cdots \wedge \underline{\omega}^n. \tag{3.114}$$

where $|g|$ is the absolute value of the determinant made by the components of the metric tensor. The tensor components of the volume form are

$$V_{\mu_1...\mu_n} = \sqrt{|g|}\varepsilon_{\mu_1...\mu_n}. \tag{3.115}$$

The volume form represents an invariant volume-element. The corresponding invariant distance in the μ-direction is

$$L_\mu = \sqrt{|g_{\mu\mu}|}\underline{\omega}^\mu. \tag{3.116}$$

3.4.2 Dual Forms

The *dual* of a p-form $\underline{\alpha}$ in an n-dimensional space is denoted by $\star\underline{\alpha}$, where $\star$ is called Hodge's star operator, and is defined by

$$\star\underline{\alpha} = \frac{1}{p!(n-p)!} V_{\nu_1...\nu_p\mu_1...\mu_{n-p}}\alpha^{\nu_1...\nu_p}\underline{\omega}^{\mu_1} \wedge \cdots \wedge \underline{\omega}^{\mu_{n-p}}. \tag{3.117}$$

The dual of an orthogonal basis p-form is

$$\star\left(\underline{\omega}^{\nu_1} \wedge \cdots \wedge \underline{\omega}^{\nu_p}\right) = \frac{1}{(n-p)!}\frac{\sqrt{|g|}}{g_p}\varepsilon_{\nu_1...\nu_p\mu_1...\mu_{n-p}}\underline{\omega}^{\mu_1} \wedge \cdots \wedge \underline{\omega}^{\mu_{n-p}}, \tag{3.118}$$

where g_p is the determinant of the metric tensor associated with the space of the p-form $\underline{\alpha}$, and g is the determinant made up of the components of the metric tensor in the n-dimensional space.

Example 3.4.3 (*Duals of basis forms in a spherical coordinate system in Euclidean 3-space*) The transformation from spherical coordinates (r, θ, ϕ) to Cartesian coordinates (x, y, z) is

$$x = r\cos\phi\,\sin\theta, \quad y = r\sin\phi\,\sin\theta, \quad z = r\cos\theta. \tag{3.119}$$

The coordinate basis vectors $(\vec{e}_r = \partial/\partial r,\ \vec{e}_\theta = \partial/\partial\theta,\ \vec{e}_\phi = \partial/\partial\phi)$ are

$$\vec{e}_r = \sin\theta\cos\phi\vec{e}_x + \sin\theta\sin\phi\vec{e}_y + \cos\theta\vec{e}_z,$$
$$\vec{e}_\theta = r\cos\theta\cos\phi\vec{e}_x + r\cos\theta\sin\phi\vec{e}_y - r\sin\theta\vec{e}_z,$$
$$\vec{e}_\phi = -r\sin\theta\sin\phi\vec{e}_x + r\sin\theta\cos\phi\vec{e}_y, \tag{3.120}$$

Inserting these expressions into Eq. (3.60) we find the non-vanishing components of the metric tensor

$$g_{rr} = 1,\quad g_{\theta\theta} = r^2,\quad g_{\phi\phi} = r^2\sin^2\theta. \tag{3.121}$$

The line-element takes the form

$$ds^2 = dr^2 + r^2 d\theta^2 + r^2\sin^2\theta d\phi^2. \tag{3.122}$$

The volume form is

$$\underline{V} = r^2\sin\theta\underline{\omega}^r \wedge \underline{\omega}^\theta \wedge \underline{\omega}^\phi. \tag{3.123}$$

The dual of a basis form is

$$\star\underline{\omega}^\nu = \frac{1}{2}\frac{\sqrt{g}}{g_\nu}\varepsilon_{\nu\mu_1\mu_2}\underline{\omega}^{\mu_1} \wedge \underline{\omega}^{\mu_2}, \tag{3.124}$$

where

$$g = r^4\sin^2\theta,\quad g_r = 1,\quad g_\theta = r^2,\quad g_\phi = r^2\sin\theta. \tag{3.125}$$

This gives

$$\star\underline{\omega}^r = r^2\sin\theta\varepsilon_{123}\underline{\omega}^2 \wedge \underline{\omega}^3 = r^2\sin\theta\underline{\omega}^\theta \wedge \underline{\omega}^\phi, \tag{3.126}$$

$$\star\underline{\omega}^\theta = \sin\theta\underline{\omega}^\phi \wedge \underline{\omega}^r, \tag{3.127}$$

$$\star\underline{\omega}^\phi = \underline{\omega}^r \wedge \underline{\omega}^\theta. \tag{3.128}$$

The double dual is given by

$$\star\star\underline{\alpha} = \hat{g}(-1)^{p(n-p)}\underline{\alpha}. \tag{3.129}$$

Hence the double dual is the identity up to a sign. The dual of the volume form is

$$\star\underline{V} = \frac{1}{n!}\varepsilon_{\mu_1...\mu_n}\varepsilon^{\mu_1...\mu_n} = \hat{g} = \pm\underline{1}. \tag{3.130}$$

Table 3.1 Right column shows the components of the forms dual to the forms in 3-space with components shown in the left column

0-form: $\underline{\phi} = \phi$	3-form: $\star\underline{\phi} = \sqrt{g}\phi$
1-form: $\underline{E} : [E_1, E_2, E_3]$	2-form: $\star\underline{E} : \sqrt{g}\begin{bmatrix} 0 & E^3 & -E^2 \\ -E^3 & 0 & E^1 \\ E^2 & -E^1 & 0 \end{bmatrix}$
2-form: $\underline{B} : \begin{bmatrix} 0 & B_{12} & -B_{31} \\ -B_{12} & 0 & B_{23} \\ B_{31} & -B_{23} & 0 \end{bmatrix}$	1-form: $\star\underline{B} : \sqrt{g}\left[B^{23}, B^{31}, B^{12}\right]$
3-form: $\underline{G} : G_{123} = G$	0-form: $\star\underline{G} : g^{-1/2}G$

Let $\underline{\alpha}$ and $\underline{\beta}$ be p-forms with corresponding vectors **A** and **B**, respectively. Then

$$(\star\underline{\alpha}) \wedge \underline{\beta} = \frac{1}{p!}\alpha^{\mu_1\ldots\mu_p}\beta_{\mu_1\ldots\mu_p}\varepsilon_{1\ldots n}\underline{\omega}^1 \wedge \cdots \wedge \underline{\omega}^n = (\mathbf{A} \cdot \mathbf{B})\underline{V}. \qquad (3.131)$$

Furthermore

$$(\star\underline{\alpha}) \wedge \underline{\beta} = \underline{\alpha}\wedge(\star\underline{\beta}). \qquad (3.132)$$

The following relationship valid for $n = 3$ between the wedge product of 1-forms and the vector product of vectors should be noted

$$\star\left(\underline{\alpha} \wedge \underline{\beta}\right) = \frac{1}{2}\varepsilon_{\nu\lambda\mu}(\mathbf{A} \wedge \mathbf{B})^{\nu\lambda}\underline{\omega}^\mu = (\mathbf{A} \times \mathbf{B})_\mu\underline{\omega}^\mu. \qquad (3.133)$$

Examples with dual forms in 3-space and 4-dimensional spacetime are shown in Tables 3.1 and 3.2.

Table 3.2 Right column shows the components of the forms dual to the forms in 4-dimensional spacetime with components shown in the left column

0-form: $\underline{\phi} = \phi$	4-form: $\star\underline{\phi} = \sqrt{-g}\phi$
1-form: $\underline{A} : [A_0, \ A_1, \ A_2, \ A_3]$	3-form: $\star\underline{A} : (\star\underline{A})_{012} = -\sqrt{-g}A^3$ etc.
2-form: $\underline{F} : \begin{bmatrix} 0 & F_{01} & F_{02} & F_{03} \\ -F_{01} & 0 & F_{12} & F_{13} \\ -F_{02} & -F_{12} & 0 & F_{23} \\ -F_{03} & -F_{13} & -F_{23} & 0 \end{bmatrix}$	2-form: $\star\underline{F} : \sqrt{-g} \begin{bmatrix} 0 & F^{23} & -F^{13} & F^{12} \\ -F^{23} & 0 & F^{03} & -F^{02} \\ F^{13} & -F^{03} & 0 & F^{01} \\ -F^{12} & F^{02} & -F^{01} & 0 \end{bmatrix}$
3-form: $\underline{G} : G_{\alpha\beta\gamma}$	1-form: $\star\underline{G} : \sqrt{-g}\left[-G^{123}, G^{230}, -G^{301}, G^{012}\right]$
4-form: $\underline{H} : H_{0123} = H$	0-form: $\star\underline{H} = -(-g)^{-1/2}H$

Exercises

3.1. *Four-vectors*

(a) Given three four-vectors

$$\vec{A} = 4\vec{e}_t + 3\vec{e}_x + 2\vec{e}_y + \vec{e}_x, \quad \vec{B} = 5\vec{e}_t + 4\vec{e}_x + 3\vec{e}_y, \quad \vec{C} = \vec{e}_t + 2\vec{e}_x + 3\vec{e}_y + 4\vec{e}_x$$
$$\vec{e}_t \cdot \vec{e}_t = -1, \ \vec{e}_x \cdot \vec{e}_x = \vec{e}_y \cdot \vec{e}_y = \vec{e}_z \cdot \vec{e}_z = 1 \qquad (3.134)$$

Show that $\vec{A}$ is time-like $\left(\vec{A} \cdot \vec{A} < 0\right)$, $\vec{B}$ is light-like $\left(\vec{B} \cdot \vec{B} = 0\right)$ and $\vec{C}$ is space-like $\left(\vec{C} \cdot \vec{C} > 0\right)$.

(b) Assume that $\vec{A}$ and $\vec{B}$ are two non-vanishing orthogonal vectors, $\vec{A} \cdot \vec{B} = 0$. Show the following

- If $\vec{A}$ is light-like, then $\vec{B}$ is space-like or light-like.
- If $\vec{A}$ and $\vec{B}$ are light-like, then they are proportional.
- If $\vec{A}$ is space-like, then $\vec{B}$ is time-like, light-like or space-like.

Illustrate this in a 3-dimensional Minkowski diagram.

(c) A change of basis is given by

$$\vec{e}_{t'} = \cosh\theta\vec{e}_t + \sinh\theta\vec{e}_x, \quad \vec{e}_{x'} = \sinh\theta\vec{e}_t + \cosh\theta\vec{e}_x, \quad \vec{e}_{y'} = \vec{e}_y, \quad \vec{e}_{z'} = \vec{e}_z \qquad (3.135)$$

Show that this describes a Lorentz transformation along the *x*-axis, where the relative velocity *v* between the reference frames is given by $v = \tanh\theta$. Draw the vectors in a 2-dimensional Minkowski diagram and find what type of curves $\vec{e}_{t'}$ and $\vec{e}_{x'}$ describe as θ varies.

(d)   The 3-vector $\vec{v}$ describing the velocity of a particle is defined *with respect to an observer*. Explain why the 4-velocity $\vec{u}$ is defined *independent* of any observer. The 4-momentum of a particle, with rest mass m, is defined by $\vec{p} = m\vec{u} = m d\vec{r}/d\tau$, where τ is the co-moving time of the particle. Show that $\vec{p}$ is timelike and that $\vec{p} \cdot \vec{p} = -m^2$. Draw, in a Minkowski diagram, the curve to which $\vec{p}$ must be tangent to and explain how this is altered as $m \to 0$. Assume that the energy of the particle is being observed by an observer with 4-velocity $\vec{u}$. Show that the energy he measures is given by

$$E = -\vec{p} \cdot \vec{u}. \tag{3.136}$$

This is an expression which is very useful when one wants to calculate the energy of a particle in an arbitrary reference frame.

3.2.   *The tensor product*

(a)   Given two 1-forms $\underline{\alpha} = (1, 1, 0, 0)$ and $\underline{\beta} = (-1, 0, 1, 0)$. Show—by using the vectors $\vec{e}_0$ and $\vec{e}_1$ as arguments—that $\underline{\alpha} \otimes \underline{\beta} \neq \underline{\beta} \otimes \underline{\alpha}$.

(b)   Find the components of the symmetric and antisymmetric parts of $\underline{\alpha} \otimes \underline{\beta}$.

3.3.   *Symmetric and antisymmetric tensors*

(a)   A tensor **T** of rank 2 in four-dimensional spacetime with Minkowski metric $\eta_{\alpha\beta} = \mathrm{diag}(-1, 1, 1, 1)$ has contravariant components

$$T^{\alpha\beta} = \begin{pmatrix} 0 & 1 & 0 & 0 \\ 1 & -1 & 0 & 2 \\ 2 & 0 & 0 & 1 \\ 1 & 0 & -2 & 0 \end{pmatrix}.$$

Find

1.   The components of the symmetric tensor $T^{(\alpha\beta)}$ and the antisymmetric tensor $T^{[\alpha\beta]}$.
2.   The mixed components T^{α}_{β}.
3.   The covariant components $T_{\alpha\beta}$.

(b)   Does it make sense to talk about the symmetric and the antisymmetric parts of a mixed tensor, i.e. a tensor with both vector- and form-arguments? Explain!

3.4.   *Contractions of tensors with different symmetries*

Let **A** be an antisymmetric tensor of rank $\begin{pmatrix} 2 \\ 0 \end{pmatrix}$, **B** a symmetric tensor of rank $\begin{pmatrix} 0 \\ 2 \end{pmatrix}$, **C** an arbitrary tensor of rank $\begin{pmatrix} 0 \\ 2 \end{pmatrix}$ and **D** an arbitrary tensor of rank $\begin{pmatrix} 2 \\ 0 \end{pmatrix}$.

Show that

$$A^{\alpha\beta} B_{\alpha\beta} = 0, \quad A^{\alpha\beta} C_{\alpha\beta} = A^{\alpha\beta} C_{[\alpha\beta]}, \quad B_{\alpha\beta} D^{\alpha\beta} = B_{\alpha\beta} D^{(\alpha\beta)}. \tag{3.137}$$

3.5. *Coordinate transformation in an Euclidean plane*

In this exercise we shall consider vectors in an Euclidean plane. Let $\{\vec{e}_x, \vec{e}_y\}$ be an orthonormal basis in the plane,

$$\vec{e}_i \cdot \vec{e}_j = \delta_{ij}.$$

A position vector as decomposed in this basis is

$$\vec{x} = x^i \vec{e}_i = x\vec{e}_x + y\vec{e}_y$$

A new coordinate system $\{x', y'\}$ is related to the $\{x, y\}$-system by the transformation

$$x' = 2x - y, \quad y' = x + y$$

(a) Find $\vec{e}_{x'}$ and $\vec{e}_{y'}$ expressed in terms of $\vec{e}_x$ and $\vec{e}_y$
(b) Find the basis vectors in the $\{x', y'\}$-system in terms of $\{\vec{e}_x, \vec{e}_y\}$.
(c) Find the components of the metric tensor in the $\{x', y'\}$-system.
(d) Calculate the line-element in the $\{x', y'\}$-system.
(e) We now define a set of basis vectors $\vec{\omega}^i$ by $\vec{\omega}^i = M^{i'}_i \vec{e}_i$ with summation over i. The scalar products of these vectors define the contravariant components of the metric tensors.

    Use this to find the contravariant components of the metric tensor in the $\{x', y'\}$-system.

Chapter 4
Accelerated Reference Frames

Abstract This chapter begins with an introduction to the formalism used to project four-dimensional spacetime into a 3-dimensional spatial 3-space. Then we apply this formalism to deduce the spatial geometry in a rotating reference frame and discuss Ehrenfest's paradox. Also we show that it is impossible to Einstein synchronize clocks around a closed path in a rotating frame because this leads to a contradiction in a non-rotating frame. Gravitational time dilation and frequency shift, and also the Sagnac experiment are discussed. Finally we give an introduction to special relativistic kinematic in a uniformly accelerated reference frame in flat spacetime. It is pointed out that an observer experiences an acceleration of gravity in such a frame.

4.1 The Spatial Metric Tensor

Let $\vec{e}_{\hat{0}}$ ($x^0 = ct$) be the 4-velocity field of the reference particles in a reference frame R. We are going to find the metric tensor γ_{ij} in a tangent space orthogonal to $\vec{e}_0$ expressed by the metric tensor $g_{\mu\nu}$ of spacetime.

The spatial basis vectors $\{\vec{e}_i\}$ are not in general orthogonal to $\vec{e}_0$ in an arbitrary coordinate basis $\{\vec{e}_\mu\}$. We choose $\vec{e}_0 \| \vec{e}_{\hat{0}}$. Let $\vec{e}_{i\perp}$ be the component of $\vec{e}_i$ orthogonal to $\vec{e}_0$, that is $\vec{e}_{i\perp} \cdot \vec{e}_0 = 0$. *The spatial metric tensor* is defined by

$$\gamma_{ij} = \vec{e}_{i\perp} \cdot \vec{e}_{j\perp}, \quad \gamma_{00} = \gamma_{i0} = \gamma_{0i} = 0. \tag{4.1}$$

It follows that the spatial metric tensor is symmetric. It is the projection of the metric tensor of spacetime onto the surface orthogonal to the 4-velocity field of the reference particles in the given reference frame. Hence it describes the spatial geometry in the simultaneity 3-space of the reference frame.

The component of $\vec{e}_i$ along $\vec{e}_0$ is

$$\vec{e}_{i\|} = \frac{\vec{e}_i \cdot \vec{e}_0}{\vec{e}_0 \cdot \vec{e}_0} \vec{e}_0 = \frac{g_{i0}}{g_{00}} \vec{e}_0. \tag{4.2}$$

Hence, the component of $\vec{e}_i$ orthogonal to $\vec{e}_0$ is

© Springer Nature Switzerland AG 2020

Ø. Grøn, *Introduction to Einstein's Theory of Relativity*,
Undergraduate Texts in Physics, https://doi.org/10.1007/978-3-030-43862-3_4

$$\vec{e}_{i\perp} = \vec{e}_i - \vec{e}_{i\parallel} = \vec{e}_i - \frac{g_{i0}}{g_{00}}\vec{e}_0. \tag{4.3}$$

Thus, the non-vanishing components of the spatial metric tensor are

$$
\begin{aligned}
\gamma_{ij} &= \left(\vec{e}_i - \frac{g_{i0}}{g_{00}}\vec{e}_0\right)\cdot\left(\vec{e}_j - \frac{g_{j0}}{g_{00}}\vec{e}_0\right) = \vec{e}_i\cdot\vec{e}_j \\
&\quad - \frac{g_{j0}}{g_{00}}\vec{e}_0\cdot\vec{e}_i - \frac{g_{i0}}{g_{00}}\vec{e}_0\cdot\vec{e}_j + \frac{g_{i0}g_{j0}}{g_{00}^2}\vec{e}_0\cdot\vec{e}_0 \\
&= g_{ij} - \frac{g_{i0}g_{j0}}{g_{00}} - \frac{g_{i0}g_{j0}}{g_{00}} + \frac{g_{i0}g_{j0}}{g_{00}} = g_{ij} - \frac{g_{i0}g_{j0}}{g_{00}}. \tag{4.4}
\end{aligned}
$$

The spatial line-element is

$$dl^2 = \gamma_{ij}dx^i dx^j = \left(g_{ij} - \frac{g_{i0}g_{j0}}{g_{00}}\right)dx^i dx^j. \tag{4.5}$$

This line-element gives the distance between simultaneous events in a reference frame where the metric tensor of spacetime in a co-moving coordinate system is $g_{\mu\nu}$. Consider a transformation of the form

$$x^0 = x^0\left(x^{\mu'}\right), \quad x^i = x^i\left(x^{i'}\right) \tag{4.6}$$

Applying the transformation Eq. (3.113) to each term in Eq. (4.4) and noting that $M_{0'}^i = 0$ for the transformation (4.6) we get

$$\gamma_{i'j'} = M_{i'}^i M_{j'}^j \gamma_{ij}. \tag{4.7}$$

This shows that γ_{ij} transform as tensor components under a transformation of the form (4.6).

From the transformation (4.6) we have

$$\frac{\partial}{\partial x^{0'}} = \frac{\partial x^\mu}{\partial x^{0'}}\frac{\partial}{\partial x^\mu}. \tag{4.8}$$

Since $\partial x^i/\partial x^{0'} = 0$. Thus

$$\vec{e}_{0'} = \frac{\partial x^0}{\partial x^{0'}}\vec{e}_0, \tag{4.9}$$

showing that $\vec{e}_{0'}$ is parallel to $\vec{e}_0$. It follows that the 4-velocity field of particles with fixed coordinates in two coordinate systems connected by a transformation of the form (4.6) is identical. This means that Eq. (4.6) represents coordinate transformations between different coordinate systems that are both co-moving with the same reference frame. Hence transformations of the form (4.6) are called *internal coordinate transformations*.

Equation (4.7) shows that the components of the spatial metric tensor transform as tensor components and the spatial line-element are invariant under internal transformations, i.e. under transformations between different coordinate systems co-moving with one and the same reference frame.

The line-element of spacetime can be expressed as

$$ds^2 = -c^2 d\hat{t}^2 + dl^2, \tag{4.10}$$

where $d\hat{t} = 0$ represents the simultaneity defining the spatial line-element. The temporal part of the line-element may be expressed as

$$
\begin{aligned}
d\hat{t}^2 &= dl^2 - ds^2 = \left(\gamma_{\mu\nu} - g_{\mu\nu} \right) dx^\mu dx^\nu \\
&= \left(\gamma_{ij} - g_{ij} \right) dx^i dx^j + 2(\gamma_{i0} - g_{i0}) dx^i dx^0 + (\gamma_{00} - g_{00}) dx^0 dx^0 \\
&= \left(g_{ij} - \frac{g_{i0} g_{j0}}{g_{00}} - g_{ij} \right) dx^i dx^j - 2 g_{i0} dx^i dx^0 - g_{00} (dx^0)^2 \\
&= -g_{00} \left[(dx^0)^2 + 2 \frac{g_{i0}}{g_{00}} dx^0 dx^i + \frac{g_{i0} g_{j0}}{g_{00}^2} dx^i dx^j \right] \\
&= \left[(-g_{00})^{1/2} \left(dx^0 + \frac{g_{i0}}{g_{00}} dx^1 \right) \right]^2.
\end{aligned}
\tag{4.11}
$$

It follows that

$$ d\hat{t} = (-g_{00})^{1/2} \left(dx^0 + \frac{g_{i0}}{g_{00}} dx^i \right). \tag{4.12} $$

The 3-space orthogonal to the world lines of the reference particles in R, defined by $d\hat{t} = 0$, corresponds to a coordinate time interval

$$ dt = -\frac{g_{i0}}{g_{00}} dx^i. \tag{4.13} $$

This is not an exact differential, that is, dt is not integrable, which means that one cannot in general define a 3-space orthogonal to the world lines of the reference particles, i.e. a "simultaneity space", in an arbitrary reference frame. We must also conclude that unless g_{i0}/g_{00} is constant, it is not possible to Einstein synchronize clocks around closed curves.

4.2   Einstein Synchronization of Clocks in a Rotating Reference Frame

We shall show that it is not possible to Einstein synchronize clocks around a closed curve in a rotating reference frame. If this is attempted, contradictory boundary conditions in the non-rotating laboratory frame will arise, due to the relativity of simultaneity (See Fig. 4.1).

The distance in the laboratory frame between two points is

$$\Delta x = \frac{2\pi r}{n}. \tag{4.14}$$

We now make a Lorentz transformation from the instantaneous rest frame (t', x') of the circumference of the rotating frame R to the non-rotating laboratory frame

$$\Delta t = \gamma\left(\Delta t' + \frac{r\omega}{c^2}\Delta x'\right), \quad \Delta x = \gamma\left(\Delta x' + r\omega\Delta t'\right), \quad \gamma = \frac{1}{\sqrt{1 - \frac{r^2\omega^2}{c^2}}}, \tag{4.15}$$

where r is the radios of the circumference and ω is the angular velocity of R.

The proper distance between two points with distance (4.14) in the laboratory frame is $\Delta x' = \gamma \Delta x$. Hence we get a time difference

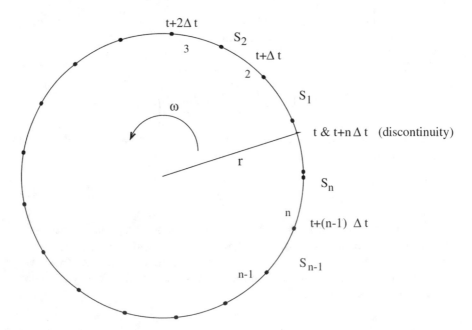

Fig. 4.1 Simultaneity in a rotating frame. Events that are simultaneous in the rotating frame are not simultaneous in the stationary frame

$$\Delta t = \gamma^2 \frac{r\omega}{c^2}\Delta x = \gamma^2 \frac{r\omega}{c^2}\frac{2\pi r}{n} \qquad (4.16)$$

In the laboratory frame for simultaneous events in the rotating frame separated by a proper distance $\Delta x'$. This is due to the relativity of simultaneity. Around the circumference the time difference is accumulated to

$$n\Delta t = \gamma \frac{2\pi r^2 \omega}{c^2}, \qquad (4.17)$$

and we get a discontinuity of simultaneity as shown in Fig. 4.2.

Let IF be an inertial frame with cylinder coordinates (T, R, Θ, Z). The line-element has then the form

$$ds^2 = -c^2 dT^2 + dR^2 + R^2 d\Theta^2 + dZ^2. \qquad (4.18)$$

In a reference frame RF rotating with constant angular velocity ω we have co-moving cylinder coordinates (t, r, θ, z). The two coordinate systems are related by the coordinate transformation

$$t = T, \quad r = R, \quad \theta = \Theta - \omega T, \quad z = Z. \qquad (4.19)$$

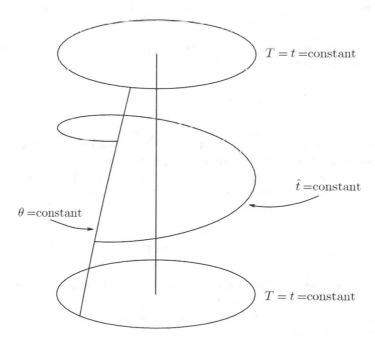

Fig. 4.2 Discontinuity of simultaneity in rotating frame

The first equation means that the coordinate clocks in RF go with a position independent rate equal to that of the clocks in the non-rotating frame IF, i.e. equal to that of the inertial rest frame of the rotational axis.

The line-element in the co-moving coordinates in RF is

$$\begin{aligned}
\mathrm{d}s^2 &= -c^2\,\mathrm{d}t^2 + \mathrm{d}r^2 + r^2(\mathrm{d}\theta + \omega\,\mathrm{d}t)^2 + \mathrm{d}z^2\\
&= -\left(1 - r^2\omega^2/c^2\right)\mathrm{d}t^2 + \mathrm{d}r^2 + r^2\mathrm{d}\theta^2 + \mathrm{d}z^2 + 2r^2\omega\,\mathrm{d}\theta\,\mathrm{d}t.
\end{aligned} \tag{4.20}$$

The metric tensor has the following components

$$g_{tt} = -\left(1 - r^2\omega^2/c^2\right),\, g_{rr} = g_{zz} = 1,\, g_{\theta\theta} = r^2,\, g_{\theta t} = g_{t\theta} = r^2\omega. \tag{4.21}$$

Putting $\mathrm{d}t = 0$ in Eq. (4.20) gives

$$\mathrm{d}s^2 = \mathrm{d}r^2 + r^2\mathrm{d}\theta^2 + \mathrm{d}z^2. \tag{4.22}$$

This represents the Euclidean geometry of the 3-space in IF.

The non-vanishing components of the spatial metric tensor (4.4) in the co-moving coordinate system of the rotating frame are

$$\gamma_{rr} = \gamma_{zz} = 1,\, \gamma_{\theta\theta} = g_{\theta\theta} - \frac{g_{\theta0}^2}{g_{00}} = r^2 - \frac{r^2\omega^2/c^2}{-\left(1 - r^2\omega^2/c^2\right)} = \frac{r^2}{1 - r^2\omega^2/c^2}. \tag{4.23}$$

Inserting this into Eq. (4.5) gives the spatial line-element in the co-moving coordinate system of the rotating frame

$$\mathrm{d}l^2 = \mathrm{d}r^2 + \frac{r^2\mathrm{d}\theta^2}{1 - r^2\omega^2/c^2} + \mathrm{d}z^2. \tag{4.24}$$

It describes the geometry of a local 3-space orthogonal to the world line of a reference particle in RF. This 3-space cannot be extended to a finite 3-dimensional space in RF since Einstein synchronization is not integrable in RF. From the line element (4.20) it is seen that the geometry of this local simultaneity space in RF is non-Euclidean. The circumference of a circle with radius r is

$$l_\theta = \frac{2\pi r}{\sqrt{1 - r^2\omega^2/c^2}} > 2\pi r. \tag{4.25}$$

This means that the spatial geometry is hyperbolic in the rotating frame.

We shall now explain this result first from the point of view of observers at rest in the non-rotating frame F, and then from the point of view of observes co-moving with the "rotating" frame, R.

We shall first define the concept *standard measuring rod*. A standard measuring rod has by definition a constant rest length even if it is accelerated. It is not allowed by a standard measuring rod to be compressed or strained. Hence a standard measuring rod will have a Lorentz contraction according to its velocity.

As observed from F the measuring rods along a circle about the origin have a velocity $v = r\omega$. Hence they will be Lorentz contracted by the factor $\sqrt{1 - r^2\omega^2/c^2}$. Hence there is place for more standard measuring rods around the circle the faster the frame R rotates. Therefore the measured length of the circle will be larger by this factor. This is the reason for the result (4.25) from the point of view of an F-observer. Hence according to the F-observers there is no question of a non-Euclidean geometry. The result (4.25) is explained by the Lorentz contraction of the standard measuring rods.

It may further be noted that since the material of a rotating disc cannot Lorentz contract an engraved scale on the disc cannot be used as a set of standard measuring rods. When the disc is put into rotation the material tries to Lorentz contract in the tangential direction, but is not allowed to do so. Hence a tangential strain will develop in the material of a disc that is put into rotation.

We shall now assume the validity of the principle of relativity for rotating motion. Then the observers in R can think of themselves as at rest and the environment as rotating. From this point of view the standard measuring rods are not Lorentz contracted. Hence the explanation of the F-observers does not work for the R-observers.

According to Einstein's interpretation of the general theory of relativity the explanation of the R-observes is as follows. The R-observer experiences what in Newton's theory is called a centrifugal force field. According to the principle of equivalence this is reckoned as a gravitational field in the theory of relativity. The R-observer will say that there is a non-Euclidean spatial geometry in the R-frame, and that this is connected with the gravitational field which is present in this frame.

In general an experimental result—in the present case that the measured length of a rotating disc with radius r is larger than $2\pi r$—is independent of the reference frame that the experiment is described from, but the *explanation* of the result depends upon the motion of the observer's reference frame.

4.3 Angular Acceleration in the Rotating Frame

We will now investigate what happens when we give RF an angular acceleration. Then we consider a rotating circle made of standard measuring rods, as shown in Fig. 4.3.

All points on a circle are accelerated simultaneously in IF (the laboratory system). We let the angular velocity increase from ω to $\omega + d\omega$, measured in IF. Lorentz transformation to an instantaneous rest frame for a point on the circumference then gives an increase in velocity in this system:

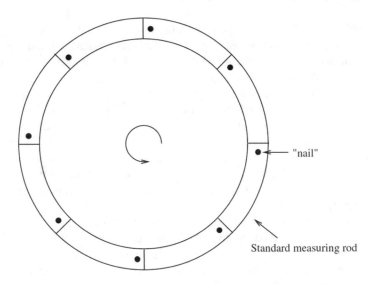

Fig. 4.3 Nonrotating disc with measuring rods. The standard measuring rods are fastened with nails at one end so that they are free to contract when the disc is put into rotation

$$r d\omega' = \frac{r d\omega}{1 - r^2\omega^2/c^2}. \tag{4.26}$$

Due to the relativity of simultaneity the points on the disc at the front end of the rods are accelerated before the points on the disc at the rear ends, in the rotating frame. The time difference of the accelerations at the front and rear ends is

$$\Delta t' = \frac{r\omega L_0/c^2}{\sqrt{1 - r^2\omega^2/c^2}}, \tag{4.27}$$

where $L_0 = 2\pi r/n$ is the rest length of the rods. In IF all points of the circumference are accelerated simultaneously. However, in RF the front points are accelerated before the rear points, so the distances between the points will increase, i.e. the length of the circumference of the disc will increase.

It is a defining property of standard measuring rods that they shall move in such a way that their proper length is preserved. They are not allowed to be stretched or compressed. The standard measuring rods must therefore move under a different acceleration program than the points on the disc. All the points of the standard rods are accelerated simultaneously in RF. Hence the measuring rods will separate from each other during a period with angular acceleration of the disc as shown in Fig. 4.4.

In IF the separation of the rods is interpreted as a consequence of the Lorentz contraction of the standard rods.

As observed in RF the distance between the rods is increased by

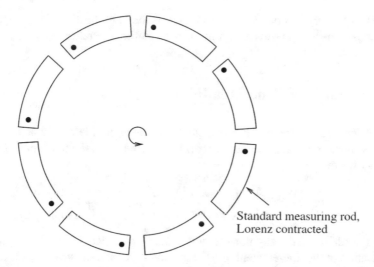

Fig. 4.4 Lorentz contacted measuring rods on a rotating disc. Separation of the measuring rods has different explanations as referred to IF and RF. The standard measuring rods have been Lorentz contracted as observed in IF, while the length of the periphery has increased as observed in RF

$$ds' = rd\omega' \Delta t' = \frac{r^2 \omega L_0 d\omega / c^2}{\left(1 - r^2\omega^2/c^2\right)^{3/2}}. \tag{4.28}$$

The increase of distance during the acceleration period is

$$s' = \frac{r^2 L_0}{c^2} \int_0^\omega \frac{\omega d\omega}{\left(1 - r^2\omega^2/c^2\right)^{3/2}} = \left(\frac{1}{\sqrt{1 - r^2\omega^2/c^2}} - 1\right) L_0. \tag{4.29}$$

Hence, after the acceleration period there is a proper distance s' between the rods. In the laboratory system, IF, the distance between the rods is

$$s = \sqrt{1 - r^2\omega^2/c^2} \, s' = \sqrt{1 - r^2\omega^2/c^2} \left(\frac{1}{\sqrt{1 - r^2\omega^2/c^2}} - 1\right) L_0$$

$$= L_0 - L_0\sqrt{1 - r^2\omega^2/c^2}. \tag{4.30}$$

We now have the situation shown in Fig. 4.4. There is room for more standard rods around the periphery the faster the disc rotates. This means that as measured with measuring rods at rest in the rotating frame the measured length of the periphery (number of standard rods) gets larger with increasing angular velocity. This is how an inertial observer would explain the measuring result of the rotating observer. According to the rotating observer, however, the disc material has been stretched in the tangential direction. Note that as measured by the inertial observer the length of the periphery is $2\pi r$ independent of the angular velocity of the disc, since the inertial observer uses measuring rods at rest in the non-rotating reference frame.

4.4 Gravitational Time Dilation

We consider a standard clock moving along a circle about the rotational axis. This clock has constant r and z. Along the world line of the clock the line element (4.20) can be written as

$$ds^2 = c^2 dt^2 \left[-\left(1 - \frac{r^2\omega^2}{c^2}\right) + \frac{r^2}{c^2}\omega_c^2 + 2\frac{r^2\omega}{c^2}\omega_c \right]. \tag{4.31}$$

where $\omega_c = d\theta/dt$ is the angular velocity of the clock in RF. Utilizing the physical interpretation of the line-element (3.91) for a time-like interval we conclude that the proper time interval $d\tau$ of the clock is related to the coordinate time interval in RF by

$$d\tau = dt\sqrt{1 - \frac{r^2\omega^2}{c^2} - \frac{r^2}{c^2}\omega_c^2 - 2\frac{r^2\omega}{c^2}\omega_c}. \tag{4.32}$$

which may be written

$$d\tau = dt\sqrt{1 - r^2(\omega + \omega_c)^2/c^2}. \tag{4.33}$$

Here t represents proper time in RF at the axis, $r = 0$, which is equal to the proper time in IF, and τ represents proper time at an arbitrary point in RF. Since the rate of coordinate time is position independent, it follows that the rate of proper time in RF decreases with increasing distance from the axis. Also it decreases with increasing angular velocity ω of RF relative to IF, and it depends upon the angular velocity ω_c of the clock in RF, both its magnitude and sign. The rate of proper time of a clock moving in RF compared to the time in IF, $d\tau/dt$, is maximal for $\omega_c = -\omega$. Such a clock is at rest in IF which is non-rotating relative to the large scale cosmic masses. For this clock $d\tau = dt$. As considered in RF such a clock moves together with the large-scale cosmic masses. Hence *a clock at rest relative to the large scale cosmic masses goes at a maximal rate.*

A standard clock at rest in RF has $\omega_c = 0$. The proper time interval of these clocks is

$$d\tau = dt\sqrt{1 - \frac{r^2\omega^2}{c^2}}. \tag{4.34}$$

Seen from IF, the non-rotating laboratory system, Eq. (4.34) represents the *velocity-dependent time dilation* from the special theory of relativity.

But how is Eq. (4.34) interpreted in RF? The clock does not move relative to an observer in this system, hence what happens cannot be interpreted as a velocity-dependent phenomenon. According to Einstein, the fact that standard clocks slow down the farther away from the axis of rotation they are is due to a *gravitational effect*.

We will now find the gravitational potential at a distance r from the axis. The centripetal acceleration is $r\omega^2$ so

$$\Phi = -\int_0^r g(r)dr = -\int_0^r r\omega^2 dr = -\frac{1}{2}r^2\omega^2. \tag{4.35}$$

We then get

$$d\tau = dt\sqrt{1 - \frac{r^2\omega^2}{c^2}} = dt\sqrt{1 + \frac{2\Phi}{c^2}}. \tag{4.36}$$

In RF the position-dependent time dilation is interpreted as a *gravitational time dilation*: Time flows slower further down in a gravitational field.

4.5   Path of Photons Emitted from the Axis in a Rotating Reference Frame

Let us first describe the photon paths in the inertial frame IF. In this frame the photon paths are radial. Consider a photon path with $\Theta = 0, R = T$ with light source at $R = 0$.

Transforming to RF the equation of the path is,

$$r = t, \quad \theta = -\omega t. \tag{4.37}$$

The orbit equation is thus $\theta = -\omega r$ which is the equation for an Archimedean spiral. The time used by a photon out to distance r from axis is $t = r/c$.

4.6   The Sagnac Effect

The Sagnac effect appears when a beam of light is split and the two beams are made to follow the same path, but in opposite directions. Returning to the point of entry the two light beams are allowed to undergo interference. The phases of the two beams, and thus the position of the interference fringes, are shifted depending upon the angular velocity of the apparatus.

Let us first describe the Sagnac effect in the inertial rest frame of the axis of the apparatus. In this frame the velocity of light is isotropic, but the emitter/receiver

Fig. 4.5 The Sagnac effect. Experiment demonstrates the anisotropic velocity of light as observed by a non-local measurement in a rotating reference frame

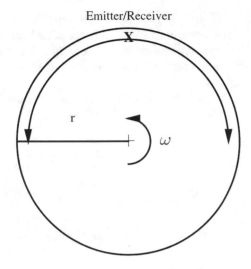

moves due to the disc's rotation as shown in Fig. 4.5. Let t_1 be the travel time around the disc of photons that move *with* the rotation.

Then

$$2\pi r + r\omega t_1 = ct_1, \tag{4.38}$$

giving

$$t_1 = \frac{2\pi r}{c - r\omega}. \tag{4.39}$$

For light travelling in the opposite direction the travel time is shorter because the light meets the emitter/receiver,

$$t_2 = \frac{2\pi r}{c + r\omega}. \tag{4.40}$$

The difference in travel time is

$$\Delta t = t_1 - t_2 = 2\pi r \left(\frac{1}{c - r\omega} - \frac{1}{c + r\omega} \right) = \frac{2\pi r 2r\omega}{c^2 - r^2\omega^2} = \gamma^2 \frac{4A\omega}{c^2}, \tag{4.41}$$

where A is the area enclosed by the photon path or orbit.

We shall now describe the same experiment from the point of view of an observer at rest in a frame rotating together with the apparatus. In this case the line-element is given by Eq. (4.31) with ω_C replaced by the angular velocity of the light, ω_L. Since $ds^2 = 0$ along the world line of the light, the equation of motion of the light takes the form

$$r^2(\omega_L + \omega)^2 - c^2 = 0, \tag{4.42}$$

which has the solutions

$$\omega_L = -\omega \pm c/r. \tag{4.43}$$

Hence, the speed of the light is

$$v_{L\pm} = -r\omega \pm c. \tag{4.44}$$

We see that in the rotating frame RF the measured round trip velocity of light is not isotropic. The difference in travel time of the two beams is

$$\Delta t = t_1 - t_2 = 2\pi r \left(\frac{1}{c - r\omega} - \frac{1}{c + r\omega} \right) = \gamma^2 \frac{4A\omega}{c^2}. \tag{4.45}$$

In agreement with the time difference (4.41) observed in IF. In IF this time difference is explained as a consequence of the motion of the apparatus, while in RF, where the apparatus is at rest, it is explained as due to the anisotropic velocity of light.

4.7 Non-integrability of a Simultaneity Curve in a Rotating Frame

We shall here give some supplementary remarks to the treatment of Einstein synchronization in a rotating reference frame in Sect. 4.1. There we made a separation of the spacetime line-element, ds^2, in a spatial part, dl^2, and a temporal part, $c^2 d\hat{t}^2$, according to $ds^2 = dl^2 - c^2 d\hat{t}^2$, where

$$dl^2 = \left(g_{ij} - \frac{g_{i0}g_{j0}}{g_{00}} \right) dx^i dx^j, \quad d d\hat{t} = \sqrt{-g_{00}} \left(dx_0 + \frac{g_{i0}}{g_{00}} dx^i \right), \quad x_0 = ct. \tag{4.46}$$

As applied to the rotating reference frame R this gives

$$dl^2 = dr^2 + \frac{r^2}{1 - r^2\omega^2/c^2} d\theta^2 + dz^2, \quad d\hat{t} = \sqrt{1 - \frac{r^2\omega}{c^2}} \left(dt - \frac{r^2\omega}{1 - r^2\omega^2/c^2} d\theta \right). \tag{4.47}$$

Here $dt = 0$ means simultaneity in the non-rotating laboratory system, IF, and $d\hat{t} = 0$ simultaneity in the rotating frame, RF. The simultaneity of the laboratory frame is defined globally, but simultaneity in the rotating frame, RF, is only defined locally. With $d\hat{t} = 0$ we get

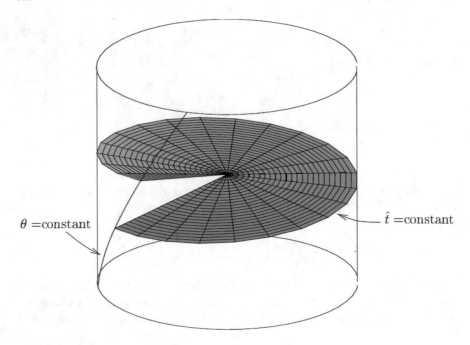

Fig. 4.6 Discontinuous simultaneity surface in a rotating frame

$$\mathrm{d}t = \frac{r^2\omega}{1 - r^2\omega^2/c^2}\mathrm{d}\theta \tag{4.48}$$

which is not a total differential. This means that simultaneity in the rotating frame
RF cannot be defined around a closed curve about the axis. If define simultaneous
events in RF along a circle about the axis, we come to progressively later events in
IF as given by Eq. (4.48). Going around the circle we arrive at the point of departure
at a later event than the one we started from. This means that the 3-space defined
by simultaneity in RF does not represent a simultaneity space in IF. In a Minkowski
diagram with reference to IF the 3-space is shaped as shown in Fig. 4.6. It has a
discontinuity.

4.8   Orthonormal Basis Field in a Rotating Frame

We saw in Sect. 4.1.1 how the spatial metric representing a simultaneity space of an
observer with 4-velocity $\vec{u}$ was defined in terms of orthogonal basis vectors, where
the time-like basis vector was chosen to be the 4-velocity of the observer. It has a
magnitude c.

Let us define an *orthonormal basis vector field*, also called a *tetrad field*, co-
moving with an observer at rest in an arbitrary reference frame. The 4-velocity of

the observer is $\vec{u}$. As time-like unit basis vector we choose

$$\vec{e}_{\hat{0}} = \frac{1}{c}\vec{u} = \frac{dt}{d\tau}\vec{e}_0 = \frac{1}{\sqrt{-g_{00}}}\vec{e}_0. \tag{4.49}$$

We shall express the spatial vectors of this co-moving orthonormal basis in terms of the coordinate basis vectors in a coordinate basis $\{\vec{e}_0, \vec{e}_1, \vec{e}_2, \vec{e}_3\}$ where $\vec{e}_0$ is parallel to $\vec{u}$, and the spatial coordinate vectors need not be orthogonal to the time-like basis vector.

As shown in Sect. 4.1.1 a space-like basis vector $\vec{e}_i$ may be separated in one component

$$\vec{e}_{i\parallel} = \frac{g_{i0}}{g_{00}}\vec{e}_0 \tag{4.50}$$

along $\vec{e}_0$ and one $\vec{e}_{i\perp} = \vec{e}_i - \vec{e}_{i\parallel}$ orthogonal to $\vec{e}_0$, i.e.

$$\vec{e}_{i\perp} = \vec{e}_i - \frac{g_{i0}}{g_{00}}\vec{e}_0. \tag{4.51}$$

Since this vector has a magnitude $|\vec{e}_{i\perp}| = \sqrt{\vec{e}_{i\perp} \cdot \vec{e}_{i\perp}} = \sqrt{\gamma_{ii}}$, the corresponding unit vector is

$$\vec{e}_{\hat{i}} = (\gamma_{ii})^{-1/2}\left(\vec{e}_i - \frac{g_{i0}}{g_{00}}\vec{e}_0\right). \tag{4.52}$$

The second and third space-like vectors in the orthonormal basis are then given by

$$\vec{e}_{\hat{j}} \cdot \vec{e}_{\hat{i}} = \vec{e}_{\hat{j}} \cdot \vec{e}_{\hat{0}} = 0, \vec{e}_{\hat{k}} = \vec{e}_{\hat{i}} \times \vec{e}_{\hat{j}}. \tag{4.53}$$

Let us now consider the rotating reference frame, RF. The coordinate transformation is

$$T = t, R = r, \Theta = \theta + \omega t, Z = z. \tag{4.54}$$

Hence the transformation from the coordinate basis vectors in IF to those in RF is

$$\vec{e}_t = \frac{\partial T}{\partial t}\vec{e}_T + \frac{\partial \Theta}{\partial t}\vec{e}_\Theta = \vec{e}_T + \omega\vec{e}_\Theta, \vec{e}_r = \vec{e}_R, \vec{e}_\theta = \vec{e}_\Theta, \vec{e}_z = \vec{e}_Z. \tag{4.55}$$

Note that even if $T = t$ the basis vectors $\vec{e}_T$ and $\vec{e}_t$ have different directions. The vector field $\vec{e}_T$ is directed along the world lines of the reference particles in IF that are parallel to the cylinder axis in the figure above while the vector field $\vec{e}_t$ is directed along the world lines of the reference particles in RF which has the spiral shape given by $\theta = $ constant shown in Fig. 4.6. The simultaneity space in IF is the horizontal planes orthogonal to $\vec{e}_T$, and the simultaneity space in RF is a succession of simultaneity spaces locally orthogonal to $\vec{e}_t$.

In order to find the orthonormal basis carried by an observer in RF by means of the formulae above, we must first find the components of the 4-velocity in the co-moving coordinate system in RF. Since the observer is at rest in RF, the time component is the only non-vanishing component. It follows from the line-element in RF as applied to an observer at rest that the 4-velocity is

$$\vec{u} = c\frac{dt}{d\tau}\vec{e}_t = \frac{c}{\sqrt{1 - r^2\omega^2/c^2}}\vec{e}_t. \tag{4.56}$$

Inserting Eq. (4.56) into Eq. (4.49) and the expressions (4.21) and (4.23) for the components of the metric tensor and the spatial metric tensor in RF into Eq. (4.52) then give the orthonormal basis carried by an observer in RF

$$\vec{e}_{\hat{t}} = \frac{1}{\sqrt{c^2 - r^2\omega^2}}\vec{e}_t, \ \vec{e}_{\hat{r}} = \vec{e}_r, \ \vec{e}_{\hat{\theta}} = \frac{\sqrt{1 - r^2\omega^2/c^2}}{r}\vec{e}_\theta + \frac{r\omega/c^2}{\sqrt{1 - r^2\omega^2/c^2}}\vec{e}_t. \tag{4.57}$$

Example 4.8.1 (*The acceleration of a velocity field representing rigid rotation*) The velocity field is

$$\vec{v} = r\omega\vec{e}_{\hat{\theta}} = \omega\vec{e}_\theta. \tag{4.58}$$

Note that the coordinate component of the velocity is not equal to the physical velocity component. *The physical velocity components are those in an orthonormal basis* (Fig. 4.7).

We shall calculate the acceleration field, $\vec{a} = \frac{d\vec{v}}{dt}$. Using the chain rule of differentiation we get

Fig. 4.7 Rigid rotation

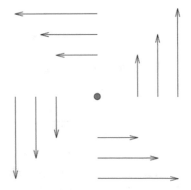

$$\vec{a} = \frac{\partial \vec{v}}{\partial x^\mu}\frac{dx^\mu}{dt} = v^\mu \frac{\partial \vec{v}}{\partial x^\mu}. \tag{4.59}$$

Since the only non-vanishing velocity component is $v^\theta = \omega$ we get

$$\vec{a} = \omega \frac{\partial \vec{v}}{\partial \theta}. \tag{4.60}$$

This gives

$$\vec{a} = \omega \frac{\partial(\omega \vec{e}_\theta)}{\partial \theta} = \omega^2 \frac{\partial \vec{e}_\theta}{\partial \theta}. \tag{4.61}$$

We have found earlier that

$$\vec{e}_r = \cos\theta \vec{e}_x + \sin\theta \vec{e}_y, \ \vec{e}_\theta = -r\sin\theta \vec{e}_x + r\cos\theta \vec{e}_y. \tag{4.62}$$

Differentiation gives

$$\frac{\partial \vec{e}_\theta}{\partial \theta} = -r\cos\theta \vec{e}_x - r\sin\theta \vec{e}_y = -r\vec{e}_r = -r\vec{e}_{\hat{r}}. \tag{4.63}$$

Hence the acceleration is

$$\vec{a} = -r\omega^2 \vec{e}_{\hat{r}}. \tag{4.64}$$

This is the centripetal acceleration for circular motion.

4.9   Uniformly Accelerated Reference Frame

Consider a particle moving along a straight line with velocity u and acceleration $a = du/dT$. The acceleration in the instantaneous inertial rest frame of the particle, its proper acceleration, is denoted by $\hat{a}$ and is given by

$$a = \left(1 - u^2/c^2\right)^{3/2}\hat{a}. \tag{4.65}$$

Assume that the particle has constant proper acceleration $\hat{a} = g$, that is

$$\frac{du}{dt} = \left(1 - u^2/c^2\right)^{3/2}g, \tag{4.66}$$

which on integration with $u(0) = 0$ gives

$$u = \frac{dX}{dT} = \frac{gT}{(1 + \frac{g^2 T^2}{c^2})^{1/2}}. \tag{4.67}$$

Integrating once more gives

$$X = \frac{c^2}{g}(1 + \frac{g^2 T^2}{c^2})^{1/2} + k, \tag{4.68}$$

where k is a constant of integration. This may be written

$$(X - k)^2 - c^2 T^2 = c^4/g^2. \tag{4.69}$$

This describes a hyperbola in the Minkowski diagram as shown in Fig. 4.8. The proper time interval as measured by a clock which follows the particle is

$$d\tau = \left(1 - \frac{u^2}{c^2}\right)^{1/2} dT. \tag{4.70}$$

Substitution for $u(T)$ and integration with $\tau(0) = 0$ give

$$\tau = \frac{c}{g}\text{arcsinh}\left(\frac{gT}{c}\right), \tag{4.71}$$

or

Fig. 4.8 Hyperbolic motion.
Graph is the world line of a
reference particle in a
uniformly accelerated
reference frames as drawn
with reference to the inertial
frame in which the particle is
at rest at the point of time
$T = 0$

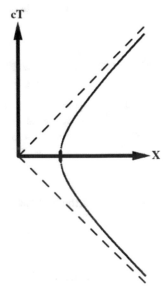

$$T = \frac{c}{g} \sinh\left(\frac{g\tau}{c}\right). \tag{4.72}$$

Inserting this into Eq. (4.68) gives

$$X = \frac{c^2}{g} \cosh\left(\frac{g\tau}{c}\right) + k. \tag{4.73}$$

We now use this particle as the origin of space in an uniformly accelerated reference frame, AF. With the initial condition $X(0) = 0$ we get $k = -c^2/g$ and

$$1 + \frac{gX}{c^2} = \cosh\frac{g\tau}{c}. \tag{4.74}$$

Example 4.9.1 (*Uniformly accelerated motion through the Milky Way*) Let a traveller traverse the Milky Way, say a distance of a hundred thousand years, with a constant proper acceleration equal to 9.8 m/s^2, to the centre and then reversing the acceleration so that the traveller stops at the other side. The travel would take approximately a hundred thousand years as measured by the stationary observer. It has been said that the traveller would only become around 20 years older during the travel.

Let us see how this comes about. Let X be the distance and T the time as measured by the stationary observer, τ the proper time of the traveller and $g = 9.8$ m/s^2 the proper acceleration. For uniformly accelerated motion, i.e. motion with constant proper acceleration, we saw in Eqs. (4.72) and (4.74) that

$$1 + \frac{gX}{c^2} = \cosh\frac{g\tau}{c}, \quad \frac{gT}{c} = \sinh\frac{g\tau}{c} \tag{4.75}$$

From the identity $\cosh^2(g\tau/c) - \sinh^2(g\tau/c) = 1$ it then follows that

$$\left(1 + \frac{gX}{c^2}\right)^2 - \left(\frac{gT}{c}\right)^2 - 1 \tag{4.76}$$

or

$$\frac{gT}{c} = \sqrt{\left(1 + \frac{gX}{c^2}\right)^2 - 1} = \sqrt{2\frac{gX}{c^2} + \left(\frac{gX}{c^2}\right)^2}. \tag{4.77}$$

In the present case we have 1 year $= 3.16 \times 10^7$ s, 1 light year $= ct = 9.48 \times 10^{15}$ m, and the distance from the initial position of the traveller to the centre of the Milky Way is $X = 5 \times 10^4$ light years $= 4.74 \times 10^{20}$ m. Hence $gX/c^2 = 5.16 \cdot 10^4 \gg 1$, so we can approximate the expression (4.73) by $T \approx X/c$. The reason that this is a good approximation in the present case is that the traveller travels with approximately the velocity of light during nearly all the time. This can be seen by calculating the velocity as measured by the stationary observer by taking the differentials of the expressions (4.75)

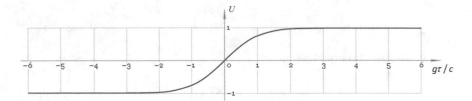

Fig. 4.9 The velocity of a uniformly accelerated particle

$$U = \frac{dX}{dT} = c \tanh \frac{g\tau}{c}. \tag{4.78}$$

The velocity as a function of the proper time is shown in Fig. 4.9.

We see that already at $g\tau/c = 2$ corresponding to $gX/c^2 = 2.8$ or a travelled distance of about 2.7 light years, the velocity of the traveller is close to that of light. Then the travel time to the centre of the Milky Way as measured by the stationary observer is close to $T = 50$ thousand years, giving $gT/c = 5.16 \times 10^4$. Inserting this into the second expression (4.75) written as

$$\tau = \frac{c}{g} \text{arsinh} \frac{gT}{c} \tag{4.79}$$

gives $\tau = 11.2$ years. So the whole travel takes a hundred thousand years as measured by the stationary observer, while the traveller only gets 22.4 years older. The traveller ages extremely slowly because she travels with a velocity which is so close to the velocity of light.

From her point of view she ages by only 22.4 years because the Milky Way moves with nearly the velocity of light and is shaped like a disc which is only 22.4 light years thick due to the Lorentz contraction.

Definition 4.9.1 (*Born-rigid Motion*) Born-rigid motion of a system is a motion such that every element of the system has constant rest length.

The uniformly accelerated reference frame is required to move in a Born-rigid way. Let the inertial frame has coordinates (T, X, Y, Z) and the accelerated frame has coordinates (t, x, y, z). We now denote the X-coordinate of the "origin particle" by X_0. From Eq. (4.73) we then have

$$1 + \frac{gX_0}{c^2} = \cosh \frac{g\tau_0}{c}, \tag{4.80}$$

where τ_0 is the proper time for this particle and k is set to $-c^2/g$. (These are Møller coordinates. Putting $k = 0$ gives Rindler coordinates).

Fig. 4.10 Simultaneity in a uniformly accelerated reference frame. The vector $\widehat{X}$ lies along the "simultaneity line" which makes the same angle with the x-axis as does $\vec{e}_{\hat{t}}$ with the cT-axis

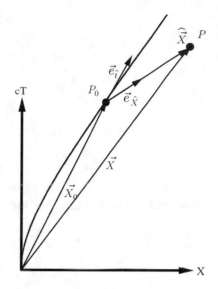

The coordinate time at an arbitrary point in AF is defined by $t = \tau_0$. That is coordinate clocks in AF run identically with the standard clock at the "origin particle". Hence, the coordinate time has a position independent rate. Let $\vec{X}_0$ be the position 4-vector of the "origin particle". Decomposed in the laboratory frame, this becomes

$$\vec{X}_0 = \left\{ \frac{c^2}{g} \sinh \frac{gt}{c}, \frac{c^2}{g} \left(\cosh \frac{gt}{c} - 1 \right), 0, 0 \right\}. \tag{4.81}$$

P is chosen such that P and P_0 are simultaneous in the accelerated frame AF. The distance (see Fig. 4.10) vector from P_0 to P, decomposed into an orthonormal co-moving basis of the "origin particle", is $\widehat{X} = (0, \hat{x}, \hat{y}, \hat{z})$ where $\hat{x}, \hat{y}$ and $\hat{z}$ are physical distances measured simultaneously in AF. The space coordinates in AF are defined by

$$x \equiv \hat{x}, \quad y \equiv \hat{y}, \quad z \equiv \hat{z}. \tag{4.82}$$

The position vector of P is $\vec{X} = \vec{X}_0 + \widehat{X}$. The relationship between basis vectors in IF and the co-moving orthonormal basis is given by a Lorentz transformation in the x-direction,

$$\vec{e}_{\hat{\mu}} = \vec{e}_\mu \frac{\partial x^\mu}{\partial x^{\hat{\mu}}} = (\vec{e}_T, \vec{e}_X, \vec{e}_Y, \vec{e}_Z,) \begin{pmatrix} \cosh \eta & \sinh \eta & 0 & 0 \\ \sinh \eta & \cosh \eta & 0 & 0 \\ 0 & 0 & 1 & 0 \\ 0 & 0 & 0 & 1 \end{pmatrix}, \tag{4.83}$$

where η is the rapidity defined by

$$\tanh \eta \equiv \frac{U_0}{c}. \tag{4.84}$$

Here U_0 is the velocity of the "origin particle",

$$U_0 = \frac{dX_0}{dT_0} = c \tanh \frac{gt}{c}. \tag{4.85}$$

Hence, in the present case the rapidity is

$$\eta = \frac{gt}{c}. \tag{4.86}$$

So the basis vectors can be written as follows,

$$\vec{e}_{\hat{t}} = \vec{e}_T \cosh \frac{gt}{c} + \vec{e}_X \sinh \frac{gt}{c},$$
$$\vec{e}_{\hat{x}} = \vec{e}_T \sinh \frac{gt}{c} + \vec{e}_X \cosh \frac{gt}{c},$$
$$\vec{e}_{\hat{y}} = \vec{e}_Y, \vec{e}_{\hat{z}} = \vec{e}_Z. \tag{4.87}$$

The equation $\vec{X} = \vec{X}_0 + \widehat{\vec{X}}$ can now be decomposed in IF,

$$cT\vec{e}_T + X\vec{e}_X + Y\vec{e}_Y + Z\vec{e}_Z$$
$$= \frac{c^2}{g} \sinh \frac{gt}{c} \vec{e}_T + \frac{c^2}{g}\left(\cosh \frac{gt}{c} - 1\right)\vec{e}_X + x \sinh \frac{gt}{c}\vec{e}_T$$
$$+ x \cosh \frac{gt}{c}\vec{e}_X + y\vec{e}_Y + z\vec{e}_Z. \tag{4.88}$$

This then gives the coordinate transformations:

$$T = \frac{c}{g} \sinh \frac{gt}{c} + \frac{x}{c} \sinh \frac{gt}{c},$$
$$X = \frac{c^2}{g}\left(\cosh \frac{gt}{c} - 1\right) + x \cosh \frac{gt}{c}. \tag{4.89}$$
$$Y = y, Z = z$$

The transformations of T and X may be written

$$\frac{gT}{c} = \left(1 + \frac{gx}{c^2}\right)\sinh \frac{gt}{c}, \quad 1 + \frac{gX}{c^2} = \left(1 + \frac{gx}{c^2}\right)\cosh \frac{gt}{c}. \tag{4.90}$$

It follows from these equations that

$$\frac{gT}{c} = (1 + \frac{gX}{c^2})\tanh \frac{gt}{c}, \tag{4.91}$$

showing that the coordinate curves t = constant are straight lines in the T, X-frame passing through the point $T = 0, X = -c^2/g$.

Using the identity $\cosh^2 \theta - \sinh^2 \theta = 1$ we get

$$\left(1 + \frac{gX}{c^2}\right)^2 - \left(\frac{gT}{c}\right)^2 = \left(1 + \frac{gx}{c^2}\right)^2, \tag{4.92}$$

showing that the coordinate curves $x =$ constant are hyperbolae in the T, X-diagram (Fig. 4.11).

Taking the differentials of the coordinates in Eq. (4.89) gives the form of the line-element in the co-moving coordinates of the uniformly accelerated reference frame,

$$ds^2 = -c^2 dT^2 + dX^2 + dY^2 + dZ^2 = -\left(1 + \frac{gx}{c^2}\right)^2 c^2 dt^2 + dx^2 + dy^2 + dz^2. \tag{4.93}$$

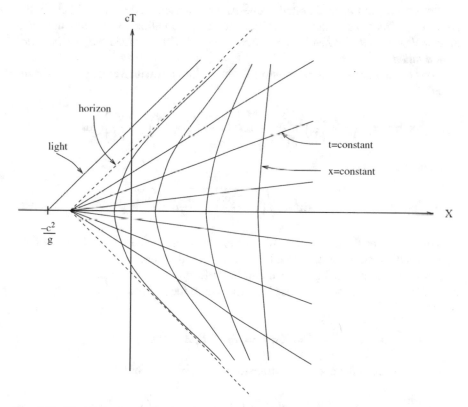

Fig. 4.11 World lines and simultaneity lines of a uniformly accelerated reference system. Minkowski diagram of the world lines and simultaneity lines of the uniformly accelerated reference frame AF with reference to the inertial frame IF in which AF is at rest at $T = 0$

The line-element is an invariant quantity, but looks different in different coordinate systems. Note that the metric is diagonal, i.e. only components with equal indices are different from zero when the basis vectors are orthogonal.

We now consider standard clocks at rest in the accelerated reference frame, i.e. $dx = dy = dz = 0$ along the world lines of the clocks. Then the line-element (4.93) reduces to

$$ds^2 = -\left(1 + \frac{gx}{c^2}\right)^2 c^2 dt^2. \tag{4.94}$$

Utilizing the physical interpretation (3.91) of the line element for a time-like interval we obtain the relationship between the proper time and the coordinate time

$$d\tau = \left(1 + \frac{gx}{c^2}\right) dt. \tag{4.95}$$

Since the rate of coordinate time is position independent, this equation tells how the rate of proper time depends upon the position.

An observer in the accelerated frame AF experiences a gravitational field in the negative x-direction. When $x < 0$ then $d\tau < dt$. Equation (4.95) says that *time passes slower further down in a gravitational field*. This is called *the gravitational time dilation*.

Consider a standard clock moving in the x-direction with velocity $v = dx/dt$. Then

$$-c^2 d\tau^2 = -\left(1 + \frac{gx}{c^2}\right)^2 c^2 dt^2 + dx^2 = -\left[\left(1 + \frac{gx}{c^2}\right)^2 - \frac{v^2}{c^2}\right]c^2 dt^2 \tag{4.96}$$

Hence

$$d\tau = \sqrt{\left(1 + \frac{gx}{c^2}\right)^2 - \frac{v^2}{c^2}}\, dt. \tag{4.97}$$

This expresses the combined effect of the position dependent gravitational time dilation and the velocity-dependent kinematic time dilation.

Let us consider how light moves in the uniformly accelerated reference frame. As a simple example we consider light moving in the y-direction in the laboratory frame,

$$X = X_0, Y = Y_0 + cT, Z = 0. \tag{4.98}$$

Inserting this in the coordinate transformation above we obtain

$$\frac{gT}{c} = \frac{g}{c^2}(y - y_0), \left(1 + \frac{gx}{c^2}\right)^2 + \left[\frac{g}{c^2}(y - y_0)\right]^2 = \left(1 + \frac{gX_0}{c^2}\right)^2. \tag{4.99}$$

The last equation shows that the light moves along a circular path with radius $X_0 + c^2/g$. Differentiating the equation of the trajectory with respect to x we obtain

$$\frac{dy}{dx} = \frac{x + c^2/g}{y - y_0}. \tag{4.100}$$

Hence $dy/dx = 0$ at the horizon. In other word the light moves in the vertical direction at the horizon. At that position the light has no motion in the y-direction. The reason is that the time does not progress at the horizon.

Note also from the line-element that

$$\frac{dx}{dt} = \left(1 + \frac{gx}{c^2}\right)c \tag{4.101}$$

for light moving in the x-direction. Light moves slower the further down it is in the gravitational field, and the coordinate velocity of the light approaches zero as the light approaches the horizon.

4.10 The Projection Tensor

An observer moves through spacetime with a 4-velocity $\vec{u}$. An arbitrary 4-vector $\vec{a}$ shall be decomposed into one component, $\vec{a}_\parallel$, along $\vec{u}$ and one, $\vec{a}_\perp$, orthogonal to $\vec{u}$, so that $\vec{a} = \vec{a}_\parallel + \vec{a}_\perp$. Using units so that $c = 1$ the 4-velocity is then a time-like unit vector, $\vec{u} \cdot \vec{u} = -1$, and

$$\vec{a}_\parallel = \frac{(\vec{a} \cdot \vec{u})\vec{u}}{\vec{u} \cdot \vec{u}} = -(\vec{a} \cdot \vec{u})\vec{u}, \vec{a}_\perp = \vec{a} + (\vec{a} \cdot \vec{u})\vec{u}. \tag{4.102}$$

In the following we shall need the corresponding 4-velocity form,

$$\underline{u} = u_\mu \underline{\omega}^\mu. \tag{4.103}$$

Definition 4.10.1 (*The Projection Tensor*) The projection tensor, P, is a mixed tensor of rank 2 which is given in terms of the unit tensor $\mathbf{I} = \underline{\omega}^\mu \otimes \vec{e}_\mu$ of rank 2, the 4-vector $\vec{u}$ and the corresponding 4-form $\underline{u}$ as

$$\mathbf{P} = \mathbf{I} + \underline{u} \otimes \vec{u}. \tag{4.104}$$

The mixed components of the projection tensor are

$$P^\nu_\mu = \delta^\nu_\mu + u_\mu u^\nu. \tag{4.105}$$

Note that the last term is *not* a scalar product. The covariant components of the projection tensor are

$$P_{\mu\nu} = g_{\mu\nu} + u_\mu u_\nu. \tag{4.106}$$

Applying the projection tensor to the vector $\vec{a}$ and noting from the line below Eq. (3.73), that $\mathbf{I}(\vec{a}) = \vec{a}$, we get

$$\mathbf{P}(\vec{a}) = \mathbf{I}(\vec{a}) + \left(\underline{u} \otimes \vec{u}\right)(\vec{a}) = \vec{a} + \underline{u}(\vec{a})\vec{u} = \vec{a} + (\vec{a} \cdot \vec{u})\vec{u} = \vec{a}_\perp. \tag{4.107}$$

This means that when one applies the projection tensor to a vector $\vec{a}$ one gets out the component $\vec{a}_\perp$ of the vector orthogonal to the vector $\vec{u}$ of the projection tensor. In other words one gets the projection onto the simultaneity space orthogonal to the time direction of the observer with 4-velocity $\vec{u}$.

Example 4.10.1 (*Covariant condition for uniformly accelerated motion*) The covariant condition for uniformly accelerated motion of a particle is that the proper acceleration of the particle is constant, which means that the projection of the rate of change of the acceleration vector vanishes in an instantaneous rest system of the particle. The covariant mathematical expression of this is: $\mathbf{P}(d\vec{a}/d\tau) = 0$, where τ is the proper time of the particle. The component form of this condition is: $P^\mu_\nu da^\nu/d\tau = 0$.

Example 4.10.2 (*Spatial metric and the projection tensor*) We shall here generalize Formula (4.4) for the spatial metric tensor. Consider a coordinate interval dx^μ. It has a component orthogonal to a 4-velocity $\vec{u}$ given by $dl^\mu = P^\mu_\alpha dx^\alpha$. The invariant spatial line-element is

$$dl^2 = dl_\alpha dl^\alpha = g_{\alpha\beta} dl^\alpha dl^\beta = g_{\alpha\beta} P^\alpha_\mu P^\beta_\nu dx^\mu dx^\nu = P_{\mu\beta} P^\beta_\nu dx^\mu dx^\nu. \tag{4.108}$$

The components of the spatial metric tensor in a simultaneity space orthogonal to $\vec{u}$ may be defined by

$$dl^2 = \gamma_{\mu\nu} dx^\mu dx^\nu. \tag{4.109}$$

This gives

$$\gamma_{\mu\nu} = P_{\mu\beta} P^\beta_\nu = \left(g_{\mu\beta} + u_\mu u_\beta\right)\left(\delta^\beta_\nu + u_\nu u^\beta\right) = g_{\mu\nu} + u_\nu u_\mu + u_\mu u_\nu$$
$$+ u_\mu u_\nu u_\beta u^\beta = g_{\mu\nu} + u_\mu u_\nu. \tag{4.110}$$

The right-hand side is the covariant components of the projection tensor. Hence the spatial metric tensor is equal to the projection tensor,

$$\gamma_{\mu\nu} = P_{\mu\nu}. \tag{4.111}$$

We shall now consider the simultaneity space of an observer at rest in the coordinate system. Then the contravariant components of the 4-velocity are

$$u^\mu = (-g_{00})^{-1/2}(1, 0, 0, 0).$$ (4.112)

The covariant components are

$$u_0 = -(-g_{00})^{1/2}, u_i = (-g_{00})^{-1/2}g_{i0}.$$ (4.113)

Substituting these expressions into Eq. (4.110) gives

$$\gamma_{00} = \gamma_{i0} = 0, \gamma_{ij} = g_{ij} - \frac{g_{i0}g_{j0}}{g_{00}},$$ (4.114)

in agreement with Eq. (4.4).

Exercises

4.1. *Relativistic rotating disc*

A disc rotates with constant angular velocity in ω in its own plane and around a fixed axis A. The axis is chosen to be the origin in a non-rotating Cartesian coordinate system (x, y) with coordinate clocks showing t. (The z-coordinate is kept constant from now on). The motion of a given point on the disc can be expressed as

$$x = r\cos(\omega t + \phi), y = r\sin(\omega t + \phi),$$ (4.115)

where (r, ϕ) are coordinates specifying the point at the disc.

(a) An observer is able to move on the disc and performs measurements of distance between neighbouring points at different locations on the disc. The measurements are performed when the observer is stationary with respect to the disc. The result is assumed to be the same as that measured in an inertial frame with the same velocity as the observer at the time of the measurement. The lengths measured by the observer are given by

$$d\ell^2 = f_1(r, \phi)dr^2 + f_2(r, \phi)d\phi^2.$$ (4.116)

Find $f_1(r, \phi)$ and $f_2(r, \phi)$.

We now assume that the observer measures the distance from the axis A to a point $(R, 0)$ along the line $\phi = 0$, by adding the result of measurements between neighbouring points. What is the result the observer finds? Furthermore the observer measures the distance around the circle $r = R$. What is then found? In what way, based on this result, is it possible to deduce that the metric considered by the observer is non-Euclidean? Will the observer find a negative or positive curvature of the disc?

We introduce coordinates $(\tilde{t}, \tilde{x}, \tilde{y})$ that follow the rotating disc. They are given by

$$\tilde{t} = t, \quad \tilde{x} = r \cos \phi, \quad \tilde{y} = r \sin \phi. \tag{4.117}$$

(b) Find the invariant interval $ds^2 = dx^2 + dy^2 - c^2 dt^2$ in terms of the coordinates $(\tilde{t}, \tilde{x}, \tilde{y})$.

(c) Light signals are sent from the axis A. How will the paths of the light signals be as seen from the $(\tilde{x}, \tilde{y})$ system? Draw a figure that illustrates this. A light signal with the frequency ν_0 is received by the observer in $r = R, \phi = 0$. Which frequency ν will be measured by the observer?

(d) We now assume that standard clocks measuring proper time are tightly packed around the circle $r = R$. The clocks are at rest on the disc. We now want to synchronize the clocks and start out with a clock at the point $(R, 0)$. The clocks are then synchronized in the direction of increasing ϕ in the following way: When a clock is tuned at a point ϕ, the clock at the neighbouring point $\phi + d\phi$ is also tuned so that it shows the same time at simultaneity in the instantaneous rest frame of the two clocks.

Show that there is a problem with synchronization when this process is performed around the entire circle, by the fact that the clock we started out with is no longer synchronous with the neighbouring clock which is tuned according to the synchronization process. Find the time difference between these two clocks.

(e) Locally around a point (r, ϕ, t) we can define an inertial system being an instantaneous rest frame of the point (r, ϕ) on the disc. We introduce an orthonormal set of basis vectors $\vec{e}_{\hat{\lambda}}$, $\vec{e}_{\hat{\eta}}$ and $\vec{e}_{\hat{\xi}}$ in this frame. The vector $\vec{e}_{\hat{\lambda}}$ points along the time axis of the system, $\vec{e}_{\hat{\xi}}$ points radially and $\vec{e}_{\hat{\eta}}$ tangentially. Find the vectors expressed by $\vec{e}_t$, $\vec{e}_x$ and $\vec{e}_y$.

4.2. Uniformly accelerated system of reference

We will now study a coordinate system (t, x) co-moving with a uniformly accelerated reference frame, AF, in a 2-dimensional Minkowski space. The connection with a Cartesian coordinate system (T, X) co-moving with an instantaneous inertial rest frame, IF, of AF at the point of time $T = 0$ is given by the coordinate transformation

$$T = x \sinh(at), \quad X = x \cosh(at) \tag{4.118}$$

where a is a constant.

(a) Draw the coordinate lines $t = $ constant and $x = $ constant in a (T, X)-diagram.

(b) Find the line-element $ds^2 = -dT^2 + dX^2$ expressed by t and x.

(c) We now assume that a particle has a path in spacetime so that it follows one of the curves $x = $ constant. Such a motion is called hyperbolic motion. Why?

Show that the particle has constant acceleration when the acceleration is measured in the instantaneous rest frame of the particle. Find the acceleration of the rod has constant acceleration and the rod g. Find also the velocity and acceleration of the particle in the system (T, X).

(d) Show that at any point on the trajectory of a reference particle in AF the direction of the coordinate axes in the (T, X)-system will overlap with the time and spatial axis of the instantaneous rest frame of the particle. Explain why it is possible to see from the line element that the X-coordinate measures length along the spatial axis, whereas the T-coordinate, which is the coordinate time, is in general not the proper time of the particle. For what value of X is the coordinate time equal to the proper time?

The (t, x)-coordinate system can be considered as an attempt to construct, from the instantaneous rest frames along the path, a coordinate system covering the entire spacetime. Explain why this is not possible for the entire space. (Hint: There is a coordinate singularity at a certain distance from the trajectory of the particle).

(e) A rod is moving in the direction of its own length. At the time $T = 0$ the rod is at rest, but still accelerated. The length of the rod measured in the stationary system is L at this time. The rod moves so that the forwards point of the rod has constant rest acceleration measured in the instantaneous rest frame.

We assume that the acceleration of the rod finds place so that the infinitesimal distance $d\ell$ between neighbouring points on the rod measured in the instantaneous rest frame is constant. Find the motion of the rear point of the rod in the stationary reference system. Why is there a maximal length of the rod, L_{max} ?

If the rear point of the rod has constant acceleration and the rod is accelerated as previously in this exercise, then is there a maximal value of L ?

4.3. Uniformly accelerated space ship

(a) A spaceship leaves the Earth at the time $T = 0$ and moves with a constant acceleration g, equal to the gravitational constant at the Earth, into space. Find how far the ship has travelled during 10 years of proper time of the ship.

(b) Radio signals are sent from the Earth towards the spaceship. Show that signals that are sent after a given time T will never reach the ship (even if the signals travel with the speed of light). Find T. At what time are the signals sent from the Earth if they reach the ship after 10 years (proper time of the ship)?

(c) Calculate the frequency of the radio signals received by the ship, given by the frequency ν_0 (emitter frequency) and the time t_0 (emitter time). Investigate the behaviour of the frequency when the proper time on the space ship $\tau \to \infty$.

4.4. *Light emitted from a point source in a gravitational field*

A point-like light source is at the position $x = x_1$, $y = 0$ in a uniformly accelerated reference frame AF. A photon is emitted from the source at a point of time $t = 0$. It is emitted in the $(x, y)-$ plane in a direction making an angle θ_0 with the x-axis.

Find the equation for the path followed by the photon, and identify the nature of the trajectory.

4.5. *Geometrical optics in a gravitational field*

It was shown in Eq. (4.99) that light does not move along a straight path in the gravitational field experienced in a uniformly accelerated reference frame AF, but along a circular trajectory.

(a) *A sphere seen from above.* The observer P is at a height $x = x_1$. The centre of the sphere is at a distance b vertically beneath P. Light is emitted from the surface of the body. It moves along a circular path with radius $R = x_1/\sin\theta$, where θ is the angle between the light path and the x-axis at the position of the observer. The corresponding angle without a gravitational field would be θ_0 given by $\sin\theta_0 = r/b$, where r is the radius of the sphere. The acceleration of gravity at the position of the observer is $g/c^2 = 1/x_1$.

Calculate how the angle θ depends upon g, r and b.

(b) *A sphere seen from below.* Same as (a) but with the observer below the sphere.

(c) *An experiment.* Place a camera one metre above the centre of a sphere of radius 10 cm and another one a metre below in the gravitational field of the Earth. Calculate the difference of θ in photographs taken with the two cameras.

Chapter 5
Covariant Differentiation

Abstract The theory of forms is a theory of antisymmetric tensors. In such a theory we need an antisymmetric version of the covariant derivative such that the derivative of a form is a form. Hence in this chapter we first introduce the covariant derivative and then the antisymmetric exterior derivative. The relativistic Euler–Lagrange equations are introduced and applied to deduce the equation of motion of free particles in curved spacetime—the geodesic equation. Since gravity is not reckoned as a force in the general theory of relativity "free particles" in this theory correspond to particles acted upon by gravity in Newton's theory. The fundamental equations of electromagnetism in form language are also presented in this chapter.

5.1 Differentiation of Forms

We must have a method of differentiation which maintains the antisymmetry so that differentiating a form gives out a form.

5.1.1 Exterior Differentiation

The *exterior derivative* of a 0-form, i.e. a scalar function, f, is given by

$$\underline{d}f = \frac{\partial f}{\partial x^\mu}\underline{\omega}^\mu = f_\mu\underline{\omega}^\mu, \tag{5.1}$$

where $\underline{\omega}^\mu$ are coordinate basis forms

$$\underline{\omega}^\mu\left(\frac{\partial}{\partial x^\nu}\right) = \delta^\mu_\nu. \tag{5.2}$$

We then get

© Springer Nature Switzerland AG 2020
Ø. Grøn, *Introduction to Einstein's Theory of Relativity*,
Undergraduate Texts in Physics, https://doi.org/10.1007/978-3-030-43862-3_5

$$\underline{\omega}^{\mu} = \delta^{\mu}_{\nu}\underline{\omega}^{\nu} = \frac{\partial x^{\mu}}{\partial x^{\nu}}\underline{\omega}^{\nu} = \underline{d}x^{\mu}. \tag{5.3}$$

This shows that we can write the basis forms as exterior derivatives of the coordinates in an arbitrary *coordinate* basis. The differential dx^{μ} is given by

$$\underline{d}x^{\mu}(\vec{dr}) = dx^{\mu}, \tag{5.4}$$

where $\vec{dr}$ is an infinitesimal position vector. Note that $\underline{d}x^{\mu}$ are *not* infinitesimal quantities. It follows that in a coordinate basis the exterior derivative of a p-form

$$\underline{\alpha} = \frac{1}{p!}\alpha_{\mu_1\dots\mu_p}\underline{d}x^{\mu_1} \wedge \cdots \wedge \underline{d}x^{\mu_p} \tag{5.5}$$

has the following component form:

$$\underline{d}\,\underline{\alpha} = \frac{1}{p!}\alpha_{\mu_1\dots\mu_p,\mu_0}\underline{d}x^{\mu_0} \wedge \underline{d}x^{\mu_1} \wedge \cdots \wedge \underline{d}x^{\mu_p}, \tag{5.6}$$

where $\mu_0 \equiv \frac{\partial}{\partial x_0^{\mu}}$. This shows that *the exterior derivative of a p-form is a $(p+1)$-form.*

Let $(\underline{d\alpha})_{\mu_0\dots\mu_p}$ be the form components of $\underline{d\alpha}$. They must, by definition, be antisymmetric under an arbitrary interchange of indices:

$$\underline{d\alpha} = \frac{1}{(p+1)!}(\underline{d\alpha})_{\mu_0\dots\mu_p}\underline{d}x^{\mu_0} \wedge \cdots \wedge \underline{d}x^{\mu_p} = \frac{1}{p!}\alpha_{[\mu_1\dots\mu_p,\mu_0]}\underline{d}x^{\mu_0} \wedge \cdots \wedge \underline{d}x^{\mu_p}. \tag{5.7}$$

Hence

$$(\underline{d\alpha})_{\mu_0\dots\mu_p} = (p+1)\alpha_{[\mu_1\dots\mu_p,\mu_0]}. \tag{5.8}$$

The component form of the form equation $\underline{d\alpha} = 0$ is

$$\alpha_{[\mu_1\dots\mu_p,\mu_0]} = 0. \tag{5.9}$$

Example 5.1.1 (*Relationship between exterior derivative and curl*) We consider a 1-form in a 3-space,

$$\underline{\alpha} = \alpha_i\underline{d}x^i, \quad x^i = (x, y, z). \tag{5.10}$$

Its exterior derivative is

$$\underline{d\alpha} = \alpha_{i,j}\underline{d}x^j \wedge \underline{d}x^i. \tag{5.11}$$

Also, assume that $\underline{d\alpha} = 0$. The corresponding component equation is $\alpha_{[i,j]} = 0$ or $\alpha_{i,j} - \alpha_{j,i} = 0$. Writing it out we have

$$\frac{\partial \alpha_y}{\partial x} - \frac{\partial \alpha_x}{\partial y} = 0, \qquad \frac{\partial \alpha_z}{\partial x} - \frac{\partial \alpha_x}{\partial z} = 0, \qquad \frac{\partial \alpha_y}{\partial z} - \frac{\partial \alpha_z}{\partial y} = 0. \qquad (5.12)$$

which corresponds to

$$\nabla \times \vec{\alpha} = 0. \qquad (5.13)$$

The outer product of an outer product is

$$\underline{d^2 \alpha} \equiv \underline{d(d\alpha)}, \qquad (5.14)$$

with component form

$$\underline{d^2 \alpha} = \frac{1}{p!} \alpha_{\mu_1 \ldots \mu_p, \nu_1 \nu_2} \underline{dx}^{\nu_2} \wedge \underline{dx}^{\nu_1} \wedge \cdots \wedge \underline{dx}^{\mu_p}, \quad \nu_1 \nu_2 \equiv \frac{\partial^2}{\partial x^{\nu_1} \partial x^{\nu_2}}. \qquad (5.15)$$

Since

$$_{,\nu_1 \nu_2} \equiv \frac{\partial^2}{\partial x^{\nu_1} \partial x^{\nu_2}} = _{,\nu_2 \nu_1} \equiv \frac{\partial^2}{\partial x^{\nu_2} \partial x^{\nu_1}} \qquad (5.16)$$

the quantities $\alpha_{\mu_1 \ldots \mu_p, \nu_1 \nu_2}$ are symmetric in ν_1 and ν_2. On the other hand, the basis is antisymmetric in ν_1 and ν_2. We saw in Eq. (3.107) that the product of a symmetric and an antisymmetric quantity vanishes. Hence, we have

$$\underline{d^2 \alpha} = 0, \qquad (5.17)$$

which means that *the second exterior derivative of a p-form* (with scalar components, see later) *vanishes*. This is sometimes called *Poincaré's lemma*. For a 1-form, it corresponds to the vector equation

$$\nabla \cdot (\nabla \times \vec{A}) = 0. \qquad (5.18)$$

Let $\underline{\alpha}$ be a p-form and $\underline{\beta}$ a q-form. Then

$$\underline{d(\alpha \wedge \beta)} = \underline{d\alpha} \wedge \underline{\beta} + (-1)^p \underline{\alpha} \wedge \underline{d\beta}. \qquad (5.19)$$

5.1.2 Covariant Derivative

The general theory of relativity contains a *covariance principle* which states that all equations expressing laws of nature must have the same form irrespective of the coordinate system in which they are derived. This is achieved by writing all equations in terms of tensors. Let us see if the partial derivative of vector components transform as tensor components.

Consider a vector $\vec{A} = A^\mu \vec{e}_\mu = A^{\mu'} \vec{e}_{\mu'}$. The basis transforms as

$$\frac{\partial}{\partial x^{\nu'}} = \frac{\partial x^\nu}{\partial x^{\nu'}} \frac{\partial}{\partial x^\nu}. \tag{5.20}$$

Hence

$$A^{\mu'}_{,\nu'} \equiv \frac{\partial}{\partial x^{\nu'}} \left(A^{\mu'} \right) = \frac{\partial x^\nu}{\partial x^{\nu'}} \frac{\partial}{\partial x^\nu} \left(A^{\mu'} \right) = \frac{\partial x^\nu}{\partial x^{\nu'}} \frac{\partial}{\partial x^\nu} \left(\frac{\partial x^{\mu'}}{\partial x^\mu} A^\mu \right). \tag{5.21}$$

Performing the differentiation of the product gives

$$A^{\mu'}_{,\nu'} = \frac{\partial x^\nu}{\partial x^{\nu'}} \frac{\partial x^{\mu'}}{\partial x^\mu} A^\mu_{,\nu} + \frac{\partial x^\nu}{\partial x^{\nu'}} A^\mu \frac{\partial^2 x^{\mu'}}{\partial x^\nu \partial x^\mu}. \tag{5.22}$$

The first term corresponds to a tensor transformation. However, the presence of the last term shows that $A^\mu_{,\nu}$ does not, in general, transform as the components of a tensor. Note that $A^\mu_{,\nu}$ will transform as a components of a tensor under linear transformations such as the Lorentz transformations.

In order to obtain a proper tensor formalism, the partial derivative must be generalized so as to ensure that when its generalization is applied to tensor components it produces tensor components.

5.2 The Christoffel Symbols

The covariant derivative of the components of a vector field was introduced by Elwin Christoffel to be able to differentiate tensor fields. It is defined in coordinate basis by generalizing the partial derivative $A^\mu_{,\nu}$ to a derivative written as $A^\mu_{;\nu}$ and which transforms tensorially,

$$A^{\mu'}_{;\nu'} \equiv \frac{\partial x^{\mu'}}{\partial x^\mu} \cdot \frac{\partial x^\nu}{\partial x^{\nu'}} A^\mu_{;\nu}. \tag{5.23}$$

Definition 5.2.1 (*Christoffel symbols*) The *Christoffel symbols* are defined by writing the covariant derivative of the vector components in the form

$$A^{\mu}_{;\nu} \equiv A^{\mu}_{,\nu} + A^{\alpha}\Gamma^{\mu}_{\alpha\nu}. \qquad (5.24)$$

The Christoffel symbols $\Gamma^{\mu}_{\alpha\nu}$ are also called the "connection coefficients in coordinate basis". From the transformation formulae of the left-hand side and the first term on the right-hand side it follows that the Christoffel symbols transform as

$$\Gamma^{\alpha'}_{\mu'\nu'} = \frac{\partial x^{\nu}}{\partial x^{\nu'}} \frac{\partial x^{\mu}}{\partial x^{\mu'}} \frac{\partial x^{\alpha'}}{\partial x^{\alpha}} \Gamma^{\alpha}_{\mu\nu} + \frac{\partial x^{\alpha'}}{\partial x^{\alpha}} \frac{\partial^2 x^{\alpha}}{\partial x^{\mu'}\partial x^{\nu'}}. \qquad (5.25)$$

Due to the last term, the Christoffel symbols do not transform as tensor components. It is possible to make *all* Christoffel symbols vanish by transforming into a locally Cartesian coordinate system which is co-moving in a locally non-rotating reference frame in free fall. Such coordinates are known as *Gaussian coordinates*.

As discussed in Sect. 1.4 according to the general theory of relativity an *inertial frame* is a non-rotating frame in free fall. The Christoffel symbols are 0 (zero) in a locally Cartesian coordinate system which is co-moving in a local inertial frame Local Gaussian coordinates are indicated with a hat over the indices since the coordinate vector basis is orthonormal in such a coordinate system, giving

$$\Gamma^{\hat{\alpha}}_{\hat{\mu}\hat{\nu}} = 0. \qquad (5.26)$$

A transformation from local Gaussian coordinates to any coordinates leads to

$$\Gamma^{\alpha'}_{\mu'\nu'} = \frac{\partial x^{\alpha'}}{\partial x^{\hat{\alpha}}} \frac{\partial^2 x^{\hat{\alpha}}}{\partial x^{\mu'}\partial x^{\nu'}}. \qquad (5.27)$$

This equation shows that the Christoffel symbols are symmetric in the two lower indices, i.e.

$$\Gamma^{\alpha'}_{\mu'\nu'} = \Gamma^{\alpha'}_{\nu'\mu'}. \qquad (5.28)$$

Example 5.2.1 (*The Christoffel symbols in plane polar coordinates*) The transformation between plane polar coordinates and Cartesian coordinates is

$$\begin{aligned} x = r\cos\theta, \quad y = r\sin\theta, \\ r = \sqrt{x^2 + y^2}, \; \theta = \arctan\tfrac{y}{x}, \end{aligned} \qquad (5.29)$$

We need the derivatives of the coordinates with respect to each other

$$\begin{aligned} \tfrac{\partial x}{\partial r} = \cos\theta, \; \tfrac{\partial x}{\partial \theta} = -r\sin\theta, \; \tfrac{\partial r}{\partial x} = \tfrac{x}{r} = \cos\theta, \; \tfrac{\partial r}{\partial y} = \sin\theta, \\ \tfrac{\partial y}{\partial r} = \sin\theta, \; \tfrac{\partial y}{\partial \theta} = r\cos\theta, \; \tfrac{\partial \theta}{\partial x} = -\tfrac{\sin\theta}{r}, \; \tfrac{\partial \theta}{\partial y} = \tfrac{\cos\theta}{r}, \end{aligned} \qquad (5.30)$$

Inserting these expressions into Eq. (5.27) gives the Christoffel symbols in the plane polar coordinates

$$\Gamma^r_{\theta\theta} = \frac{\partial r}{\partial x}\frac{\partial^2 x}{\partial\theta^2} + \frac{\partial r}{\partial y}\frac{\partial^2 y}{\partial\theta^2} = \cos\theta(-r\cos\theta) + \sin\theta(-r\sin\theta)$$

$$= -r(\cos\theta^2 + \sin\theta^2) = -r, \tag{5.31}$$

$$\Gamma^\theta_{r\theta} = \Gamma^\theta_{\theta r} = \frac{\partial\theta}{\partial x}\frac{\partial^2 x}{\partial\theta\partial r} + \frac{\partial\theta}{\partial y}\frac{\partial^2 y}{\partial\theta\partial r} = -\frac{\sin\theta}{r}(-\sin\theta) + \frac{\cos\theta}{r}(\cos\theta) = \frac{1}{r}. \tag{5.32}$$

The geometrical interpretation of the covariant derivative was given by Levi-Civita. Consider a curve S in any (e.g. curved) space. It is parameterized by λ, i.e. $x^\mu = x^\mu(\lambda)$. The parameter λ is invariant and chosen to represent the arc length. The tangent vector field of the curve is

$$\vec{u} = (dx^\mu/d\lambda)\vec{e}_\mu. \tag{5.33}$$

The curve passes through a vector field $\vec{A}$.

Definition 5.2.2 (*Covariant directional derivative*) The *covariant directional derivative* of the vector field along the curve is defined as

$$\nabla_{\vec{u}}\vec{A} = \frac{d\vec{A}}{d\lambda} \equiv A^\mu_{;\nu}\frac{dx^\nu}{d\lambda}\vec{e}_\mu = A^\mu_{;\nu}u^\nu\vec{e}_\mu. \tag{5.34}$$

Definition 5.2.3 (*Parallel transport*) The vectors are said to be connected by *parallel transport* along the curve if

$$A^\mu_{;\nu}u^\nu = 0. \tag{5.35}$$

According to the geometrical interpretation of Levi-Civita, the covariant directional derivative is

$$\nabla_{\vec{u}}\vec{A} = A^\mu_{;\nu}u^\nu\vec{e}_\mu = \lim_{\Delta\lambda\to 0}\frac{\vec{A}_\|(\lambda + \Delta\lambda) - \vec{A}(\lambda)}{\Delta\lambda}, \tag{5.36}$$

where $\vec{A}_\|(\lambda + \Delta\lambda)$ means the vector $\vec{A}$ parallel transported from Q to P (Fig. 5.1).

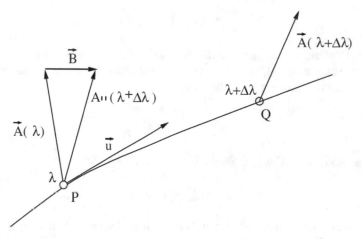

Fig. 5.1 Parallel transport from P to Q. The vector $\vec{B} = A^{\mu}_{;\nu}u^{\nu}\Delta\lambda\vec{e}_{\mu}$

5.3 Geodesic Curves

Definition 5.3.1 (*Geodesic curves*) A *geodesic curve* is defined by the requirement that the vectors of the tangent vector field of the curve are connected by parallel transport.

This definition says that geodesic curves are "as straight as possible". If vectors in a vector field $\vec{A}(\lambda)$ are connected by parallel transport by a displacement along a vector $\vec{u}$, we have $A^{\mu}_{;\nu}u^{\nu} = 0$. For geodesic curves, we then have

$$u^{\mu}_{;\nu}u^{\nu} = 0 \tag{5.37}$$

which is the *geodesic equation*

$$(u^{\mu}_{,\nu} + \Gamma^{\mu}_{\alpha\nu}u^{\alpha})u^{\nu} = 0. \tag{5.38}$$

Then we are using that

$$\frac{d}{d\lambda} \equiv \frac{dx^{\nu}}{d\lambda}\frac{\partial}{\partial x^{\nu}} = u^{\nu}\frac{\partial}{\partial x^{\nu}}, \tag{5.39}$$

which gives

$$\frac{du^{\mu}}{d\lambda} = u^{\nu}\frac{\partial u^{\mu}}{\partial x^{\nu}} = u^{\nu}u^{\mu}_{,\nu}. \tag{5.40}$$

Hence, the geodesic equation can be written as

$$\frac{du^\mu}{d\lambda} + \Gamma^\mu_{\alpha\nu} u^\alpha u^\nu = 0. \tag{5.41}$$

A usual notation is to represent the derivative with respect to an invariant curve parameter by an over dot, $\cdot = \frac{d}{d\lambda}$. Then the expression for the components of the tangent vector of the curve takes the form

$$u^\mu = \frac{dx^\mu}{d\lambda} = \dot{x}^\mu, \tag{5.42}$$

and *the geodesic equation is written as*

$$\ddot{x}^\mu + \Gamma^\mu_{\alpha\nu} \dot{x}^\alpha \dot{x}^\nu = 0. \tag{5.43}$$

Geodesic curves on a flat surface and on a spherical surface are shown in Figs. 5.2 and 5.3, respectively.

Fig. 5.2 Geodesic curve on a flat surface. On a flat surface, the geodesic curve is the minimal distance between P and Q

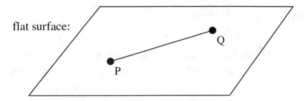

Fig. 5.3 Geodesic curves on a sphere

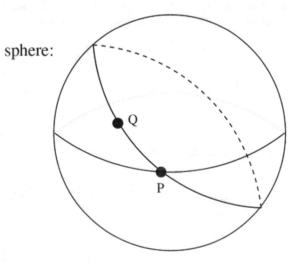

5.4 The Covariant Euler–Lagrange Equations

Let a particle have a world line between two points in spacetime (events) P_1 and P_2. The curve is described by an invariant parameter λ (proper time τ is used for particles with a rest mass, i.e. for time-like world lines).

The Lagrange function is a function of the coordinates and their derivatives,

$$L = L(x^\mu, \dot{x}^\mu), \quad \dot{x}^\mu \equiv \frac{\mathrm{d}x^\mu}{\mathrm{d}\lambda}. \tag{5.44}$$

Note that if $\lambda = \tau$ then $\dot{x}^\mu$ are the 4-velocity components.

The action integral is $S = \int L(x^\mu, \dot{x}^\mu)\mathrm{d}\lambda$. The *principle of extremal action* (*Hamiltons principle*) says that the world line of a particle is determined by the condition that S shall be extremal for all infinitesimal variations of curves which keep P_1 and P_2 fixed, i.e.

$$\delta \int_{\lambda_1}^{\lambda_2} L(x^\mu, \dot{x}^\mu)\mathrm{d}\lambda = 0, \tag{5.45}$$

where λ_1 and λ_2 are the parameter values at P_1 and P_2 (Fig. 5.4).

For all the variations the following condition applies,

$$\delta x^\mu(\lambda_1) = \delta x^\mu(\lambda_2) = 0. \tag{5.46}$$

We write Eq. (5.45) as

Fig. 5.4 Neighboring geodesics in a Minkowski diagram

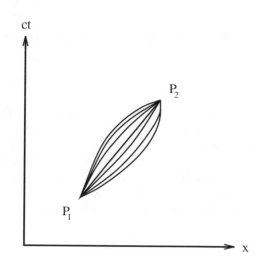

$$\delta \int_{\lambda_1}^{\lambda_2} L d\lambda = \int_{\lambda_1}^{\lambda_2} \left[\frac{\partial L}{\partial x^\mu} \delta x^\mu + \frac{\partial L}{\partial \dot{x}^\mu} \delta \dot{x}^\mu \right] d\lambda. \tag{5.47}$$

Partial integration of the last term gives

$$\int_{\lambda_1}^{\lambda_2} \frac{\partial L}{\partial \dot{x}^\mu} \delta \dot{x}^\mu d\lambda = \left[\frac{\partial L}{\partial \dot{x}^\mu} \delta x^\mu \right]_{\lambda_1}^{\lambda_2} - \int_{\lambda_1}^{\lambda_2} \frac{d}{d\lambda} \left(\frac{\partial L}{\partial \dot{x}^\mu} \right) \delta x^\mu d\lambda. \tag{5.48}$$

Due to the conditions $\delta x^\mu(\lambda_1) = \delta x^\mu(\lambda_2) = 0$, the first term becomes zero. Then we have

$$\delta S = \int_{\lambda_1}^{\lambda_2} \left[\frac{\partial L}{\partial x^\mu} - \frac{d}{d\lambda} \left(\frac{\partial L}{\partial \dot{x}^\mu} \right) \right] \delta x^\mu d\lambda. \tag{5.49}$$

The world line the particle follows is determined by the condition $\delta S = 0$ for any variation δx^μ. Hence, the world line of the particle must be given by

$$\frac{\partial L}{\partial x^\mu} - \frac{d}{d\lambda} \left(\frac{\partial L}{\partial \dot{x}^\mu} \right) = 0. \tag{5.50}$$

These are the covariant *Euler–Lagrange* equations.

The canonical momentum p_μ conjugated to a coordinate x^μ is defined as

$$p_\mu \equiv \frac{\partial L}{\partial \dot{x}^\mu}. \tag{5.51}$$

The Lagrange equations can now be written as

$$\frac{dp_\mu}{d\lambda} = \frac{\partial L}{\partial x^\mu} \quad \text{or} \quad \dot{p}_\mu = \frac{\partial L}{\partial x^\mu}. \tag{5.52}$$

A coordinate which the Lagrange function does not depend on is known as a *cyclic coordinate*. Hence, $\frac{\partial L}{\partial x^\mu} = 0$ for a cyclic coordinate. From Eq. (5.52) then follows: *The canonical momentum conjugated to a cyclic coordinate is a constant of motion.* That is, $p_\mu = $ constant if x^μ is cyclic. Note that only the covariant component of the momentum is constant, not the corresponding contravariant component since raising an index introduces metric functions.

A free particle in spacetime has the Lagrange function

$$L = \frac{1}{2} \vec{u} \cdot \vec{u} = \frac{1}{2} \dot{x}_\mu \dot{x}^\mu = \frac{1}{2} g_{\mu\nu} \dot{x}^\mu \dot{x}^\nu. \tag{5.53}$$

An integral of the Lagrange equations is obtained readily from *the 4-velocity identity*:

$$\begin{cases} \dot{x}_\mu \dot{x}^\mu = -c^2 & \text{for a particle with rest mass} \\ \dot{x}_\mu \dot{x}^\mu = 0 & \text{for light} \end{cases} \qquad (5.54)$$

The line element is

$$ds^2 = g_{\mu\nu}dx^\mu dx^\nu = g_{\mu\nu}\dot{x}^\mu \dot{x}^\nu d\lambda^2 = 2L d\lambda^2. \qquad (5.55)$$

Thus the Lagrange function of a free particle is obtained from the line element.

5.5 Application of the Lagrange Formalism to Free Particles

To describe the motion of a free particle, we start by setting up the line element of the spacetime in the chosen coordinate system. There are coordinates on which the metric does not depend. For example, given axial symmetry we may choose the angle θ which is a cyclic coordinate here and the conjugate (covariant) impulse P_θ is a constant of the motion (the orbital spin of the particle). If, in addition, the metric is time independent (*stationary metric*) then t is also cyclic and p_t is a constant of the motion (the mechanical energy of the particle).

A *static metric* is time independent and unchanged under time reversal (i.e. t › $-t$). A *stationary metric* is independent of time, but changes under time reversal. Examples of static metrics are the Minkowski metric and the metric in a uniformly accelerated reference frame. The rotating cylindrical coordinate system is stationary.

5.5.1 Equation of Motion from Lagrange's Equations

The Lagrange function for a free particle is

$$L = \frac{1}{2} g_{\mu\nu} \dot{x}^\mu \dot{x}^\nu, \qquad (5.56)$$

where $g_{\mu\nu} = g_{\mu\nu}(x^\lambda)$, and the Lagrange equations are

$$\frac{\partial L}{\partial x^\beta} - \frac{d}{d\tau}\left(\frac{\partial L}{\partial \dot{x}^\beta}\right) = 0. \qquad (5.57)$$

Differentiation of L with respect to the coordinates and velocity components give

$$\frac{\partial L}{\partial x^\beta} = \frac{1}{2} g_{\mu\nu,\beta} \dot{x}^\mu \dot{x}^\nu, \quad \frac{\partial L}{\partial \dot{x}^\beta} = g_{\beta\nu} \dot{x}^\nu. \tag{5.58}$$

Differentiation of the latter quantities with respect to the proper time leads to

$$\frac{d}{d\tau}\left(\frac{\partial L}{\partial \dot{x}^\beta}\right) \equiv \left(\frac{\partial L}{\partial \dot{x}^\beta}\right)^{\cdot} = \dot{g}_{\beta\nu}\dot{x}^\nu + g_{\beta\nu}\ddot{x}^\nu = g_{\beta\nu,\mu}\dot{x}^\mu\dot{x}^\nu + g_{\beta\nu}\ddot{x}^\nu \tag{5.59}$$

Equations (5.58) and (5.59) give

$$\frac{1}{2} g_{\mu\nu,\beta} \dot{x}^\mu \dot{x}^\nu - g_{\beta\nu,\mu} \dot{x}^\mu \dot{x}^\nu - g_{\beta\nu}\ddot{x}^\nu = 0. \tag{5.60}$$

The second term on the left-hand side of Eq. (5.60) may be rewritten making use of the fact that $\dot{x}^\mu \dot{x}^\nu$ is symmetric in $\mu\nu$, as follows:

$$g_{\beta\nu,\mu}\dot{x}^\mu\dot{x}^\nu = \frac{1}{2}(g_{\beta\mu,\nu} + g_{\beta\nu,\mu})\dot{x}^\mu\dot{x}^\nu. \tag{5.61}$$

Hence, the equation of motion of the free particle may be written,

$$g_{\beta\nu}\ddot{x}^\nu + \frac{1}{2}(g_{\beta\mu,\nu} + g_{\beta\nu,\mu} - g_{\mu\nu,\beta})\dot{x}^\mu\dot{x}^\nu = 0. \tag{5.62}$$

Finally, since we are free to multiply (5.62) through by $g^{\alpha\beta}$, we can isolate $\ddot{x}^\alpha$ to get the equation of motion in the form

$$\ddot{x}^\alpha + \frac{1}{2}g^{\alpha\beta}(g_{\beta\mu,\nu} + g_{\beta\nu,\mu} - g_{\mu\nu,\beta})\dot{x}^\mu\dot{x}^\nu = 0. \tag{5.63}$$

This may be written as

$$\ddot{x}^\alpha + \Gamma^\alpha_{\mu\nu}\dot{x}^\mu\dot{x}^\nu = 0, \tag{5.64}$$

where the *symbols* $\Gamma^\alpha_{\mu\nu}$ in (5.64) are given by

$$\Gamma^\alpha_{\mu\nu} \equiv \frac{1}{2}g^{\alpha\beta}(g_{\beta\mu,\nu} + g_{\beta\nu,\mu} - g_{\mu\nu,\beta}). \tag{5.65}$$

Comparison with Eq. (5.43) shows that Eq. (5.64) describes a geodesic curve and that $\Gamma^\alpha_{\mu\nu}$ are Christoffel symbols. Hence, *free particles follow geodesic curves in spacetime*.

The equation of a time-like geodesic curve has been deduced from a variational principle which says that if there are two fixed points P_1 and P_2 in spacetime, then there exists an open subset of spacetime containing these two points such that among all curves contained in this open subset, the geodesic will be the curve of longest length between these two points. Note that for time-like curves, the length of a curve between the two events P_1 and P_2 which the curve passes through is the proper time used by a particle travelling from P_1 to P_2 along the curve.

So, *the variational principle says that time-like geodesics maximizes the proper time along the curve among all curves in spacetime close to the geodesic.*

There exist non-geodesic curves between two events far away from the geodesic curve between the events, along which a particle following the curve may have larger proper time between the events than a particle following the geodesic curve. Consider for example a clock at rest outside the Earth compared to a clock moving freely along a circular path, and calculate the proper time between two meetings of the clocks. It turns out that a standard clock on the non-geodesic particle at rest measures a larger proper time between departure and arrival of a clock following the particle which moves freely along a circular path in 3-space. (An exact calculation requires the Schwarzschild spacetime. See Chap. 9) Hence in general the variational principle has a local character.

Example 5.5.1 (*Vertical free fall in a uniformly accelerated reference frame*) The Lagrange function of the particle is

$$L = -\frac{1}{2}\left(1 + \frac{gx}{c^2}\right)^2 \dot{t}^2 + \frac{1}{2}\frac{\dot{x}^2}{c^2}, \tag{5.66}$$

where the dot denotes differentiation with respect to the proper time τ of the freely falling particle. This gives

$$\frac{\partial L}{\partial x} = -\frac{g}{c^2}\left(1 + \frac{gx}{c^2}\right)\dot{t}^2, \quad \frac{\partial L}{\partial \dot{x}} = \frac{\dot{x}}{c^2}. \tag{5.67}$$

Hence the Euler–Lagrange equation

$$\frac{\partial L}{\partial x} - \left(\frac{\partial L}{\partial \dot{x}}\right)^{\cdot} = 0 \tag{5.68}$$

takes the form

$$\ddot{x} + g\left(1 + \frac{gx}{c^2}\right)\dot{t}^2 = 0. \tag{5.69}$$

Furthermore we have

$$\frac{\partial L}{\partial t} = 0, \quad \frac{\partial L}{\partial \dot{t}} = -\left(1 + \frac{gx}{c^2}\right)^2 \dot{t}, \quad \left(\frac{\partial L}{\partial \dot{t}}\right)^{\cdot} = -2\left(1 + \frac{gx}{c^2}\right)\frac{g}{c^2}\dot{x}\dot{t} - \left(1 + \frac{gx}{c^2}\right)^2 \ddot{t}. \tag{5.70}$$

Hence, the Lagrange equation

$$\frac{\partial L}{\partial t} - \left(\frac{\partial L}{\partial \dot{t}}\right)^{\cdot} = 0 \tag{5.71}$$

takes the form

$$\frac{2g}{c^2}\dot{x}\dot{t} + \left(1 + \frac{gx}{c^2}\right)\ddot{t} = 0. \tag{5.72}$$

Combining Eqs. (5.69) and (5.72) leads to

$$-c^2\left(1 + \frac{gx}{c^2}\right)\frac{g}{c^2}\dot{x}\dot{t}^2 - c^2\left(1 + \frac{gx}{c^2}\right)^2\dot{t}\ddot{t} + \dot{x}\ddot{x} = 0. \tag{5.73}$$

Integrating this equation gives the 4-velocity identity

$$-c^2\left(1 + \frac{gx}{c^2}\right)^2\dot{t}^2 + \dot{x}^2 = -c^2. \tag{5.74}$$

Since the metric is static the momentum

$$p_t = \partial L/\partial \dot{t} = -\left(1 + \frac{gx}{c^2}\right)^2\dot{t} \tag{5.75}$$

is a constant of motion. The value of p_t is determined by the initial condition. Inserting the expression for $\dot{t}$ into the 4-velocity identity gives

$$-\frac{p_t^2}{\left(1 + \frac{gx}{c^2}\right)^2} + \frac{\dot{x}^2}{c^2} = -1. \tag{5.76}$$

Assume that the particle is falling from rest at an initial position $x = x_0$, i.e. $\dot{x}(x_0) = 0$. Then

$$p_t = -\left(1 + \frac{gx_0}{c^2}\right). \tag{5.77}$$

Inserting this into the 4-velocity identity leads after a short calculation to

$$\int_{x_0}^{x} \frac{1 + \frac{gx}{c^2}}{\sqrt{\left(1 + \frac{gx_0}{c^2}\right)^2 - \left(1 + \frac{gx}{c^2}\right)^2}}dx = \int_{0}^{\tau} cd\tau. \tag{5.78}$$

Performing the integration gives

$$\sqrt{\left(1 + \frac{gx_0}{c^2}\right)^2 - \left(1 + \frac{gx}{c^2}\right)^2} = \frac{g\tau}{c}. \tag{5.79}$$

This expression shows that the particle uses a finite proper time

$$\tau_{\max} = \frac{c}{g}\left(1 + \frac{gx_0}{c^2}\right) \tag{5.80}$$

to arrive at the horizon at $x = -c^2/g$. The position of the particle as a function of its proper time is

$$x = \frac{c^2}{g}\left(\sqrt{\left(1 + \frac{g x_0}{c^2}\right)^2 - \left(\frac{g\tau}{c}\right)^2} - 1\right).$$

(5.81)

5.5.2 Geodesic World Lines in Spacetime

Consider two time-like curves between two events in Minkowski spacetime. In Fig. 5.5 they are drawn in a Minkowski diagram which refers to an inertial reference frame.

The general interpretation of the line element for a time-like interval is: The spacetime distance between O and P in Fig. 5.5 equals the proper time interval between two events O and P measured on a clock moving in a such way that it is present both at O and P.

$$ds^2 = -c^2 d\tau^2$$

(5.82)

The proper time interval between two events at coordinate times T_0 and T_1 are

$$\tau_{0\ 1} = \int_{T_0}^{T_1} \sqrt{1 - \frac{v^2(T)}{c^2}}\, dT.$$

(5.83)

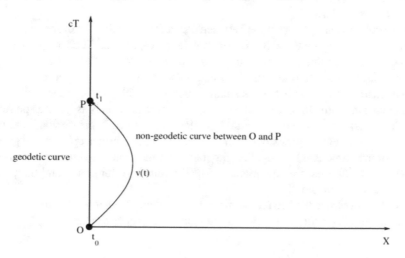

Fig. 5.5 Time like geodesics

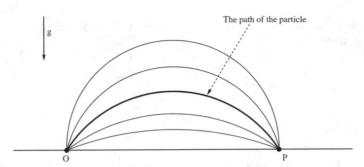

Fig. 5.6 Projectiles in 3-space. The particle moves between two events O and P at fixed points in time

We can see that τ_{0-1} is maximal along the geodesic curve with $v(T) = 0$. Time-like geodesic curves in spacetime have maximal distance between two points.

Example 5.5.2 (*How geodesics in spacetime can give parabolas in space*) A geodesic curve between two events O and P has maximal proper time. The proper time interval of a particle with position x and velocity v in a gravitational field with acceleration of gravity g is

$$d\tau = dt\sqrt{\left(1 + \frac{gx}{c^2}\right)^2 - \frac{v^2}{c^2}}. \tag{5.84}$$

This expression shows that the proper time of the particle proceeds faster the higher up in the field the particle is, and it proceeds slower the faster the particle moves.

In Fig. 5.6, we have drawn several paths between to events O and P with the same height in a gravitational field.

The path chosen by the particle between O and P is such that the proper time taken by the particle between these two events is as large as possible. Thus the particle will follow a path such that its co-moving standard clock goes as fast as possible. If the particle follows the horizontal line between O and P it goes as slowly as possible and the kinematical time dilation is as small as possible. Then the slowing down of its co-moving standard clock due to the kinematical time dilation is as small as possible, favoring a fast rate of the clock, but the particle is far down in the gravitational field and its standard clock goes slowly for that reason. Paths further up lead to a greater rate of proper time due to a smaller gravitational time dilation. But above the curve drawn as a thick line, the kinematical time dilation will dominate, and the proper time proceeds more slowly.

We shall now deduce the mathematical expression of what has been said above. Time-like geodesic curves are curves with maximal proper time, i.e.

$$\tau = \int_0^{\tau_1} \sqrt{-g_{\mu\nu}\dot{x}^\mu \dot{x}^\nu}\,d\tau \tag{5.85}$$

is maximal for a geodesic curve. However, the action

$$S = -2\int_0^{\tau_1} L\,d\tau = -\int_0^{\tau_1} g_{\mu\nu}\dot{x}^\mu \dot{x}^\nu\,d\tau \tag{5.86}$$

is maximal for the same curves, and this gives an easier calculation.

In the case of a vertical curve in a hyperbolically accelerated reference frame, the Lagrangian is

$$L = \frac{1}{2}\left(-\left(1 + \frac{gx}{c^2}\right)^2 \dot{t}^2 + \frac{\dot{x}^2}{c^2}\right). \tag{5.87}$$

Using the Euler–Lagrange equations now gives Eq. (5.69). Since spacetime is flat, the equation represents straight lines in spacetime. The projection of such curves into the three space of arbitrary inertial frames gives straight paths in 3-space, in accordance with Newton's 1st law. However, projecting it into an accelerated frame where the particle also has a horizontal motion, and taking the Newtonian limit, one finds the parabolic path of projectile motion.

5.5.3 Acceleration of Gravity

A free particle has vanishing 4-acceleration and moves along a time-like geodesic curve. The i-component of the geodesic equation is

$$\ddot{x}^i + \Gamma^i_{\mu\nu}\dot{x}^\mu \dot{x}^\nu = 0. \tag{5.88}$$

We define *the acceleration of gravity* as the 3-acceleration of a free particle instantaneously at rest. Since the spatial components of the 4-velocity of a particle at rest vanish, the acceleration of gravity is given by

$$\ddot{x}^i = -\Gamma^i_{tt}\dot{t}^2. \tag{5.89}$$

Hence the acceleration of gravity is given by the Christoffel symbols Γ^i_{tt}. They vanish in a local inertial reference frame, i.e. in a freely falling non-rotating reference frame. *There is an acceleration of gravity in any non-inertial laboratory independent of the geometrical properties of spacetime.*

In the Newtonian limit $d\tau = dt$, $\dot{t} = 1$ and the components of the acceleration of gravity are written $\ddot{x}^i = g^i$. It follows that

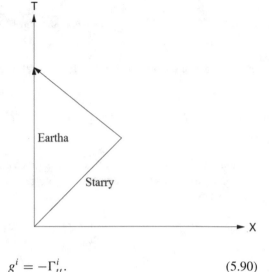

Fig. 5.7 The twin paradox. The twins Eartha and Starry each travel between two fixed events in spacetime

$$g^i = -\Gamma^i_{tt}. \tag{5.90}$$

Example 5.5.3 (*The Twin Paradox*) Consider two twins, Starry and Eartha. Starry travels to Proxima Centauri, four light years from the Earth, with a velocity $v = 0.8c$ so that $\gamma = \left(1 - v^2/c^2\right)^{-1/2} = 5/3$. The trip takes five years out and five years back. This means that Eartha, who remains at the Earth, is ten years older when she meets Starry at the end of her journey. Starry, on the other hand, is $10/\gamma = 6$ years older (Fig. 5.7).

According to the general principle of relativity, Starry can consider herself as being at rest and Eartha as the one whom undertakes the long journey. In this picture it seems that Starry and Eartha must be ten and six years older, respectively, upon their return.

Let us accept the principle of general relativity as applied to accelerated reference frames and review the twin "paradox" in this light.

Starry's description of the trip when she sees herself as stationary is as follows. She perceives a Lorentz-contracted distance between the Earth and Proxima Centauri, namely, four light years $\times 1/\gamma = 2.4$ light years. The Earth and Eartha travel with $v = 0.8c$. Her travel time in one direction is then $\frac{2.4 \text{ light years}}{0.8c} = 3$ years. So the round trip takes six years according to Starry. This means that Starry is six years older when they meet again. This is in accordance with the result arrived at by Eartha. According to Starry, Eartha ages by only six years $\times 1/\gamma = 3.6$ years during the round trip, not ten years as Eartha found. It is this conflict which constitutes the twin paradox. Note that the formulation of the twin paradox makes use of the general principle of relativity.

On turning about Starry experiences a force which reduces her velocity and accelerates her towards the Earth and Eartha. This means that she experiences a gravitational force directed *away* from the Earth. Eartha is higher up in this gravitational field and ages *faster* than Starry, because of the gravitational time dilation. We assume that

Starry has constant proper acceleration and is stationary in a uniformly accelerated frame as she turns about. Let us follow Starry's calculation of the ageing of Eartha while Starry experiences this gravitation field.

The canonical momentum p_t for Eartha is then (see Eq. (5.75))

$$p_t = -\left(1 + \frac{gx}{c^2}\right)^2 c\dot{t}. \tag{5.91}$$

Inserting this into the 4-velocity identity gives

$$p_t^2 - c^2\left(1 + \frac{gx}{c^2}\right)^2 = \left(1 + \frac{gx}{c^2}\right)^2 \dot{x}^2 \tag{5.92}$$

or

$$d\tau = \frac{1 + \frac{gx}{c^2}}{\sqrt{p_t^2 - c^2\left(1 + \frac{gx}{c^2}\right)^2}} dx. \tag{5.93}$$

Since $\dot{x} = 0$ for $x = x_2$ (x_2 is Eartha's turning point according to Starry), it follows that

$$p_t = c\left(1 + \frac{gx_2}{c^2}\right). \tag{5.94}$$

Let x_1 be Eartha's position according to Starry at the moment that Starry begins to notice the gravitational field, that is when Eartha begins to slow down in Starry's frame. Integration from x_1 to x_2 and inserting the value of p_t gives

$$\tau_{1-2} = \frac{c}{g}\sqrt{\left(1 + \frac{gx_2}{c^2}\right)^2 - \left(1 + \frac{gx_1}{c^2}\right)^2}. \tag{5.95}$$

Eartha neglected the time used by Starry to change from an outwards to a return velocity at Proxima Centauri when she calculated the aging of herself and Starry. This means that she took the limit of an infinitely large acceleration. Hence Starry must do the same. In this limit Starry gets

$$\lim_{g \to \infty} \tau_{1-2} = \frac{1}{c}\sqrt{x_2^2 - x_1^2}. \tag{5.96}$$

Now setting $x_2 = 4$ years and $x_1 = 2.4$ light years, respectively, we get $\lim_{g \to \infty} \tau_{1-2} = 3.2$ years. Hence, Eartha's aging as she turns about is, according to Starry,

$$\Delta\tau_{\text{Eartha}} = 2 \lim_{g \to \infty} \tau_{1-2} = 6.4 \text{ years}.$$

So all in all Eartha has aged by $\tau_{\text{Eartha}} = 3.6 + 6.4 = 10$ years, according to Starry, which is just what Eartha herself found.

We have now seen: The formulation of the twin paradox makes use of the general principle of relativity for accelerated motion. The paradox arises if one tries to describe the ageing of both twins from the point of view of each twin without taking into account the effect of gravity upon the rate of time. Hence, both the formulation and the solution of the twin paradox involves the general theory of relativity.

5.5.4 Gravitational Shift of Wavelength

We shall consider the shift of wavelength of light moving up or down in a gravitational field. The 4-momentum of a particle with relativistic energy E and spatial velocity $\vec{v}$ (3-velocity) is given by

$$\vec{P} = \frac{E}{c^2}(c, \vec{v}). \tag{5.97}$$

Let $\vec{U}$ be the 4-velocity of an observer. In a co-moving orthonormal basis of the observer we have $\vec{U} = (c, 0, 0, 0)$. This gives

$$\vec{U} \cdot \vec{P} = -\hat{E}. \tag{5.98}$$

Hence, the energy of a particle with 4-momentum $\vec{P}$ measured by an observer with 4-velocity $\vec{U}$ is

$$\hat{E} = -\vec{U} \cdot \vec{P}. \tag{5.99}$$

Let $E_e = -(\vec{U} \cdot \vec{P})_e$ and $E_0 = -(\vec{U} \cdot \vec{P})_0$ be the energy of a photon, measured locally by observers at rest in the emitter and observer positions, respectively. This gives[1]

$$\frac{E_e}{(\vec{U} \cdot \vec{P})_e} = \frac{E_0}{(\vec{U} \cdot \vec{P})_0}. \tag{5.100}$$

Let the wavelength of the light, measured by the emitter and observer, be λ_e and λ_0, respectively. We then have

$$\lambda_e = \frac{hc}{E_e}, \quad \lambda_0 = \frac{hc}{E_0} \tag{5.101}$$

---

[1] $\vec{A} \cdot \vec{B} = A_0 B^0 + A_1 B^1 + \cdots = g_{00} A^0 B^0 + g_{11} A^1 B^1 + \cdots$, an orthonormal basis gives $\vec{A} \cdot \vec{B} = -A^0 B^0 + A^1 B^1 + \cdots$.

which gives

$$\lambda_0 = \frac{(\vec{U} \cdot \vec{P})_e}{(\vec{U} \cdot \vec{P})_0} \lambda_e. \tag{5.102}$$

This formula may be applied to observers with arbitrary motion.

We shall here restrict ourselves to an observer at rest in a coordinate system with time-independent diagonal metric. Then we have

$$\vec{U} \cdot \vec{P} = U^t P_t = c \frac{dt}{d\tau} P_t, \tag{5.103}$$

where P_t is a constant of motion (since t is a cyclic coordinate) for photons and hence has the same value in emitter and observer positions. The line element is

$$ds^2 = g_{tt} c^2 dt^2 + g_{ii} (dx^i)^2. \tag{5.104}$$

Using the physical interpretation (3.82) of the line element for a time-like interval, we obtain for the proper time of an observer at rest

$$d\tau^2 = -g_{tt} dt^2 \Rightarrow d\tau = \sqrt{-g_{tt}} dt. \tag{5.105}$$

Hence

$$\frac{dt}{d\tau} = \frac{1}{\sqrt{-g_{tt}}}, \tag{5.106}$$

which gives

$$\vec{U} \cdot \vec{P} = \frac{c}{\sqrt{-g_{tt}}} P_t. \tag{5.107}$$

Inserting this into the expression for the wavelength (5.102) gives the formula for the gravitational shift of the wavelength of light emitted and observed by an emitter and an observer at rest in the reference frame,

$$\lambda_0 = \sqrt{\frac{(g_{tt})_0}{(g_{tt})_e}} \lambda_e. \tag{5.108}$$

Example 5.5.4 (*Gravitational redshift or blueshift of light*) Inserting the metric of a uniformly accelerated reference frame with

$$g_{tt} = -\left(1 + \frac{gx}{c^2}\right)^2 \tag{5.109}$$

gives

$$\lambda_0 = \frac{1 + \frac{gx_0}{c^2}}{1 + \frac{gx_e}{c^2}} \lambda_e. \tag{5.110}$$

The gravitational redshift (light moving upwards in a gravitational field) or blueshift (light moving downwards) is

$$z_G = \frac{\lambda_0 - \lambda_e}{\lambda_e} = \frac{1 + \frac{gx_0}{c^2}}{1 + \frac{gx_e}{c^2}} - 1 = \frac{\frac{g}{c^2}(x_0 - x_e)}{1 + \frac{gx_e}{c^2}} \approx \frac{gh}{c^2}, \tag{5.111}$$

where $h = x_0 - x_e$ is the difference in height between transmitter and receiver. In the case of motion upwards the observed wavelength increases with the height. By motion of radiation downwards there is a blueshift.

This effect was first measured in 1959 in the Pound–Rebka experiment with a height difference between emitter and receiver of $h = 22.5\,\mathrm{m}$ giving $z_G = 2.5 \times 10^{-15}$.

5.6 Connection Coefficients

The covariant directional derivative of a scalar field f in the direction of a vector $\vec{u}$ is defined as

$$\nabla_{\vec{u}} f \equiv \vec{u}(f). \tag{5.112}$$

Here the vector $\vec{u}$ should be taken as a differential operator. (In coordinate basis, $\vec{u} = u^\mu \frac{\partial}{\partial x^\mu}$.) The directional derivative along a basis vector $\vec{e}_\nu$ is written as

$$\nabla_\nu \equiv \nabla_{\vec{e}_\nu}. \tag{5.113}$$

Hence $\nabla_\mu(\ ) = \nabla_{\vec{e}_\mu}(\ ) = \vec{e}_\mu(\ )$.

Definition 5.6.1 (*Koszul's connection coefficients in an arbitrary basis.*) In an arbitrary basis the *Koszul connection coefficients*, $\Gamma^\alpha_{\mu\nu}$, are defined by

$$\nabla_\nu \vec{e}_\mu \equiv \Gamma^\alpha_{\mu\nu} \vec{e}_\alpha, \tag{5.114}$$

which may also be written $\vec{e}_\nu(\vec{e}_\mu) = \Gamma^\alpha_{\mu\nu} \vec{e}_\alpha$. In coordinate basis $\Gamma^\alpha_{\mu\nu}$ is reduced to Christoffel symbols and one often writes $\vec{e}_{\mu,\nu} = \Gamma^\alpha_{\mu\nu} \vec{e}_\alpha$. In an arbitrary basis $\Gamma^\alpha_{\mu\nu}$ has no symmetry.

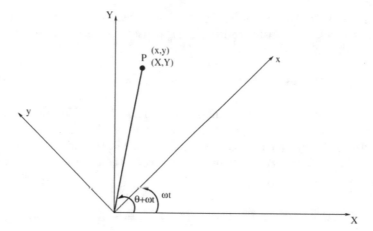

Fig. 5.8 Rotating coordinate system. The non-rotating coordinate system (X, Y) and the rotating system (x, y), rotating with angular velocity ω

Example 5.6.1 (*The connection coefficients in a rotating reference frame*) In an inertial reference frame, there is a co-moving Cartesian coordinate system with coordinates X, Y and corresponding plane polar coordinates R, Θ. In a reference frame rotating with angular velocity ω, the corresponding co-moving coordinates are x, y and r, θ (Fig. 5.8). The coordinate clocks in the rotating frame shows the same time as those in the inertial frame, $t = T$.

The transformation between the coordinates is

$$t = T, \quad r = R, \quad \theta = \Theta - \omega T,$$
$$X = R \cos \Theta, \quad Y = R \sin \Theta,$$
$$X = r \cos(\theta + \omega t), \quad Y = r \sin(\theta + \omega t). \tag{5.115}$$

The time-like coordinate basis vector in the rotating frame is calculated from

$$\vec{e}_t = \frac{\partial}{\partial t} = \frac{\partial X}{\partial t} \frac{\partial}{\partial X} + \frac{\partial Y}{\partial t} \frac{\partial}{\partial Y} + \frac{\partial T}{\partial t} \frac{\partial}{\partial T}, \tag{5.116}$$

Using this formula together with the transformation (5.115), and corresponding formulae for the spatial basis vectors, give

$$\vec{e}_t = -r\omega \sin(\theta + \omega t)\vec{e}_X + r\omega \cos(\theta + \omega t)\vec{e}_Y + \vec{e}_T,$$
$$\vec{e}_r = \frac{\partial X}{\partial r} \frac{\partial}{\partial X} + \frac{\partial Y}{\partial r} \frac{\partial}{\partial Y} = \cos(\theta + \omega t)\vec{e}_X + \sin(\theta + \omega t)\vec{e}_Y,$$
$$\vec{e}_\theta = \frac{\partial X}{\partial \theta} \frac{\partial}{\partial X} + \frac{\partial Y}{\partial \theta} \frac{\partial}{\partial Y} = -r \sin(\theta + \omega t)\vec{e}_X + r \cos(\theta + \omega t)\vec{e}_Y. \tag{5.117}$$

We are going to find the Christoffel symbols, which involves differentiation of the basis vectors. The transformation (5.117) makes this easy, since $\vec{e}_X, \vec{e}_Y, \vec{e}_T$ are constant. Differentiation gives

$$\nabla_t \vec{e}_t = -r\omega^2 \cos(\theta + \omega t)\vec{e}_X - r\omega^2 \sin(\theta + \omega t)\vec{e}_Y. \tag{5.118}$$

The connection coefficients are calculated from Eq. (5.114). Note that to calculate $\Gamma^\alpha_{\mu\nu}$, the right-hand side of Eq. (5.118) has to be expressed by the basis which we are differentiating. Comparing the right-hand side of Eq. (5.118) with the expression for $\vec{e}_r$ we see that $\nabla_t \vec{e}_t = -r\omega^2 \vec{e}_r$ giving $\Gamma^r_{tt} = -r\omega^2$. The other nonzero Christoffel symbols are

$$\Gamma^\theta_{rt} = \Gamma^\theta_{tr} = \frac{\omega}{r}, \quad \Gamma^\theta_{\theta r} = \Gamma^\theta_{r\theta} = \frac{1}{r}, \quad \Gamma^r_{\theta t} = \Gamma^r_{t\theta} = -r\omega, \quad \Gamma^r_{\theta\theta} = -r. \tag{5.119}$$

Example 5.6.2 (*Acceleration in a non-rotating reference frame*) The covariant expression for the acceleration is

$$\ddot{\vec{r}} = \dot{\vec{v}} = v^i_{;j} v^j \vec{e}_i = (\dot{v}^i + \Gamma^i_{\alpha\beta} v^\alpha v^\beta)\vec{e}_i, \tag{5.120}$$

where $\dot{} \equiv d/dt$. Here i, j and k are space indices. Inserting the Christoffel symbols (5.31), (5.32) for plane polar coordinates gives

$$\vec{a}_{\text{inert}} = (\ddot{r} - r\dot{\theta}^2)\vec{e}_r + \left(\ddot{\theta} + \frac{2}{r}\dot{r}\dot{\theta}\right)\vec{e}_\theta = (\ddot{r} - r\dot{\theta}^2)\vec{e}_{\hat{r}} + (r\ddot{\theta} + 2\dot{r}\dot{\theta})\vec{e}_{\hat{\theta}}. \tag{5.121}$$

Example 5.6.3 (*Acceleration in a rotating reference frame*) Inserting the Christoffel symbols from Example 5.6.1 into Eq. (5.119) gives

$$\begin{aligned}
\vec{a}_{\text{rot}} &= (\ddot{r} - r\dot{\theta}^2 + \Gamma^r_{tt}\dot{t}^2 + \Gamma^r_{\theta t}\dot{\theta}\dot{t} + \Gamma^r_{t\theta}\dot{t}\dot{\theta})\vec{e}_r + \left(\ddot{\theta} + \frac{2}{r}\dot{r}\dot{\theta} + \Gamma^\theta_{rt}\dot{r}\dot{t} + \Gamma^\theta_{tr}\dot{t}\dot{r}\right)\vec{e}_\theta \\
&= (\ddot{r} - r\dot{\theta}^2 - r\omega^2 - 2r\omega\dot{\theta})\vec{e}_{\hat{r}} + (r\ddot{\theta} + 2\dot{r}\dot{\theta} + 2\dot{r}\omega)\vec{e}_{\hat{\theta}} \\
&= \vec{a}_{\text{inert}} - (r\omega^2 + 2r\omega\dot{\theta})\vec{e}_{\hat{r}} + 2\dot{r}\omega\vec{e}_{\hat{\theta}}.
\end{aligned} \tag{5.122}$$

The angular velocity of the reference frame, is $\vec{\omega} = \omega\vec{e}_z$. We also introduce $\vec{r} = r\vec{e}_r$. The velocity in a rotating reference frame is then

$$\dot{\vec{r}} = \dot{r}\vec{e}_r + r\dot{\vec{e}}_r. \tag{5.123}$$

Furthermore

$$\dot{\vec{e}}_r = \frac{d\vec{e}_r}{dt} = \frac{\partial\vec{e}_r}{\partial x^i}\frac{dx^i}{dt} = v^i\vec{e}_{r,i}. \tag{5.124}$$

Using Definition 5.6.1 in a coordinate basis, this may be written as

$$\dot{\vec{e}}_r = v^i \Gamma^j_{ri} \vec{e}_j. \tag{5.125}$$

Applying the expressions for the Christoffel symbols in Example 5.6.1 and introducing orthonormal basis, we get

$$\dot{\vec{e}}_r = v^\theta \Gamma^\theta_{r\theta} \vec{e}_\theta = \dot{\theta} \frac{1}{r} \vec{e}_\theta = \dot{\theta} \vec{e}_{\hat{\theta}}. \tag{5.126}$$

Hence

$$\vec{v} = \dot{\vec{r}} = \dot{r} \vec{e}_{\hat{r}} + r\dot{\theta} \vec{e}_{\hat{\theta}}. \tag{5.127}$$

Inserting this into the expression for the acceleration gives

$$\ddot{\vec{r}}_{\text{rot}} = \ddot{\vec{r}}_{\text{inert}} + \vec{\omega} \times (\vec{\omega} \times \vec{r}) + 2\vec{\omega} \times \vec{v}. \tag{5.128}$$

We see that the centrifugal acceleration (the term in the middle) and the Coriolis acceleration (last term) are contained in the expression for the covariant derivative.

5.6.1 Structure Coefficients

The commutator of two vectors, $\vec{u}$ and $\vec{v}$, expressed by covariant directional derivatives is given by

$$[\vec{u}, \vec{v}] = \nabla_{\vec{u}} \vec{v} - \nabla_{\vec{v}} \vec{u}. \tag{5.129}$$

Let $\vec{u} = \vec{e}_\mu$ and $\vec{v} = \vec{e}_\nu$. We then have

$$[\vec{e}_\mu, \vec{e}_\nu] = \nabla_\mu \vec{e}_\nu - \nabla_\nu \vec{e}_\mu. \tag{5.130}$$

Using the definitions of the connection and structure coefficients we get

$$c^\alpha_{\mu\nu} \vec{e}_\alpha = (\Gamma^\alpha_{\nu\mu} - \Gamma^\alpha_{\mu\nu}) \vec{e}_\alpha. \tag{5.131}$$

Thus (in a torsion free space)

$$c^\alpha_{\mu\nu} = \Gamma^\alpha_{\nu\mu} - \Gamma^\alpha_{\mu\nu}. \tag{5.132}$$

In *coordinate basis* we have

$$\vec{e}_\mu = \frac{\partial}{\partial x^\mu}, \quad \vec{e}_\nu = \frac{\partial}{\partial x^\nu}, \tag{5.133}$$

and therefore

$$[\vec{e}_\mu, \vec{e}_\nu] = \left[\frac{\partial}{\partial x^\mu}, \frac{\partial}{\partial x^\nu}\right] = \frac{\partial}{\partial x^\mu}\left(\frac{\partial}{\partial x^\nu}\right) - \frac{\partial}{\partial x^\nu}\left(\frac{\partial}{\partial x^\mu}\right) = \frac{\partial^2}{\partial x^\mu \partial x^\nu} - \frac{\partial^2}{\partial x^\nu \partial x^\mu} = 0.$$

(5.134)

Equation (5.134) shows that $c^\alpha_{\mu\nu} = 0$, and Eq. (5.132) then implies that the connection coefficients in Eq. (5.98) are symmetrical in a coordinate basis:

$$\Gamma^\alpha_{\nu\mu} = \Gamma^\alpha_{\mu\nu}.$$

(5.135)

5.7 Covariant Differentiation of Vectors, Forms and Tensors

5.7.1 Covariant Differentiation of Vectors

The covariant directional derivative of a vector field was defined in Eq. (5.34). It should be noted, however, that there are several ways of defining the covariant derivative that are consistent with each other. Here we shall formulate some definitions that are alternative to those given in Sect. 5.2.

Definition 5.7.1 (*Covariant derivative of a vector field*) The covariant derivative of a vector in an arbitrary basis is defined by

$$\nabla_\nu \vec{A} = \nabla_\nu (A^\mu \vec{e}_\mu) = \nabla_\nu A^\mu \vec{e}_\mu + A^\alpha \nabla_\nu \vec{e}_\alpha.$$

(5.136)

Using the definition (5.114) of the connection coefficients, this may be written

$$\nabla_\nu \vec{A} = [\vec{e}_\nu(A^\mu) + A^\alpha \Gamma^\mu_{\alpha\nu}]\vec{e}_\mu.$$

(5.137)

The covariant derivative of the components of a vector field was introduced in Eq. (5.23) by the requirement that the derivative of a tensor component shall transform as a tensor component. There exists, however, another point of view. The ordinary partial derivative of the vector components does only describe the change of the components of a vector, not of the vector itself. The covariant derivatives of the vector components, on the other hand, describe the change of the vector itself. Let us see how that comes about.

Definition 5.7.2 (*Covariant derivative of a vector component*) The covariant derivative of a vector component A^μ is written as $A^\mu_{;\nu}$ and is defined by

$$\nabla_\nu \vec{A} = A^\mu_{;\nu} \vec{e}_\mu.$$

(5.138)

Comparison with Eq. (5.137) shows that

$$A^{\mu}_{;\nu} = \vec{e}_{\nu}(A^{\mu}) + A^{\alpha}\Gamma^{\mu}_{\alpha\nu}. \tag{5.139}$$

Note that the term with the connection coefficients come from the second term in Eq. (5.136). Hence, it represents the change of the basis vector field with position and time. The first term describes the changes of the vector component with position and time. The sum describes the change of the vector itself, as seen from the definition (5.138).

5.7.2 Covariant Differentiation of Forms

Definition 5.7.3 (*Covariant directional derivative of a 1-form field*) Given a vector field $\vec{A}$ and a 1-form field $\underline{\alpha}$, the covariant directional derivative of $\underline{\alpha}$ in the direction of the vector $\vec{u}$ is defined by

$$(\nabla_{\vec{u}}\underline{\alpha})(\vec{A}) \equiv \nabla_{\vec{u}}[\underbrace{\underline{\alpha}(\vec{A})}_{\alpha_{\mu}A^{\mu}}] - \underline{\alpha}(\nabla_{\vec{u}}\vec{A}). \tag{5.140}$$

Let the 1-form $\underline{\alpha}$ be a basis field, $\underline{\alpha} = \underline{\omega}^{\mu}$. Then $\underline{\omega}^{\mu}(\vec{e}_{\nu}) = \delta^{\mu}_{\nu}$. Furthermore let $\vec{A} = \vec{e}_{\nu}$ and $\vec{u} = \vec{e}_{\lambda}$. We then have

$$(\nabla_{\lambda}\underline{\omega}^{\mu})(\vec{e}_{\nu}) = \nabla_{\lambda}[\underbrace{\underline{\omega}^{\mu}(\vec{e}_{\nu})}_{\delta^{\mu}_{\nu}}] - \underline{\omega}^{\mu}(\nabla_{\lambda}\vec{e}_{\nu}). \tag{5.141}$$

The covariant directional derivative ∇_{λ} of a constant scalar field is zero, $\nabla_{\lambda}\delta^{\mu}_{\nu} = 0$. We therefore get

$$(\nabla_{\lambda}\underline{\omega}^{\mu})(\vec{e}_{\nu}) = -\underline{\omega}^{\mu}(\nabla_{\lambda}\vec{e}_{\nu}) = -\underline{\omega}^{\mu}(\Gamma^{\alpha}_{\nu\lambda}\vec{e}_{\alpha}) = -\Gamma^{\alpha}_{\nu\lambda}\underline{\omega}^{\mu}(\vec{e}_{\alpha}) = -\Gamma^{\alpha}_{\nu\lambda}\delta^{\mu}_{\alpha} = -\Gamma^{\mu}_{\nu\lambda}. \tag{5.142}$$

The contraction of a 1-form and a basis vector gives the components of the 1-form, $\underline{\alpha}(\vec{e}_{\nu}) = \alpha_{\nu}$. Equation (5.142) gives the directional derivatives of the basis forms and tells that the ν-component of $\nabla_{\lambda}\underline{\omega}^{\mu}$ is equal to $-\Gamma^{\mu}_{\nu\lambda}$. Hence

$$\nabla_{\lambda}\underline{\omega}^{\mu} = -\Gamma^{\mu}_{\nu\lambda}\underline{\omega}^{\nu}. \tag{5.143}$$

Using the rule for differentiation of a product gives

$$\nabla_{\lambda}\underline{\alpha} = \nabla_{\lambda}(\alpha_{\mu}\underline{\omega}^{\mu}) = \nabla_{\lambda}(\alpha_{\mu})\underline{\omega}^{\mu} + \alpha_{\mu}\nabla_{\lambda}\underline{\omega}^{\mu} = \vec{e}_{\lambda}(\alpha_{\mu})\underline{\omega}^{\mu} - \alpha_{\mu}\Gamma^{\mu}_{\nu\lambda}\underline{\omega}^{\nu}. \tag{5.144}$$

This motivates the following definition.

Definition 5.7.4 (*Covariant derivative of a 1-form*) The covariant derivative of a
1-form $\underline{\alpha} = \alpha_\mu \underline{\omega}^\mu$ is given by

$$\nabla_\lambda \underline{\alpha} = [\vec{e}_\lambda(\alpha_\nu) - \alpha_\mu \Gamma^\mu_{\nu\lambda}]\underline{\omega}^\nu. \tag{5.145}$$

Definition 5.7.5 (*Covariant derivative of a 1-form component*) The covariant
derivative of the 1-form components α_μ is denoted by $\alpha_{\nu;\lambda}$ and defined by

$$\nabla_\lambda \underline{\alpha} \equiv \alpha_{\nu;\lambda} \underline{\omega}^\nu. \tag{5.146}$$

It follows that

$$\alpha_{\nu;\lambda} = \vec{e}_\lambda(\alpha_\nu) - \alpha_\mu \Gamma^\mu_{\nu\lambda}. \tag{5.147}$$

Note that $\Gamma^\mu_{\nu\lambda}$ in Eq. (5.147) are not in general the same as the Christoffel symbols.
However, in coordinate basis we get

$$\alpha_{\nu;\lambda} = \alpha_{\nu,\lambda} - \alpha_\mu \Gamma^\mu_{\lambda\nu}, \tag{5.148}$$

where $\Gamma^\mu_{\lambda\nu} = \Gamma^\mu_{\nu\lambda}$ are Christoffel symbols.

5.7.3   Covariant Differentiation of Tensors of Arbitrary Rank

Definition 5.7.6 (*Covariant derivative of tensors*) Let A and B be two tensors of
arbitrary rank. The covariant directional derivative along a basis vector $\vec{e}_\lambda$ of a tensor
$A \otimes B$ of arbitrary rank is defined by

$$\nabla_\lambda(A \otimes B) \equiv (\nabla_\lambda A) \otimes B + A \otimes (\nabla_\lambda B). \tag{5.149}$$

We will use Eq. (5.149) to find the formula for the covariant derivative of the
components of a tensor of rank 2:

$$\begin{aligned}
\nabla_\alpha S &= \nabla_\alpha(S_{\mu\nu}\underline{\omega}^\mu \otimes \underline{\omega}^\nu) \\
&= (\nabla_\alpha S_{\mu\nu})\underline{\omega}^\mu \otimes \underline{\omega}^\nu + S_{\mu\nu}(\nabla_\alpha \underline{\omega}^\mu) \otimes \underline{\omega}^\nu + S_{\mu\nu}\underline{\omega}^\mu \otimes (\nabla_\alpha \underline{\omega}^\nu) \\
&= (S_{\mu\nu,\alpha} - S_{\beta\nu}\Gamma^\beta_{\mu\alpha} - S_{\mu\beta}\Gamma^\beta_{\nu\alpha})\underline{\omega}^\mu \otimes \underline{\omega}^\nu,
\end{aligned} \tag{5.150}$$

where $S_{\mu\nu,\alpha} = \vec{e}_\alpha(S_{\mu\nu})$.

Definition 5.7.7 (*Covariant derivative of tensor components*) The covariant derivative $S_{\mu\nu;\alpha}$ of the covariant components of a tensor of rank 2 is defined by

$$\nabla_\alpha S = S_{\mu\nu;\alpha}\underline{\omega}^\mu \otimes \underline{\omega}^\nu \tag{5.151}$$

we get

$$S_{\mu\nu;\alpha} = S_{\mu\nu,\alpha} - S_{\beta\nu}\Gamma^{\beta}_{\mu\alpha} - S_{\mu\beta}\Gamma^{\beta}_{\nu\alpha}. \tag{5.152}$$

For the metric tensor we get

$$g_{\mu\nu;\alpha} = g_{\mu\nu,\alpha} - g_{\beta\nu}\Gamma^{\beta}_{\mu\alpha} - g_{\mu\beta}\Gamma^{\beta}_{\nu\alpha}. \tag{5.153}$$

From

$$g_{\mu\nu} = \vec{e}_{\mu} \cdot \vec{e}_{\nu} \tag{5.154}$$

we get

$$\begin{aligned} g_{\mu\nu,\alpha} = \nabla_{\alpha}g_{\mu\nu} &= \nabla_{\alpha}(\vec{e}_{\mu} \cdot \vec{e}_{\nu}) = (\nabla_{\alpha}\vec{e}_{\mu}) \cdot \vec{e}_{\nu} + \vec{e}_{\mu}(\nabla_{\alpha}\vec{e}_{\nu}) \\ &= \Gamma^{\beta}_{\mu\alpha}\vec{e}_{\beta} \cdot \vec{e}_{\nu} + \vec{e}_{\mu} \cdot \Gamma^{\beta}_{\nu\alpha}\vec{e}_{\beta} = g_{\beta\nu}\Gamma^{\beta}_{\mu\alpha} + g_{\mu\beta}\Gamma^{\beta}_{\nu\alpha}. \end{aligned} \tag{5.155}$$

This means that

$$g_{\mu\nu;\alpha} = 0. \tag{5.156}$$

So the metric tensor is a constant tensor.

5.8 The Cartan Connection

Definition 5.8.1 (*Exterior derivative of a basis vector*)

$$\underline{d}\vec{e}_{\mu} = \Gamma^{\nu}_{\mu\alpha}\vec{e}_{\nu} \otimes \underline{\omega}^{\alpha}. \tag{5.157}$$

Note that this can be thought of as a 1-form with vector components $\Gamma^{\nu}_{\mu\alpha}\vec{e}_{\nu}$, a *vectorial form*. The exterior derivative of a vector field is

$$\underline{d}\vec{A} = \underline{d}(\vec{e}_{\mu}A^{\mu}) = \vec{e}_{\nu} \otimes \underline{d}A^{\nu} + A^{\mu}\underline{d}\vec{e}_{\mu}. \tag{5.158}$$

In arbitrary basis the exterior derivative of the vector components are

$$\underline{d}A^{\nu} = \vec{e}_{\lambda}(A^{\nu})\underline{\omega}^{\lambda} \tag{5.159}$$

In coordinate basis $\vec{e}_{\lambda}(A^{\nu}) = \frac{\partial}{\partial x^{\lambda}}(A^{\nu}) = A^{\nu}_{,\lambda}$, giving

$$\underline{d}\vec{A} = \vec{e}_{\nu} \otimes [\vec{e}_{\lambda}(A^{\nu})\underline{\omega}^{\lambda}] + A^{\mu}\Gamma^{\nu}_{\mu\lambda}\vec{e}_{\nu} \otimes \underline{\omega}^{\lambda} = (\vec{e}_{\lambda}(A^{\nu}) + A^{\mu}\Gamma^{\nu}_{\mu\lambda})\vec{e}_{\nu} \otimes \underline{\omega}^{\lambda}. \tag{5.160}$$

Together with Eq. (5.139), this gives

$$\underline{d}\vec{A} = A^{\nu}_{,\lambda}\vec{e}_{\nu} \otimes \underline{\omega}^{\lambda}. \tag{5.161}$$

Definition 5.8.2 (*Connection forms* $\underline{\Omega}^{\nu}_{\mu}$) The connection forms $\underline{\Omega}^{\nu}_{\mu}$ are 1-forms, defined by

$$\underline{d}\vec{e}_{\mu} \equiv \vec{e}_{\nu} \otimes \underline{\Omega}^{\nu}_{\mu}. \tag{5.162}$$

From Eqs. (5.157) and (5.162) we have

$$\underline{d}\vec{e}_{\mu} = \Gamma^{\nu}_{\mu\alpha}\vec{e}_{\nu} \otimes \underline{\omega}^{\alpha} = \vec{e}_{\nu} \otimes \Gamma^{\nu}_{\mu\alpha}\underline{\omega}^{\alpha} = \vec{e}_{\nu} \otimes \underline{\Omega}^{\nu}_{\mu}. \tag{5.163}$$

This shows that

$$\underline{\Omega}^{\nu}_{\mu} = \Gamma^{\nu}_{\mu\alpha}\underline{\omega}^{\alpha}. \tag{5.164}$$

The exterior derivatives of the components of the metric tensor:

$$\underline{d}g_{\mu\nu} = \underline{d}(\vec{e}_{\mu} \cdot \vec{e}_{\nu}) = \vec{e}_{\mu} \cdot \underline{d}\vec{e}_{\nu} + \vec{e}_{\nu} \cdot \underline{d}\vec{e}_{\mu}, \tag{5.165}$$

where the meaning of the dot is defined as follows.

Definition 5.8.3 (*Scalar product between vector and 1-form*) The scalar product between a vector $\vec{u}$ and a vectorial 1-form $\underline{A} = A^{\mu}_{\nu}\vec{e}_{\mu} \otimes \underline{\omega}^{\nu}$ is defined by

$$\vec{u} \cdot \underline{A} \equiv u^{\alpha}A^{\mu}_{\nu}(\vec{e}_{\alpha} \cdot \vec{e}_{\mu})\underline{\omega}^{\nu}. \tag{5.166}$$

Using this definition, we get

$$\underline{d}g_{\mu\nu} = (\vec{e}_{\mu} \cdot \vec{e}_{\lambda})\underline{\Omega}^{\lambda}_{\nu} + (\vec{e}_{\nu} \cdot \vec{e}_{\gamma})\underline{\Omega}^{\gamma}_{\mu} = g_{\mu\lambda}\underline{\Omega}^{\lambda}_{\nu} + g_{\nu\gamma}\underline{\Omega}^{\gamma}_{\mu}. \tag{5.167}$$

Lowering an index gives

$$\underline{d}g_{\mu\nu} = \underline{\Omega}_{\mu\nu} + \underline{\Omega}_{\nu\mu}. \tag{5.168}$$

In an orthonormal basis field there is Minkowski metric: $g_{\hat{\mu}\hat{\nu}} = \eta_{\hat{\mu}\hat{\nu}}$ which is constant. Then we have $\underline{d}g_{\hat{\mu}\hat{\nu}} = 0$ which implies

$$\underline{\Omega}_{\hat{\nu}\hat{\mu}} = -\underline{\Omega}_{\hat{\mu}\hat{\nu}}. \tag{5.169}$$

The connection forms with lower indices are antisymmetric in an orthonormal basis. It follows from (5.164) and (5.169) that

$$\Gamma_{\hat{\nu}\hat{\mu}\hat{\alpha}} = -\Gamma_{\hat{\mu}\hat{\nu}\hat{\alpha}}. \tag{5.170}$$

Note also that

$$\Gamma^{\hat{i}}_{\hat{i}\hat{j}} = -\Gamma_{\hat{i}\hat{i}\hat{j}} = \Gamma_{\hat{i}\hat{i}\hat{j}} = \Gamma^{\hat{i}}_{\hat{i}\hat{j}}, \quad \Gamma^{i}_{\hat{j}k} = -\Gamma^{\hat{j}}_{ik}. \tag{5.171}$$

We shall now consider Cartan's 1 structure equation which gives a relationship between the structure coefficients and the connection coefficients. As a preparation for deducing Cartan's 1 structure equation we shall first deduce a mathematical identity valid for 1-forms.

The commutator of two vectors is itself a vector, and the contraction of a 1-form with a vector is essentially the same quantity as a scalar product of two vectors. Hence for a 1-form $\underline{\alpha}$ we have

$$\underline{\alpha}([\vec{u}, \vec{v}]) = \alpha_v [\vec{u}, \vec{v}]^v = \alpha_v (u^\mu v^\nu{}_{,\mu} - v^\mu u^\nu{}_{,\mu}). \tag{5.172}$$

From this we get

$$\vec{u}(\underline{\alpha}(\vec{v})) = u^\mu \vec{e}_\mu (\alpha_v v^\nu) = u^\mu (\alpha_{v,\mu} v^\nu + \alpha_v v^\nu{}_{,\mu}), \tag{5.173}$$

and

$$\vec{v}(\underline{\alpha}(\vec{u})) = v^\mu (\alpha_{v,\mu} u^\nu + \alpha_v u^\nu{}_{,\mu}), \tag{5.174}$$

Furthermore

$$\underline{d}\underline{\alpha}(\vec{u}, \vec{v}) = (\alpha_{\mu,\nu} - \alpha_{\nu,\mu}) u^\nu v^\mu. \tag{5.175}$$

It follows from these expressions that

$$\underline{d}\underline{\alpha}(\vec{u}, \vec{v}) = \vec{u}(\underline{\alpha}(\vec{v})) - \vec{v}(\underline{\alpha}(\vec{u})) - \underline{\alpha}([\vec{u}, \vec{v}]). \tag{5.176}$$

This identity is valid in an arbitrary basis.
Let $\underline{\alpha} = \underline{\omega}^\rho$, $\vec{u} = \vec{e}_\mu$, $\vec{v} = \vec{e}_\nu$. Then

$$\begin{aligned}
\underline{d}\,\underline{\omega}^\rho(\vec{e}_\mu, \vec{e}_\nu) &= \vec{e}_\mu(\underline{\omega}^\rho(\vec{e}_\nu)) - \vec{e}_\nu(\underline{\omega}^\rho(\vec{e}_\mu)) - \underline{\omega}^\rho([\vec{e}_\mu, \vec{e}_\nu]) \\
&= \vec{e}_\mu(\delta^\rho_\nu) - \vec{e}_\nu(\delta^\rho_\mu) - \underline{\omega}^\rho(c^\alpha_{\mu\nu}\vec{e}_\alpha) = -c^\alpha_{\mu\nu}\delta^\rho_\alpha = -c^\rho_{\mu\nu}. \tag{5.177}
\end{aligned}$$

A 2-form applied to two basis vectors is equal to its components. Hence we get Cartan's 1 structure equation

$$\underline{d}\underline{\omega}^\rho = -\frac{1}{2}c^\rho_{\mu\nu}\underline{\omega}^\mu \wedge \underline{\omega}^\nu. \tag{5.178}$$

Combining this with Eq. (5.132) and utilizing the antisymmetry of the wedge product we have

$$\underline{d\omega}^\rho = -\frac{1}{2}(\Gamma^\rho_{\nu\mu} - \Gamma^\rho_{\mu\nu})\underline{\omega}^\mu \wedge \underline{\omega}^\nu = \Gamma^\rho_{\mu\nu}\underline{\omega}^\mu \wedge \underline{\omega}^\nu. \tag{5.179}$$

Together with Eq. (5.164) this gives

$$\underline{d\omega}^\rho = -\underline{\Omega}^\rho_\nu \wedge \underline{\omega}^\nu. \tag{5.180}$$

In coordinate basis, we have $\underline{\omega}^\rho = \underline{d}x^\rho$. Thus,

$$\underline{d\omega}^\rho = \underline{d}^2 x^\rho = 0. \tag{5.181}$$

The exterior derivatives of the basis forms vanish in coordinate basis. We also have $c^\rho_{\mu\nu} = 0$, and Cartan's 1 structure equation is reduced to an identity. This formalism cannot be used in coordinate basis! But due to the antisymmetry (5.169) *the Cartan formalism is particularly useful in orthonormal basis.*

Example 5.8.1 (*Cartan connection in an orthonormal basis field in plane polar coordinates.*) Here the line-element is

$$ds^2 = dr^2 + r^2 d\theta^2.$$

Introducing basis forms in an orthonormal basis field (where the metric is $g_{\hat{r}\hat{r}} = g_{\hat{\theta}\hat{\theta}} = 1$), the metric tensor takes the form

$$\mathbf{g} = g_{\hat{r}\hat{r}}\underline{\omega}^{\hat{r}} \otimes \underline{\omega}^{\hat{r}} + g_{\hat{\theta}\hat{\theta}}\underline{\omega}^{\hat{\theta}} \otimes \underline{\omega}^{\hat{\theta}} = \underline{\omega}^{\hat{r}} \otimes \underline{\omega}^{\hat{r}} + \underline{\omega}^{\hat{\theta}} \otimes \underline{\omega}^{\hat{\theta}}.$$

Hence

$$\underline{\omega}^{\hat{r}} = \underline{d}r, \quad \underline{\omega}^{\hat{\theta}} = r\underline{d}\theta$$

Exterior differentiation gives

$$\underline{d\omega}^{\hat{r}} = \underline{d}^2 r = 0, \quad \underline{d\omega}^{\hat{\theta}} = \underline{d}r \wedge \underline{d}\theta = \frac{1}{r}\underline{\omega}^{\hat{r}} \wedge \underline{\omega}^{\hat{\theta}}.$$

From Eq. (5.180) we then have

$$\underline{d\omega}^{\hat{\mu}} = -\underline{\Omega}^{\hat{\mu}}_{\hat{\nu}} \wedge \underline{\omega}^\nu = -\underline{\Omega}^{\hat{\mu}}_{\hat{r}} \wedge \underline{\omega}\hat{r} - \underline{\Omega}^{\hat{\mu}}_{\hat{\theta}} \wedge \underline{\omega}^{\hat{\theta}}.$$

Furthermore $\underline{d\omega}^{\hat{r}} = 0$, which gives

$$\underline{\Omega}^{\hat{r}}_{\hat{\theta}} = \Gamma^{\hat{r}}_{\hat{\theta}\hat{\theta}}\underline{\omega}^{\hat{\theta}} \tag{5.182}$$

since $\underline{\omega}^{\hat{\theta}} \wedge \underline{\omega}^{\hat{\theta}} = 0$. Note that $\underline{\Omega}^{\hat{r}}_{\hat{r}} = 0$ because of the antisymmetry $\underline{\Omega}_{\hat{\nu}\hat{\mu}} = -\underline{\Omega}_{\hat{\mu}\hat{\nu}}$. We also have

$$\underline{d\omega}^{\hat{\theta}} = -\frac{1}{r}\underline{\omega}^{\hat{\theta}} \wedge \underline{\omega}^{\hat{r}},$$

and from Eq. (5.180) it follows that

$$\underline{d\omega}^{\hat{\theta}} = -\underline{\Omega}^{\hat{\theta}}_{\hat{r}} \wedge \underline{\omega}^{\hat{r}} - \underbrace{\underline{\Omega}^{\hat{\theta}}_{\hat{\theta}}}_{=0} \wedge \underline{\omega}^{\hat{\theta}},$$

From Eq. (5.164) we have

$$\underline{\Omega}^{\hat{\theta}}_{\hat{r}} - \Gamma^{\hat{\theta}}_{\hat{r}\hat{\theta}}\underline{\omega}^{\hat{\theta}} + \Gamma^{\hat{\theta}}_{\hat{r}\hat{r}}\underline{\omega}^{\hat{r}}, \qquad\qquad (5.183)$$

giving

$$\Gamma^{\hat{\theta}}_{\hat{r}\hat{\theta}} = \frac{1}{r}.$$

Furthermore $\underline{\Omega}^{\hat{r}}_{\hat{\theta}} = -\underline{\Omega}^{\hat{\theta}}_{\hat{r}}$. Using Eqs. (5.182) and (5.183) we get

$$\Gamma^{\hat{\theta}}_{\hat{r}\hat{r}} = 0, \quad \Gamma^{\hat{r}}_{\hat{\theta}\hat{\theta}} = -\frac{1}{r}$$

giving

$$\underline{\Omega}^{\hat{r}}_{\hat{\theta}} = \underline{\Omega}^{\hat{\theta}}_{\hat{r}} = \frac{1}{r}\underline{\omega}^{\hat{\theta}}.$$

5.9 Covariant Decomposition of a Velocity Field

I order to develop some intuition about the kinematics of fluids we shall first consider an ordinary 3-velocity field in Newtonian hydrodynamics.

5.9.1 Newtonian 3-Velocity

The *total* or *material derivative* of the velocity is

$$\frac{D\vec{v}}{Dt} = \frac{\partial\vec{v}}{\partial t} + (\vec{v}\cdot\nabla)\vec{v}. \qquad\qquad (5.184)$$

Here $\partial \vec{v}/\partial t$ is the *local derivative* which represents the change of the velocity field at a certain position, for example the increase with time of the velocity at certain place in a river due to rain. The term $(\vec{v} \cdot \nabla)\,\vec{v}$ is called the *convective derivative* and is sometimes written $d\vec{v}/dt$. It represents the change of velocity with position at a fixed point of time, for example the slowing down of the velocity further downwards due to a widening of the river. The total derivative represents the change of the velocity field following a fluid particle, due to both local changes with time and the inhomogeneity of the velocity field. The component version of Eq. (5.184) is

$$\frac{Dv^i}{Dt} = \frac{\partial v^i}{\partial t} + v^j \frac{\partial v^i}{\partial x^j}. \tag{5.185}$$

Introducing a Cartesian coordinate system with coordinates (x, y, z), the convective derivative may be written in matrix form as

$$\begin{pmatrix} \frac{dv^x}{dt} \\ \frac{dv^y}{dt} \\ \frac{dv^z}{dt} \end{pmatrix} = \begin{pmatrix} \frac{\partial v^x}{\partial x} & \frac{\partial v^x}{\partial y} & \frac{\partial v^x}{\partial z} \\ \frac{\partial v^y}{\partial x} & \frac{\partial v^y}{\partial y} & \frac{\partial v^y}{\partial z} \\ \frac{\partial v^z}{\partial x} & \frac{\partial v^z}{\partial y} & \frac{\partial v^z}{\partial z} \end{pmatrix} \begin{pmatrix} v^x \\ v^y \\ v^z \end{pmatrix}. \tag{5.186}$$

The 3×3 matrix may be separated into three parts: The trace

$$\theta = \frac{\partial v^x}{\partial x} + \frac{\partial v^y}{\partial y} + \frac{\partial v^z}{\partial z}, \tag{5.187}$$

The antisymmetric part

$$\omega_{ij} = \frac{1}{2} \begin{pmatrix} 0 & \frac{\partial v^x}{\partial y} - \frac{\partial v^y}{\partial x} & \frac{\partial v^x}{\partial z} - \frac{\partial v^z}{\partial x} \\ & 0 & \frac{\partial v^y}{\partial z} - \frac{\partial v^z}{\partial y} \\ \text{minus the same} & & 0 \end{pmatrix}, \tag{5.188}$$

and the symmetric, trace-free part,

$$\sigma_{ij} = \begin{pmatrix} \frac{1}{3}\left(2\frac{\partial v^x}{\partial x} - \frac{\partial v^y}{\partial y} - \frac{\partial v^z}{\partial z} \right) & \frac{\partial v^x}{\partial y} + \frac{\partial v^y}{\partial x} & \frac{\partial v^x}{\partial z} + \frac{\partial v^z}{\partial x} \\ & \frac{1}{3}\left(2\frac{\partial v^y}{\partial y} - \frac{\partial v^x}{\partial x} - \frac{\partial v^z}{\partial z} \right) & \frac{\partial v^y}{\partial z} + \frac{\partial v^z}{\partial x} \\ & & \frac{1}{3}\left(2\frac{\partial v^z}{\partial z} - \frac{\partial v^x}{\partial x} - \frac{\partial v^y}{\partial y} \right) \end{pmatrix}. \tag{5.189}$$

Here θ *represents the volume expansion (or contraction)*, ω_{ij} *rotation and* σ_{ij} *the shear*. The component form of these expressions is

$$v_{i,j} = \theta_{ij} + \omega_{ij} + \sigma_{ij}, \tag{5.190}$$

$$\theta_{ij} = \frac{1}{3}v^k{}_{,k}\,\delta_{ij}, \tag{5.191}$$

$$\omega_{ij} = \frac{1}{2}\left(v_{i,j} - v_{j,i}\right) = \frac{1}{2}(\nabla \times \vec{v})_{ij}, \tag{5.192}$$

$$\sigma_{ij} = \frac{1}{2}\left(v_{i,j} + v_{j,i}\right) - \frac{1}{2}v^k{}_{,k}\,\delta_{ij}. \tag{5.193}$$

5.9.2 Relativistic 4-Velocity

We shall now find the relativistic generalizations of these expressions, i.e. we shall find covariant expressions that represent expansion, rotation and shear as measured by an observer following the fluid.

The covariant derivative may be thought of as a generalization of the total derivative. We first separate the covariant directional derivative of a 4-velocity field in a component along the 4-velocity and a component orthogonal to it. The directional derivative of the 4-velocity along itself is the 4-acceleration. It is orthogonal to the 4-velocity and has covariant components

$$\dot{u}_\alpha = u_{\alpha;\mu}u^\mu = \left(u_{\alpha,\mu} - u_\nu\Gamma^\nu_{\alpha\mu}\right)u^\mu. \tag{5.194}$$

Putting them equal to zero gives the geodesic equation describing the world line of a freely falling particle.

The projection of the tensor with components $u_{\alpha;\beta}$ into the spatial simultaneity space orthogonal to the 4-velocity, is

$$\left(u_{\alpha;\beta}\right)_\perp = u_{\mu;\nu}P^\mu_\alpha P^\nu_\beta. \tag{5.195}$$

where P^μ_α are the mixed components of the projection tensor defined in Eq. (4.104). In the same way as we did in the Newtonian case, this may be separated in three parts.
Expansion:

$$\theta_{\alpha\beta} = \frac{1}{3}\theta P_{\alpha\beta}, \quad \theta = u^\mu{}_{;\mu}. \tag{5.196}$$

Shear:

$$\sigma_{\alpha\beta} = \frac{1}{2}\left(u_{\mu;\nu} + u_{\nu;\mu}\right)P^\mu_\alpha P^\nu_\beta - \frac{1}{3}u^\mu{}_{;\mu}P_{\alpha\beta}. \tag{5.197}$$

Rotation:

$$\omega_{\alpha\beta} = \frac{1}{2}\left(u_{\mu;\nu} - u_{\nu;\mu}\right)P_\alpha^\mu P_\beta^\nu. \tag{5.198}$$

The covariant derivative of a 4-velocity field can now the be separated in expansion, shear, rotation and a 4-acceleration-term as follows

$$u_{\alpha;\beta} = \theta_{\alpha\beta} + \sigma_{\alpha\beta} + \omega_{\alpha\beta} - \dot{u}_\alpha u_\beta \tag{5.199}$$

In the relativistic literature, the expressions for shear and rotation are written in different ways that are found by inserting the components of the projection tensor from Eq. (4.106) into Eq. (5.195),

$$u_{\mu;\nu}P_\alpha^\mu P_\beta^\nu = u_{\mu;\nu}\left(\delta_\alpha^\eta + u^\mu u_\alpha\right)\left(\delta_\beta^\nu + u^\nu u_\beta\right). \tag{5.200}$$

Using that

$$u_{\mu;\nu}\delta_\alpha^\mu \delta_\beta^\nu = u_{\alpha;\beta}, \quad u_{\mu;\nu}\delta_\alpha^\mu u^\nu u_\beta = u_{\alpha;\nu}u^\nu u_\beta = \dot{u}_\alpha u_\beta, \tag{5.201}$$

we get

$$u_{\mu;\nu}P_\alpha^\mu P_\beta^\nu = u_{\alpha;\beta} + \dot{u}_\alpha u_\beta + u_{\mu;\beta}u^\mu u_\alpha + \dot{u}_\mu u^\mu u_\alpha u_\beta. \tag{5.202}$$

It follows from the 4-velocity identity, $u_\mu u^\mu = -1$, that the 4-acceleration is orthogonal to the 4-velocity, $\dot{u}_\mu u^\mu = 0$, and that $\left(u_\mu u^\mu\right)_{;\beta} = 0$, giving $u_{\mu;\beta}u^\mu = 0$. Hence Eq. (5.202) simplifies to

$$u_{\mu;\nu}P_\alpha^\mu P_\beta^\nu = u_{\alpha;\beta} + \dot{u}_\alpha u_\beta. \tag{5.203}$$

Furthermore

$$u_{\alpha;\mu}P_\beta^\mu = u_{\alpha;\mu}\delta_\beta^\mu + u_{\alpha;\mu}u^\mu u_\beta = u_{\alpha;\beta}\dot{u}_\alpha + u_\beta, \tag{5.204}$$

or

$$u_{\alpha;\beta} = u_{\alpha;\mu}P_\beta^\mu - \dot{u}_\alpha u_\beta. \tag{5.205}$$

We then get

$$u_{\alpha;\mu}P_\beta^\mu = \theta_{\alpha\beta} + \sigma_{\alpha\beta} + \omega_{\alpha\beta}, \tag{5.206}$$

with the alternative expressions for shear and rotation

$$\sigma_{\alpha\beta} = \frac{1}{2}\left(u_{\alpha;\mu}P_\beta^\mu + u_{\beta;\mu}P_\alpha^\mu\right) - \frac{1}{3}u_{;\mu}^\mu P_{\alpha\beta}, \tag{5.207}$$

$$\omega_{\alpha\beta} = \frac{1}{2}\left(u_{\alpha;\mu}P_{\beta}^{\mu} - u_{\beta;\mu}P_{\alpha}^{\mu}\right). \tag{5.208}$$

5.10 Killing Vectors and Symmetries

Killing vectors are useful for describing the symmetry properties of space in an invariant way.

Definition 5.10.1 (*Killing vectors*) We will here define a *Killing vector* as a vector $\vec{\xi}$ which obeys *Killing's equation*

$$\xi_{\mu;\nu} + \xi_{\nu;\mu} = 0. \tag{5.209}$$

In this form Killing's equation is valid in an arbitrary basis. In a coordinate basis the equation reduces to

$$\xi_{\mu,\nu} + \xi_{\nu,\mu} = 2\xi_{\alpha}\Gamma_{\mu\nu}^{\alpha}. \tag{5.210}$$

One may show that if $\vec{\xi}^{(1)}$ and $\vec{\xi}^{(2)}$ are two Killing vectors, and a, b are constants, then $a\vec{\xi}^{(1)} + b\vec{\xi}^{(2)}$ is a Killing vector. Furthermore, the commutator $\left[\vec{\xi}^{(1)}, \vec{\xi}^{(2)}\right]$ between two Killing vectors is also a Killing vector.

In an n-dimensional space, there are maximally $n(n + 1)/2$ linearly independent Killing vectors. In four-dimensional spacetime, there may be up to ten such vectors. A metric and the corresponding space that admit the maximum number of Killing vectors is said to be *maximally symmetric*.

Example 5.10.1 (*Killing vectors of an Euclidean plane*) We consider a two-dimensional Euclidean plane with Cartesian coordinates. Then the line-element has the form

$$dl^2 = dx^2 + dy^2, \tag{5.211}$$

and the Christoffel symbols vanish. Hence, the Killing equations reduce to

$$\xi_{x,x} = \xi_{y,y} = 0, \quad \xi_{x,y} + \xi_{y,x} = 0. \tag{5.212}$$

These equations have the following solutions,

$$\vec{\xi}_1 = \frac{\partial}{\partial x}, \quad \vec{\xi}_2 = \frac{\partial}{\partial y}, \quad \vec{\xi}_1 = x\frac{\partial}{\partial y} - y\frac{\partial}{\partial x}. \tag{5.213}$$

The first vector represents invariance by translation in the x-direction, the second in the y-direction, and the third vector represents invariance by rotation about an axis orthogonal to the plane. A two-dimensional space with three Killing vectors is maximally symmetric. Hence the Euclidean space is maximally symmetric.

Definition 5.10.2 (*Invariant basis*) An *invariant basis* is defined as a basis where the basis vectors commute with a Killing vector,

$$\left[\vec{\xi}, \vec{e}_\mu \right] = 0. \tag{5.214}$$

Since the commutator of two coordinate basis vectors vanish, a consequence of this definition is that if a basis vector of a coordinate basis is a Killing vector, then this coordinate basis is invariant.

The commutator of two vectors, $\left[\vec{u}, \vec{v} \right]$ is also called the *Lie derivative* of $\vec{v}$ with respect to $\vec{u}$, which is writte

$$\pounds_{\vec{u}} \vec{v} = \left[\vec{u}, \vec{v} \right]. \tag{5.215}$$

The Lie derivative of a scalar function with respect to a vector is the directional derivative of the scalar function along the vector,

$$\pounds_{\vec{u}} f = \vec{u}(f). \tag{5.216}$$

We have the rule

$$\pounds_{\vec{u}} \left(\vec{A} \cdot \vec{B} \right) = \left(\pounds_{\vec{u}} \vec{A} \right) \cdot \vec{B} + \vec{A} \cdot \left(\pounds_{\vec{u}} \vec{B} \right). \tag{5.217}$$

From this we get

$$\pounds_{\vec{u}} \left(g_{\mu\nu} \right) = \pounds_{\vec{u}} \left(\vec{e}_\mu \cdot \vec{e}_\nu \right) = \left(\pounds_{\vec{u}} \vec{e}_\mu \right) \cdot \vec{e}_\nu + \vec{e}_\mu \cdot \left(\pounds_{\vec{u}} \vec{e}_\nu \right). \tag{5.218}$$

Equations (5.214) and (5.215) give

$$\pounds_{\vec{\xi}} \vec{e}_\mu = 0 \tag{5.219}$$

It follows from Eq. (5.215) for an invariant basis, together with Eqs. (5.218) and (5.219), that

$$\vec{\xi} \left(g_{\mu\nu} \right) = 0 \tag{5.220}$$

for an invariant basis. Hence the components of the metric tensor are constants along Killing vectors in a space with an invariant basis field.

We shall finally find the relation between the Killing vectors of a space and the constants of motion for a particle moving freely in the space. A free particle moves along a geodesic curve with the equation

$$\nabla_{\vec{u}}\,\vec{u} = 0 \tag{5.221}$$

where $\vec{u}$ is the 4-velocity of the particle. Consider the scalar product $\vec{u} \cdot \vec{\xi}$, where $\vec{\xi}$ is a Killing vector field. The covariant directional derivative of this product along the geodesic curve is

$$\nabla_{\vec{u}}\left(\vec{u} \cdot \vec{\xi}\right) = u^{\sigma}\left(u^{\mu}\xi_{\mu}\right)_{;\sigma} = u^{\sigma}u^{\mu}_{;\sigma}\xi_{\mu} + u^{\sigma}u^{\mu}\xi_{\mu;\nu}. \tag{5.222}$$

Here the first term vanishes because of Eq. (5.221), and the second vanishes since $u^{\sigma}u^{\mu}$ is symmetric in σ and μ while $\xi_{\mu;\nu}$ is antisymmetric in σ and μ due to the Killing equation (5.209). Thus

$$\nabla_{\vec{u}}\left(\vec{u} \cdot \vec{\xi}\right) = 0. \tag{5.223}$$

We then have the result that $\vec{u} \cdot \vec{\xi}$ is constant along a geodesic curve. For a particle with constant rest mass, this may also be expressed as: $\vec{p} \cdot \xi$ is constant along a geodesic curve, where $\vec{p}$ is the 4 momentum of the particle.

Assume that $\vec{\xi}_{\alpha}$ is a time-like Killing vector associated with a cyclic coordinate x^{α}. Then a coordinate system can be chosen such that $\vec{\xi}_{\alpha} = (-1, 0, 0, 0)$, i.e. $\vec{\xi}_{\alpha} = -\delta^{0}_{\alpha}\vec{e}_{0}$. In this case

$$\vec{p} \cdot \vec{\xi}_{\alpha} = -p_{\mu}\delta^{0}_{\alpha} = -p_{\alpha 0} = \text{constant}. \tag{5.224}$$

which is the energy of the particle. Correspondingly, if $\vec{\xi}_{\alpha}$ is a space-like Killing vector associated with a cyclic coordinate, and one chooses a coordinate system so that $\vec{\xi}_{\alpha} = (0, 1, 0, 0)$, one obtains

$$p_{\alpha 1} = \text{constant}. \tag{5.225}$$

This represents conservation of momentum in the x-direction for a free particle and is in accordance with the Lagrangian dynamics that the covariant momenta conjugate to cyclic coordinates are constants of motion for freely falling particles. The Killing vectors describe symmetries of spacetime. Hence the equation $\vec{p} \cdot \vec{\xi}_{\alpha} = \text{constant}$ relates constants of motion of a free particle to symmetries of spacetime.

5.11 Covariant Expressions for Gradient, Divergence, Curl, Laplacian and D'Alembert's Wave Operator

The exterior derivative $\underline{d}$ of a p-form was defined in Eq. (5.6). It is an antisymmetric derivative giving a $(p + 1)$-form. The so-called *Hodge dual* of the exterior derivative $\underline{\delta}$ is called the *codifferential*. In an n-dimensional space, it is defined by

$$\underline{\delta} = (-1)^{n(p+1)+1} \star \underline{d} \star. \tag{5.226}$$

where the Hodge star operator acting on a form gives the dual form defined in Eq. (3.118). The codifferential acting on a p-form gives out a $(p-1)$-form. For example the codifferential of a 1-form is a scalar function. If the dimension of the space is even, for example in four-dimensional spacetime, then

$$\underline{\delta} = -\star \underline{d} \star. \tag{5.227}$$

Since there only exist forms with $p - 1 \geq 0$ the codifferential of a scalar function f is not defined as a form equation, but according to a separate definition the codifferential of a scalar function vanishes,

$$\underline{\delta} f = 0. \tag{5.228}$$

In the same manner as the exterior derivative as applied to a p-form with scalar components, satisfies Poincare's lemma, $\underline{d}^2 = 0$, the codifferential satisfies

$$\underline{\delta}\,\underline{\delta} = \underline{\delta}^2 = 0. \tag{5.229}$$

A generalization of the Laplacian operator in 3-space

$$\Delta = \frac{\partial^2}{\partial x^2} + \frac{\partial^2}{\partial y^2} + \frac{\partial^2}{\partial z^2}, \tag{5.230}$$

is the form operator called the *Hodge Laplacian* or alternatively the *Laplace–Beltrami operator*, defined by

$$\underline{\Delta} = \underline{\delta}\,\underline{d} + \underline{d}\,\underline{\delta}. \tag{5.231}$$

Due to Eq. (5.228) the Hodge Laplacian as applied to a scalar function takes the form

$$\underline{\Delta}\phi = \underline{\delta}\,\underline{d}\phi = \star \underline{d} \star \underline{d}\phi. \tag{5.232}$$

The Hodge Laplacian of a 1-form is

$$\underline{\Delta}\,\underline{\alpha} = (\underline{d}\star\underline{d}\star + \star\underline{d}\star\underline{d})\underline{\alpha}. \tag{5.233}$$

Let us find the covariant form of the Hodge Laplacian in four-dimensional spacetime as applied to a scalar function f. Using Eq. (3.118) and the definition (5.6) of the exterior derivative, Eq. (5.232) gives

$$\underline{\Delta}f = \frac{1}{4!} V^{\alpha\beta\gamma\delta} 4 \frac{\partial}{\partial x^\alpha}\left(V_{\lambda\beta\gamma\delta}\frac{\partial f}{\partial x_\lambda}\right)$$

$$= \frac{1}{3!} \frac{-1}{\sqrt{-g}} \varepsilon^{\alpha\beta\gamma\delta} \frac{\partial}{\partial x^\alpha} \left(\sqrt{-g} \varepsilon_{\lambda\beta\gamma\delta} g^{\lambda\rho} \frac{\partial f}{\partial x^\rho} \right)$$

$$= -\frac{3!}{3!} \frac{1}{\sqrt{-g}} \delta^\alpha_\lambda \frac{\partial}{\partial x^\alpha} \left(\sqrt{-g} g^{\lambda\rho} \frac{\partial f}{\partial x^\rho} \right)$$

$$= -\frac{1}{\sqrt{-g}} \frac{\partial}{\partial x^\lambda} \left(\sqrt{-g} g^{\lambda\rho} \frac{\partial f}{\partial x^\rho} \right). \tag{5.234}$$

where $V_{\alpha\beta\gamma\delta} = \sqrt{-g}\varepsilon_{\alpha\beta\gamma\delta}$, are the components of the volume form as given in Eq. (3.116), and the minus sign comes from

$$V^{\mu\beta\gamma\delta} = \frac{1}{g} V_{\alpha\beta\gamma\delta} = \frac{1}{g} \sqrt{-g} \varepsilon_{\alpha\beta\gamma\delta} = -\frac{1}{\sqrt{-g}} \varepsilon_{\alpha\beta\gamma\delta} = -\frac{1}{\sqrt{-g}} \varepsilon^{\alpha\beta\gamma\delta}. \tag{5.235}$$

In this case, the Hodge Laplacian reduces to minus the d'Lambertian wave operator, $\Delta = -\square$. Hence we have the covariant form of the wave operator

$$\square f = \frac{1}{\sqrt{-g}} \frac{\partial}{\partial x^\lambda} \left(\sqrt{-g} g^{\lambda\rho} \frac{\partial f}{\partial x^\rho} \right). \tag{5.236}$$

Performing the same calculation for a scalar field in 3-space, we get the covariant form of the Laplacian

$$\Delta f = \frac{1}{\sqrt{g}} \frac{\partial}{\partial x^i} \left(\sqrt{g} g^{ij} \frac{\partial f}{\partial x^j} \right). \tag{5.237}$$

We shall finally look at the relationship between the exterior derivative and the gradient, divergence and curl. The gradient is defined as the vector corresponding to the 1-form $\underline{d}\phi$. Hence we get the covariant expression for the gradient as decomposed in a coordinate basis

$$\vec{\nabla} f = g^{ij} \frac{\partial f}{\partial x^j} \vec{e}_i. \tag{5.238}$$

Considering an orthogonal coordinate basis, the physical components of the gradient are obtained by transforming to an orthonormal basis according to $\vec{e}_i = \sqrt{g_{ii}}\vec{e}_{\hat{i}}$. Hence the physical components of the gradient are given by

$$\nabla f = \frac{1}{\sqrt{g_{ii}}} \frac{\partial f}{\partial x^i} \vec{e}_{\hat{i}}. \tag{5.239}$$

The divergence of a vector is given by

$$|g| \text{div} \, \vec{A} = \star \underline{d} \star \underline{A}, \tag{5.240}$$

where the components of the 1-form $\underline{A}$ are found by lowering the components of the vector $\vec{A}$. From Eqs. (3.116) and (3.118) we have

$$\star \underline{A} = \frac{1}{(n-1)!} \sqrt{|g|} \varepsilon_{\nu \cdot \mu_1 \ldots \mu_{n-1}} A^\nu \underline{dx}^{\nu_1} \wedge \underline{dx}^{\mu_1} \wedge \cdots \wedge \underline{dx}^{\mu_{n-1}}. \tag{5.241}$$

The exterior derivative of this form is

$$\underline{d} \star \underline{A} = \frac{1}{(n-1)!} \varepsilon_{\nu \mu_1 \ldots \mu_{n-p}} \frac{\partial \left(\sqrt{|g|} A^\nu \right)}{\partial x^\alpha} \underline{dx}^\alpha \wedge \underline{dx}^{\mu_1} \wedge \cdots \wedge \underline{dx}^{\mu_{n-1}}. \tag{5.242}$$

To avoid multicounting we use the notation $\varepsilon_{|\nu_1 \cdot \mu_1 \ldots \mu_{n-1}|}$ meaning that the summation over the indices is performed under the condition $\nu < \mu_1 < \cdots < \mu_{n-1}$. Then

$$\underline{d} \star \underline{A} = \varepsilon_{|\nu \cdot \mu_1 \ldots \mu_{n-1}|} \frac{\partial \left(\sqrt{|g|} A^\nu \right)}{\partial x^\alpha} \underline{dx}^\alpha \wedge \underline{dx}^{\mu_1} \wedge \cdots \wedge \underline{dx}^{\mu_{n-1}}. \tag{5.243}$$

The dual of this form is

$$\star \underline{d} \star \underline{A} = \sqrt{|g|} \frac{\partial \left(\sqrt{|g|} A^\nu \right)}{\partial x^\nu}. \tag{5.244}$$

Inserting this into Eq. (5.240) gives the covariant expression for the divergence of a vector

$$\operatorname{div} \vec{A} = \frac{1}{\sqrt{|g|}} \frac{\partial \left(\sqrt{|g|} A^\nu \right)}{\partial x^\nu}. \tag{5.245}$$

We shall find the covariant expressions for the components of the curl of a vector $\vec{A}$ in the special case of a coordinate system with an orthogonal vector basis. The corresponding 1-form is $\underline{A}$. The covariant expressions of the components of the curl $\vec{\nabla} \times \vec{A}$ is obtained from

$$\vec{\nabla} \times \vec{A} = (\star \underline{d} \, \underline{A})^\#. \tag{5.246}$$

This involves the following operations: The vector is originally given in an orthonormal basis associated with a chosen coordinate system. We call the corresponding vector components for the *physical components* of the vector. One wants to express the physical components of $\vec{\nabla} \times \vec{A}$ in an arbitrary orthogonal coordinate system in terms of the physical components of the vector and the components of the metric tensor in the coordinate system. Since the expression (5.246) involves the exterior derivative of the 1-form $\underline{A}$, which is most easily performed in a coordinate basis because then the exterior derivatives of the basis form vanish due to Poincare's lemma, one first transforms to the coordinate basis of the coordinate system. Then the (contravariant) indices of the components of the vector $\vec{A}$ is lowered, and one obtains the (covariant) components of the 1-form $\underline{A}$ in the arbitrary coordinate basis in terms of the physical components of the vector and the components of the metric tensor (Eq. 5.250 below). Then one takes the exterior derivative of this form and gets the 2-form $\underline{d}\,\underline{A}$ as decomposed in the coordinate basis. One now transforms to the form basis corresponding to the orthonormal vector basis. Further one takes the dual of this form and gets a 1-form $\star\underline{d}\,\underline{A}$. The symbol # means that one converts this 1-form to a vector by raising the indices of the components of $\star\underline{d}\,\underline{A}$. This is automatically taken account of when we decompose $\underline{d}\,\underline{A}$ in a 1-form basis corresponding to an orthonormal vector basis where the covariant and contravariant components are equal. Hence we get the vector as decomposed in the orthonormal basis. In this way we have found the physical components of $\vec{\nabla}\times\vec{A}$.

We shall demonstrate this procedure in the case of an arbitrary orthogonal vector basis, so that the metric tensor is diagonal, i.e. the only non-vanishing components of the metric tensor are those with equal indices.

In a coordinate system with orthogonal basis vectors, the line element is

$$dl^2 = g_{ii}\,dx^i\,dx^i. \tag{5.247}$$

The coordinate basis vectors are $\vec{e}_i = \partial/\partial x^i$, and the corresponding coordinate basis forms are $\underline{dx}^i$. Since the metric is diagonal, the transformation of the basis vectors to a coordinate basis has the form $\vec{e}_{\hat{i}} = (g_{ii})^{-1/2}\vec{e}_i$. The corresponding basis forms dual to the orthonormal vector basis are $\underline{\omega}^{\hat{i}} = \sqrt{g_{ii}}\underline{dx}^i$. In this space we have a vector $\vec{A} = A^{\hat{i}}\vec{e}_{\hat{i}}$, $\hat{i} = 1, 2, 3$ as decomposed in an orthonormal basis. As decomposed in the coordinate basis the vector components are

$$A^i = (g_{ii})^{-1/2}A^{\hat{i}}. \tag{5.249}$$

So that

$$\vec{A} = A^i\vec{e}_i = (g_{ii})^{-1/2}A^{\hat{i}}\vec{e}_i. \tag{5.249}$$

The corresponding 1-form is

$$\underline{A} = g_{ii}A^i\underline{dx}^i = \sqrt{g_{ii}}A^{\hat{i}}\underline{dx}^i. \tag{5.250}$$

The exterior derivative of this form is

$$\begin{aligned}
\underline{d}\,\underline{A} &= \frac{1}{2}\left[\left(\sqrt{g_{ii}}A^{\hat{i}}\right)_{,j} - \left(\sqrt{g_{jj}}A^{\hat{j}}\right)_{,i}\right]\underline{dx}^j \wedge \underline{dx}^i \\
&= \frac{1}{2}\frac{1}{\sqrt{g_{ii}g_{jj}}}\left[\left(\sqrt{g_{ii}}A^{\hat{i}}\right)_{,j} - \left(\sqrt{g_{jj}}A^{\hat{j}}\right)_{,i}\right]\underline{\omega}^j \wedge \underline{\omega}^i.
\end{aligned} \tag{5.251}$$

The dual of this form is the 1-form

$$\star \underline{d}\,\underline{A} = \frac{1}{\sqrt{g_{ii}g_{jj}}}\Big[\big(\sqrt{g_{ii}}A^i\big)_{,j} - \big(\sqrt{g_{jj}}A^j\big)_{,i}\Big]\omega^{\hat{k}}, \quad \hat{k} \neq \hat{i},\hat{j}. \tag{5.252}$$

Thus the curl is

$$\nabla \times \vec{A} = \frac{1}{\sqrt{g_{ii}g_{jj}}}\Big[\big(\sqrt{g_{ii}}A^i\big)_{,j} - \big(\sqrt{g_{jj}}A^j\big)_{,i}\Big]\vec{e}_{\hat{k}}. \tag{5.253}$$

Example 5.11.1 (Differential operators in spherical coordinates) The components of the metric tensor in spherical coordinates in Euclidean 3-space are given in Eq. (3.122), $g_{rr} = 1, g_{\theta\theta} = r^2, g_{\phi\phi} = r^2 \sin^2 \theta$. Inserting this into Eq. (5.239) we get the expression for the gradient in spherical coordinates

$$\nabla f = \frac{\partial f}{\partial r}\vec{e}_{\hat{r}} + \frac{1}{r}\frac{\partial f}{\partial \theta}\vec{e}_{\hat{\theta}} + \frac{1}{r\sin\theta}\frac{\partial f}{\partial \phi}\vec{e}_{\hat{\phi}}. \tag{5.254}$$

The curl is

$$\nabla \times \vec{A} = \frac{1}{r\sin\theta}\left[\frac{\partial}{\partial \theta}\big(\sin\theta A^{\hat{\phi}}\big) - \frac{\partial A^{\hat{\theta}}}{\partial \phi}\right]\vec{e}_{\hat{r}}$$

$$+ \frac{1}{r\sin\theta}\left[\frac{\partial A^{\hat{r}}}{\partial \phi} - \sin\theta\frac{\partial}{\partial r}\big(r A^{\hat{\phi}}\big)\right]\vec{e}_{\hat{\theta}}$$

$$+ \frac{1}{r}\left[\frac{\partial}{\partial r}\big(r A^{\hat{\theta}}\big) - \frac{\partial A^{\hat{r}}}{\partial \theta}\right]\vec{e}_{\hat{\phi}}. \tag{5.255}$$

The divergence is

$$\nabla \cdot \vec{A} = \frac{1}{r^2}\frac{\partial(r^2 A^{\hat{r}})}{\partial r} + \frac{1}{r\sin\theta}\frac{\partial\big(\sin\theta A^{\hat{\theta}}\big)}{\partial \theta} + \frac{1}{r\sin\theta}\frac{\partial A^{\hat{\phi}}}{\partial \phi}. \tag{5.256}$$

The Laplacian is

$$\nabla^2 f = \frac{1}{r^2}\frac{\partial}{\partial r}\left(r^2\frac{\partial f}{\partial r}\right) + \frac{1}{r^2\sin\theta}\frac{\partial}{\partial \theta}\left(\sin\theta\frac{\partial f}{\partial \theta}\right) + \frac{1}{r\sin^2\theta}\frac{\partial^2 f}{\partial \phi^2}. \tag{5.257}$$

The d'Alembertian wave operator is

$$\Box f = \frac{1}{r^2}\frac{\partial}{\partial r}\left(r^2\frac{\partial f}{\partial r}\right) + \frac{1}{r^2\sin\theta}\frac{\partial}{\partial \theta}\left(\sin\theta\frac{\partial f}{\partial \theta}\right) + \frac{1}{r\sin^2\theta}\frac{\partial^2 f}{\partial \phi^2} - \frac{1}{c^2}\frac{\partial^2 f}{\partial t^2}. \tag{5.258}$$

5.12   Electromagnetism in Form Language

We shall consider electromagnetism in flat spacetime decomposing the vectors and forms in a coordinate basis. In this section it will be shown that the simplest possible form equations for electromagnetism in four-dimensional spacetime leads by means of Poincare's lemma to Maxwell's equations.

Let us introduce an electromagnetic potential form

$$\underline{A} = A_\nu \underline{dx}^\nu = -\phi c \underline{dt} + c A_i \underline{dx}^i. \tag{5.259}$$

Hence

$$A^0 = -A_0 = \phi, \tag{5.260}$$

where ϕ is the electric scalar potential. The spatial component of $\underline{A}$ is the electromagnetic vector potential.

The exterior derivative of $\underline{A}$ is

$$\underline{F} = \underline{d}\,\underline{A}, \tag{5.261}$$

and is called the electromagnetic field form. In coordinate basis it is written in component form as

$$\underline{F} = \frac{1}{2} F_{\mu\nu} \underline{dx}^\mu \wedge \underline{dx}^\nu = \underline{d}\underline{A} = \frac{1}{2}\left(\frac{\partial A_\nu}{\partial x^\mu} - \frac{\partial A_\mu}{\partial x^\nu}\right)\underline{dx}^\mu \wedge \underline{dx}^\nu. \tag{5.262}$$

Hence, the components of the electromagnetic field form are

$$F_{\mu\nu} = \frac{\partial A_\nu}{\partial x^\mu} - \frac{\partial A_\mu}{\partial x^\nu}. \tag{5.263}$$

The electric field strength $\vec{E}$ and the magnetic flux density $\vec{B}$ are given by

$$\underline{F} = E_i \underline{dx}^i \wedge c\underline{dt} + \frac{1}{2} c B_{ij} \underline{dx}^i \wedge \underline{dx}^j, \quad B_{|ij|} = B_k, \quad k \neq i, j. \tag{5.264}$$

Here $|ij|$ means that $i < j$. The components of the electromagnetic field form is given in terms of the electric field strength and the magnetic flux density by

$$F_{\mu\nu} = \begin{pmatrix} 0 & -E_x & -E_y & -E_z \\ E_x & 0 & cB_z & -cB_y \\ E_y & -cB_z & 0 & cB_x \\ E_z & cB_y & -cB_x & 0 \end{pmatrix}. \tag{5.265}$$

The electromagnetic field form can now be written as

$$\underline{F} = F_{i\,t}\underline{dx}^i \wedge \underline{dt} + \frac{1}{2}c\left(\frac{\partial A_j}{\partial x^i} - \frac{\partial A_i}{\partial x^j}\right)\underline{dx}^i \wedge \underline{dx}^j$$

$$= c\left(\frac{\partial A_t}{\partial x^i} - \frac{\partial A_i}{\partial t}\right)\underline{dx}^i \wedge \underline{dt} + \frac{1}{2}c\left(\frac{\partial A_j}{\partial x^i} - \frac{\partial A_i}{\partial x^j}\right)\underline{dx}^i \wedge \underline{dx}^j$$

$$= -c\left((\nabla\phi)_i - \frac{\partial A_i}{\partial t}\right)\underline{dx}^i \wedge \underline{dt} + \frac{1}{2}c\left(\frac{\partial A_j}{\partial x^i} - \frac{\partial A_i}{\partial x^j}\right)\underline{dx}^i \wedge \underline{dx}^j. \qquad (5.266)$$

It follows from Eqs. (5.264) and (5.266) that the electric field strength and the magnetic flux density is given in terms of the electric scalar potential and the electromagnetic vector potential by

$$\vec{E} = -\vec{\nabla}\phi - \frac{\partial \vec{A}}{\partial t}, \quad \vec{B} = \vec{\nabla} \times \vec{A}. \qquad (5.267)$$

Consider a transformation of the form

$$\underline{A}' = \underline{A} + c\underline{d}\Lambda, \qquad (5.268)$$

where Λ is a scalar function. Then, using Poincare's lemma, we get

$$\underline{F}' = \underline{d}\,\underline{A}' = \underline{d}\,\underline{A} + c\underline{d}\,\underline{d}\Lambda = \underline{d}\,\underline{A} = \underline{F}. \qquad (5.269)$$

Hence the field form $\underline{F}$, and thereby the electromagnetic field strengths $\vec{E}$ and $\vec{B}$ are invariant against a transformation of the form (5.268). This transformation is called a *gauge transformation*. It has the component form

$$-\phi'c\underline{dt} + cA_i'\underline{dx}^i = -\phi c\underline{dt} + cA_i\underline{dx}^i + c\frac{\partial \Lambda}{\partial t}\underline{dt} + c\frac{\partial \Lambda}{\partial x^i}\underline{dx}^i. \qquad (5.270)$$

Hence the gauge transformation takes the form

$$\phi' = \phi - \frac{\partial \Lambda}{\partial t}, \quad \vec{A}' = \vec{A} + \vec{\nabla}\phi. \qquad (5.271)$$

As a consequence of Poincare's lemma the electromagnetic field is invariant against this transformation. This freedom of gauge has been utilized to introduce a condition upon the potentials as follows. It follows from Eq. (5.268) that

$$\delta\,\underline{A} = \delta\,\underline{A}' - c\delta\,\underline{d}f = \delta\,\underline{A}' - c\Box\,f, \qquad (5.272)$$

where $\Box$ is the d'Alembertian wave operator in Cartesian coordinates, i.e.

$$\Box = -\frac{1}{c^2}\frac{\partial^2}{\partial t^2} + \frac{\partial^2}{\partial x^2} + \frac{\partial^2}{\partial y^2} + \frac{\partial^2}{\partial z^2}, \qquad (5.273)$$

and f is a scalar gauge function. Introducing a gauge function which satisfies the Lorenz condition

$$c\,\Box\,f = \delta\,\underline{A}',$$
(5.274)

Equation (5.272) reduces to

$$\delta\,\underline{A} = 0.$$
(5.275)

This is called the Lorenz gauge (Lorenz is a Danish physicist who died in 1892). In this gauge the 4-divergence of $\underline{A}$ vanishes in vacuum. This corresponds to the equation

$$\vec{\nabla}\cdot\vec{A} + \frac{1}{c^2}\frac{\partial\phi}{\partial t} = 0.$$
(5.276)

It follows from Poincare's lemma that

$$\underline{d}\,\underline{F} = \underline{d}\,\underline{d}\,A = 0.$$
(5.277)

The component form of the field form is given in Eq. (5.264). Writing it out gives

$$\underline{F} = E_x\underline{dx}\wedge c\underline{dt} + E_y\underline{dy}\wedge c\underline{dt} + E_z\underline{dz}\wedge c\underline{dt} + cB_z\underline{dx}\wedge\underline{dy}$$
$$+ cB_y\underline{dz}\wedge\underline{dx} + cB_x\underline{dy}\wedge\underline{dz}.$$
(5.278)

The exterior derivative of $\underline{F}$ is the 3-form

$$\underline{d}\,\underline{F} = \frac{\partial F_{|ij|}}{\partial x^k}\underline{dx}^k\wedge\underline{dx}^i\wedge\underline{dx}^j$$
$$= c\left(\frac{\partial B_x}{\partial x} + \frac{\partial B_y}{\partial y} + \frac{\partial B_z}{\partial z}\right)\underline{dx}\wedge\underline{dy}\wedge\underline{dz}$$
$$= c\left(\frac{\partial B_y}{\partial t} + \frac{\partial E_x}{\partial z} - \frac{\partial E_z}{\partial x}\right)\underline{dt}\wedge\underline{dy}\wedge\underline{dz}$$
$$= c\left(\frac{\partial B_x}{\partial t} + \frac{\partial E_z}{\partial y} - \frac{\partial E_y}{\partial z}\right)\underline{dt}\wedge\underline{dz}\wedge\underline{dx}$$
$$= c\left(\frac{\partial B_z}{\partial t} + \frac{\partial E_y}{\partial x} - \frac{\partial E_x}{\partial y}\right)\underline{dt}\wedge\underline{dx}\wedge\underline{dy}$$
$$= c\left(\vec{\nabla}\cdot\vec{B}\right)\underline{dx}\wedge\underline{dy}\wedge\underline{dz} + c\left(\frac{\partial\vec{B}}{\partial t} + \vec{\nabla}\times\vec{E}\right)_k\underline{dt}\wedge\underline{dx}^i\wedge\underline{dy}^j, k\neq i, j.$$
(5.279)

This shows that the form equation

$$\underline{d}F = 0 \tag{5.280}$$

corresponds to Maxwell's source free equations

$$\vec{\nabla} \cdot \vec{B} = 0 \tag{5.281}$$

and

$$\frac{\partial \vec{B}}{\partial t} + \nabla \times \vec{E} = 0. \tag{5.282}$$

The electromagnetic current form $\underline{J}$ is the 1-form

$$\underline{J} = -\rho c \underline{d}t + (1/c) j_i \underline{d}x^i. \tag{5.283}$$

where ρ is the charge density and j_i the i-component of the current density. The source of the electromagnetic field form is the current form

$$\delta \underline{F} = -\frac{1}{\varepsilon_0} \underline{J}, \tag{5.284}$$

where $\underline{\delta}$ is the codifferential defined in Eq. (5.227), and ε_0 is the permittivity of empty space. Writing out this equation gives

$$-\underline{\delta} \underline{F} = \star \underline{d} \star \underline{F} = \partial^\nu F_{\nu\lambda} \underline{d}x^\lambda = \left(\partial^0 F_{00} + \partial^i F_{i0}\right) c \underline{d}t + \left(\partial^0 F_{0j} + \partial^i F_{ij}\right) \underline{d}x^j. \tag{5.285}$$

Using the component form (5.278) of $\underline{F}$ we obtain

$$-\underline{\delta} \underline{F} = \left(\vec{\nabla} \cdot \vec{E}\right) c \underline{d}t + \left(\frac{1}{c}\frac{\partial \vec{E}}{\partial t} - c\vec{\nabla} \times \vec{B}\right)_j \underline{d}x^j. \tag{5.286}$$

Equations (5.283), (5.284) and (5.285) give

$$\vec{\nabla} \cdot \vec{E} = \frac{\rho}{\varepsilon_0} \tag{5.287}$$

and

$$\vec{\nabla} \times \vec{B} - \frac{1}{c^2}\frac{\partial \vec{E}}{\partial t} = \mu_0 \vec{j}. \tag{5.288}$$

where $\mu_0 = 1/\varepsilon_0 c^2$ is the permeability of empty space. Equations (5.287) and (5.288) are the Maxwell source equations. Hence the form equation (5.284) contains the Maxwell source equations. The displacement current $\partial \vec{E}/\partial t$ is automatically included in the formulation of Maxwell's equation in the form language.

Let $\underline{\alpha}$ be a p-form. In Minkowski spacetime

$$\star\star\underline{\alpha} = -(-1)^{p(4-p)}\underline{\alpha}. \tag{5.289}$$

Since $\underline{d\star F}$ is a 3-form we have

$$\star\star\underline{d\star F} = -(-1)^3\underline{d\star F} = \underline{d\star F}. \tag{5.290}$$

Hence, taking the dual of Eq. (5.284) gives

$$\star\underline{J} = -\varepsilon_0\underline{d\star F}. \tag{5.291}$$

Taking the exterior derivative we get

$$\underline{d\star J} = -\varepsilon_0\underline{d d\star F} = 0 \tag{5.292}$$

due to Poincare's lemma. Taking the dual of this equation we get

$$\underline{\delta J} = 0. \tag{5.293}$$

The divergence of the current form vanishes. This is the mathematical expression of the conservation of charge in the form language. Using Eq. (5.283) the vector version of this equation is

$$\frac{\partial\rho}{\partial t} + \vec{\nabla}\cdot\vec{j} = 0, \tag{5.294}$$

which is the equation of continuity in the electromagnetism.

It follows from Eqs. (5.261) and (5.284) that

$$\underline{\delta d A} = -\frac{1}{\varepsilon_0}\underline{J}. \tag{5.295}$$

Applying the Laplace–Betrami operator defined in Eq. (5.231) to $\underline{A}$ gives

$$\underline{\Delta A} = \underline{\delta d A} + \underline{d\delta A}. \tag{5.296}$$

It follows that

$$\underline{\Delta A} = -\frac{1}{\varepsilon_0}\underline{J} + \underline{d\delta A}. \tag{5.297}$$

Asuming that $\underline{A}$ fulfills the Lorenz gauge condition (5.275) the last term in this equation vanishes, so that

$$\Delta\,\underline{A} = -\frac{1}{\varepsilon_0}\underline{J}. \tag{5.298}$$

This is the equation for electromagnetic waves. The time component of this equation is

$$\Box\phi = -\frac{\rho}{\varepsilon_0}, \tag{5.299}$$

and the space component is

$$\Box\vec{A} = -\mu_0\vec{j}. \tag{5.300}$$

The equation of motion and the energy equation of a charged particle in an electromagnetic field is combined in the following equation

$$\underline{M} = \frac{q}{c}\underline{F}(\vec{u}). \tag{5.301}$$

Here $\underline{F}(\vec{u})$ is the contraction as defined in Eq. (3.46) of the field form $\underline{F}$ with the 4-velocity $\vec{u}$ of the charge, and $\underline{M}$ is the Minkowski force form,

$$\underline{M} = -\gamma\frac{dE}{dt}\underline{dt} + \gamma\frac{dp_i}{dt}\underline{dx^i}, \tag{5.302}$$

where E and $\vec{p}$ are the energy and momentum of the particle, and $\vec{u} = \gamma(c,\vec{v})$. Writing out Eq. (5.301) gives

$$\begin{aligned}
-\gamma\frac{dE}{dt}\underline{dt} + \gamma\frac{dp_i}{dt}\underline{dx^i} &= \left(-E_xu^x - E_yu^y - E_zu^z\right)c\underline{dt} \\
&\quad + \left(E_xu^t\ \ cB_zu^y - cB_yu^z\right)\underline{dx} \\
&\quad + \left(E_xu^t\ \ -cB_zu^x\ \ cB_xu^z\right)\underline{dy} \\
&\quad + \left(E_zu^t\ \ cB_yu^x - cB_xu^y\right)\underline{dz} \\
&= -\gamma\left(\vec{E}\cdot\vec{v}\right)c\underline{dt} + \gamma c\left(\vec{E} + \vec{v}\times\vec{B}\right)_i\underline{dx^i}. \tag{5.303}
\end{aligned}$$

The time component of this equation is the equation representing energy conservation

$$\frac{dE}{dt} = q\vec{E}\cdot\vec{v}, \tag{5.304}$$

and the space component is the equation of motion of a charge in an electromagnetic field

$$\frac{d\vec{p}}{dt} = q\left(\vec{E} + \vec{v} \times \vec{B}\right). \tag{5.305}$$

The covariant form of the equation of motion is found from Eq. (5.301),

$$\frac{dp^{\mu}}{d\tau} = q F^{\mu}_{\nu} u^{\nu}. \tag{5.306}$$

For a particle with constant rest mass m_0, this gives

$$m_0 a^{\mu} = q F^{\mu}_{\nu} u^{\nu}, \tag{5.307}$$

where a^{μ} is the μ-component of the particle's 4-acceleration.

Exercises

5.1 Dual forms

Let $\{\vec{e}_i\}$ be a Cartesian basis in the three-dimensional Euclidean space. Using a vector $\vec{a} = a^i \vec{e}_i$ there are two ways of constructing a form:

(i) By constructing a 1-form from its covariant components $a_j = g_{ji} a^i$:

$$\underline{A} = a_i \underline{dx}^i.$$

(ii) By constructing a 2-form from its dual components, defined by $a_{ij} = \varepsilon_{ijk} a^k$:

$$\underline{a} = \frac{1}{2} a_{ij} \underline{dx}^i \wedge \underline{dx}^j.$$

We write this form as $\underline{a} = \star\underline{A}$ where $\star$ means to take the dual form.

(a) Given the vectors $\vec{a} = \vec{e}_x + 2\vec{e}_y - \vec{e}_z$ and $\vec{b} = 2\vec{e}_x - 3\vec{e}_y + \vec{e}_z$.
      Find the corresponding 1-forms $\underline{A}$ and $\underline{B}$, and the dual 2-forms $\underline{a} = \star\underline{A}$ and $\underline{b} = \star\underline{B}$, and also the dual form θ to the 1-form $\underline{\sigma} = \underline{dx} - 2\underline{dy}$.

(b) Take the exterior product $\underline{A} \wedge \underline{B}$ and show that

$$\theta_{ij} = \varepsilon_{ijk} C^k,$$

where $\underline{\theta} = \underline{A} \wedge \underline{B}$ and $\vec{C} = \vec{a} \times \vec{b}$.

(c)   Show that the exterior product $\underline{A} \wedge \star\underline{B}$ is given by the 3-form

$$\underline{A} \wedge \star\underline{B} = (\vec{a} \cdot \vec{b})\underline{dx} \wedge \underline{dy} \wedge \underline{dz}.$$

(d)   Show that the exterior derivative of a 1-form, $\underline{dA}$, corresponds to the curl $\nabla \times \vec{A}$ of the corresponding vector.

(e)   Finally show that

$$\underline{d} * \underline{df} = \nabla^2 f \underline{dx} \wedge \underline{dy} \wedge \underline{dz}$$

for a scalar field f.

5.2   Differential operators in spherical coordinates

We consider an Euclidean three-dimensional space with Cartesian coordinates (x, y, z) and spherical coordinates (r, θ, ϕ). The transformation from the spherical to the Cartesian coordinates is

$$x = r \sin\theta \cos\phi, \quad y = r \sin\theta \sin\phi, \quad z = r \cos\theta$$

(a)   Find the components of the metric tensor and the form of the line-element in spherical coordinates.

(b)   Let f be a scalar field. The gradient of f is given by

$$\nabla f = \frac{\partial f}{\partial r}\vec{e}_{\hat{r}} + \frac{\partial f}{\partial \theta}\vec{e}_{\hat{\theta}} + \frac{\partial f}{\partial \phi}\vec{e}_{\hat{\phi}},$$

where $\vec{e}_{\hat{i}}$ are the orthonormal basis vectors formed from the coordinate basis vectors in the spherical coordinate system.

Find the expressions for the gradient of f in coordinate basis.

(c)   In a coordinate system with orthogonal coordinate basis vectors, the curl of a vector field is given by

$$\nabla \times \vec{A} = \frac{1}{\sqrt{g_{22}g_{33}}}\left(\frac{\partial A^{\hat{3}}}{\partial x^2} - \frac{\partial A^{\hat{2}}}{\partial x^3}\right)\vec{e}_1 + \frac{1}{\sqrt{g_{11}g_{33}}}\left(\frac{\partial A^{\hat{1}}}{\partial x^3} - \frac{\partial A^{\hat{3}}}{\partial x^1}\right)\vec{e}_2$$

$$+ \frac{1}{\sqrt{g_{11}g_{22}}}\left(\frac{\partial A^{\hat{2}}}{\partial x^1} - \frac{\partial A^{\hat{1}}}{\partial x^2}\right)\vec{e}_3.$$

Find an expression for the curl in spherical coordinates. (The division by the factors $\sqrt{g_{ii}g_{jj}}$ is a normalization of the area of a surface element normal to the basis vector $\vec{e}_k$, $k \neq i, j$.)

(d)   The divergence of a vector field can in general (in an arbitrary basis) be defined by

$$\left(\nabla \cdot \vec{A}\right)\underline{\varepsilon} = \underline{d} * \underline{A},$$

where

$$\underline{\varepsilon} = \sqrt{|g|}\underline{\omega}^1 \wedge \underline{\omega}^2 \wedge \underline{\omega}^3,$$

Is the volume form, and $|g|$ is the determinant of the matrix formed by the components of the metric tensor. The volume form represents an invariant volume element.

Find the expression for the divergence of $\vec{A}$ in spherical coordinate.

Finally find the expression for the Laplacian of f in the spherical coordinate system.

5.3 Spatial geodesics in a rotating frame of reference

Our point of departure is the line-element (4.20) for 3-space in a rotating reference frame,

$$d\ell^2 = dr^2 + \frac{r^2 d\theta^2}{1 - r^2\omega^2/c^2} + dz^2.$$

We shall consider geodesics in the two-dimensional surface with $z = $ constant. The task is to calculate the shortest curve between two points with the same distance from the axis using the Lagrangian equations with the Lagrange function $L = (1/2)\dot{\ell}^2$, where the dot denotes differentiation with respect to an invariant parameter representing the arc length along the curve.

(a) Find the form of the 2-vector identity for the tangent vectors of the curve.
(b) Find an expression for the momentum p_θ conjugate to the cyclic coordinate θ of L.
(c) Find the differential equation for the geodesic curves.
(d) Use the boundary condition that the point on the curve closest to the axis has a distance r_0 from the axis, to show that

$$p_\theta = \frac{r_0}{\sqrt{1 - r_0^2\omega^2/c^2}}.$$

(e) Show that the differential equation of the curve can be written as

$$\frac{dr}{r\sqrt{r^2 - r_0^2}} - \frac{\omega^2}{c^2}\frac{r\,dr}{\sqrt{r^2 - r_0^2}} = \frac{d\theta}{r_0}.$$

Integrate this equation and find the equation of the curve. Finally draw the curve.

5.4 *Christoffel symbols in a uniformly accelerated reference frame*

(a) Use the coordinate transformation (4.80)–(4.82) and the formula (5.27) to calculate the non-vanishing Christoffel symbols in the coordinate system of Chap. 4 co-moving with a uniformly accelerated reference frame.

(b) Use Eq. (5.65) to calculate the same Christoffel symbols as in (a).

5.5 *Relativistic vertical projectile motion*

A particle is thrown vertically upwards with velocity v from the origin of the coordinate system in the gravitational field of a uniformly accelerated reference frame.

Calculate the maximal height of the particle.

5.6 *The geodesic equation and constants of motion*

(a) Show that the geodesic equation can be written in the following form: $\frac{du_\alpha}{ds} - \frac{1}{2}\frac{\partial g_{\beta\gamma}}{\partial x^\alpha}u^\beta u^\gamma = 0$.

(b) Assume that the metric is static and the space is cylindrically symmetric with cylindrical coordinates (r, θ, z). What constants of motion are there then for a free particle?

Chapter 6
Curvature

Abstract The Riemann curvature tensor is introduced, and the expression of its components in terms of the derivatives of the metric and the structure coefficients is deduced. Tidal forces are discussed in a relativistic context, and it is pointed out that the relativistic gravitational field has both a non-tidal component due to the motion of the reference frame and a tidal component due to spacetime curvature.

6.1 The Riemann Curvature Tensor

The covariant directional derivative of a vector field $\vec{A}$ along a vector $\vec{v}$ was defined and interpreted geometrically in Sect. 5.2 as follows:

$$\nabla_{\vec{v}}\vec{A} = \frac{d\vec{A}}{d\lambda} = A^{\mu}_{;\nu}v^{\nu}\vec{e}_{\mu} - \lim_{\Delta\lambda \to 0}\frac{\vec{A}_{QP}(\lambda + \Delta\lambda) - \vec{A}(\lambda)}{\Delta\lambda}. \tag{6.1}$$

Let $\vec{A}_{QP}$ be the vector $\vec{A}$ parallel transported from Q to P in Fig. 6.1.

In Fig. 6.2 we have illustrated that a vector parallel transported around a closed curve on a curved surface changes direction during the round trip.

Then to first order in $\Delta\lambda$ we have $\vec{A}_{QP} = \vec{A}_P + (\nabla_{\vec{v}}\vec{A})_P\Delta\lambda$ and

$$\vec{A}_{PQ} = \vec{A}_Q - (\nabla_{\vec{v}}\vec{A})_Q\Delta\lambda. \tag{6.2}$$

To second order in $\Delta\lambda$ we have

$$\vec{A}_{PQ} = \left(1 - \nabla_{\vec{v}}\Delta\lambda + \frac{1}{2}\nabla_{\vec{v}}\nabla_{\vec{v}}(\Delta\lambda)^2\right)\vec{A}_Q. \tag{6.3}$$

We shall now parallel transport a vector around the closed polygon shown in Fig. 6.3.

If $\vec{A}_{PQ}$ is parallel transported further onto R we get

$$\vec{A}_{PQR} = \left(1 - \nabla_{\vec{u}}\Delta\lambda + \frac{1}{2}\nabla_{\vec{u}}\nabla_{\vec{u}}(\Delta\lambda)^2\right)\left(1 - \nabla_{\vec{v}}\Delta\lambda + \frac{1}{2}\nabla_{\vec{v}}\nabla_{\vec{v}}(\Delta\lambda)^2\right)\vec{A}_R, \tag{6.4}$$

© Springer Nature Switzerland AG 2020

Ø. Grøn, *Introduction to Einstein's Theory of Relativity*,
Undergraduate Texts in Physics, https://doi.org/10.1007/978-3-030-43862-3_6

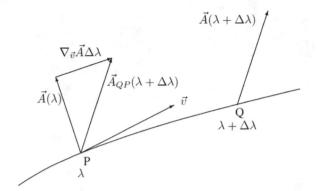

Fig. 6.1 Parallel transport of $\vec{A}$

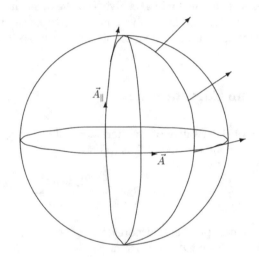

Fig. 6.2 Parallel transport of a vector around a triangle

where $\vec{A}_Q$ is replaced by $\vec{A}_R$ because the differential operator always shall be applied to the vector in the first position. If we parallel transport $\vec{A}$ around the whole polygon in Fig. 6.3 we get

$$\vec{A}_{PQRSTP} = \left(1 + \nabla_{\vec{u}}\Delta\lambda + \frac{1}{2}\nabla_{\vec{u}}\nabla_{\vec{u}}(\Delta\lambda)^2\right)\left(1 + \nabla_{\vec{v}}\Delta\lambda + \frac{1}{2}\nabla_{\vec{v}}\nabla_{\vec{v}}(\Delta\lambda)^2\right)$$

$$\cdot \left(1 - \nabla_{[\vec{u},\vec{v}]}(\Delta\lambda)^2\right)\left(1 - \nabla_{\vec{u}}\Delta\lambda + \frac{1}{2}\nabla_{\vec{u}}\nabla_{\vec{u}}(\Delta\lambda)^2\right)$$

$$\left(1 - \nabla_{\vec{v}}\Delta\lambda + \frac{1}{2}\nabla_{\vec{v}}\nabla_{\vec{v}}(\Delta\lambda)^2\right)\vec{A}_P. \tag{6.5}$$

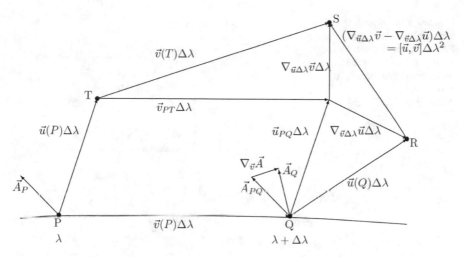

Fig. 6.3 Curvature and parallel transport

Calculating to second order in $\Delta\lambda$ gives

$$\vec{A}_{PQRSTP} = \vec{A}_P + \left([\nabla_{\vec{u}}, \nabla_{\vec{v}}] - \nabla_{[\vec{u},\vec{v}]}\right)(\Delta\lambda)^2\vec{A}_P. \tag{6.6}$$

There is a variation of the vector under parallel transport around the closed polygon,

$$\delta\vec{A} - \vec{A}_{PQRSTP} - \vec{A}_P = \left([\nabla_{\vec{u}}, \nabla_{\vec{v}}] - \nabla_{[\vec{u},\vec{v}]}\right)\vec{A}_P(\Delta\lambda)^2. \tag{6.7}$$

Definition 6.1.1 (*The Riemann curvature tensor*) The *Riemann's curvature tensor is defined* as

$$R(\ \ ,\vec{A}, \vec{u}, \vec{v}) = \left([\nabla_{\vec{u}}, \nabla_{\vec{v}}] - \nabla_{[\vec{u},\vec{v}]}\right)(\vec{A}). \tag{6.8}$$

The components of the Riemann curvature tensor are defined by applying the tensor on basis vectors

$$R^{\mu}_{\nu\alpha\beta}\vec{e}_{\mu} \equiv \left([\nabla_{\alpha}, \nabla_{\beta}] - \nabla_{[\vec{e}_{\alpha},\vec{e}_{\beta}]}\right)(\vec{e}_{\nu}). \tag{6.9}$$

It follows from Definition (6.8) that the Riemann tensor is antisymmetric in its last two arguments. Hence the components are antisymmetric in their last two indices,

$$R^{\mu}_{\nu\beta\alpha} = -R^{\mu}_{\nu\alpha\beta}. \tag{6.10}$$

The expression for the change of $\vec{A}$ under parallel transport around the polygon, Eq. (6.7), can now be written as

Fig. 6.4 Curl as area of a
vectorial parallelogram.
Parallelogram defined by the
vectors $\vec{u}\Delta\lambda$ and $\vec{v}\,\Delta\lambda$ is
represented mathematically
by the vector
$\Delta\vec{S} = \vec{u} \times \vec{v}(\Delta\lambda)^2$

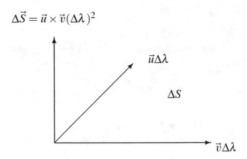

$\Delta\vec{S} = \vec{u} \times \vec{v}(\Delta\lambda)^2$

$$
\begin{aligned}
\Delta\vec{A} &= R(\ ,\vec{A},\vec{u},\vec{v})(\Delta\lambda)^2 \\
&= R(\ ,A^\nu\vec{e}_\nu, u^\alpha\vec{e}_\alpha, v^\beta\vec{e}_\beta)(\Delta\lambda)^2 \\
&= \vec{e}_\mu R^\mu{}_{\nu\alpha\beta}A^\nu u^\alpha v^\beta \cdot (\Delta\lambda)^2 \\
&= \frac{1}{2}\vec{e}_\mu R^\mu{}_{\nu\alpha\beta}A^\nu (u^\alpha v^\beta - u^\beta v^\alpha)(\Delta\lambda)^2 .
\end{aligned}
\tag{6.11}
$$

The area of the parallelogram defined by the vectors $\vec{u}\Delta\lambda$ and $\vec{v}\,\Delta\lambda$ is

$$
\Delta\vec{S} = \vec{u} \times \vec{v}\,(\Delta\lambda)^2 .
\tag{6.12}
$$

These vectors are shown in Fig. 6.4.
Using that

$$
(\vec{u} \times \vec{v})^{\alpha\beta} = u^\alpha v^\beta - u^\beta v^\alpha ,
\tag{6.13}
$$

we can write Eq. (6.11) as

$$
\Delta\vec{A} = \frac{1}{2}A^\nu R^\mu{}_{\nu\alpha\beta}\Delta S^{\alpha\beta}\vec{e}_\mu .
\tag{6.14}
$$

The components of the Riemann tensor expressed by the connection coefficients
(5.114) and structure coefficients (3.40) are given by

$$
\begin{aligned}
\vec{e}_\mu R^\mu{}_{\nu\alpha\beta} &= [\nabla_\alpha, \nabla_\beta]\vec{e}_\nu - \nabla_{[\vec{e}_\alpha,\vec{e}_\beta]}\vec{e}_\nu \\
&= (\nabla_\alpha\nabla_\beta - \nabla_\beta\nabla_\alpha - c^\rho_{\alpha\beta}\nabla_\rho)\vec{e}_\nu \\
&= \nabla_\alpha\nabla_\beta\vec{e}_\nu - \nabla_\beta\nabla_\alpha\vec{e}_\nu - c^\rho_{\alpha\beta}\nabla_\rho\vec{e}_\nu \\
&= \nabla_\alpha\Gamma^\mu_{\nu\beta}\vec{e}_\mu - \nabla_\beta\Gamma^\mu_{\nu\alpha}\vec{e}_\mu - c^\rho_{\alpha\beta}\Gamma^\mu_{\nu\rho}\vec{e}_\mu \\
&= (\nabla_\alpha\Gamma^\mu_{\nu\beta})\vec{e}_\mu + \Gamma^\mu_{\nu\beta}\nabla_\alpha\vec{e}_\mu - (\nabla_\beta\Gamma^\mu_{\nu\alpha})\vec{e}_\mu - \Gamma^\mu_{\nu\alpha}\nabla_\beta\vec{e}_\mu - c^\rho_{\alpha\beta}\Gamma^\mu_{\nu\rho}\vec{e}_\mu \\
&= \vec{e}_\alpha(\Gamma^\mu_{\nu\beta})\vec{e}_\mu + \Gamma^\rho_{\nu\beta}\Gamma^\mu_{\rho\alpha}\vec{e}_\mu - \vec{e}_\beta(\Gamma^\mu_{\nu\alpha})\vec{e}_\mu - \Gamma^\rho_{\nu\alpha}\Gamma^\mu_{\rho\beta}\vec{e}_\mu - c^\rho_{\alpha\beta}\Gamma^\mu_{\nu\rho}\vec{e}_\mu .
\end{aligned}
\tag{6.15}
$$

This gives the components of the Riemann curvature tensor in an arbitrary basis

$$R^{\mu}_{\nu\alpha\beta} = \vec{e}_{\alpha}(\Gamma^{\mu}_{\nu\beta}) - \vec{e}_{\beta}(\Gamma^{\mu}_{\nu\alpha}) + \Gamma^{\rho}_{\nu\beta}\Gamma^{\mu}_{\rho\alpha} - \Gamma^{\rho}_{\nu\alpha}\Gamma^{\mu}_{\rho\beta} - c^{\rho}_{\alpha\beta}\Gamma^{\mu}_{\nu\rho}. \tag{6.16}$$

In coordinate basis Eq. (6.16) is reduced to

$$R^{\mu}_{\nu\alpha\beta} = \Gamma^{\mu}_{\nu\beta,\alpha} - \Gamma^{\mu}_{\nu\alpha,\beta} + \Gamma^{\rho}_{\nu\beta}\Gamma^{\mu}_{\rho\alpha} - \Gamma^{\rho}_{\nu\alpha}\Gamma^{\mu}_{\rho\beta}, \tag{6.17}$$

where $\Gamma^{\mu}_{\nu\beta} = \Gamma^{\mu}_{\beta\nu}$ are the Christoffel symbols.

Since the basis vectors are derivative operators, the first two terms is a linear combination of derivatives of the connection coefficients. In a local Cartesian coordinate system co-moving with a local inertial reference frame all the connection coefficients vanish, and only the first two terms in the expression of the components of the Riemann curvature tensor remain. As we have seen in Sect. 5.5.3 this means that in such a system there is no acceleration of gravity. But in general the derivatives of the connections coefficients will not vanish. Hence in general spacetime is curved. This shows that the acceleration of gravity does not depend upon the curvature of spacetime. It depends instead upon the motion of the reference frame. The curvature of spacetime is given by a tensor and is an invariant property of spacetime at the considered position. The acceleration of gravity is, however, not an invariant property of spacetime since it is given by certain connection coefficients that are not tensor components. They can be transformed away. This is the mathematical expression of the fact that you can transform away the acceleration of gravity locally by going into a local inertial frame.

Definition 6.5.1 (*Contraction of a tensor component*) Contraction of a tensor component is an operation defined by

$$R_{\nu\beta} \equiv R^{\mu}_{\nu\mu\beta}. \tag{6.18}$$

The summation is over μ. In this way a new tensor is constructed from another tensor with a rank 2 lower than the original tensor.

The tensor with components $R_{\nu\beta}$ is called *the Ricci curvature tensor*. Another contraction gives *the Ricci curvature scalar* $R = R^{\mu}_{\mu}$.

Due to the antisymmetry (6.10) we can define a matrix of *curvature forms*,

$$\underline{R}^{\mu}_{\nu} = \frac{1}{2}R^{\mu}_{\nu\alpha\beta}\underline{\omega}^{\alpha} \wedge \underline{\omega}^{\beta}. \tag{6.19}$$

Inserting the components of the Riemann tensor from Eq. (6.14) gives

$$\underline{R}^{\mu}_{\nu} = \left(\vec{e}_{\alpha}(\Gamma^{\mu}_{\nu\beta}) + \Gamma^{\rho}_{\nu\beta}\Gamma^{\mu}_{\rho\alpha} - \frac{1}{2}c^{\rho}_{\alpha\beta}\Gamma^{\mu}_{\nu\rho}\right)\underline{\omega}^{\alpha} \wedge \underline{\omega}^{\beta}. \tag{6.20}$$

We shall now introduce the curvature form. Then we need the connection forms,

$$\underline{\Omega}^{\mu}_{\nu} = \Gamma^{\mu}_{\nu\alpha}\underline{\omega}^{\alpha}. \tag{6.21}$$

The exterior derivatives of the basis forms are,

$$d\underline{\omega}^\rho = -\frac{1}{2}c^\rho_{\alpha\beta}\underline{\omega}^\alpha \wedge \underline{\omega}^\beta = -\underline{\Omega}^\rho_\alpha \wedge \underline{\omega}^\alpha \tag{6.22}$$

The exterior derivatives of the connection forms can now be written

$$d\underline{\Omega}^\mu_\nu = d\Gamma^\mu_{\nu\beta} \wedge \underline{\omega}^\beta + \Gamma^\mu_{\nu\rho}d\underline{\omega}^\rho$$

$$= \vec{e}_\alpha(\Gamma^\mu_{\nu\beta})\underline{\omega}^\alpha \wedge \underline{\omega}^\beta - \frac{1}{2}c^\rho_{\alpha\beta}\Gamma^\mu_{\nu\rho}\underline{\omega}^\alpha \wedge \underline{\omega}^\beta. \tag{6.23}$$

Combining Eqs. (6.20)–(6.22) the curvature forms take the form

$$R^\mu_\nu = d\underline{\Omega}^\mu_\nu + \underline{\Omega}^\mu_\lambda \wedge \underline{\Omega}^\lambda_\nu. \tag{6.24}$$

This is Cartans 2. structure equation.

Example 6.1.1 *The Riemann curvature tensor of a spherical surface calculated from Cartan's structure equations*

Let $r = R$ be the radius of the spherical surface. The calculation is performed in 5 steps.

1. Write down the metric tensor and introduce a form basis dual to an orthonormal vector basis.

$$\mathbf{g} = \underline{\omega}^{\hat\theta} \otimes \underline{\omega}^{\hat\theta} + \underline{\omega}^{\hat\varphi} \otimes \underline{\omega}^{\hat\varphi} = R^2 d\underline{\theta} \otimes d\underline{\theta} + R^2 \sin^2\theta d\underline{\varphi} \otimes d\underline{\varphi},$$

giving

$$\underline{\omega}^{\hat\theta} = Rd\underline{\theta}, \quad \underline{\omega}^{\hat\varphi} = R\sin\theta d\underline{\varphi}.$$

2. Use Cartan's 1. structure equation, $d\underline{\omega}^\mu = \underline{\omega}^\nu \wedge \underline{\Omega}^\mu_\nu$, to calculate the structure forms. Since R is constant and using Poincare's lemma and the antisymmetry of the connection forms, exterior differentiation of the basis forms lead to

$$d\underline{\omega}^{\hat\theta} = 0 = \underline{\omega}^{\hat\varphi} \wedge \underline{\Omega}^{\hat\theta}_{\hat\varphi}, \quad d\underline{\omega}^{\hat\varphi} = R\cos\theta d\underline{\theta} \wedge d\underline{\varphi} = \underline{\omega}^{\hat\theta} \wedge \frac{1}{R}\frac{\cos\theta}{\sin\theta}\underline{\omega}^{\hat\varphi} = \underline{\omega}^{\hat\theta} \wedge \underline{\Omega}^{\hat\varphi}_{\hat\theta}$$

giving

$$\underline{\Omega}^{\hat\theta}_{\hat\varphi} = f(\theta, \varphi)\underline{\omega}^{\hat\varphi}, \quad \underline{\Omega}^{\hat\varphi}_{\hat\theta} = g(\theta, \varphi)\underline{\omega}^{\hat\theta} + \frac{1}{R}\frac{\cos\theta}{\sin\theta}\underline{\omega}^{\hat\varphi}$$

where the functions $f(\theta, \varphi)$ and $g(\theta, \varphi)$ are determined from the antisymmetry of the connection forms. Using that $\underline{\Omega}^{\hat\theta}_{\hat\varphi} = \underline{\Omega}_{\hat\theta\hat\varphi} = -\underline{\Omega}_{\hat\varphi\hat\theta} = -\underline{\Omega}^{\hat\varphi}_{\hat\theta}$, we get

$$f(\theta, \varphi) = \frac{1}{R} \frac{\cos \theta}{\sin \theta}, \quad g(\theta, \varphi) = 0.$$

Hence,

$$\underline{\Omega}_{\hat{\theta}}^{\hat{\varphi}} = -\underline{\Omega}_{\hat{\varphi}}^{\hat{\theta}} = \frac{1}{R} \frac{\cos \theta}{\sin \theta} \underline{\omega}^{\hat{\varphi}} = \cos \theta \underline{d\varphi}.$$

The reason for going back to coordinate basis here is that then it is easier to calculate the exterior derivative $\underline{d\Omega}_{\hat{\theta}}^{\hat{\varphi}}$.

3. Calculate the Riemann curvature forms from Cartan's 2. structure equation,

$$\underline{R}_{\hat{\varphi}}^{\hat{\theta}} = \underline{d\Omega}_{\hat{\varphi}}^{\hat{\theta}} + \underline{\Omega}_{\hat{\varphi}}^{\hat{\theta}} \wedge \underline{\Omega}_{\hat{\varphi}}^{\hat{\varphi}} = \underline{d\Omega}_{\hat{\varphi}}^{\hat{\theta}} = \underline{d}(\cos \theta \underline{d\varphi}) = -\sin \theta \underline{d\theta} \wedge \underline{d\varphi} = -\frac{1}{R^2} \underline{\omega}^{\hat{\theta}} \wedge \underline{\omega}^{\hat{\varphi}} = -\underline{R}_{\hat{\theta}}^{\hat{\varphi}}.$$

4. Calculate the non-vanishing components of the Riemann tensor from

$$\underline{R}_{\hat{\nu}}^{\hat{\mu}} = (1/2) R^{\mu}_{\nu \alpha \beta} \underline{\omega}^{\alpha} \wedge \underline{\omega}^{\beta}.$$

This gives

$$R^{\hat{\theta}}_{\hat{\varphi}\hat{\theta}\hat{\varphi}} = R^{\hat{\varphi}}_{\hat{\theta}\hat{\varphi}\hat{\theta}} = -R^{\hat{\theta}}_{\hat{\varphi}\hat{\varphi}\hat{\theta}} = -R^{\hat{\varphi}}_{\hat{\theta}\hat{\theta}\hat{\varphi}} = \frac{1}{R^2}.$$

5. Calculate the components of the Ricci curvature tensor and the Ricci curvature scalar,

$$R_{\hat{\theta}\hat{\theta}} = R_{\hat{\varphi}\hat{\varphi}} = \frac{1}{R^2}, \quad R = R^{\hat{\theta}}_{\hat{\theta}} + R^{\hat{\varphi}}_{\hat{\phi}} = \frac{2}{R^2}.$$

6.2 Differential Geometry of Surfaces

Imagine an arbitrary surface embedded in an Euclidean 3-dimensional space (Fig. 6.5).

The coordinate basis vectors on the surface are

$$\vec{e}_u = \frac{\partial}{\partial u}, \quad \vec{e}_v = \frac{\partial}{\partial v}, \tag{6.25}$$

where u and v are coordinates on the surface. The directional derivatives of the basis vectors are written as

$$\vec{e}_{\mu,\nu} = \Gamma^{\alpha}_{\mu\nu} \vec{e}_{\alpha} + K_{\mu\nu} \vec{N}, \quad \alpha = 1, 2. \tag{6.26}$$

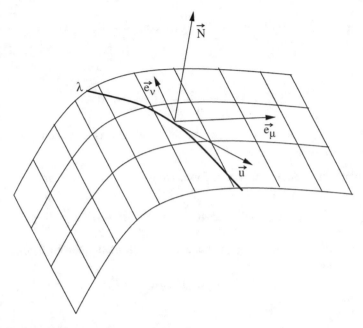

Fig. 6.5 Surface with tangent vectors and normal vector

Greek indices run through the surface coordinates, and $\vec{N}$ is a unit vector orthogonal to the surface. It follows from Eq. (6.26) that

$$K_{\mu\nu} = \vec{e}_{\mu,\nu} \cdot \vec{N}. \tag{6.27}$$

Equation (6.26) is called Gauss' equation. In coordinate basis we have $\vec{e}_{\mu,\nu} = \frac{\partial^2}{\partial x^\mu \partial x^\nu} = \frac{\partial^2}{\partial x^\nu \partial x^\mu} = \vec{e}_{\nu,\mu}$. Hence

$$K_{\mu\nu} = K_{\nu\mu}. \tag{6.28}$$

Let $\vec{u}$ be the unit tangent vector to a curve on the surface, parameterized by λ. Differentiating $\vec{u}$ along the curve, we get

$$\frac{d\vec{u}}{d\lambda} = u^\mu_{;\nu} u^\nu \vec{e}_\mu + \underbrace{K_{\mu\nu} u^\mu u^\nu}_{\text{2nd fundamental form}} \vec{N}. \tag{6.29}$$

We define two scalar quantities κ_g and κ_N by

$$\frac{d\vec{u}}{d\lambda} = \kappa_g \vec{e} + \kappa_N \vec{N}. \tag{6.30}$$

Here $\vec{e} = \vec{N} \times \vec{u}$ is a unit vector in the surface which is orthogonal both to $\vec{N}$ and to $\vec{u}$. κ_g is called the *geodesic curvature*, and κ_N the *normal curvature* (external curvature). Note that $\kappa_g = 0$ for geodesic curves on the surface. The geodesic curvature is given by the covariant directional derivative of the tangent vector field of a curve along the curve,

$$\kappa_g \vec{e} = u^\mu_{;\nu} u^\nu \vec{e}_\mu = \nabla_{\vec{u}} \vec{u}. \tag{6.31}$$

It follows from Eqs. (6.29) that

$$\frac{d\vec{u}}{d\lambda} \cdot \vec{N} = K_{\mu\nu} u^\mu u^\nu, \tag{6.32}$$

and from Eq. (6.30) that

$$\kappa_N = \frac{d\vec{u}}{d\lambda} \cdot \vec{N}. \tag{6.33}$$

We also have that $\vec{u} \cdot \vec{N} = 0$ along the whole curve. Differentiation gives

$$\frac{d\vec{u}}{d\lambda} \cdot \vec{N} + \vec{u} \cdot \frac{d\vec{N}}{d\lambda} = 0. \tag{6.34}$$

It follows from the last three equations that

$$\kappa_N = K_{\mu\nu} u^\mu u^\nu = -\vec{u} \cdot \frac{d\vec{N}}{d\lambda}, \tag{6.35}$$

which is called *Weingarten's equation*.

The geodesic curvature κ_g and normal curvature κ_N together give a complete description of the geometry of a surface in a flat 3-dimensional space. We are now going to consider geodesic curves through a point on the surface. The point of departure is the tangent vector $\vec{u} = u^\mu \vec{e}_\mu$ with $\vec{u} \cdot \vec{u} = g_{\mu\nu} u^\mu u^\nu = 1$. The directions with maximum and minimum values for the normal curvatures are found by extremalizing κ_N under the condition $g_{\mu\nu} u^\mu u^\nu = 1$. We then solve the variation problem $\delta F = 0$ for arbitrary u^μ where $F = K_{\mu\nu} u^\mu u^\nu - k(g_{\mu\nu} u^\mu u^\nu - 1)$. Here k is the Lagrange multiplicator. Variation with respect to u^μ gives

$$\delta F = 2(K_{\mu\nu} - k g_{\mu\nu}) u^\nu \delta u^\mu. \tag{6.36}$$

The requirement that $\delta F = 0$ for arbitrary u^μ demands that

$$(K_{\mu\nu} - k g_{\mu\nu}) u^\nu = 0. \tag{6.37}$$

For this system of equations to have nonzero solutions, we must have

$$\det(K_{\mu\nu} - kg_{\mu\nu}) = 0, \tag{6.38}$$

or

$$\begin{vmatrix} K_{11} - kg_{11} & K_{12} - kg_{12} \\ K_{21} - kg_{21} & K_{22} - kg_{22} \end{vmatrix} = 0. \tag{6.39}$$

This gives the following quadratic equation for k:

$$k^2 \det(g_{\mu\nu}) - (g_{11}K_{22} - 2g_{12}K_{12} + g_{22}K_{11})k + \det(K_{\mu\nu}) = 0$$
$$(K \text{ symmetric } K_{12} = K_{21}). \tag{6.40}$$

The equation has two solutions, k_1 and k_2. These are the extremal values of k. In order to find the meaning of k, we multiply Eq. (6.37) by u^μ, which gives

$$\begin{aligned} 0 &= (K_{\mu\nu} - kg_{\mu\nu})u^\mu u^\nu \\ &= K_{\mu\nu}u^\mu u^\nu - kg_{\mu\nu}u^\mu u^\nu \\ &= \kappa_N - k \Rightarrow k = \kappa_N. \end{aligned} \tag{6.41}$$

The extremal values of κ_N are called the *principal curvatures* of the surface. Let the directions of the geodesics with extreme normal curvature be given by the tangent vectors $\vec{u}$ and $\vec{v}$. Equation (6.37) gives

$$K_{\mu\nu}u^\nu = kg_{\mu\nu}u^\nu. \tag{6.42}$$

Multiplying by v^μ we get

$$\begin{aligned} K_{\mu\nu}u^\nu v^\mu &= k_1 g_{\mu\nu}u^\nu v^\mu = k_1 u_\mu v^\mu = k_1(\vec{u} \cdot \vec{v}), \\ K_{\mu\nu}v^\nu u^\mu &= k_2 g_{\mu\nu}v^\nu u^\mu = k_2(\vec{u} \cdot \vec{v}), \end{aligned} \tag{6.43}$$

which gives

$$(k_1 - k_2)(\vec{u} \cdot \vec{v}) = K_{\mu\nu}(u^\nu v^\mu - v^\nu u^\mu) = 2K_{\mu\nu}u^{[\nu}v^{\mu]}. \tag{6.44}$$

$K_{\mu\nu}$ is symmetric in μ and ν. So we get $(k_1 - k_2)(\vec{u} \cdot \vec{v}) = 0$. For $k_1 \neq k_2$ we have to demand $\vec{u} \cdot \vec{v} = 0$. So the geodesics with extremal normal curvature are orthogonal to each other.

The *Gaussian curvature* (at a point) is defined as

$$K = \kappa_{N1} \cdot \kappa_{N2} \tag{6.45}$$

Since κ_{N1} and κ_{N2} are solutions of the quadratic equation above, we get

$$K = \frac{\det(K_{\mu\nu})}{\det(g_{\mu\nu})}. \tag{6.46}$$

6.2.1 Surface Curvature Using the Cartan Formalism

When we use the Cartan formalism, we introduce an orthonormal set of basis vectors at each point of the surface. Greek indices run through the surface coordinates (2-dimensional) and Latin indices through the space coordinates (3-dimensional):

$$\vec{e}_{\hat{a}} = (\vec{e}_{\hat{1}}, \vec{e}_{\hat{2}}, \vec{N}), \quad \vec{e}_{\hat{\mu}} = \{\vec{e}_{\hat{1}}, \vec{e}_{\hat{2}}\}. \tag{6.47}$$

where $\vec{N}$ is a unit vector field orthogonal to the surface. Using the exterior derivative and form formalism, we find how the unit vectors on the surface change:

$$\underline{d}\vec{e}_{\hat{\nu}} = \vec{e}_{\hat{a}} \otimes \underline{\Omega}_{\hat{\nu}}^{\hat{a}} = \vec{e}_{\hat{\alpha}} \otimes \underline{\Omega}_{\hat{\nu}}^{\hat{\alpha}} + \vec{N} \otimes \underline{\Omega}_{\hat{\nu}}^{\hat{3}}, \tag{6.48}$$

where $\underline{\Omega}_{\hat{\nu}}^{\hat{\mu}} = \Gamma_{\hat{\nu}\hat{\alpha}}^{\mu}\underline{\omega}^{\hat{\alpha}}$ are the connection forms on the surface, i.e. the intrinsic connection forms. The extrinsic connection forms are

$$\underline{\Omega}_{\hat{\nu}}^{\hat{3}} = K_{\hat{\nu}\hat{\alpha}}\underline{\omega}^{\hat{\alpha}}, \quad \underline{\Omega}_{\hat{3}}^{\hat{\mu}} = K_{\hat{\alpha}}^{\hat{\mu}}\underline{\omega}^{\hat{\alpha}}. \tag{6.49}$$

We let the surface be embedded in an Euclidean (flat) 3-dimensional space. This means that the curvature forms of the 3-dimensional space are zero:

$$\underline{R}_{\hat{3}\hat{b}}^{\hat{a}} = 0 = \underline{d}\underline{\Omega}_{\hat{b}}^{\hat{a}} + \underline{\Omega}_{\hat{k}}^{\hat{a}} \wedge \underline{\Omega}_{\hat{b}}^{\hat{k}}, \tag{6.50}$$

which gives

$$\underline{R}_{\hat{3}\hat{\nu}}^{\hat{\mu}} = 0 = \underline{d}\underline{\Omega}_{\hat{\nu}}^{\hat{\mu}} + \underline{\Omega}_{\hat{\alpha}}^{\hat{\mu}} \wedge \underline{\Omega}_{\hat{\nu}}^{\hat{\alpha}} + \underline{\Omega}_{\hat{3}}^{\hat{\mu}} \wedge \underline{\Omega}_{\hat{\nu}}^{\hat{3}} = \underline{R}_{\hat{\nu}}^{\hat{\mu}} + \underline{\Omega}_{\hat{3}}^{\hat{\mu}} \wedge \underline{\Omega}_{\hat{\nu}}^{\hat{3}}, \tag{6.51}$$

where $\underline{R}_{\hat{\nu}}^{\hat{\mu}}$ are the *curvature forms of the surface*. We then have

$$\frac{1}{2}R_{\hat{\nu}\hat{\alpha}\hat{\beta}}^{\hat{\mu}}\underline{\omega}^{\hat{\alpha}} \wedge \underline{\omega}^{\hat{\beta}} = -\underline{\Omega}_{\hat{3}}^{\hat{\mu}} \wedge \underline{\Omega}_{\hat{\nu}}^{\hat{3}}. \tag{6.52}$$

Inserting the components of the extrinsic connection forms, we get (using the antisymmetry of α and β in $\underline{R}^\mu_{\nu\alpha\beta}$)

$$R^{\hat{\mu}}_{\hat{\nu}\hat{\alpha}\hat{\beta}} = K^{\hat{\mu}}_{\hat{\alpha}} K_{\hat{\nu}\hat{\beta}} - K^{\hat{\mu}}_{\hat{\beta}} K_{\hat{\nu}\hat{\alpha}}. \tag{6.53}$$

We now lower the first index:

$$R_{\hat{\mu}\hat{\nu}\hat{\alpha}\hat{\beta}} = K_{\hat{\mu}\hat{\alpha}} K_{\hat{\nu}\hat{\beta}} - K_{\hat{\mu}\hat{\beta}} K_{\hat{\nu}\hat{\alpha}}. \tag{6.54}$$

$R_{\hat{\mu}\hat{\nu}\hat{\alpha}\hat{\beta}}$ are the components of a curvature tensor which *only* refer to the dimensions of the surface. In particular

$$R_{\hat{1}\hat{2}\hat{1}\hat{2}} = K_{\hat{1}\hat{1}} K_{\hat{2}\hat{2}} - K_{\hat{1}\hat{2}} K_{\hat{2}\hat{1}} = \det K. \tag{6.55}$$

We then have the following connection between this component of the Riemann curvature tensor of the surface and the Gaussian curvature of the surface:

$$K = \kappa_{N1}\kappa_{N2} = \frac{\det K_{\hat{\mu}\hat{\nu}}}{\det g_{\hat{\mu}\hat{\nu}}} = \frac{R_{\hat{1}\hat{2}\hat{1}\hat{2}}}{\det g_{\hat{\mu}\hat{\nu}}} = R^{\hat{1}}_{\hat{2}\hat{1}\hat{2}}, \tag{6.56}$$

where we have used that $\det g_{\hat{\mu}\hat{\nu}} = 1$ in orthonormal basis. Since the right-hand side refers to the intrinsic curvature and the metric on the surface, we have proved that the Gaussian curvature of a surface is an intrinsic quantity. It can be measured by observers on the surface without embedding the surface in a 3-dimensional space. This is the contents of Gauss' *theorema egregium*.

6.3   The Ricci Identity

Applying the Riemann tensor to a vector gives

$$\vec{e}_\mu R^\mu_{\nu\alpha\beta} A^\nu = (\nabla_\alpha \nabla_\beta - \nabla_\beta \nabla_\alpha - \nabla_{[\vec{e}_\alpha, \vec{e}_\beta]})(\vec{A}). \tag{6.57}$$

In coordinate basis this is reduced to

$$\vec{e}_\mu R^\mu_{\nu\alpha\beta} A^\nu = (A^\mu_{;\beta\alpha} - A\mu_{;\alpha\beta})\vec{e}_\mu, \tag{6.58}$$

where

$$A^\mu_{;\alpha\beta} \equiv (A^\mu_{;\beta})_{;\alpha}. \tag{6.59}$$

The *Ricci identity* on component form is

$$A^\nu R^\mu_{\nu\alpha\beta} = A^\mu_{;\beta\alpha} - A^\mu_{;\alpha\beta}. \tag{6.60}$$

We can write this as

$$\underline{d}^2\vec{A} = \frac{1}{2}R^\mu_{\nu\alpha\beta}A^\nu \vec{e}_\mu \otimes \underline{\omega}^\alpha \wedge \underline{\omega}^\beta. \tag{6.61}$$

This shows that the 2. exterior derivative of a vector is equal to zero only in a *flat* space. Equations (6.60) and (6.61) both represent the Ricci identity.

6.4 Bianchi's 1. Identity

We shall here need Cartan's 1. structure equation,

$$\underline{d}\omega^\mu = -\underline{\Omega}^\mu_\nu \wedge \underline{\omega}^\nu, \tag{6.62}$$

and Cartan's 2. structure equation,

$$\underline{R}^\mu_\nu = \underline{d}\Omega^\mu_\nu + \underline{\Omega}^\mu_\lambda \wedge \underline{\Omega}^\lambda_\nu. \tag{6.63}$$

Exterior differentiation of Eq. (6.62) and use of Poincaré's lemma (5.16) give $d^2\underline{\omega}^\mu = 0$. Hence,

$$0 = \underline{d}\Omega^\mu_\nu \wedge \underline{\omega}^\nu - \underline{\Omega}^\mu_\lambda \wedge \underline{d}\omega^\lambda. \tag{6.64}$$

Use of (6.62) gives

$$\underline{d}\Omega^\mu_\nu \wedge \underline{\omega}^\nu + \Omega^\mu_\lambda \wedge \underline{\Omega}^\lambda_\nu \wedge \underline{\omega}^\nu = 0. \tag{6.65}$$

From this we see that

$$(\underline{d}\Omega^\mu_\nu + \underline{\Omega}^\mu_\lambda \wedge \underline{\Omega}^\lambda_\nu) \wedge \underline{\omega}^\nu = 0. \tag{6.66}$$

We now get *Bianchi's 1. identity*,

$$\underline{R}^\mu_\nu \wedge \underline{\omega}^\nu = 0 \tag{6.67}$$

The component form of Bianchi's 1. identity is

$$\underbrace{\frac{1}{2}R^\mu_{\nu\alpha\beta}\underline{\omega}^\alpha \wedge \underline{\omega}^\beta}_{\underline{R}^\mu_\nu} \wedge \underline{\omega}^\nu = 0. \tag{6.68}$$

The component equation is

$$R^{\mu}_{[\nu\alpha\beta]} = 0 \tag{6.69}$$

or

$$R^{\mu}_{\nu\alpha\beta} + R^{\mu}_{\alpha\beta\nu} + R^{\mu}_{\beta\nu\alpha} = 0, \tag{6.70}$$

where the antisymmetry $R^{\mu}_{\nu\alpha\beta} = -R^{\mu}_{\nu\beta\alpha}$ has been used.

6.5 Bianchi's 2. Identity

Exterior differentiation of Eq. (6.63) gives

$$\begin{aligned}
\underline{\mathrm{d}R^{\mu}_{\nu}} &= \underline{R^{\mu}_{\lambda}} \wedge \underline{\Omega^{\lambda}_{\nu}} - \underline{\Omega^{\mu}_{\rho}} \wedge \underline{\Omega^{\rho}_{\lambda}} \wedge \underline{\Omega^{\lambda}_{\nu}} - \underline{\Omega^{\mu}_{\lambda}} \wedge \underline{R^{\lambda}_{\nu}} + \underline{\Omega^{\mu}_{\lambda}} \wedge \underline{\Omega^{\lambda}_{\rho}} \wedge \underline{\Omega^{\rho}_{\nu}} \\
&= \underline{R^{\mu}_{\lambda}} \wedge \underline{\Omega^{\lambda}_{\nu}} - \underline{\Omega^{\mu}_{\lambda}} \wedge \underline{R^{\lambda}_{\nu}}.
\end{aligned} \tag{6.71}$$

We now have *Bianchi's 2. identity* as a form equation:

$$\underline{\mathrm{d}\, R^{\mu}_{\nu}} + \underline{\Omega^{\mu}_{\lambda}} \wedge \underline{R^{\lambda}_{\nu}} - \underline{R^{\mu}_{\lambda}} \wedge \underline{\Omega^{\lambda}_{\nu}} = 0 \tag{6.72}$$

As a component equation Bianchi's 2. identity is given by

$$R^{\mu}_{\nu[\alpha\beta;\gamma]} = 0. \tag{6.73}$$

The Riemann curvature tensor has four symmetries.

I.   The definition of the Riemann tensor implies that: $R^{\mu}_{\nu\alpha\beta} = -R^{\mu}_{\nu\beta\alpha}$.
II.   Bianchi's 1. identity: $R^{\mu}_{[\nu\alpha\beta]} = 0$.
III. From Cartan's 2. structure equation follows:

$$R_{\mu\nu\alpha\beta} = -R_{\nu\mu\alpha\beta}. \tag{6.74}$$

By choosing a locally Cartesian coordinate system in an inertial frame we get the following expression for the components of the Riemann curvature tensor:

$$R_{\mu\nu\alpha\beta} = \frac{1}{2}(g_{\mu\beta,\nu\alpha} - g_{\mu\alpha,\nu\beta} + g_{\nu\alpha,\mu\beta} - g_{\nu\beta,\mu\alpha}), \tag{6.75}$$

from which we get:

IV.   The fourth symmetry of the Riemann curvature tensor is $R_{\mu\nu\alpha\beta} = R_{\alpha\beta\mu\nu}$.

These four symmetries reduce the number of independent components of the Riemann tensor in 4-dimensional spacetime from 256 to 20. Contraction of μ and α leads to

$$R_{\nu\beta} = R_{\beta\nu}, \tag{6.76}$$

that is the Ricci tensor is symmetric. In 4-dimensional spacetime the Ricci tensor has 10 independent components.

6.6 Torsion

Definition 6.6.1 (*The torsion 2-form*) The torsion 2-form is defined by

$$\underline{T}(\vec{u} \wedge \vec{v}) = \nabla_{\vec{u}}\vec{v} - \nabla_{\vec{v}}\vec{u} - [\vec{u}, \vec{v}], \tag{6.77}$$

where $\vec{u}$ and $\vec{v}$ are arbitrary vectors. It is a vectorial form, which means that the torsion form has vector components. The contraction of the torsion with $\vec{u} \wedge \vec{v}$ has the component form,

$$\underline{T}(\vec{u} \wedge \vec{v}) = -\left(\Gamma^\rho_{\mu\nu} - \Gamma^\rho_{\nu\mu} + c^\rho_{\mu\nu}\right)u^\mu v^\nu \vec{e}_\rho. \tag{6.78}$$

It follows that the torsion form has the component form

$$\underline{T} = \frac{1}{2}\left(\Gamma^\rho_{\nu\mu} - \Gamma^\rho_{\mu\nu} - c^\rho_{\mu\nu}\right)\vec{e}_\rho \otimes \underline{\omega}^\mu \wedge \underline{\omega}^\nu. \tag{6.79}$$

Introducing the scalar torsion components $T^\rho_{\mu\nu}$ by

$$\underline{T}(\vec{u} \wedge \vec{v}) = T^\rho_{\mu\nu}u^\mu v^\nu \vec{e}_\rho, \tag{6.80}$$

we get

$$T^\rho_{\mu\nu} = \Gamma^\rho_{\nu\mu} - \Gamma^\rho_{\mu\nu} - c^\rho_{\mu\nu}, \tag{6.81}$$

So that

$$\underline{T} = \frac{1}{2}T^\rho_{\mu\nu}\vec{e}_\rho \otimes \underline{\omega}^\mu \wedge \underline{\omega}^\nu .. \tag{6.82}$$

In coordinate basis $c^\rho_{\mu\nu} = 0$, giving

$$T^\rho_{\mu\nu} = \Gamma^\rho_{\nu\mu} - \Gamma^\rho_{\mu\nu}. \tag{6.83}$$

Hence torsion induces an antisymmetric part of the connection coefficients in a coordinate basis.

The spacetime of the general theory of relativity is assumed to be torsion free. This is called a Riemannian space. Then the connection coefficients are related to the structure coefficients by

$$c^{\rho}_{\mu\nu} = \Gamma^{\rho}_{\nu\mu} - \Gamma^{\rho}_{\mu\nu}. \tag{6.84}$$

This shows that the structure coefficients represent the antisymmetric parts of the connection coefficients. In a coordinate basis in a Riemannian space the connection coefficients are symmetric and the structure coefficients vanish.

It follows from Eqs. (5.164), (5.78) and (6.79) that the vectorial torsion form (form with vector components) may be written

$$\underline{T} = \vec{e}_{\rho} \otimes \left(\underline{d\omega}^{\rho} + \underline{\Omega}^{\mu}_{\nu} \wedge \underline{\omega}^{\rho} \right). \tag{6.85}$$

The torsion two forms, $\underline{T}^{\rho}$, with scalar components are defined by

$$\underline{T} = \vec{e}_{\rho} \otimes \underline{T}^{\rho}, \tag{6.86}$$

hence,

$$\underline{T}^{\rho} = \underline{d\omega}^{\rho} + \underline{\Omega}^{\mu}_{\nu} \wedge \underline{\omega}^{\rho}. \tag{6.87}$$

6.7   The Equation of Geodesic Deviation

Consider two nearby geodesic curves (Fig. 6.6), both parametrized by a parameter λ. Let $\vec{s}$ be a vector connecting two curves with the same value of λ. The connecting vector $\vec{s}$ is said to measure the *geodesic deviation* of the curves

In order to deduce an equation describing how the geodesic deviation varies along the curves, we consider the covariant directional derivative of $\vec{s}$ along the curves, $\nabla_{\vec{u}}\vec{s}$, where $\vec{u}$ is the tangent vector field of the curves.

Let $\vec{u}$ and $\vec{s}$ be coordinate basis vectors of a coordinate system. Then $\left[\vec{s}, \vec{u} \right] = 0$ so that

$$\nabla_{\vec{u}}\vec{s} = \nabla_{\vec{s}}\vec{u}, \tag{6.88}$$

giving

$$\nabla_{\vec{u}}\nabla_{\vec{u}}\vec{s} = \nabla_{\vec{u}}\nabla_{\vec{s}}\vec{u}. \tag{6.89}$$

Fig. 6.6 Geodesic deviation. Two neighbouring geodesic curves with a vector $\vec{s}$ connecting points on the curves with the same parameter value

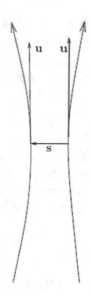

Furthermore

$$\underline{R}(\vec{u}, \vec{s})\,\vec{u} = \left([\nabla_{\vec{u}}, \nabla_{\vec{s}}] - \nabla_{[\vec{u}, \vec{s}]}\right)\vec{u} = [\nabla_{\vec{u}}, \nabla_{\vec{s}}]\vec{u}. \tag{6.90}$$

Thus

$$\nabla_{\vec{u}}\nabla_{\vec{u}}\,\vec{s} = \nabla_{\vec{s}}\nabla_{\vec{u}}\vec{u} + \underline{R}(\vec{u}, \vec{s})\,\vec{u}. \tag{6.91}$$

Since the curves are geodesics $\nabla_{\vec{u}}\vec{u} = 0$, and $\underline{R}(\vec{u}, \vec{s}) = -\underline{R}(\vec{s}, \vec{u})$ due to the antisymmetry of the Riemann tensor, the equation reduces to

$$\nabla_{\vec{u}}\nabla_{\vec{u}}\,\vec{s} + \underline{R}(\vec{s}, \vec{u})\,\vec{u} = 0. \tag{6.92}$$

This is *the equation of geodesic deviation*. The component form of the equation is

$$\left(\frac{D^2 s}{d\lambda^2}\right)^{\mu} + R^{\mu}_{\alpha\nu\beta}u^{\alpha}s^{\nu}u^{\beta} = 0. \tag{6.93}$$

where $D/d\lambda$ is the covariant derivative with respect to an invariant curve parameter λ. In co-moving geodesic normal coordinates with $\vec{u} = (1, 0, 0, 0)$ and vanishing Christoffel symbols, the covariant derivative reduces to the ordinary derivative, and the equation for geodesic deviation reduces to

$$\left(\frac{d^2 s}{d\lambda^2}\right)^{i} + R^{i}_{0j0}s^{j} = 0. \tag{6.94}$$

6.8 Tidal Acceleration and Spacetime Curvature

In Chap. 1 we found Eq. (1.50) for the *tidal acceleration*, i.e. the relative acceleration between two nearby particles,

$$\frac{d^2 \zeta^k}{dt^2} = -\zeta^i \frac{\partial^2 \phi}{\partial x_i \partial x_k}, \tag{6.95}$$

where ζ^j is the j-component of the separation vector, and ϕ is the Newtonian gravitational potential. Comparing these equations we see that in the Newtonian limit the non-vanishing components of the Riemann curvature tensor of spacetime are the second derivatives of the Newtonian potential,

$$R^i_{0j0} = \frac{\partial^2 \phi}{\partial x_i \partial x^j}. \tag{6.96}$$

In Newtonian physics the acceleration of gravity is given by

$$\vec{g} = -\nabla \phi, \tag{6.97}$$

or in component form

$$g^i = -\frac{\partial \phi}{\partial x_i}. \tag{6.98).}$$

Comparing with Eq. (5.90) we see that with a locally Cartesian coordinate system the non-vanishing Christoffel symbols are

$$\Gamma^i_{00} = \frac{\partial \phi}{\partial x_i}. \tag{6.99}$$

The Christoffel symbols are the first derivatives of the Newtonian gravitational potential. According to Eq. (6.96) the second derivatives are the components of the Riemann curvature tensor. Hence in the Newtonian approximation the non-vanishing components of the curvature tensor are

$$R^i_{0j0} = \frac{\partial \Gamma^i_{00}}{\partial x^j}. \tag{6.100}$$

6.9 The Newtonian Tidal Tensor

There are several definitions of the Newtonian tidal tensor that are mathematically equivalent. One is as follows.

Definition 6.9.1 (*Newtonian tidal tensor*) The Newtonian tidal tensor is a symmetrical tensor of rank 2 with components

$$E_{ij} = -\frac{\partial g_i}{\partial x^j}, \tag{6.101}$$

i.e. E_{ij} is minus the change of the i-component of the acceleration of gravity due to a displacement in the j-direction. Since

$$g_i = -\frac{\partial \phi}{\partial x^i}, \tag{6.102}$$

the components of the Newtonian tidal tensor may be written

$$E_{ij} = \frac{\partial^2 \phi}{\partial x^i \partial x^j}. \tag{6.103}$$

It follows that the Newtonian tidal tensor is symmetrical.
The Newtonian gravitational field equation

$$\nabla^2 \phi = \frac{\partial^2 \phi}{\partial x_i \partial x^i} = 4\pi G \rho \tag{6.104}$$

can now be written

$$E^i_i = 4\pi G \rho. \tag{6.105}$$

Also it follows that the equation of tidal acceleration can be written

$$\frac{d^2 \zeta^k}{dt^2} = -E^k_i \zeta^i, \tag{6.106}$$

and that in the Newtonian limit the tidal tensor is related to the Riemann curvature tensor of spacetime by

$$R^i_{0j0} = E^i_j. \tag{6.107}$$

6.10   The Tidal and Non-tidal Components
   of a Gravitational Field

In Newton's theory of gravitation all gravitational fields are caused by masses. A gravitational field may be described mathematically in terms of a gravitational potential ϕ. The acceleration of gravity can be expressed as the negative gradient of the gravitational potential as in Eq. (6.97). The gravitational field is an acceleration field which has a gravitational field strength equal to the acceleration of gravity at each point in space. The gravitational field strength is a local quantity, while the gravitational field itself is a global concept.

Einstein generalized Newton's 1 law to the form: The 4-acceleration of a free particle vanishes when no non-gravitational forces act upon it,

$$A^\mu = \frac{\mathrm{d}U^\mu}{\mathrm{d}\tau} + \Gamma^\mu_{\alpha\beta} U^\alpha U^\beta = 0. \tag{6.108}$$

This equation holds in arbitrary coordinate systems in all frames of reference, inertial or accelerated. But only in coordinate systems in which the coefficients of the metric tensor are constant, do the Christoffel symbols vanish, so that the equation takes the Newtonian three-vector form $\vec{a} = 0$. In an arbitrary reference frame the term $\mathrm{d}U^i/\mathrm{d}\tau$ represents the i-component of the acceleration of the particle relative to the reference frame, and according to Eq. (6.108) it is given by

$$\frac{\mathrm{d}U^i}{\mathrm{d}\tau} = -\Gamma^i_{\alpha\beta} U^\alpha U^\beta. \tag{6.109}$$

The acceleration of gravity may be defined in two mathematically identical ways, either in terms of the 3-acceleration of a free particle instantaneously at rest, or in terms of the 4-acceleration

$$A^\mu = \frac{\mathrm{d}U^\mu}{\mathrm{d}\tau} + \Gamma^\mu_{\alpha\beta} U^\alpha U^\beta \tag{6.110}$$

of a particle permanently at rest in a reference frame. Let the gravitational field point in the i-direction at a point. Then the quantity $\Gamma^i_{\alpha\beta} U^\alpha U^\beta$ determines the acceleration of gravity.

This has an important conceptual consequence: *Experiencing a gravitational field strength, i.e. that there is a non-vanishing acceleration of gravity, in other words that a free particle falls, has nothing to do with the curvature of spacetime.* It depends upon certain Christoffel symbols. The fact that the Christoffel symbols are not tensor components implies that all of them may be transformed away at a point by introducing locally Cartesian coordinates co-moving with a freely falling local reference frame. This is the mathematical expression of one of the properties of gravity expressed in connection with the principle of equivalence, that gravity may be transformed away locally by going into a freely falling room. So when do we experience a gravitational field strength? The answer is: *We experience a gravitational field strength when we are in a room which is not freely falling.* Note that this is valid whether spacetime is curved of flat.

As seen from Eq. (6.93) of geodesic deviation spacetime curvature is connected to inhomogeneity in a gravitational field—the difference between acceleration at two nearby points. In Newton's theory this is associated with the phenomenon of tidal forces.

Inertial effects upon physical phenomena may be defined as those effects that depend upon the state of acceleration or rotation of the reference frame. In accelerated or rotating reference frames there are non-vanishing Christoffel symbols that represent inertial effects, such as the Coriolis and centrifugal acceleration in a rotating reference frame. We shall take a closer look at this connection.

It should be noted that not all of the Christoffel symbols represent a deviation from uniform motion of free particles. For example, there are non-vanishing Christoffel symbols in a system of polar coordinates in an inertial reference frame that only tell about the geometrical properties of the coordinate system, but do not have any kinematical significance. On the other hand the Christoffel symbols Γ^i_{00} represent the acceleration of a free particle instantaneously at rest, and Γ^i_{j0} represent the Coriolis acceleration.

The gravitational field strength depends upon the chosen frame of reference and vanishes in a local freely falling reference frame, i.e., an inertial reference frame. Generally a gravitational field will have both a non-tidal tidal component which can be transformed away, and a tidal component which cannot be transformed away. We shall now consider the mathematical representation of these components.

Let us start by considering gravity in Newton's theory. Introducing the gravitational potential ϕ and the separation vector $\vec{s}$ between P_0 and a nearby point P, the acceleration of gravity at P is given by the first two terms of a Taylor expansion about P_0,

$$g_i = -\left(\frac{\partial \phi}{\partial x^i}\right)_P = -\left(\frac{\partial \phi}{\partial x^i}\right)_{P_0} - \left(\frac{\partial^2 \phi}{\partial x^i \partial x^j}\right) s^j = (g_{NT})_i + (g_T)_i, \qquad (6.111)$$

where the non-tidal component of the gravitational field strength is

$$(g_{NT})_i = -\left(\frac{\partial \phi}{\partial x^i}\right)_{P_0}, \qquad (6.112)$$

and the tidal component is

$$(g_T)_i = -\left(\frac{\partial^2 \phi}{\partial x^i \partial x^j}\right)_{P_0} s^j. \qquad (6.113)$$

Noting that $(g_T)_i = d^2s_i/dt^2$ and comparing with Eq. (6.94) we see that in the weak field limit there is a simple connection between the tidal component of the gravitational field strength and spacetime curvature,

$$(g_T)_i = -R^i_{0j0}s^j = -\left(\frac{\partial^2\phi}{\partial x^i \partial x^j}\right)_{P_0} s^j. \tag{6.114}$$

This shows that the relativistic counterpart to inhomogeneity of a gravitational field, i.e. to a tidal gravitational field, is spacetime curvature.

We now proceed to consider gravity according to Einstein's theory. Making a Taylor expansion of Eq. (6.109) for a free particle instantaneously at rest in a reference frame with a stationary metric, we find the relativistic expression for the i-component of the gravitational field strength

$$g^i_R = -\left(\Gamma^i_{00}\right)_{P_0} - \left(\Gamma^i_{00,j}\right)_{P_0} s^j. \tag{6.115}$$

In this case Eq. (6.17) gives

$$\Gamma^i_{00,j} = R^i_{0j0} + \Gamma^\alpha_{0j}\Gamma^i_{\alpha 0} - \Gamma^k_{00}\Gamma^i_{kj}. \tag{6.116}$$

Inserting this into Eq. (6.115) leads to

$$g^i_R = -\left(\Gamma^i_{00}\right)_{P_0} + \left(\Gamma^k_{00}\Gamma^i_{kj} - \Gamma^\alpha_{0j}\Gamma^i_{\alpha 0}\right)_{P_0} s^j - \left(R^i_{0j0}\right)_{P_0} s^j. \tag{6.117}$$

The first term of Eq. (6.117) represents the acceleration of gravity at point P_0, i.e., it represents the uniform part of the gravitational field. The second term represents the non-uniform part of the gravitational field which is also present in a non-inertial reference frame in flat spacetime, for example, the non-uniformity of the centrifugal field in a rotating reference frame. The last term represents the tidal effects, which in the general theory is proportional to the spacetime curvature. This suggests the following separation of a gravitational field into a non-tidal part and a tidal part:

$$g^i_R = g^i_{NT} + g^i_T, \tag{6.118}$$

where the non-tidal component of the gravitational field is given by

$$g^i_{NT} = -\left(\Gamma^i_{00}\right)_{P_0} + \left(\Gamma^k_{00}\Gamma^i_{kj} - \Gamma^\alpha_{0j}\Gamma^i_{\alpha 0}\right)_{P_0} s^j, \tag{6.119}$$

and the tidal part by

$$g^i_T = -\left(R^i_{0j0}\right)_{P_0} s^j. \tag{6.120}$$

As in the Newtonian case, the non-tidal part of the gravitational field can be transformed away locally by going into a local inertial frame. The tidal part cannot be transformed away.

Example 6.10.1 *Non-tidal gravitational field*. Let us as a simple illustrationdp consider the gravitational field strength in a rotating reference frame in flat spacetime. In this case the Riemann curvature tensor vanishes, so the gravitational field has no tidal component. The non-vanishing Christoffel symbols are given in Eq. (5.119). Inserting these into Eq. (6.119) gives for the non-tidal component of the gravitational field

$$g^r_{NT} = r\omega^2 + \omega^2 s^r, \quad g^\theta_{NT} = \omega^2 s^\theta. \tag{6.121}$$

The term $r\omega^2$ is the centrifugal acceleration at the point P_0, and the other terms are due to the inhomogeneity of the centrifugal gravitational field.

Exercises

6.1   Parallel transport and curvature

(a)   A curve $P(\lambda)$ runs through a point $P = P(0)$, and a vector $\vec{A}$ is defined at this point. The vector is parallel transported along the curve so that in each point $P(\lambda)$ there is a well-defined vector $\vec{A}(\lambda)$. Express the condition that the vectors along the curve are parallel as an equation of the components of the vector $A^\mu(\lambda)$. Show that the change of the components of the vector by an infinitesimal displacement dx^μ is

$$dA^\mu = -\Gamma^\mu_{\lambda\nu}(x)A^\lambda dx^\nu.$$

(b)   A closed curve has the shape of a parallelogram with the sides $d\vec{a}$ and $d\vec{b}$. The corners of the parallelogram are denoted by A, B, C and D, respectively. A vector $\vec{A}$ is parallel transported from A and C along the two curves ABC and ADC. Show that the result in these two cases is in general not the same. Then use this fact to show that the change of $\vec{A}$, by parallel transporting it along the closed curve $ABCDA$, is

$$\delta A^\alpha = -R^\alpha_{\beta\gamma\delta}A^\beta da^\gamma db^\delta,$$

where $R^\alpha_{\beta\gamma\delta}$ is the Riemann curvature tensor.

6.2   Curvature of the simultaneity space in a rotating reference frame

Calculate the curvature scalar R of a 2-dimensional simultaneity space in a rotating reference frame with the line-element

Fig. 6.7 Spacetime
curvature and the tidal force
pendulum. Here
$a = R_{\hat\phi\hat t\hat\phi\hat t}\ell^{\hat\phi}\cos\theta -$
$R_{\hat r\hat t\hat r\hat t}\ell^{\hat r}\sin\theta,\quad \vec{\ell} =$
$\ell\sin\theta\,\vec{e}_{\hat\phi} + \ell\cos\theta\,\vec{e}_{\hat r}$

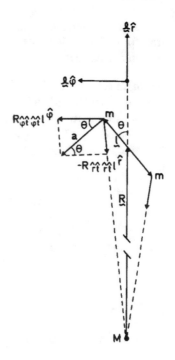

$$dl^2 = dr^2 + \frac{r^2 d\theta^2}{1 - r^2\omega^2/c^2}.$$

6.3   The tidal force pendulum and the curvature of space

We will again consider the tidal force pendulum. Here we shall use the equation for
geodesic deviation to find the period of the pendulum.

(a)   Why can the equation for geodesic equation be used to find the period of the
pendulum in spite of the fact that the particles do not move along geodesics?
Assume that the centre of the pendulum is fixed at a distance R from the centre
of mass of the Earth. Introduce an orthonormal basis $\{\mathbf{e}_{\hat a}\}$ with the origin at the
centre of the pendulum (see Fig. 6.7).

(b)   Show to first order in v/c and ϕ/c^2, where v is the 3-velocity of the masses and
ϕ the gravitational potential at the position of the pendulum, that the equation
of geodesic deviation takes the form

$$\frac{d^2\ell^{\hat i}}{dt^2} + R^{\hat i}_{\hat 0\hat j\hat 0}\ell^{\hat j} = 0.$$

(c)   Find the period of the pendulum expressed in terms of the components of
Riemann's curvature tensor.

Chapter 7
Einstein's Field Equations

Abstract This chapter starts with a hydrodynamical description of energy–momentum conservation in a Newtonians context in order to give some intuition about the relativistic formulation of energy–momentum conservation as represented by a vanishing divergence of the energy–momentum tensor. Einstein demanded that energy–momentum conservation should follow from the field equations, and hence he needed a divergence-free curvature tensor. This is deduced from Bianchi's 2. identity. It is shown that one need not postulate that free particles follow geodesic curves, but that it follows from the field equations.

7.1 Newtonian Fluid

We shall begin this chapter by giving a mathematical formulation of the law of energy–momentum conservation. Again (like we did in the Sect. 5.9.1) we shall start by considering a Newtonian fluid.

The total derivative of a velocity field was presented in Sect. 5.9.1 and is

$$\frac{D\vec{v}}{Dt} \equiv \frac{\partial \vec{v}}{\partial t} + (\vec{v} \cdot \vec{\nabla})\vec{v}. \tag{7.1}$$

The component notation for the expression of the total derivative of the velocity takes the form

$$\frac{D\vec{v}^i}{Dt} \equiv \frac{\partial v^i}{\partial t} + v^j \frac{\partial v^i}{\partial x^j}. \tag{7.2}$$

In Newtonian hydrodynamics mass conservation is represented mathematically by the *continuity equation*,

© Springer Nature Switzerland AG 2020

Ø. Grøn, *Introduction to Einstein's Theory of Relativity*,
Undergraduate Texts in Physics, https://doi.org/10.1007/978-3-030-43862-3_7

$$\frac{\partial \rho}{\partial t} + \nabla \cdot (\rho \vec{v}) = 0 \quad \text{or} \quad \frac{\partial \rho}{\partial t} + \frac{\partial (\rho v^i)}{\partial x^i} = 0. \tag{7.3}$$

Conservation of momentum is represented by the Euler's equation of motion (ignoring gravity),

$$\rho \frac{D \vec{v}}{Dt} = -\vec{\nabla} p \quad \text{or} \quad \rho \left(\frac{\partial v^i}{\partial t} + v^j \frac{\partial v^i}{\partial x^j} \right) = -\frac{\partial p}{\partial x^i}. \tag{7.4}$$

The *energy–momentum tensor* is a symmetric tensor of rank 2 which describes material properties,

$$T^{\mu\nu} = \begin{pmatrix} T^{00} & T^{01} & T^{02} & T^{03} \\ T^{10} & T^{11} & T^{12} & T^{13} \\ T^{20} & T^{21} & T^{22} & T^{23} \\ T^{30} & T^{31} & T^{32} & T^{33} \end{pmatrix}, \tag{7.5}$$

Here T^{00} represents mass–energy density, T^{i0} represents momentum density, T^{ii} represents pressure for $T^{ii} > 0$ and tension for $T^{ii} < 0$, and T^{ij} represents shear forces for $i \neq j$.

Example 7.1.1 (*Energy–momentum tensor of a Newtonian fluid*) In the case of a Newtonian fluid the components of the energy–momentum tensor are

$$T^{00} = \rho c^2, \quad T^{i0} = \rho c v^i, \quad T^{ij} = \rho v^i v^j + p \delta^{ij}, \tag{7.6}$$

where ρ is the mass density, p the pressure, assumed isotropic here and v^i the i-component of the velocity. We choose a locally Cartesian coordinate system in an inertial frame such that the covariant derivatives are reduced to partial derivatives. The divergence of the momentum–energy tensor, $T^{\mu\nu}_{;\nu}$, has four components, one for each value of μ.

The zeroth component is

$$T^{0\nu}_{;\nu} = T^{0\nu}_{,\nu} = T^{00}_{,0} + T^{0i}_{,i} = \frac{\partial \rho}{\partial t} + \frac{\partial (\rho v^i)}{\partial x^i}, \tag{7.7}$$

which in comparison with Newtonian hydrodynamics shows that

$$T^{0\nu}_{;\nu} = 0 \tag{7.8}$$

is the continuity equation. This equation represents the conservation of mass–energy.

The i- component of the divergence is

$$T^{i\nu}_{,\nu} = T^{i0}_{,0} + T^{ij}_{,j} = \frac{\partial (\rho v^i)}{\partial t} + \frac{\partial (\rho v^i v^j + p \delta^{ij})}{\partial x^j}$$

$$= \rho \frac{\partial v^i}{\partial t} + v^i \frac{\partial \rho}{\partial t} + v^i \frac{\partial \rho v^j}{\partial x^j} + \rho v^j \frac{\partial v^i}{\partial x^j} + \frac{\partial p}{\partial x^i}, \tag{7.9}$$

According to the continuity equation

$$\frac{\partial(\rho v^i)}{\partial x^i} = -\frac{\partial \rho}{\partial t}, \tag{7.10}$$

which implies that

$$T^{iv}_{,v} = \rho \frac{\partial v^i}{\partial t} + v^i \frac{\partial \rho}{\partial t} - v^i \frac{\partial \rho}{\partial t} + \rho v^j \frac{\partial v^i}{\partial x^j} + \frac{\partial p}{\partial x^i} = \rho \frac{Dv^i}{Dt} + \frac{\partial p}{\partial x^i} \tag{7.11}$$

Hence the Euler equation of motion, which represents momentum conservation, takes the form

$$T^{iv}_{;v} = 0. \tag{7.12}$$

This is Euler's equation of motion. It expresses the conservation of momentum. It follows from Eqs. (7.8) and (7.12) that the equations

$$T^{\mu v}_{;v} = 0 \tag{7.13}$$

are *the general expressions for energy and momentum conservation.*

7.2 Perfect Fluids

We now turn to the general relativistic case. A perfect fluid is a fluid with no viscosity. The components of the energy–momentum tensor of a perfect fluid are

$$T_{\mu v} = \left(\rho + \frac{p}{c^2}\right) u_\mu u_v + p g_{\mu v}, \tag{7.14}$$

where ρ and p are the mass–density and pressure of tension, respectively, measured in the rest frame of the fluid, and u_μ are the components of the 4-velocity of the fluid.

In a co-moving orthonormal basis the components of the 4-velocity are $u^{\hat{\mu}} = (c, 0, 0, 0)$. Then the energy–momentum tensor is given by

$$T_{\hat{\mu}\hat{v}} = \begin{pmatrix} \rho c^2 & 0 & 0 & 0 \\ 0 & p & 0 & 0 \\ 0 & 0 & p & 0 \\ 0 & 0 & 0 & p \end{pmatrix}, \tag{7.15}$$

where $p > 0$ is pressure and $p < 0$ is tension.

There are three different types of perfect fluids that are particularly useful:

1. *Dust* or non-relativistic gas is given by $p = 0$ and the energy–momentum tensor $T_{\mu\nu} = \rho u_\mu u_\nu$.
2. *Radiation* or ultra-relativistic gas is given by a traceless energy–momentum tensor, i.e. $T^\mu_\mu = 0$. It follows that $p = (1/3)\rho c^2$.
3. The third type is a sort of *vacuum energy* of particular significance for construction of relativistic universe models.

7.2.1 Lorentz Invariant Vacuum Energy—LIVE

Let us follow a thought presented by the Belgian cosmologist Georges Lemaître around 1935. Assume a particle is alone in the universe. It is not possible to define motion for such a particle. Hence all motion is relative.

Quantum mechanics implies that particle–antiparticle pairs are created and then annihilated again in a very short time restricted by the Heisenberg uncertainty relationships. Averaging over a macroscopic time and region in space this implies the existence of a quantum mechanical vacuum energy on a macroscopic scale. If it is possible to measure velocity relative to this energy, it would act as a sort of ether and re-establish absolute motion into the physics.

According to the special theory of relativity, which has been experimentally confirmed in several ways, this cannot be the case. Hence it must be impossible to measure velocity relative to the vacuum energy. This implies that all the components of the energy–momentum tensor of the vacuum energy must be Lorentz invariant. It is shown in Exercise 7.2 that Lorentz invariance of all the components of an energy–momentum tensor implies that the energy–momentum tensor is proportional to the metric tensor.

Assume now that the Lorentz invariant vacuum energy, LIVE, can be described as a perfect fluid.

$$T_{\mu\nu} = \left(\rho_{\text{LIVE}} + p_{\text{LIVE}}/c^2\right)u_\mu u_\nu + \rho_{\text{LIVE}}g_{\mu\nu} \tag{7.16}$$

Lorentz invariance then requires that the vacuum energy obeys the equation of state

$$p_{\text{LIVE}} = -\rho_{\text{LIVE}}c^2. \tag{7.17}$$

Hence LIVE is in a state of strain.

7.2.2 Energy–Momentum Tensor of an Electromagnetic Field

Given an electric field $\vec{E} = E^i \vec{e}_i$ and a magnetic field $\vec{B} = B^i \vec{e}_i$. The electromagnetic field tensor is an antisymmetric tensor of rank 2 given by (using units so that $c = 1$)

$$F_{\eta\nu} = \begin{bmatrix} 0 & E^1 & E^2 & E^3 \\ -E^1 & 0 & -B^3 & B^2 \\ -E^2 & B^3 & 0 & -B^1 \\ -E^3 & -B^2 & B^1 & 0 \end{bmatrix}. \tag{7.18}$$

The electromagnetic energy–momentum tensor is a symmetric tensor of rank 2 with components

$$T_{\mu\nu} = F_{\mu\alpha} F^{\alpha}_{\ \nu} - \frac{1}{4} g_{\mu\nu} F_{\alpha\beta} F^{\alpha\beta}. \tag{7.19}$$

This tensor is used when one is going to find solutions of Einstein's field equations for spacetimes with electromagnetic fields.

7.3 Einstein's Curvature Tensor

Einstein assumed that the field equations representing the relativistic generalization of Newton's law of gravitation have the form: *spacetime curvature* $\propto$ *momentum–energy tensor*. Also, he demanded that energy and momentum conservation should follow as a consequence of the field equation. This puts the following constraints on the curvature tensor: It must be a symmetric, divergence-free tensor of rank 2.

A good candidate is the Ricci tensor introduced in Eq. (6.18). It is a symmetric curvature tensor of rank 2. Let us see whether it is divergence free. In order to calculate its divergence we start with Bianchi's 2. identity which was deduced in Sect. 6.5,

$$R^{\mu}_{\ \nu\alpha\beta;\sigma} + R^{\mu}_{\ \nu\sigma\alpha;\beta} + R^{\mu}_{\ \nu\beta\sigma;\alpha} = 0. \tag{7.20}$$

Contraction of μ and α and using Eq. (6.73) gives

$$R_{\nu\beta;\sigma} - R_{\nu\sigma;\beta} + R^{\mu}_{\ \nu\beta\sigma;\mu} = 0. \tag{7.21}$$

Further contraction of ν and σ leads to

$$2R^{\sigma}_{\ \beta;\sigma} = R_{;\beta}. \tag{7.22}$$

Thus, the divergence of the Ricci tensor is

$$R^{\sigma}_{\beta;\sigma} = \frac{1}{2} R_{;\beta}.$$

(7.23)

It is not vanishing. Hence, the Ricci tensor is not the curvature tensor to be put into the left-hand side of the field equations. However, we can use this expression to construct a new divergence-free curvature tensor.

Since the metric tensor is covariant divergence free, we have that $(g^{\sigma}_{\beta} R)_{;\sigma} = g^{\sigma}_{\beta} R_{;\sigma}$. Now we multiply Eq. (7.23) by g^{β}_{α} to get

$$\left(g^{\beta}_{\alpha} R^{\sigma}_{\beta}\right)_{;\sigma} - \frac{1}{2}\left(g^{\beta}_{\alpha} R\right)_{;\beta} = 0.$$

(7.24)

Interchanging σ and β in the first term and using that the mixed components of the metric tensor are $g^{\beta}_{\alpha} = \delta^{\beta}_{\alpha}$, we get

$$\left(R^{\beta}_{\alpha} - \frac{1}{2}\delta^{\beta}_{\alpha} R\right)_{;\beta} = 0.$$

(7.25)

Hence the tensor $R^{\beta}_{\alpha} - (1/2)\delta^{\beta}_{\alpha} R$ is divergence-free. This tensor is called the *Einstein tensor*, and its covariant components are denoted by $E_{\alpha\beta}$, that is

$$E_{\alpha\beta} = R_{\alpha\beta} - \frac{1}{2} g_{\alpha\beta} R.$$

(7.26)

In 4. dimensional spacetime the metric tensor has ten independent components. Due to the identities

$$E^{\mu\nu}_{;\nu} = 0,$$

(7.27)

which represent four equations that any Einstein tensor fulfils as a consequence of Bianchi's 2. identity, the field equations give only six independent equations to determine the components of the metric tensor. Since there are ten independent components, this leaves four free metric functions. This secures a free choice of coordinate system.

7.4 Einstein's Field Equations

Einstein's field equations are

$$E_{\mu\nu} = \kappa T_{\mu\nu}$$

(7.28)

where κ is a constant called *Einstein's constant of gravity*. Inserting the expression (7.26) for the components of the Einstein tensor we have

$$R_{\mu\nu} - \frac{1}{2}g_{\mu\nu}R = \kappa T_{\mu\nu}. \tag{7.29}$$

Contraction and using that $g^\mu_\mu = \delta^\mu_\mu = 4$ gives

$$R = -\kappa T, \tag{7.30}$$

where $T = T^\mu_\mu$. Thus the field equations may be written in the form

$$R_{\mu\nu} = \kappa\left(T_{\mu\nu} - \frac{1}{2}g_{\mu\nu}T\right). \tag{7.31}$$

In the Newtonian limit the metric may be written as

$$ds^2 = -\left(1 + \frac{2\phi}{c^2}\right)c^2 dt^2 + (1 + h_{ii})(dx^2 + dy^2 + dz^2), \tag{7.32}$$

where the Newtonian potential $|\phi| \ll c^2$, and h_{ii} is a perturbation of the metric satisfying $|h_{ii}| \ll 1$. We also have $T_{00} \gg T_{kk}$ and $T \approx -T_{00}$. Then the 00 component of the field equations may be approximated by

$$R_{00} \approx \frac{\kappa}{2}T_{00}. \tag{7.33}$$

Furthermore we have

$$R_{00} = R^\mu_{0\mu 0} = R^i_{0i0} = \Gamma^i_{00,i} - \Gamma^i_{0i,0} = \frac{\partial \Gamma^k_{00}}{\partial x^k} = \frac{1}{c^2}\nabla^2\phi. \tag{7.34}$$

Since $T_{00} \approx \rho c^2$, Eq. (7.33) can be written as

$$\nabla^2\phi = \frac{1}{2}\kappa c^4 \rho. \tag{7.35}$$

Comparing this equation with the Newtonian law of gravitation on local form,

$$\nabla^2\phi = 4\pi G\rho, \tag{7.36}$$

we see that Einstein's constant of gravity is

$$\kappa = \frac{8\pi G}{c^4}. \tag{7.37}$$

It has the value $\kappa = 2.077 \times 10^{-43} \mathrm{s}^2/\mathrm{m\,kg}$.

In classical *empty space* we have $T_{\mu\nu} = 0$, which gives

$$E_{\mu\nu} = 0, \tag{7.38}$$

or

$$R_{\mu\nu} = 0. \tag{7.39}$$

These are the field equations for empty space. Note that $R_{\mu\nu} = 0$ does *not* imply $R_{\mu\nu\alpha\beta} = 0$. In general curvature is non-vanishing in empty space. In the words of J. A. Wheeler: "Mass there curves spacetime here".

It was shown by D. Hilbert that the field equations may be deduced from a variational principle with action

$$\int R\sqrt{-g}\,\mathrm{d}^4 x, \tag{7.40}$$

where $R\sqrt{-g}$ is the Lagrange density. One may also include a so-called cosmological constant Λ:

$$\int (R + 2\Lambda)\sqrt{-g}\,\mathrm{d}^4 x. \tag{7.41}$$

The field equations with cosmological constant are

$$R_{\mu\nu} - \frac{1}{2} g_{\mu\nu} R + \Lambda g_{\mu\nu} = \kappa T_{\mu\nu}. \tag{7.42}$$

The field equations of empty space with a cosmological constant are

$$R_{\mu\nu} - \frac{1}{2} g_{\mu\nu} R + \Lambda g_{\mu\nu} = 0. \tag{7.43}$$

Solutions of these equations are sometimes called *Einstein spaces*.

7.5 The "Geodesic Postulate" as a Consequence of the Field Equations

The principle that free particles follow geodesic curves has been called the "geodesic postulate". We shall now show that the "geodesic postulate" follows as a consequence of the field equations.

Consider a system of free particles in curved spacetime. This system can be regarded as a pressure-free gas. Such a gas is called *dust*. It is described by an energy–momentum tensor

$$T^{\mu\nu} = \rho u^{\mu} u^{\nu}, \tag{7.44}$$

where ρ is the rest density of the dust as measured by an observer at rest in the dust, and u^{μ} are the components of the 4-velocity of the dust particles.

Einstein's field equations as applied to spacetime filled with dust take the form

$$R^{\mu\nu} - \frac{1}{2} g^{\mu\nu} R = \kappa \rho u^{\mu} u^{\nu}. \tag{7.45}$$

Because the divergence of the left-hand side is zero, the divergence of the right-hand side must be zero, too,

$$(\rho u^{\mu} u^{\nu})_{;\nu} = 0 \tag{7.46}$$

or

$$(\rho u^{\nu} u^{\mu})_{;\nu} = 0. \tag{7.47}$$

We now regard the quantity in the parenthesis as a product of ρu^{ν} and u^{μ}. By the rule for differentiating a product we get

$$(\rho u^{\nu})_{;\nu} u^{\mu} + \rho u^{\nu} u^{\mu}_{;\nu} = 0. \tag{7.48}$$

Since the 4-velocity of any object has a magnitude equal to the velocity of light we have

$$u_{\mu} u^{\mu} = -c^2. \tag{7.49}$$

Differentiation gives

$$(u_{\mu} u^{\mu})_{;\nu} = 0. \tag{7.50}$$

Using, again, the rule for differentiating a product, we get

$$u_{\mu;\nu} u^{\mu} + u_{\mu} u^{\mu}_{;\nu} = 0. \tag{7.51}$$

From the rule for raising an index and the freedom of changing a summation index from α to μ, say, we get

$$u_{\mu;\nu} u^{\mu} = u^{\mu} u_{\mu;\nu} = g^{\mu\alpha} u_{\alpha} u_{\mu;\nu} = u_{\alpha} g^{\mu\alpha} u_{\mu;\nu} = u_{\alpha} u^{\alpha}_{;\nu} = u_{\mu} u^{\mu}_{;\nu}. \tag{7.52}$$

Thus the two terms of Eq. (7.51) are equal. It follows that each of them are equal to zero. So we have

$$u_\mu u^\mu_{;\nu} = 0. \tag{7.53}$$

Multiplying Eq. (7.48) by u_μ, we get

$$(\rho u^\nu)_{;\nu} u_\mu u^\mu + \rho u^\nu u_\mu u^\mu_{;\nu} = 0. \tag{7.54}$$

Using Eq. (7.47) in the first term and Eq. (7.51) in the last term, which then vanishes, we get

$$(\rho u^\nu)_{;\nu} = 0. \tag{7.55}$$

Thus the first term in Eq. (7.48) vanishes and we get

$$\rho u^\nu u^\mu_{;\nu} = 0. \tag{7.56}$$

Since $\rho \neq 0$ we must have

$$u^\nu u^\mu_{;\nu} = 0. \tag{7.57}$$

This is just the geodesic equation. Hence, *it follows from Einstein's field equations that free particles follow geodesic curves of spacetime.*

7.6 Einstein's Field Equations Deduced from a Variational Principle

It was shown by David Hilbert how Einstein's field equations can be deduced from a variational principle. A detailed discussion of this is given in for example Ø. Grøn and S. Hervik: "Einstein's General Theory of Relativity", Chap. 8. We shall here only give a brief summary as a preparation for the presentation of the Kaluza–Klein theory in the Appendix.

Hilbert's variational principle has the form

$$\delta I = 0, \quad I = \int L\sqrt{-g}\, \mathrm{d}^4 x, \tag{7.58}$$

where I is the action, L the Lagrange function—also called the Lagrangian—and g the determinant of the metric tensor. Einstein's field equations for empty space are obtained from the Lagrange function

$$L = R, \tag{7.59}$$

where R is the Ricci curvature scalar. This gives

$$\delta I = \int \left(R^{\mu\nu} - \frac{1}{2} R \, g^{\mu\nu} \right) \delta \, g_{\mu\nu} \sqrt{-g} \, \mathrm{d}^4 x. \tag{7.60}$$

Requiring that $\delta I = 0$ for arbitrary variations of $g_{\mu\nu}$ gives the field equations for empty space

$$R^{\mu\nu} - \frac{1}{2} R \, g^{\mu\nu} = 0. \tag{7.61}$$

Einstein's field equations (7.29) for a spacetime filled with energy and matter, described by an energy–momentum tensor with mixed components T^{μ}_{ν}, are now written

$$R^{\mu}_{\nu} - \frac{1}{2} R \, \delta^{\mu}_{\nu} = \kappa \, T^{\mu}_{\nu}. \tag{7.62}$$

According to the Lagrangian formalism the energy momentum tensor of an electromagnetic field is given by

$$\mu_0 \bar{T}^{\mu}_{\nu} = \frac{\partial L}{\partial A_{\rho, \mu}} A_{\rho, \nu} - L \delta^{\mu}_{\nu}, \tag{7.63}$$

where A_{ρ} are the components of the electromagnetic vector potential, and μ_0 is the permeability of empty space. The electromagnetic field tensor is given in terms of the electromagnetic vector potential in Eq. (5.263) which is here written in the form

$$F_{\mu\nu} = A_{\nu, \mu} - A_{\mu, \nu}. \tag{7.64}$$

The Lagrange function of the electromagnetic field is

$$L = \frac{1}{4} F^{\mu\nu} F_{\mu\nu}. \tag{7.65}$$

In order to calculate the first term at the right-hand side of Eq. (7.3) this is written as

$$L = \frac{1}{4} F^{\mu\nu} \left(A_{\nu, \mu} - A_{\mu, \nu} \right). \tag{7.66}$$

Due to the antisymmetry of $F^{\mu\nu}$ this can be written as

$$L = \frac{1}{2} F^{\mu\nu} A_{\nu, \mu}. \tag{7.67}$$

Hence

$$\frac{\partial L}{\partial A_{\rho,\mu}} = F^{\mu\rho}, \tag{7.68}$$

so

$$\mu_0 \bar{T}^{\mu}_{\nu} = F^{\rho\mu} A_{\rho,\nu} - \frac{1}{4} F_{\rho\sigma} F^{\rho\sigma} \delta^{\mu}_{\nu}. \tag{7.69}$$

The covariant components are

$$\mu_0 \bar{T}_{\mu\nu} = F^{\rho}_{\mu} A_{\rho,\nu} - \frac{1}{4} F_{\rho\sigma} F^{\rho\sigma} g_{\mu\nu}. \tag{7.70}$$

The energy–momentum tensor should be symmetric and divergence free in order to fit into Einstein's field equations. But the first term in the expression (7.70) is not symmetric in μ and ν. The usual procedure for obtaining a symmetric energy–momentum tensor is the following. Add a term of the form $K^{\lambda\mu}_{\nu,\lambda}$ to the expression (7.69), where $K^{\lambda\mu}_{\nu}$ is antisymmetric in the first two indices. The mathematical expression of energy–momentum conservation is that the divergence of the energy–momentum tensor vanishes. The divergence of the added term is $K^{\lambda\mu}_{\nu,\lambda\mu}$, which is symmetric in the lower indices λ and μ and antisymmetric in the upper ones. So the summation over λ and μ makes this term vanish. This means that the tensor

$$T^{\mu}_{\nu} = \bar{T}^{\mu}_{\nu} + K^{\lambda\mu}_{\nu,\lambda} \tag{7.71}$$

is equally good for describing energy–momentum conservation as $\bar{T}^{\mu}_{\nu}$.

For electromagnetism we choose

$$K^{\lambda\mu\nu} = F^{\mu\lambda} A^{\nu}. \tag{7.72}$$

Hence

$$K^{\lambda\mu}_{\nu,\lambda} = \left(F^{\mu\lambda} A_{\nu} \right)_{,\lambda} = F^{\mu\lambda}_{,\lambda} A_{\nu} + F^{\mu\lambda} A_{\nu,\lambda}. \tag{7.73}$$

The component form of Maxwell's source free equations (5.280) is

$$F^{\mu\lambda}_{,\lambda} = 0, \tag{7.74}$$

giving

$$K^{\lambda\mu}_{\nu,\lambda} = F^{\mu\lambda} A_{\nu,\lambda}. \tag{7.75}$$

From Eqs. (7.70), (7.71) and (7.75) we get

$$\mu_0 T_{\mu\nu} = F_\mu^\rho \left(A_{\nu,\,\rho} - A_{\rho,\,\nu} \right) - \frac{1}{4} F_{\rho\sigma} F^{\rho\sigma} g_{\mu\nu}. \tag{7.76}$$

Using Eq. (7.64) we finally get *the symmetric energy–momentum tensor of an electromagnetic field*

$$\mu_0 T_{\mu\nu} = F_\mu^\rho F_{\rho\nu} - \frac{1}{4} F_{\rho\sigma} F^{\rho\sigma} g_{\mu\nu}. \tag{7.77}$$

Example 7.6.1 (*The energy–momentum tensor of an electric field in a spherically symmetric spacetime*) We shall consider a spherically symmetric space with a charge at the centre of the coordinate system. Then there is a static, radial electric field in this spacetime, and the electromagnetic field tensor has only two non-vanishing components

$$F_{01} = -F_{10} = \frac{1}{c} E_r = \frac{Q}{4\pi\varepsilon_0 c r^2}. \tag{7.78}$$

In the present case the only non-vanishing components of the electromagnetic energy–momentum tensor (7.77) are

$$\mu_0 T_{\mu\,\mu} = g_{\mu 0} F_{\mu 1} F^{01} + g_{\mu 1} F_{\mu 0} F^{10} - \frac{1}{2} g_{\mu\,\mu} F_{01} F^{01}, \tag{7.79}$$

where we have summarized two equal terms in the last term due to the antisymmetry $F_{01} = -F_{10}$.

We use spherical coordinates so the angular components are

$$g_{\theta\theta} = r^2, \quad g_{\phi\phi} = r^2 \sin^2\theta \tag{7.80}$$

as given in Eq. (3.122). The non-vanishing components of the energy–momentum tensor are

$$T_{00} = \frac{1}{\mu_0} \left(g_{00} F_{01} F^{01} - \frac{1}{2} g_{00} F_{01} F^{01} \right) = \frac{1}{2\mu_0} g_{00} F_{01} F^{01} = \frac{Q^2}{32\pi^2\varepsilon_0} \frac{g_{00}}{r^4}, \tag{7.81}$$

$$T_{11} = \frac{1}{\mu_0} \left(g_{11} F_{01} F^{01} - \frac{1}{2} g_{11} F_{01} F^{01} \right) = \frac{1}{2\mu_0} g_{11} F_{01} F^{01} = \frac{Q^2}{32\pi^2\varepsilon_0} \frac{g_{11}}{r^4}, \tag{7.82}$$

$$T_{22} = \frac{1}{2\mu_0} g_{22} F_{01} F^{01} = \frac{1}{2\mu_0} r^2 F_{01} F^{01} = \frac{Q^2}{32\pi^2\varepsilon_0} \frac{1}{r^2}. \tag{7.83}$$

$$T_{33} = \frac{1}{2\mu_0} g_{33} F_{01} F^{01} = \frac{1}{2\mu_0} r^2 \sin^2\theta \, F_{01} F^{01} = \frac{Q^2}{32\pi^2\varepsilon_0} \frac{\sin^2\theta}{r^2}. \tag{7.84}$$

Exercises

7.1. *Newtonian approximation of perfect fluid*
    Let

$$T^{\alpha\beta} = p\,\eta^{\alpha\beta} + \left(\rho + p/c^2\right)u^\alpha u^\beta$$

be the components of the energy momentum tensor of a perfect fluid in flat
spacetime with Minkowski metric $\eta_{\mu\nu}$. Here p is the pressure and ρ the mass
density of the fluid, and u^α the components of its 4-velocity.

(a) Explain why the conservation law $T^{\alpha\beta}_{;\beta} = 0$ in this case reduces to $T^{\alpha\beta}_{,\beta} = 0$.
(b) We shall consider the Newtonian limit where p/c^2 can be neglected com-
     pared to ρ in the second term of $T^{\alpha\beta}$, and the components of the 4-velocity
     of the fluid are $u^\alpha \approx (c,\ \vec{v})$ where $\vec{v}$ is the ordinary velocity of a fluid
     element. Show that in this case the conservation law (a) implies mass
     conservation as represented by the equation of continuity,

$$\frac{\partial\rho}{\partial t} + \nabla\cdot(\rho\,\vec{v}) = 0,$$

(c) And momentum conservation as represented by the Euler equation of
     motion,

$$\rho\left(\frac{\partial\vec{v}}{\partial t} + (\vec{v}\cdot\nabla)\vec{v}\right) = -\nabla p.$$

7.2. *The energy–momentum tensor of LIVE*

(a) Show that the energy–momentum tensor of a Lorentz invariant medium
     is proportional to the metric tensor.
(b) Show that a Lorentz invariant perfect fluid has equation of state $p = -\rho c^2$.
(c) How is the density of Lorentz invariant vacuum energy, LIVE, related to
     the cosmological constant?

Chapter 8
Schwarzschild Spacetime

Abstract The Schwarzschild solution describing spacetime outside a spherical mass distribution is deduced. In this deduction we give a detailed prescription of how one calculates the components of Einstein's curvature tensor using differential forms as decomposed in an orthonormal basis. The predictions for the classical tests of Einstein's theory—gravitational frequency shift and time dilation, deflection of light passing the Sun and the perihelion shift of Mercury—are deduced. Finally the Reissner–Nordström solution describing spacetime outside a charged particle is deduced.

8.1 Schwarzschild's Exterior Solution

This is a solution of the vacuum field equations $E_{\mu\nu} = 0$ for a static spherically symmetric spacetime. One can then choose the following form of the line element (employing units so that $c = 1$):

$$ds^2 = -e^{2\alpha(r)}dt^2 + e^{2\beta(r)}dr^2 + r^2 d\Omega^2,$$
$$d\Omega^2 = d\theta^2 + \sin^2\theta d\phi^2. \tag{8.1}$$

These coordinates are chosen so that the area of a sphere with radius r is $4\pi r^2$. They are often called "curvature coordinates".

The physical distance in the radial direction, corresponding to a coordinate distance dr, is

$$dl_r = \sqrt{g_{rr}}dr = e^{\beta(r)}dr. \tag{8.2}$$

We shall now determine the components of the Einstein tensor by using the Cartan formalism and give the procedure in eight steps.

© Springer Nature Switzerland AG 2020
Ø. Grøn, *Introduction to Einstein's Theory of Relativity*,
Undergraduate Texts in Physics, https://doi.org/10.1007/978-3-030-43862-3_8

1. We first express the basis forms in orthonormal basis in terms of the coordinate basis forms,

$$\underline{\omega}^{\hat{t}} = e^{\alpha(r)}\underline{dt} \quad \underline{\omega}^{\hat{r}} = e^{\beta(r)}\underline{dr}, \quad \underline{\omega}^{\hat{\theta}} = r\underline{d\theta} \quad \underline{\omega}^{\hat{\phi}} = r\sin\theta\underline{d\phi}. \tag{8.3}$$

2. Then we compute the connection forms by applying Cartan's 1. structure equations

$$\underline{d\omega}^{\hat{\mu}} = -\underline{\Omega}^{\hat{\mu}}_{\hat{\nu}} \wedge \underline{\omega}^{\hat{\nu}}. \tag{8.4}$$

$$\begin{aligned}
\underline{d\omega}^{\hat{t}} &= e^{\alpha}\alpha'\underline{dr} \wedge \underline{dt} \\
&= e^{\alpha}\alpha' e^{-\beta}\underline{\omega}^{\hat{r}} \wedge e^{-\alpha}\underline{\omega}^{\hat{t}} \\
&= -e^{-\beta}\alpha'\underline{\omega}^{\hat{t}} \wedge \underline{\omega}^{\hat{r}} \\
&= -\underline{\Omega}^{\hat{t}}_{\hat{r}} \wedge \underline{\omega}^{\hat{r}} \tag{8.5}
\end{aligned}$$

Hence

$$\underline{\Omega}^{\hat{t}}_{\hat{r}} = e^{-\beta}\alpha'\underline{\omega}^{\hat{t}} + f_1\underline{\omega}^{\hat{r}}. \tag{8.6}$$

3. To determine the f-functions, we apply the antisymmetry

$$\underline{\Omega}_{\hat{\mu}\hat{\nu}} = -\underline{\Omega}_{\hat{\nu}\hat{\mu}}. \tag{8.7}$$

This gives the non-zero connection forms

$$\begin{aligned}
\underline{\Omega}^{\hat{r}}_{\hat{\phi}} &= -\underline{\Omega}^{\hat{\phi}}_{\hat{r}} = -\frac{1}{r}e^{-\beta}\underline{\omega}^{\hat{\phi}}, \\
\underline{\Omega}^{\hat{\theta}}_{\hat{\phi}} &= -\underline{\Omega}^{\hat{\phi}}_{\hat{\theta}} = -\frac{1}{r}\cot\theta\,\underline{\omega}^{\hat{\phi}}, \\
\underline{\Omega}^{\hat{t}}_{\hat{r}} &= +\underline{\Omega}^{\hat{r}}_{\hat{t}} = e^{-\beta}\alpha'\underline{\omega}^{\hat{t}}, \\
\underline{\Omega}^{\hat{r}}_{\hat{\theta}} &= -\underline{\Omega}^{\hat{\theta}}_{\hat{r}} = -\frac{1}{r}e^{-\beta}\underline{\omega}^{\hat{\theta}}.
\end{aligned} \tag{8.8}$$

4. We then proceed to determine the curvature forms by applying Cartan's 2. structure equations

$$\underline{R}^{\hat{\mu}}_{\hat{\nu}} = \underline{d\Omega}^{\hat{\mu}}_{\hat{\nu}} + \underline{\Omega}^{\hat{\mu}}_{\hat{\alpha}} \wedge \underline{\Omega}^{\hat{\alpha}}_{\hat{\nu}}, \tag{8.9}$$

which gives

$$R^{\hat{t}}_{\hat{r}} = -e^{-2\beta}(\alpha'' + \alpha'^2 - \alpha'\beta')\underline{\omega}^{\hat{t}} \wedge \underline{\omega}^{\hat{r}},$$

$$R^{\hat{t}}_{\hat{\theta}} = -\tfrac{1}{r}e^{-2\beta}\alpha'\underline{\omega}^{\hat{t}} \wedge \underline{\omega}^{\hat{\theta}},$$

$$R^{\hat{t}}_{\hat{\phi}} = -\tfrac{1}{r}e^{-2\beta}\alpha'\underline{\omega}^{\hat{t}} \wedge \underline{\omega}^{\hat{\phi}},$$

$$R^{\hat{r}}_{\hat{\theta}} = \tfrac{1}{r}e^{-2\beta}\beta'\underline{\omega}^{\hat{r}} \wedge \underline{\omega}^{\hat{\theta}}, \tag{8.10}$$

$$R^{\hat{r}}_{\hat{\phi}} = \tfrac{1}{r}e^{-2\beta}\beta'\underline{\omega}^{\hat{r}} \wedge \underline{\omega}^{\hat{\phi}},$$

$$R^{\hat{\theta}}_{\hat{\phi}} = \tfrac{1}{r^2}(1 - e^{-2\beta})\underline{\omega}^{\hat{\theta}} \wedge \underline{\omega}^{\hat{\phi}}.$$

5. By applying the relation

$$\underline{R}^{\hat{\mu}}_{\hat{\nu}} = \frac{1}{2}R^{\hat{\mu}}_{\hat{\nu}\hat{\alpha}\hat{\beta}}\underline{\omega}^{\hat{\alpha}} \wedge \underline{\omega}^{\hat{\beta}}, \tag{8.11}$$

we find the components of Riemann's curvature tensor.
6. Contraction gives the components of Ricci's curvature tensor,

$$R_{\hat{\mu}\hat{\nu}} \equiv R^{\hat{\alpha}}_{\hat{\mu}\hat{\alpha}\hat{\nu}}. \tag{8.12}$$

7. A new contraction gives Ricci's curvature scalar,

$$R \equiv R^{\hat{\mu}}_{\hat{\mu}}. \tag{8.13}$$

8. The components of the Einstein tensor can then be found,

$$E_{\hat{\mu}\hat{\nu}} = R_{\hat{\mu}\hat{\nu}} - \frac{1}{2}\eta_{\hat{\mu}\hat{\nu}}R, \tag{8.14}$$

where $\eta_{\hat{\mu}\hat{\nu}} = \mathrm{diag}(-1, 1, 1, 1)$. Hence

$$\begin{aligned}
E_{\hat{t}\hat{t}} &= \tfrac{2}{r}e^{-2\beta}\beta' + \tfrac{1}{r^2}(1 - e^{-2\beta}),\\
E_{\hat{r}\hat{r}} &= \tfrac{2}{r}e^{-2\beta}\alpha' - \tfrac{1}{r^2}(1 - e^{-2\beta}),\\
E_{\hat{\theta}\hat{\theta}} = E_{\hat{\phi}\hat{\phi}} &= e^{-2\beta}\left(\alpha'' + \alpha'^2 - \alpha'\beta' + \tfrac{\alpha'}{r} - \tfrac{\beta'}{r}\right).
\end{aligned} \tag{8.15}$$

We want to solve the equations $E_{\hat{\mu}\hat{\nu}} = 0$. In the present case there are only two independent equations,

$$E_{\hat{t}\hat{t}} = 0 \quad \text{and} \quad E_{\hat{r}\hat{r}} = 0. \tag{8.16}$$

Adding the two equations we get

$$\frac{2}{r}e^{-2\beta}(\beta' + \alpha') = 0. \tag{8.17}$$

Integration gives

$$\alpha + \beta = K_1,\tag{8.18}$$

where K_1 is an integration constant.

We now have

$$ds^2 = -e^{2\alpha}dt^2 + e^{2(K_1-\alpha)}dr^2 + r^2d\Omega^2.\tag{8.19}$$

Letting $r \to \infty$ the line element should describe flat spacetime in spherical coordinates with $\alpha = \beta = 0$. This requires that $K_1 = 0$, and hence, that $\beta = -\alpha$.

Since we have $ds^2 = -e^{2\alpha}dt^2 + e^{-2\alpha}dr^2 + r^2d\Omega^2$, this means that $g_{rr} = -1/g_{tt}$. We must solve one more equation to get the complete solution and choose the equation $E_{\hat{t}\hat{t}} = 0$, which gives

$$\frac{2}{r}e^{-2\beta}\beta' + \frac{1}{r^2}(1 - e^{-2\beta}) = 0.\tag{8.20}$$

This equation can be written as

$$\frac{1}{r^2}\frac{d}{dr}\left[r\left(1 - e^{-2\beta}\right)\right] = 0.\tag{8.21}$$

Integration gives

$$r(1 - e^{-2\beta}) = K_2.\tag{8.22}$$

If we choose $K_2 = 0$ we get $\beta = 0$ giving $\alpha = 0$ and

$$ds^2 = -dt^2 + dr^2 + r^2d\Omega^2,\tag{8.23}$$

which is the Minkowski spacetime described in spherical coordinates. In general, $K_2 \neq 0$ and $1 - e^{-2\beta} = \frac{K_2}{r} \equiv \frac{K}{r}$, giving

$$e^{2\alpha} = e^{-2\beta} = 1 - \frac{K}{r},\tag{8.24}$$

and

$$ds^2 = -\left(1 - \frac{K}{r}\right)dt^2 + \frac{dr^2}{1 - \frac{K}{r}} + r^2d\Omega^2.\tag{8.25}$$

We can find K by going to the Newtonian limit and compare with a purely Newtonian calculation. According to Newton's theory the acceleration of gravity at a distance r from a spherical mass M is

$$g = \frac{d^2 r}{dt^2} = -\frac{GM}{r^2}. \tag{8.26}$$

Let us now calculate the corresponding acceleration in the Newtonian limit of the general theory of relativity. In this limit the proper time τ of a particle will be approximately equal to the coordinate time t. The acceleration of a freely falling particle in 3-space is given by the geodesic equation,

$$\frac{d^2 x^\mu}{d\tau^2} + \Gamma^\mu_{\alpha\beta} u^\alpha u^\beta = 0. \tag{8.27}$$

For a particle instantaneously at rest in a weak field, we have $d\tau \approx dt$. Using $u^\mu = (1, 0, 0, 0)$, we get

$$g = \frac{d^2 r}{dt^2} = -\Gamma^r_{tt}. \tag{8.28}$$

This equation gives a physical interpretation of Γ^r_{tt} as the gravitational acceleration. This is a mathematical way to express the principle of equivalence: The gravitational acceleration can be transformed to 0 since the Christoffel symbols always can be transformed into 0 locally, by going into a freely falling non-rotating frame, i.e. a local inertial frame. In the Newtonian approximation we have

$$\Gamma^r_{tt} = \frac{1}{2} \underbrace{g^{r\alpha}}_{\frac{1}{g_{r\alpha}}} \left(\underbrace{\frac{\partial g_{\alpha t}}{\partial t}}_{=0} + \underbrace{\frac{\partial g_{\alpha t}}{\partial t}}_{=0} - \frac{\partial g_{tt}}{\partial r^\alpha} \right) = -\frac{1}{2 g_{rr}} \frac{\partial g_{tt}}{\partial r}. \tag{8.29}$$

Inserting

$$g_{tt} = -\left(1 - \frac{K}{r}\right), \qquad \frac{\partial g_{tt}}{\partial r} = -\frac{K}{r^2} \tag{8.30}$$

into Eq. (8.29) and then calculating the acceleration of gravity from Eq. (8.28) we get

$$g = -\Gamma^r_{tt} = -\frac{K}{2r^2}. \tag{8.31}$$

Comparing with the Newtonian expression for the acceleration of gravity, Eq. (8.26), and inserting the velocity of light, then lead to

$$K = \frac{2GM}{c^2}. \tag{8.32}$$

Then we have the line element of *the exterior Schwarzschild metric*,

$$ds^2 = -\left(1 - \frac{2GM}{c^2 r}\right)c^2 dt^2 + \frac{dr^2}{1 - \frac{2GM}{c^2 r}} + r^2 d\Omega^2. \qquad (8.33)$$

We now introduce the Schwarzschild radius of a mass M,

$$R_S \equiv \frac{2GM}{c^2}. \qquad (8.34)$$

Hence, the line element of the exterior Schwarzschild spacetime as expressed in curvature coordinates, takes the form

$$ds^2 = -\left(1 - \frac{R_S}{r}\right)c^2 dt^2 + \frac{dr^2}{1 - \frac{R_S}{r}} + r^2 d\Omega^2. \qquad (8.35)$$

The Schwarzschild radius of the Earth is $R_S \sim 0.9$ cm and of the Sun, $R_S \sim 3$ km.

Far from a localized mass distribution the gravitational field is weak. The definition is that in a region where $r \gg R_S$, there is a *weak gravitational field*.

A standard clock at rest in the Schwarzschild spacetime shows a proper time τ:

$$d\tau = \sqrt{1 - \frac{R_S}{r}} dt. \qquad (8.36)$$

It follows from the time independence of the metric that *the coordinate clocks are adjusted to go at the same rate independent of their position*. Hence Eq. (8.36) shows that *the rate of proper time is slower for decreasing value of r, i.e. farther down in the gravitational field*. Time is not running at the Schwarzschild radius.

Definition 8.1.1 (*Physical singularity*) A physical singularity is a point where the curvature is infinitely large.

Definition 8.1.2 (*Coordinate singularity*) A coordinate singularity is a point (or a surface) where at least one of the components of the metric tensor is infinitely large, but where the curvature of spacetime is finite.

The Kretschmann's curvature scalar is $R_{\mu\nu\alpha\beta} R^{\mu\nu\alpha\beta}$. From the Schwarzschild metric, we get

$$R_{\mu\nu\alpha\beta} R^{\mu\nu\alpha\beta} = 12\frac{R_S^2}{r^6}, \qquad (8.37)$$

which diverges only at the origin. Since there is no physical singularity at $r = R_S$, the singularity here is just a coordinate singularity and can be removed by a transformation to a coordinate system falling inwards (Eddington–Finkelstein coordinates, Kruskal–Szekeres analytical extension of the description of Schwarzschild spacetime to include the region inside R_S).

8.2 Radial Free Fall in Schwarzschild Spacetime

The Lagrangian function of a particle moving radially in Schwarzschild spacetime is

$$L = -\frac{1}{2}\left(1 - \frac{R_S}{r}\right)c^2\dot{t}^2 + \frac{1}{2}\frac{\dot{r}^2}{\left(1 - \frac{R_S}{r}\right)}, \quad \bullet \equiv \frac{d}{d\tau}, \tag{8.38}$$

where τ is the time measured on a standard clock which the particle is carrying. The momentum p_t conjugate to the cyclic coordinate t is a constant of motion,

$$p_t = \frac{\partial L}{\partial \dot{t}} = -\left(1 - \frac{R_S}{r}\right)c^2\dot{t}. \tag{8.39}$$

The 4-velocity identity, $u_\mu u^\mu = -c^2$, takes the form

$$-\left(1 - \frac{R_S}{r}\right)c^2\dot{t}^2 + \frac{\dot{r}^2}{1 - \frac{R_S}{r}} = -c^2. \tag{8.40}$$

Inserting the expression for $\dot{t}$ from Eq. (8.39) gives

$$\dot{r}^2 - \frac{p_t^2}{c^2} = -\left(1 - \frac{R_S}{r}\right)c^2. \tag{8.41}$$

The boundary condition is that the particle is falling from rest at $r = r_0$, giving

$$\frac{p_t^2}{c^2} = \left(1 - \frac{R_S}{r_0}\right)c^2. \tag{8.42}$$

Inserting this into Eq. (8.41) gives

$$\dot{r} = \frac{dr}{d\tau} = -\sqrt{\frac{p_t^2}{c^2} - \left(1 - \frac{R_S}{r}\right)c^2} = -\sqrt{R_S}\sqrt{\frac{1}{r} - \frac{1}{r_0}}. \tag{8.43}$$

With the initial condition $\tau(r_0) = 0$ we have

$$\int_{r_0}^{r} \sqrt{\frac{r}{r_0 - r}}\,dr = -c\sqrt{\frac{R_S}{r_0}}\,\tau. \tag{8.44}$$

Performing the integration gives

$$\tau = \frac{r_0^{3/2}}{c\sqrt{R_S}}\left[\frac{\pi}{2} + \sqrt{\frac{r}{r_0}\left(1 - \frac{r}{r_0}\right)} - \arcsin\sqrt{\frac{r}{r_0}}\right].$$
(8.45)

The proper time that a particle spends falling from r_0 to R_S is

$$\tau = \frac{r_0^{3/2}}{c\sqrt{R_S}}\left[\frac{\pi}{2} + \sqrt{\frac{R_S}{r_0}\left(1 - \frac{R_S}{r_0}\right)} - \arcsin\sqrt{\frac{R_S}{r_0}}\right],$$
(8.46)

which is finite. Assuming that $r_0 \gg R_S$ and calculating to 1. order in R_S/r_0 we obtain

$$\tau(R_S) \approx \frac{\pi}{2}\frac{r_0}{c}\sqrt{\frac{r_0}{R_S}}.$$
(8.47)

Let us calculate the corresponding travelling time as measured by a stationary observer. From Eqs. (7.39) and (7.42) we have

$$dt = \frac{\sqrt{1 - \frac{R_S}{r_0}}}{1 - \frac{R_S}{r}}d\tau.$$
(8.48)

From (8.43) we have

$$d\tau = -\sqrt{\frac{r_0 r}{R_S(r_0 - r)}}dr.$$
(8.49)

Hence the time taken by the particle to fall down to the Schwarzschild radius, as measured by the stationary observer, is

$$t(R_S) = \sqrt{\frac{1}{R_S} - \frac{1}{r_0}}\int_{R_S}^{r_0}\frac{r^{3/2}}{(r - R_S)\sqrt{r_0 - r}}dr,$$
(8.50)

which diverges. Hence it takes an infinitely long coordinate time to fall down to the Schwarzschild radius. But the proper time of the falling object is finite.

8.3 Light Cones in Schwarzschild Spacetime

The Schwarzschild line element (with $c = 1$) is

$$ds^2 = -\left(1 - \frac{R_S}{r}\right)dt^2 + \frac{dr^2}{\left(1 - \frac{R_S}{r}\right)} + r^2 d\Omega^2.$$
(8.51)

We will look at *radially moving photons*. Then $ds^2 = d\Omega^2 = 0$, giving

$$\frac{r^{\frac{1}{2}}dr}{\sqrt{r - R_S}} = \pm\frac{\sqrt{r - R_S}}{r^{\frac{1}{2}}}dt, \tag{8.52}$$

or

$$\frac{rdr}{r - R_S} = \pm dt \tag{8.53}$$

with $+$ for outward motion and $-$ for inward motion. For inwardly moving photons, integration yields

$$r + t + R_S \ln\left|\frac{r}{R_S} - 1\right| = k = \text{constant.} \tag{8.54}$$

We now introduce a new time coordinate t' such that the equation of motion for photons moving *inwards* takes the form

$$r + t' = k. \tag{8.55}$$

Hence in this coordinate system the coordinate velocity of light is equal to the invariant velocity of light,

$$\frac{dr}{dt'} = -1. \tag{8.56}$$

It follows from Eqs. (8.54) and (8.55) that

$$t' = t + R_S \ln\left|\frac{r}{R_S} - 1\right|. \tag{8.57}$$

The coordinate t' is called an *ingoing Eddington–Finkelstein coordinate*. For photons moving *outwards* we have

$$r + R_S \ln\left|\frac{r}{R_S} - 1\right| = t + k. \tag{8.58}$$

Substituting for t from Eq. (8.57) we get

$$r + 2R_S \ln\left|\frac{r}{R_S} - 1\right| = t' + k. \tag{8.59}$$

Differentiating we find the coordinate velocity of outgoing light in this coordinate system,

$$\frac{dr}{dt'} = \frac{r - R_S}{r + R_S}.$$ (8.60)

Making use of curvature coordinates we get the following coordinate velocities for inwardly and outwardly moving photons,

$$v_L = \frac{dr}{dt} = \pm\left(1 - \frac{R_S}{r}\right),$$ (8.61)

which shows how light is decelerated in a gravitational field. Figure 8.1 illustrates how this is viewed by a non-moving observer located far away from the mass. In Fig. 8.2 we have instead used the time coordinate t' of the ingoing Eddington–Finkelstein coordinate system.

Note that since the special theory of relativity is valid locally, all material particles have world lines inside the light cone formed by the light they emit. From the shape of the light cone inside $r = R_S$ we see that nothing emitted from a position inside the spherical surface $r = R_S$ can escape from this region. An observer outside this surface cannot see anything from the inside region. Hence this surface is a *horizon* for an external observer.

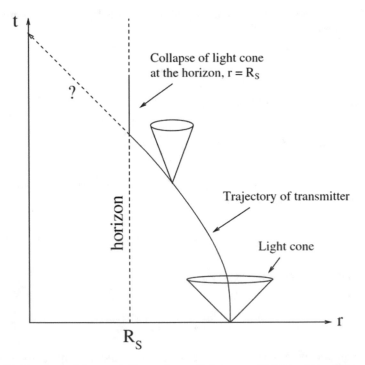

Fig. 8.1 Light cones in Schwarzschild spacetime with Schwarzschild time. At a radius $r = R_S$ the light cones collapse, and nothing can any longer escape, when we use the Schwarzschild coordinate time

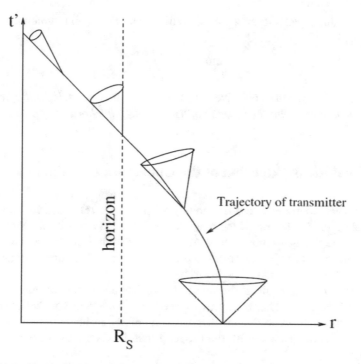

Fig. 8.2 Light cones in Schwarzschild spacetime with Eddington-Finkelstein time. Using the ingoing Eddington–Finkelstein time coordinate there is no collapse of the light cone at $r = R_S$. Instead we get a collapse at the singularity at $r = 0$. The angle between the left part of the light cone and the t'-axis is always 45°. We also see that once the emitter gets inside the horizon at $r - R_S$, nothing can escape

The region inside the horizon is called a *black hole* since it is invisible because radiation cannot come out of this region according to classical (non-quantum mechanical) general relativity. A time-reversed black hole is called a *white hole*.

Let us compare the coordinate velocity with curvature coordinates of a particle falling vertically with the velocity of light moving vertically. It follows from Eqs. (8.48) and (8.49) that a particle falling from rest at $r = r_0$ has a coordinate velocity

$$v = \frac{dr}{dt} = \frac{\dot{r}}{\dot{t}} = -\left(1 - \frac{R_S}{r}\right)\sqrt{\frac{R_S(r_0 - r)}{(r_0 - R_S)r}}. \tag{8.62}$$

Comparing with the coordinate velocity of light as given in Eq. (8.61) we see that

$$v = \sqrt{\frac{R_S(r_0 - r)}{(r_0 - R_S)r}} v_L. \tag{8.63}$$

The coordinate velocity of a particle falling from rest at an infinitely far position
is

$$\lim_{r \to \infty} v = v_L \sqrt{\frac{R_S}{r}}. \tag{8.64}$$

Both Eqs. (8.63) and (8.64) give $v(R_S) = v_L$. Hence, *a particle falling freely from
any distance moves with the velocity of light through the horizon of a black hole.*

8.4   Analytical Extension of the Curvature Coordinates

The curvature coordinates are co-moving with a static reference frame outside a
spherical mass distribution. If the mass has collapsed to a black hole, there exists a
horizon at the Schwarzschild radius. As we have seen in Sect. 8.3 there do not exist
static observers at finite radii inside the horizon. Hence, the curvature coordinates
are well defined only outside the horizon.

Also the rr-component of the metric tensor has a coordinate singularity at the
Schwarzschild radius. The curvature of spacetime is finite here. Kruskal and Szekeres
have introduced new coordinates that are well defined inside as well as outside the
Schwarzschild radius, and with the property that the metric tensor is non-singular
for all $r > 0$.

In order to arrive at these coordinates we start by considering a photon moving
radially inwards. From Eq. (8.54) we then have

$$t = -r - R_S \ln \left| \frac{r}{R_S} - 1 \right| + v, \tag{8.65}$$

where v is a constant along the world line of the photon. We introduce a new radial
coordinate

$$r^* \equiv r + R_S \ln \left| \frac{r}{R_S} - 1 \right|. \tag{8.66}$$

Then the equation of the world line of the photon takes the form

$$t + r^* = v. \tag{8.67}$$

The value of the constant v does only depend upon the point of time when the
photon was emitted. We may therefore use v as a new time coordinate.

For an outgoing photon we get in the same way

$$t - r^* = u, \tag{8.68}$$

where u is a constant of integration, which may be used as a new time coordinate for outgoing photons. The coordinates u and v are the generalization of the *light cone coordinates* of Minkowski spacetime to the Schwarzschild spacetime.

From Eqs. (8.67) and (8.68) we get

$$dt = \frac{1}{2}(dv + du),\tag{8.69}$$

$$dr^* = \frac{1}{2}(dv - du),\tag{8.70}$$

and from Eq. (8.66),

$$dr = \left(1 - \frac{R_s}{r}\right)dr^*.\tag{8.71}$$

Inserting these differentials into Eq. (8.51) we arrive at a new form of the Schwarzschild line element,

$$ds^2 = -\left(1 - \frac{R_s}{r}\right)du\,dv + r^2 d\Omega^2.\tag{8.72}$$

The metric is still not well behaved at the horizon. Kruskal and Szekeres found coordinates that are well behaved at the horizon.

Introducing the coordinates

$$U = -e^{-\frac{u}{2R_s}},\tag{8.73}$$

$$V = e^{\frac{v}{2R_s}},\tag{8.74}$$

gives

$$UV = -e^{\frac{v-u}{2R_s}} = -e^{\frac{r^*}{R_s}} = -\left|\frac{R_s}{r} - 1\right|e^{\frac{r}{R_s}}\tag{8.75}$$

and

$$du\,dv = -4R_s^2\frac{dU\,dV}{UV}.\tag{8.76}$$

The line element (8.72) then takes the form

$$ds^2 = -\frac{4R_s^3}{r}e^{-\frac{r}{R_s}}dU\,dV + r^2 d\Omega^2.\tag{8.77}$$

This is the first form of the Kruskal–Szekeres line element. Here there is no coordinate singularity, only a physical singularity at $r = 0$.

We may furthermore introduce two new coordinates:

$$T = \frac{1}{2}(V + U) = \left|\frac{r}{R_s} - 1\right|^{\frac{1}{2}} e^{\frac{r}{2R_s}} \sinh \frac{t}{2R_s}, \tag{8.78}$$

$$Z = \frac{1}{2}(V - U) = \left|\frac{r}{R_s} - 1\right|^{\frac{1}{2}} e^{\frac{r}{2R_s}} \cosh \frac{t}{2R_s}. \tag{8.79}$$

Hence

$$V = T + Z, \tag{8.80}$$

$$U = T - Z, \tag{8.81}$$

giving

$$dU dV = dT^2 - dZ^2. \tag{8.82}$$

Inserting this into Eq. (8.72) we arrive at the second form of the Kruskal–Szekeres line element

$$ds^2 = -\frac{4R_s^3}{r} e^{-\frac{r}{R_s}} \left(dT^2 - dZ^2\right) + r^2 d\Omega^2. \tag{8.83}$$

The inverse transformations of Eqs. (8.74) and (8.75) are

$$\left|\frac{r}{R_s} - 1\right| e^{\frac{r}{R_s}} = Z^2 - T^2, \tag{8.84}$$

$$\tanh \frac{t}{2R_s} = \frac{T}{Z}. \tag{8.85}$$

Note from Eq. (8.83) that with the Kruskal–Szekeres coordinates T and Z the equation of the radial null geodesics has the same form as in flat spacetime:

$$Z = \pm T + \text{constant.} \tag{8.86}$$

8.5 Embedding of the Schwarzschild Metric

We will now look at a static, spherically symmetric space. A simultaneity surface $dt = 0$ through the equatorial plane, $d\theta = 0$, has the line element

$$ds^2 = g_{rr}dr^2 + r^2 d\phi^2 \tag{8.87}$$

with a radial coordinate such that a circle with radius r has a circumference of length $2\pi r$.

Now we embed this surface in a flat three-dimensional space with cylinder coordinates (z, r, ϕ) and line element

$$ds^2 = dz^2 + dr^2 + r^2 d\phi^2. \tag{8.88}$$

The surface described by the line element in Eq. (8.87) has the equation $z = z(r)$. The line element in (8.88) can therefore be written as

$$ds^2 = \left[1 + \left(\frac{dz}{dr}\right)^2\right]dr^2 + r^2 d\phi^2. \tag{8.89}$$

Demanding that (8.89) is in agreement with (8.87) we get

$$g_{rr} = 1 + \left(\frac{dz}{dr}\right)^2, \tag{8.90}$$

or

$$\frac{dz}{dr} = \pm\sqrt{g_{rr} - 1}. \tag{8.91}$$

Choosing the positive solution gives

$$dz = \sqrt{g_{rr} - 1}\,dr. \tag{8.92}$$

In the Schwarzschild spacetime we have

$$g_{rr} = \frac{1}{1 - \frac{R_S}{r}}. \tag{8.93}$$

Making use of this we find the equation of the intersection of the simultaneity surface $dt = 0$ through the equatorial plane, $d\theta = 0$, and the paper plane,

$$z = \int_{R_S}^{r} \frac{dr}{\sqrt{\frac{r}{R_S} - 1}} = \sqrt{4R_S(r - R_S)}. \tag{8.94}$$

Fig. 8.3 Embedding of the
extended Schwarzschild
spacetime. It represents a
worm-hole connecting a
black and a white hole

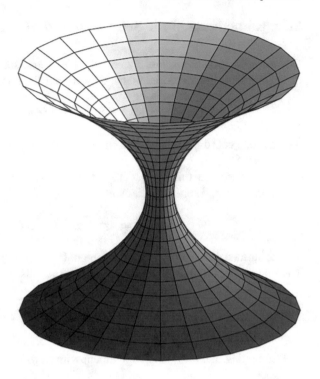

Rotating this about the symmetry axis and including negative values of z give the
surface shown in Fig. 8.3, which consists of two so-called Flamm paraboloids glued
together at the Schwarzschild radius.

8.6 The Shapiro Experiment

The radial speed of light in curvature coordinates as measured by an observer far
from a mass distribution, say the Sun, is given in Eq. (8.61). The formula shows that
the speed of light slows down farther down in a gravitational field.

To measure this effect, one can look at how long it takes for light to get from, for
example, Mercury (Shapiro used Venus in the first experiment in 1967) to the Earth
[1]. This is illustrated in Fig. 8.4.

The travel time from z_1 to z_2 is

$$\Delta t = \int_{z_1}^{z_2} \frac{dz}{1 - \frac{R_S}{r}} \approx \int_{z_1}^{z_2} \left(1 + \frac{R_S}{r}\right) dz = \int_{z_1}^{z_2} \left(1 + \frac{R_S}{\sqrt{b^2 + z^2}}\right) dz$$

$$= z_2 + |z_1| + R_S \ln \frac{\sqrt{z_2^2 + b^2} + z_2}{\sqrt{z_1^2 + b^2} - |z_1|}, \tag{8.95}$$

Fig. 8.4 The Shapiro experiment. General relativity predicts that light travelling from Mercury to the Earth will be delayed due to the effect of the Sun's gravity field on the speed of light. This effect has been measured by Shapiro et al. [1] and is therefore called the *Shapiro effect*

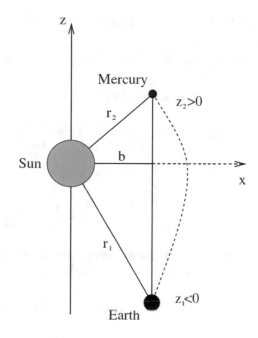

where R_S is the Schwarzschild radius of the Sun.

The deceleration is greatest when Earth and Mercury (where the light is reflected) are on nearly opposite sides of the Sun. The impact parameter b is then small. A series expansion to the lowest order of b/z gives

$$\Delta t = z_2 + |z_1| + R_S \ln \frac{4|z_1||z_2|}{b^2}. \tag{8.96}$$

The last term represents the extra travelling time due to the effect of the Sun's gravity field on the speed of light. The main effect is slowing down of the velocity of light farther down in the gravitational field of the Sun. (Also the path is a little longer because of the bending due to gravity, but the delay due to this is smaller than the velocity effect and has been neglected in the calculation). Let us insert the magnitudes of the quantities in Eq. (8.96) and calculate the magnitude of the Shapiro effect.

R_S = the Schwarzschild radius of the Sun = ~ 3 km
$|z_1|$ = the radius of Earth's orbit = 15×10^{10} m
z_2 = the radius of Mercury's orbit = 5.8×10^{10} m
$b = R_\odot = 7 \times 10^8$ m

give a delay of 1.1×10^{-4}s. In addition to this one must also take into account the effect of the Earth's atmosphere upon the travelling time of the light.

8.7 Particle Trajectories in Schwarzschild 3-Space

The Lagrange function of a free particle in the Schwarzschild spacetime is

$$L = \frac{1}{2} g_{\mu\nu} \dot{X}^{\mu} \dot{X}^{\nu}$$

$$= -\frac{1}{2}\left(1 - \frac{R_s}{r}\right)\dot{t}^2 + \frac{\frac{1}{2}\dot{r}^2}{1 - \frac{R_s}{r}} + \frac{1}{2}r^2\dot{\theta}^2 + \frac{1}{2}r^2\sin^2\theta\dot{\phi}^2. \qquad (8.97)$$

Since t is a cyclic coordinate

$$-p_t = -\frac{\partial L}{\partial \dot{t}} = \left(1 - \frac{R_s}{r}\right)\dot{t} = \text{constant} = E, \qquad (8.98)$$

where E is the particle's energy per unit rest mass as measured by an observer "far away" ($r \gg R_s$). Also ϕ is a cyclic coordinate so that

$$p_\phi = \frac{\partial L}{\partial \dot{\phi}} = r^2 \sin^2\theta \, \dot{\phi} = \text{constant}, \qquad (8.99)$$

where p_ϕ is the particle's orbital angular momentum per unit rest mass, with units so that $c = 1$ the angular velocity $\dot{\phi}$ has dimension length^{-1}, and p_ϕ has dimension length.

In the present case the 4-velocity identity $\vec{U}^2 = g_{\mu\nu}\dot{X}^{\mu}\dot{X}^{\nu} = -1$ takes the form

$$-\left(1 - \frac{R_s}{r}\right)\dot{t}^2 + \frac{\dot{r}^2}{1 - \frac{R_s}{r}} + r^2\dot{\theta}^2 + r^2\sin^2\theta\dot{\phi}^2 = -1, \qquad (8.100)$$

which on substitution for $\dot{t} = \frac{E}{1 - \frac{R_s}{r}}$ and $\dot{\phi} = \frac{p_\phi}{r^2 \sin^2\theta}$ becomes

$$-\frac{E^2}{1 - \frac{R_s}{r}} + \frac{\dot{r}^2}{1 - \frac{R_s}{r}} + r^2\dot{\theta}^2 + \frac{p_\phi^2}{r^2 \sin^2\theta} = -1. \qquad (8.101)$$

Now, referring back to the Lagrange equation

$$\frac{d}{d\tau}\left(\frac{\partial L}{\partial \dot{X}^{\mu}}\right) - \frac{\partial L}{\partial X^{\mu}} = 0 \qquad (8.102)$$

we get

$$(r^2\dot{\theta})^{\cdot} = r^2 \sin\theta \cos\theta\dot{\phi}^2 = \frac{p_\phi^2 \cos\theta}{r^2 \sin^3\theta}. \qquad (8.103)$$

Multiplying this by $r^2 \dot{\theta}$ leads to

$$(r^2 \dot{\theta})(r^2 \dot{\theta})^{\cdot} = \frac{\cos \theta \dot{\theta}}{\sin^3 \theta} p_\phi^2, \tag{8.104}$$

which, on integration, gives

$$(r^2 \dot{\theta})^2 = k - \left(\frac{p_\phi}{\sin \theta} \right)^2, \tag{8.105}$$

where k is the constant of integration.

Because of the spherical geometry we are free to choose a coordinate system such that the particle moves in the equatorial plane and along the equator at a given time $t = 0$. That is $\theta = \pi/2$ and $\dot{\theta} = 0$ at time $t = 0$. This determines the constant of integration, giving $k = p_\phi^2$ such that

$$(r^2 \dot{\theta})^2 = p_\phi^2 \left(1 - \frac{1}{\sin^2 \theta} \right). \tag{8.106}$$

The right-hand side is negative for all $\theta \neq \pi/2$. It follows that the particle cannot deviate from its original (equatorial) trajectory. Also, since this particular choice of trajectory was arbitrary, we can conclude, quite generally, that *any motion of free particles in a spherically symmetric gravitational field is planar.*

8.7.1 Motion in the Equatorial Plane

We now consider motion in the equatorial plane. With $\theta = \pi/2$ Eq. (8.101) reduces to

$$-\frac{E^2}{1 - \frac{R_S}{r}} + \frac{\dot{r}^2}{1 - \frac{R_S}{r}} + \frac{p_\phi^2}{r^2} = -1, \tag{8.107}$$

that is

$$\dot{r}^2 = E^2 - \left(1 - \frac{R_S}{r} \right) \left(1 + \frac{p_\phi^2}{r^2} \right), \tag{8.108}$$

or

$$\dot{r}^2 = E^2 - 1 + \frac{R_S}{r} - \frac{p_\phi^2}{r^2} + \frac{R_S p_\phi^2}{r^3}. \tag{8.109}$$

A particle falling from rest infinitely far from the mass distribution, $\dot{r}(\infty) = 0$, has $E = 1$, i.e. its energy at the starting point is equal to it rest mass (when the units are chosen so that $c = 1$). Hence the energy equation takes the form

$$\frac{1}{2}\dot{r}^2 + V_R = 0 \qquad (8.110)$$

with a relativistic effective potential

$$V_R = -\frac{GM}{r} + \frac{p_\phi^2}{2r^2} - \frac{GMp_\phi^2}{r^3}, \qquad (8.111)$$

where we have used that $R_S = 2\,GM$. Note that the Newtonian "mechanical energy" of a free particle falling from rest at the zero level of the effective potential vanishes, since the rest mass energy is not included in this energy. It is, however, included in the total relativistic energy E.

The Newtonian potential V_N is

$$V_N = -\frac{GM}{r} + \frac{p_\phi^2}{2r^2}. \qquad (8.112)$$

Hence the last term in Eq. (8.111) is a relativistic effect. The potential (8.112) is plotted in Fig. 8.5.

The Newtonian potential has a centrifugal barrier preventing a particle with $p_\phi \neq 0$ to arrive at the origin, $r = 0$.

The relativistic potential (8.111) is compared to the Newtonian potential in Fig. 8.6.

The centrifugal barrier is not infinitely high according to the theory of relativity as it is in Newton's theory. This means that *relativistic gravity is stronger than Newtonian gravity*. The reason is that the increase of the moving particle's kinetic energy gives it a larger mass, and hence, increases the gravity.

Fig. 8.5 Newtonian centrifugal barrier. Newtonian potential as function of the radius. Note the *centrifugal barrier*. Due to this, particles with $p_\phi \neq 0$ cannot arrive at $r = 0$

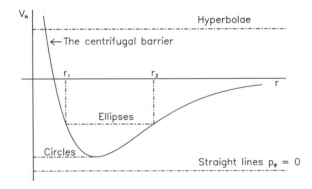

Fig. 8.6 Relativistic gravitational potential outside a spherical body. V_R and V_N plotted as a function of r. When relativistic effects are included, there is no longer an infinitely high potential barrier, and a particle with $p_\phi \neq 0$ can fall down to $r = 0$

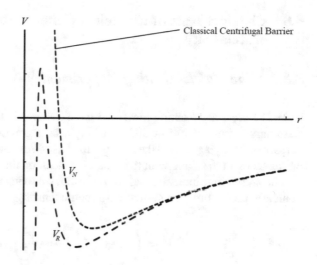

An **orbit equation** is one which connects r and ϕ. For motion in the equatorial plane for weak fields we have

$$\frac{d\phi}{dt} = \frac{p_\phi}{mr^2}, \quad \bullet \equiv \frac{d}{dt} = \frac{p_\phi}{mr^2}\frac{d}{d\phi}. \tag{8.113}$$

Introducing the new radial coordinate $u \equiv 1/r$ we get

$$\frac{du}{d\phi} = -\frac{1}{r^2}\frac{dr}{d\phi} = -\frac{1}{r^2}\frac{mr^2}{p_\phi}\frac{dr}{dt} = -\frac{m}{p_\phi}\dot{r}. \tag{8.114}$$

Hence,

$$\dot{r} - -\frac{p_\phi}{m}\frac{du}{d\phi}. \tag{8.115}$$

Substitution from Eq. (8.115) for $\dot{r}$ in the energy equation yields the orbit equation

$$\left(\frac{du}{d\phi}\right)^2 + (1 - 2GMu)\left(u^2 + \frac{1}{p_\phi^2}\right) = \frac{E^2}{p_\phi^2}. \tag{8.116}$$

Differentiating this, we find

$$\frac{d^2u}{d\phi^2} + u = \frac{GM}{p_\phi^2} + 3GMu^2 = \frac{R_S}{2p_\phi^2} + \frac{3}{2}R_Su^2. \tag{8.117}$$

The last term on the right-hand side is a relativistic correction term.

8.8 Classical Tests of Einstein's General Theory of Relativity

8.8.1 The Hafele–Keating Experiment

In 1971 Hafele and Keating measured the difference in time shown on moving and stationary atomic clocks at different height in a gravitational field [2]. This was performed by flying around the Earth in the East–West direction, comparing the time on the clock in the plane with the time on a clock on the ground.

The proper time interval measured on a clock moving with a velocity $v^i = dx^i/dt$ in an arbitrary coordinate system with metric tensor $g_{\mu\nu}$ is

$$d\tau = \left(-\frac{g_{\mu\nu}}{c^2} dx^\mu dx^\nu\right)^{\frac{1}{2}} = \left(-g_{00} - 2g_{i0}\frac{v^i}{c} - \frac{v^2}{c^2}\right)^{\frac{1}{2}} dt, \quad v^2 = g_{ij}v^i v^j. \quad (8.118)$$

For a diagonal metric tensor, $g_{i0} = 0$, we get

$$d\tau = \left(-g_{00} - \frac{v^2}{c^2}\right)^{\frac{1}{2}} dt, \quad v^2 = g_{ii}(v^i)^2. \quad (8.119)$$

We now look at an idealized situation where a plane flies at constant altitude h and with constant speed along the equator,

$$d\tau = \left(1 - \frac{R_S}{r} - \frac{v^2}{c^2}\right)^{\frac{1}{2}} dt, \quad r = R + h. \quad (8.120)$$

where R is the radius of the Earth. To lowest order in R_S/r and v^2/c^2, we get

$$d\tau = \left(1 - \frac{R_S}{2r} - \frac{1}{2}\frac{v^2}{c^2}\right) dt. \quad (8.121)$$

The speed of the moving clock is

$$v = (R + h)\Omega + u, \quad (8.122)$$

where Ω is the angular velocity of the Earth and u is the speed of the plane. A series expansion and use of the expression (8.122) for v give

$$\Delta\tau = \left(1 - \frac{GM}{Rc^2} - \frac{1}{2}\frac{R^2\Omega^2}{c^2} + \frac{gh}{c^2} - \frac{2R\Omega u + u^2}{2c^2}\right)\Delta t, \quad g = \frac{GM}{R^2} - R\Omega^2. \quad (8.123)$$

$u > 0$ when flying in the direction of the Earth's rotation, i.e. eastwards.

Let us find the velocity of a clock with maximal rate of ageing. Differentiating $\Delta\tau$ with respect to u gives

$$\frac{\mathrm{d}\Delta\tau}{\mathrm{d}u} = -\frac{R\Omega + u}{c^2}\Delta t. \tag{8.124}$$

Putting the derivative equal to zero shows that the maximal rate of ageing happens for a clock moving along the equator on the Earth with a velocity $u = -R\Omega$. This may be understood as a consequence of the fact that this clock is at rest in a non-rotating reference frame.

For the clock that was left on the airport (stationary, $h = u = 0$) we get

$$\Delta\tau_0 = \left(1 - \frac{GM}{Rc^2} - \frac{1}{2}\frac{R^2\Omega^2}{c^2}\right)\Delta t. \tag{8.125}$$

To lowest order, the relative difference in travel time is

$$k = \frac{\Delta\tau - \Delta\tau_0}{\Delta\tau_0} \cong \frac{gh}{c^2} - \frac{2R\Omega u + u^2}{2c^2}. \tag{8.126}$$

Inserting approximate values,
$g = 10$ m/s^2, $h = 10$ km, $R = 6400$ km, $\Omega = 7.27 \times 10^{-5}$ rad/s, $u = \pm 250$ m/s gives a

Travel time: $\Delta\tau_0 = 1.2 \times 10^5$ s (a little over 24 h);
For travelling eastwards: $k_e = -1.0 \times 10^{-12}$;
For travelling westwards: $k_w = 2.1 \times 10^{-12}$.

Hence, $(\Delta\tau - \Delta\tau_0)_e = -1.2 \times 10^{-7}$ s ≈ -120 ns and $(\Delta\tau - \Delta\tau_0)_w = 2.5 \times 10^{-7}$ s ≈ 250 ns.

These relativistic predictions were verified with about 15% accuracy in the experiment.

8.8.2 Mercury's Perihelion Precession

We shall now calculate Mercury's perihelion shift. The point of departure is the orbit Eq. (8.117). This will here be slightly generalized to

$$\frac{\mathrm{d}^2u}{\mathrm{d}\phi^2} + u = \frac{R_S}{2p_\phi^2} + ku^2, \tag{8.127}$$

where k is a theory-dependent or situation-dependent constant. The general theory of relativity gives $k = (3/2)R_S$, and Newton's theory of gravitation gives $k = 0$.

Here $p_\phi = r^2 d\phi/d\tau$. For light $d\tau \to 0$ and $p_\phi \to \infty$. Hence the orbit equation for light reduces to

$$\frac{d^2u}{d\phi^2} + u = ku^2. \tag{8.128}$$

The orbit equation for a material particle has a circular solution with an inverse radius fulfilling

$$u_0 = \frac{R_S}{2p_\phi^2} + ku_0^2. \tag{8.129}$$

The corresponding Newtonian radius is

$$u_{0N} = \frac{R_S}{2p_\phi^2}. \tag{8.130}$$

We shall now apply the procedure used in stability analysis to calculate the perihelion precession of Mercury. The equilibrium solution will be perturbed, and the equation of motion of the perturbation is calculated to 1. order in the perturbation. This equation will tell whether the equilibrium solution is stable or not.

Hence, the circular motion is perturbed so that $u = u_0 + u_1$ with $u_1 \ll u_0$. To 1. order in u_1, we have

$$\frac{d^2u_1}{d\phi^2} + u_0 + u_1 = \frac{R_S}{2p_\phi^2} + ku_0^2 + 2ku_0u_1. \tag{8.131}$$

Using the equation for the circular solution we get the equation of motion for the perturbation

$$\frac{d^2u_1}{d\phi^2} + (1 - 2ku_0)u_1 = 0. \tag{8.132}$$

If $1 - ku_0 > 0$, this is the equation of harmonic oscillations, and the circular motion is stable. If $1 - ku_0 < 0$, the solution of the equation is exponential functions, and the equilibrium solution is unstable. In the present case we assume that k is so small that the solution is stable. In this case, and with the initial condition $u_1(0) = eu_0$, where e is an integration constant, the solution of this equation can be written

$$u_1 = u_0 e \cos(f\phi), \quad f = \sqrt{1 - 2ku_0}. \tag{8.133}$$

The constant e is called the *eccentricity* of the orbit and tells how elongated it is. Using that

$$u = u_0 + u_1 = u_0[1 + e\cos(f\phi)], \tag{8.134}$$

we have

$$r = \frac{r_0}{1 + e\cos(f\phi)}. \tag{8.135}$$

In the Newtonian case with $k = 0$ we get $f = 1$, and then this expression describes a (non-precessing) ellipse. However in the relativistic case $k > 0$ and $f < 1$. Then ϕ has to increase by $2\pi/f > 2\pi$ in order that r shall return to the initial value. The precession angle per orbit is

$$\Delta\phi = 2\pi\left(\frac{1}{f} - 1\right) = 2\pi\left(\frac{1}{\sqrt{1 - 2ku_0}} - 1\right) \approx 2\pi k u_0. \tag{8.136}$$

Inserting the Newtonian value of u_0 we have

$$\Delta\phi = \frac{\pi k R_S}{p_\phi^2}. \tag{8.137}$$

On the other hand, the general relativistic value $k = (3/2)R_S$ leads to

$$\Delta\phi = \frac{3}{2}\pi\left(\frac{R_S}{p_\phi}\right)^2. \tag{8.138}$$

The angular momentum per unit mass can be expressed in terms of the period by means of Kepler's 2. law. This law says that the planet has a constant areal velocity, dA/dt, where dA is the area swept out by the radius vector from the Sun to the planet during a time dt. In a small time dt the planet sweeps out a small triangle with base line r and height $r\,d\phi$ and area $dA = (1/2)r^2\,d\phi$, and so the constant areal velocity is

$$\frac{dA}{dt} = \frac{1}{2}r^2\frac{d\phi}{dt}. \tag{8.139}$$

The area enclosed by the elliptical orbit is πab where a and b are the semi major and semi minor axes of the ellipse. Hence integrating over one period, T, we get

$$\frac{1}{2}r^2\frac{d\phi}{dt}T = \frac{1}{2}p_\phi T = \pi ab = \pi a^2\sqrt{1 - e^2}. \tag{8.140}$$

Hence the angular per unit mass can be written

$$p_\phi = \frac{2\pi a^2\sqrt{1 - e^2}}{T}. \tag{8.141}$$

Inserting this in the general formula for the precession angle per orbit gives

$$\Delta\phi = \frac{R_S T^2 k}{4\pi a^4 (1 - e^2)}. \tag{8.142}$$

According to Kepler's 3. law the square of the period is proportional to the cube of the semi major axis,

$$T^2 = \frac{4\pi^2 a^3}{G(M_{\text{Sun}} + m_{\text{Mercury}})} \approx \frac{4\pi^2 a^3}{GM} = \frac{8\pi^2 a^3}{R_S}. \tag{8.143}$$

Inserting this into Eq. (8.142) gives

$$\Delta\phi = \frac{2\pi k}{a(1 - e^2)}. \tag{8.144}$$

Finally, substituting the general relativistic value of k leads to

$$\Delta\phi = \frac{3\pi R_S}{a(1 - e^2)}. \tag{8.145}$$

Inserting the Schwarzschild radius of the Sun, $R_S = 3$ km, the semi major axis of Mercury's orbit, $a = 5.8 \times 10^7$ km and the eccentricity $e = 0.2$, and using that the period of Mercury's orbital motion is 88 days, give a precession angle 43 arc seconds per hundred years.

This solved an old problem, namely that observations showed that Mercury's elliptical orbit precesses by 575 arc seconds per hundred years, while only 532 arc seconds per hundred years could be accounted for by gravitational forces from the other planets in the solar system. So there was a discrepancy of 43 arc seconds per hundred years between observations and the Newtonian prediction.

8.8.3 Deflection of Light

The orbit Eq. (8.128) for light, i.e. for a free particle with mass $m = 0$, reduces to

$$\frac{d^2 u}{d\phi^2} + u = ku^2. \tag{8.146}$$

If light is not deflected, it will follow the straight line

$$\cos\phi_0 = \frac{b}{r_0} = bu_0, \tag{8.147}$$

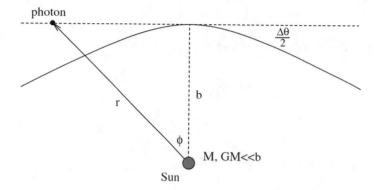

Fig. 8.7 Deflection of light. Light travelling close to a massive object is deflected

where b is the impact parameter of the path. This represents the horizontal dashed line in Fig. 8.7. The zeroth order solution (8.147) fulfils

$$\frac{d^2 u_0}{d\phi^2} + u_0 = 0. \tag{8.148}$$

Hence it is a solution of (8.146) with $k = 0$.
The perturbed solution is

$$u = u_0 + u_1, \quad |u_1| \ll u_0. \tag{8.149}$$

Inserting this into the orbit equation gives

$$\frac{d^2 u_0}{d\phi^2} + \frac{d^2 u_1}{d\phi^2} + u_0 + u_1 = k u_0^2 + 2k u_0 u_1 + k u_1^2. \tag{8.150}$$

The first and third terms at the left-hand side cancel each other due to Eq. (8.148), and the last term at the right-hand side is small to second order in u_1 and will be neglected. Hence we get

$$\frac{d^2 u_1}{d\phi^2} + u_1 = k u_0^2 + 2k u_0 u_1. \tag{8.151}$$

The last term at the right-hand side is much smaller than the first and will also be neglected. Inserting for u_0 from (8.147) we then get

$$\frac{d^2 u_1}{d\phi^2} + u_1 = \frac{k}{b^2} \cos^2 \phi. \tag{8.152}$$

This equation has a particular solution of the form

$$u_{1p} = A + B \cos^2 \phi. \tag{8.153}$$

Inserting this into (8.152) we find

$$A = \frac{2k}{3b^2}, \quad B = -\frac{k}{3b^2}. \tag{8.154}$$

Hence

$$u_{1p} = \frac{k}{3b^2} \left(2 - \cos^2 \phi\right) \tag{8.155}$$

giving

$$\frac{1}{r} = u = u_0 + u_1 = \frac{\cos \phi}{b} + \frac{k}{3b^2} \left(2 - \cos^2 \phi\right). \tag{8.156}$$

The deflection of the light $\Delta\theta$ is assumed to be small. We therefore put $\phi = \frac{\pi}{2} + \frac{\Delta\theta}{2}$ where $\Delta\theta \ll \pi$ (see Fig. 8.7). Hence

$$\cos \phi = \cos\left(\frac{\pi}{2} + \frac{\Delta\theta}{2}\right) = -\sin \frac{\Delta\theta}{2} \approx -\frac{\Delta\theta}{2}. \tag{8.157}$$

Thus, the term $\cos^2 \phi$ in (8.155) can be neglected. Furthermore, the deflection of the light is found by letting $r \to \infty$, i.e. $u \to 0$. Then we get

$$\Delta\theta = \frac{4k}{3b}. \tag{8.158}$$

For motion in the Schwarzschild spacetime outside the Sun, $k = (3/2)R_S$ where R_S is the Schwarzschild radius of the Sun, and for light passing the surface of the Sun $b = R_\odot$, where $R_\odot$ is the actual radius of the Sun. The deflection is then

$$\Delta\theta = 2\frac{R_S}{R_\odot} = 1.75''. \tag{8.159}$$

This general relativistic prediction was verified in two British expeditions utilizing a solar eclipse in 1919.

8.9   The Reissner–Nordström Spacetime

We shall now solve Einstein's field equations for a static spherically symmetric space-time with a radial electric field outside a body with mass M and charge Q. According to Eqs. (7.81)–(7.84) the non-vanishing components of the energy–momentum tensor in an orthonormal basis are

$$T_{\hat{t}\hat{t}} = T_{\hat{\theta}\hat{\theta}} = T_{\hat{\phi}\hat{\phi}} = \frac{Q^2}{32\pi^2\varepsilon_0}\frac{1}{r^4}, \quad T_{\hat{r}\hat{r}} = -\frac{Q^2}{32\pi^2\varepsilon_0}\frac{1}{r^4}. \tag{8.160}$$

From Eq. (8.15) we get by adding the $\hat{t}\hat{t}-$ and $\hat{r}\hat{r}-$field equations and integrating

$$\alpha(r) = -\beta(r). \tag{8.161}$$

Inserting this into the $\hat{t}\hat{t}$ field equation we get

$$\frac{1}{r^2}[r(1 - e^{-2\beta})]' = \frac{R_Q^2}{r^4}, \quad R_Q^2 = \frac{GQ^2}{4\pi\varepsilon_0 c^4}. \tag{8.162}$$

Integrating this equation and determining the integration constant by the requirement that the solution shall reduce to the Schwarzschild metric with vanishing charge, we obtain

$$e^{-\beta} = 1 - \frac{R_S}{r} + \frac{R_Q^2}{r^2}, \quad R_S = \frac{2GM}{c^2}. \tag{8.163}$$

Hence, the line element of the spacetime outside a massive changed body takes the form

$$ds^2 = -\left(1 - \frac{R_S}{r} + \frac{R_Q^2}{r^2}\right)c^2 dt^2 + \frac{dr^2}{1 - \frac{R_S}{r} + \frac{R_Q^2}{r^2}} + r^2 d\Omega^2. \tag{8.164}$$

This is *the Reissner–Nordström solution*. There are two coordinate singularities at

$$r_\pm = \frac{R_S}{2} \pm \sqrt{\left(\frac{R_S}{2}\right)^2 - R_Q^2}. \tag{8.165}$$

The exterior singularity is a horizon for an observer outside the mass and charge distribution.

The maximum allowed charge of the body is given by $R_Q = R_S/2$ which leads to

$$Q_{max} = 2\sqrt{\pi\varepsilon_0 G}M. \tag{8.166}$$

One may note that the "Reissner–Nordström length" as given in Eq. (8.162), corresponding to the elementary charge, is

$$R_e = \frac{e}{2c^2} \sqrt{\frac{G}{\pi \varepsilon_0}}. \tag{8.167}$$

Inserting the values of the constants gives $R_e = 1.38 \times 10^{-35}$ m, which is a little smaller than the Planck length.

Exercises

8.1. *Non-relativistic Kepler motion*

In the first part of this exercise we will consider the gravitational potential at a distance r from the Sun, $V(r) = -GM/r$, where M is the mass of the Sun.

(a) Write down the classical Lagrangian in spherical coordinates (r, θ, ϕ) for a planet with mass m moving in this field. The Sun is assumed to be stationary.

What is the physical interpretation of the canonical momenta $p_\phi = \ell$?

How is it possible, by just looking at the Lagrangian, to state that p_ϕ is a constant of motion?

Find the Euler equation for ϑ and show that it can be written into the form

$$\frac{d}{dt}\left(mr^4\dot{\theta}^2 + \frac{\ell^2}{m \sin^2 \theta}\right) = 0. \tag{8.168}$$

Based on the above equation, show that the planet moves in a plane by choosing a direction of the z-axis so that at a given time, $t = 0$, we have that $\theta = \pi/2$ and $\dot{\theta} = 0$.

(b) Write down the Euler equation for r and use this equation to find $u = 1/r$ as a function of ϕ. Show that the orbits that describe bound states are elliptic. Find the period T_0 for a circular orbit in terms of the radius R of the circle.

(c) If the Sun is not entirely spherical, but rather a little deformed (i.e. more flat near the poles), the gravitational field in the plane where the Sun has its greatest extension will be modified into

$$V(r) = -\frac{GM}{r} - \frac{S}{r^3}, \tag{8.169}$$

where S is a small constant. We now assume that the motion of the planet takes place in the plane where the expression of is $V(r)$ valid. Show that a circular motion is still possible. What is the period T now, expressed by the radius R?

We now assume that the motion deviates slightly from a pure circular orbit, that is $u = \frac{1}{R} + u_1$, where $u_1 \ll \frac{1}{R}$. Show that u_1 varies periodically around the orbit,

$$u_1 = k \sin(f\phi). \tag{8.170}$$

(d) Find f and show that the path rotates in space. What is the size of the angle $\Delta\phi$ that the planetary orbit rotates per round trip?

The constant S can be written as $S = \frac{1}{2}J_2 GM R_{Sun}^2$ where J_2 is a parameter describing the quadrupole moment and R_{Sun} is the radius of the Sun. Observational data indicate that $J_2 \lesssim 3 \cdot 10^{-5}$. Calculate how large the rotation $\Delta\phi$ of the orbit of Mercury this can cause. Is this sufficient to explain the discrepancy between the observed perihelion motion of Mercury and that predicted by Newtonian theory?

8.2. *The Schwarzschild solution in isotropic coordinates*

(a) We introduce a new radial coordinate ρ so that the Schwarzschild metric (with units so that $c = 1$),

$$ds^2 = -\left(1 - \frac{R_S}{r}\right)dt^2 + \left(1 - \frac{R_S}{r}\right)^{-1}dr^2 + r^2 d\Omega^2, \qquad (8.171)$$

gets the following form

$$ds^2 = -\left(1 - \frac{R_R}{r(\rho)}\right)dt^2 + f^2(\rho)(d\rho^2 + \rho^2 d\Omega^2), \qquad (8.172)$$

where $d\Omega^2 = d\theta^2 + \sin^2\theta d\phi^2$. Find the functions $r(\rho)$ and $f(\rho)$, and write down the explicit expression of the line element with ρ as the radial coordinate.

(b) What is the value of ρ at the Schwarzschild horizon $r = R_S$ and at the origin, $r = 0$? The Schwarzschild coordinates t and r interchange their roles as $r < R_S$. What is the behaviour of ρ inside the horizon?

8.3. *Proper radial distance in the external Schwarzschild space*

Calculate the proper radial distance from a coordinate position r to the horizon R_S in the external Schwarzschild space.

8.4. *The Schwarzschild–de Sitter metric*

The Einstein equations for empty space with a cosmological constant Λ are

$$R_{\mu\nu} - \frac{1}{2}Rg_{\mu\nu} + \Lambda g_{\mu\nu} = 0. \qquad (8.173)$$

(a) Use curvature coordinates and solve the Einstein field equations with a cosmological constant for a static spacetime with a spherically symmetric 3-space outside a spherical body with mass M.

(b) The solution of Einsteins field equations with a cosmological constant in globally empty space, i.e. with $M = 0$, is called the *de Sitter spacetime*. Introduce a de Sitter radius $R_\Lambda = \sqrt{3/\Lambda}$. Give a physical interpretation of this radius and calculate how large it is in a universe where the value of the cosmological constant corresponds to a density of LIVE equal to the average density of the masse and vacuum energy of the universe, $\Lambda = 10^{-52}$ m^{-2}.

8.5. *The perihelion precession of Mercury and the cosmological constant*

(a) Show that the orbit equation for free particles moving outside a spherically symmetric body with mass M has the form

$$\frac{d^2u}{d\phi^2} + u = \frac{M}{L^2} + 3Mu^2 - \frac{\Lambda}{3L^2u^3}, \tag{8.174}$$

where $u = 1/r$, and L is the angular momentum per unit mass for the particle.

(b) Assume that the orbit can be described as a perturbation of a circle and calculate the precession angle per round trip.

(c) Estimate the contribution to the precession of Mercury's perihelion from the cosmological constant if we assume that the value of the cosmological constant is $\Lambda \approx 10^{-52}$ m^{-2}.

8.6. *Relativistic time effects and GPS*

Calculate the magnitude of the kinematical and gravitational time effects upon the GPS satellite clocks. Are standard clocks on the GPS satellites going slower or faster than a standard clock at rest on the surface of the Earth.

To compute the position of an object by means of the GPS system with a precision of 1 m, the GPS satellite clocks must measure time with a precision of one part in 10^{13}.

Are the relativistic effects so small that they can be neglected?

8.7. *The photon sphere*

The photon sphere is defined as a spherical shell made up of light moving horizontally in the Schwarzschild spacetime. Calculate the radius of the photon sphere.

References

1. Shapiro, I.I., Ingalls, R.P., Smith, W.B., Campbell, D.B., Dyce, R.E., Jurgens, R.B., Pettengill, G.H.: Fourth test of general relativity—new radar result. Phys. Rev. Lett. **26**, 1132–1135 (1971)
2. Hafele, J., Keating, R.: Around the world atomic clocks: predicted relativistic time gains. Science **177**, 166–168 (1972)

Chapter 9
The Linear Field Approximation and Gravitational Waves

Abstract The linear field approximation of Einstein's field equations is presented. The solutions of these equations inside and outside a rotating spherical shell are deduced. It is shown that inertial dragging is a consequence of the general theory of relativity. The equations of gravitomagnetism are deduced. Gravitational waves are described and related to their detection by the LIGO-detectors.

9.1 The Linear Field Approximation

In this chapter we shall describe weak gravitational fields, i.e. fields far from a black hole, meaning that $r \gg R_S$, where R_S is the Schwarzschild radius of the black hole. Normalizing the gravitational potential, ϕ, to zero far from the black hole, this means that $|\phi| \ll c^2$, and that the curvature of spacetime is small. It is then possible to introduce a coordinate system such that the metric deviates very little from the Minkowski metric. Then it will be a good approximation in order to linearize the field equations.

We now assume that the gravitational field is weak and introduce a near-Cartesian coordinate system. The components of the metric tensor can then be written as $g_{\mu\nu} = \eta_{\mu\nu} + h_{\mu\nu}$ where $\eta_{\mu\nu}$ is the Minkowski metric, and $h_{\mu\nu} \ll 1$. Also the derivatives $h_{\mu\nu,\lambda}$ are small.

Einstein's field equations are

$$R_{\mu\nu} - \frac{1}{2} g_{\mu\nu} R = \kappa T_{\mu\nu}, \tag{9.1}$$

Here $R \equiv R_\beta^\beta$ is the Ricci curvature scalar and $R_{\mu\nu} = R_{\mu\alpha\nu}^\alpha$ the components of the Ricci curvature tensor, where $R_{\mu\alpha\nu}^\alpha$ are the components of the Riemann curvature tensor. According to Eq. (6.17) they are given by the expression

$$R_{\mu\beta\nu}^\alpha = \Gamma_{\mu\nu,\beta}^\alpha - \Gamma_{\mu\beta,\nu}^\alpha + \Gamma_{\lambda\beta}^\alpha \Gamma_{\mu\nu}^\lambda - \Gamma_{\lambda\nu}^\alpha \Gamma_{\mu\beta}^\lambda \tag{9.2}$$

© Springer Nature Switzerland AG 2020

Ø. Grøn, *Introduction to Einstein's Theory of Relativity*,
Undergraduate Texts in Physics, https://doi.org/10.1007/978-3-030-43862-3_9

in coordinate basis. Here $\Gamma^\alpha_{\mu\nu}$ are the Christoffel symbols. They are calculated from the expression (5.65),

$$\Gamma^\alpha_{\mu\nu} = \frac{1}{2} g^{\alpha\beta} \left(g_{\beta\mu,\nu} + g_{\beta\nu,\mu} - g_{\mu\nu,\beta} \right). \tag{9.3}$$

In the linear approximation we will only calculate to first order in the metric perturbation $h_{\mu\nu}$. Then the expression for the Christoffels symbols becomes

$$\Gamma^\alpha_{\mu\nu} = \frac{1}{2} \eta^{\alpha\beta} \left(h_{\beta\mu,\nu} + h_{\beta\nu,\mu} - h_{\mu\nu,\beta} \right). \tag{9.4}$$

Inserting this into Eq. (9.2) gives

$$R^\alpha_{\mu\beta\nu} = \frac{1}{2} \eta^{\alpha\gamma} \left(h_{\gamma\nu,\mu\beta} + h_{\mu\beta,\gamma\nu} - h_{\mu\nu,\gamma\beta} - h_{\gamma\beta,\mu\nu} \right). \tag{9.5}$$

Calculating the Ricci tensor by contracting the 1. and 3. index we find

$$R_{\mu\nu} = \frac{1}{2} \left(h^\alpha_{\mu\alpha,\nu} + h^\alpha_{\nu\alpha,\mu} - h^\alpha_{\mu\nu,\alpha} - h_{,\mu\nu} \right), \tag{9.6}$$

where $h \equiv h^\alpha_\alpha = \eta^{\alpha\beta} h_{\alpha\beta}$. Hence the Ricci scalar is

$$R = R^\mu_\mu = h_{\alpha\beta,}{}^{\alpha\beta} - h_{,\alpha}^{\ \alpha}. \tag{9.7}$$

Then the Einstein's field equations in the linear approximation take the form

$$h^\alpha_{\mu\alpha,\nu} + h^\alpha_{\nu\alpha,\mu} - h^\alpha_{\mu\nu,\alpha} - h_{,\mu\nu} - \eta_{\mu\nu} (h^{\alpha\beta}_{\alpha\beta,} - h^\beta_{,\beta}) = 16\pi G T_{\mu\nu}. \tag{9.8}$$

This equation can be simplified by introducing

$$\bar{h}_{\mu\nu} = h_{\mu\nu} - \frac{1}{2} \eta_{\mu\nu} h. \tag{9.9}$$

We assume that $\bar{h}$ satisfies the so-called Lorenz gauge condition

$$\bar{h}^\alpha_{\mu,\alpha} = 0. \tag{9.10}$$

Then the field equations take the form

$$\bar{h}^\alpha_{\mu\nu,\alpha} = -2\kappa T_{\mu\nu}. \tag{9.11}$$

Introducing the d'Alembert's wave operator in Minkowski spacetime

$$\Box = \eta^{\alpha\beta}\partial_\alpha\partial_\beta = -\partial^2/\partial t^2 + \partial^2/\partial x^2 + \partial^2/\partial y^2 + \partial^2/\partial z^2 = -\partial^2/\partial t^2 + \nabla^2, \tag{9.12}$$

Einstein's field equations in the linear field approximation can be written as

$$\Box \bar{h}_{\mu\nu} = -2\kappa T_{\mu\nu}. \tag{9.13}$$

Hence, the linearized field equations for empty space take the form

$$\Box \bar{h}_{\mu\nu} = 0. \tag{9.14}$$

This is d'Alembert's wave equation for waves moving with the velocity of light. It here describes gravitational waves. Hence, it follows from Einstein's theory that gravitational waves move with the velocity of light. This reveals a deep relationship between electromagnetism and gravity which is not yet fully understood. The explanation may be hidden in the Kaluza–Klein theory (see the Appendix).

In the case of a static spacetime in matter with density ρ the tt-component of Einstein's field equations takes the form

$$\nabla^2 \bar{h}_{tt} = -2\kappa\rho. \tag{9.15}$$

Comparing with the Newtonian gravitational field Eq. (1.29) we get

$$\bar{h}_{tt} = -4\phi/c^2. \tag{9.16}$$

where ϕ is the Newtonian gravitational potential.

In order to find the line-element in the linear field approximation we need the reverse of Eq. (9.9). Taking the trace of $\bar{h}_{\mu\nu}$ and utilizing that $\eta^\mu_\mu = \delta^\mu_\mu = 4$ we get $\bar{h} = -h$. Hence,

$$h_{\mu\nu} = \bar{h}_{\mu\nu} - \frac{1}{2}\eta_{\mu\nu}\bar{h}. \tag{9.17}$$

Inserting Eq. (9.16) gives

$$h_{tt} = h_{ii} = (1/2)\bar{h}_{tt} = -2\phi. \tag{9.18}$$

Hence the line element takes the form

$$ds^2 = -\left(1 + \frac{2\phi}{c^2}\right)c^2 dt^2 + \left(1 - \frac{2\phi}{c^2}\right)(dx^2 + dy^2 + dz^2), \tag{9.19}$$

or

$$ds^2 = -\left(1 + \frac{2\phi}{c^2}\right)c^2 dt^2 + \left(1 - \frac{2\phi}{c^2}\right)(dr^2 + r^2 d\Omega^2). \tag{9.20}$$

9.2 Solutions of the Linearized Field Equations

9.2.1 The Gravitational Potential of a Point Mass

The space time outside a spherical mass distribution with given mass does not depend upon the radius of the mass distribution. Hence the spacetime is the same as that outside a particle. We choose to consider spacetime outside a particle in order to simplify the calculation.

A point mass with the mass m is situated in the origin of a Cartesian coordinate system. Its energy–momenta tensor has only one non-vanishing component,

$$T_{00} = m\delta(\vec{r}), \tag{9.21}$$

where $\delta(\vec{r})$ is the 3-dimensional δ-function.

In this case Einstein's field equation (9.13) takes the form

$$\nabla^2 \bar{h}_{00} = -2\kappa m \delta(\vec{r}). \tag{9.22}$$

Using that

$$\int \delta(\vec{r}) d^3 r = 1, \tag{9.23}$$

and integrating Eq. (9.22) gives

$$\int \nabla^2 \bar{h}_{00} d^3 r = -2\kappa m. \tag{9.24}$$

Utilizing Gauss's integral theorem we can transform the left-hand side to an integral over the surface of a spherical surface with radius r and get

$$\int \nabla \bar{h}_{00} \cdot d\vec{S} = 4\pi r^2 \frac{d\bar{h}_{00}}{dr} = -2\kappa m. \tag{9.25}$$

where $r = \sqrt{x^2 + y^2 + z^2}$. Hence,

$$\frac{d\bar{h}_{00}}{dr} = -\frac{4Gm}{c^2 r^2}. \tag{9.26}$$

Integration with the boundary condition $\lim_{r \to \infty} \bar{h}_{00} = 0$ gives

$$\bar{h}_{00} = \frac{4Gm}{c^2 r}. \tag{9.27}$$

Hence,

$$\bar{h} = \bar{h}^\mu_\mu = -\frac{4Gm}{c^2 r}. \tag{9.28}$$

Inserting Eqs. (9.27) and (9.28) in Eq. (9.17) gives

$$h_{00} = h_{ii} = \frac{2GM}{c^2 r} = \frac{R_S}{r}. \tag{9.29}$$

Hence the line element of the spacetime outside the particle is

$$ds^2 = -\left(1 - \frac{R_S}{r}\right)c^2 dt^2 + \left(1 + \frac{R_S}{r}\right)(dx^2 + dy^2 + dz^2). \tag{9.30}$$

In terms of spherical coordinates the line element takes the form

$$ds^2 = -\left(1 - \frac{R_S}{r}\right)c^2 dt^2 + \left(1 + \frac{R_S}{r}\right)(dr^2 + r^2 d\Omega^2). \tag{9.31}$$

This is the same as the linearized line element, (S8.50), of the Schwarzschild spacetime as expressed in isotropic coordinates.

9.2.2 Spacetime Inside and Outside a Rotating Spherical Shell

We shall solve the linearized field equations, (9.13), inside and outside a rotating spherical shell with radius R and mass M consisting of dust particles. The energy–momentum tensor of the shell is

$$T_{\mu\nu} = \rho u_\mu u_\nu, \quad \rho = \frac{M}{4\pi R^2}\delta(r - R), \quad u_\mu = (-1, -R\omega \sin\theta \sin\phi, R\omega \sin\theta \cos\phi, 0). \tag{9.32}$$

where ω is the angular velocity of the shell. In this case Einstein's field equations (9.13) take the form

$$\nabla^2 \bar{h}_{00} = -\frac{2R_S}{R^2}\delta(r - R), \quad \nabla^2 \bar{h}_{ii} = 0$$
$$\nabla^2 \bar{h}_{0x} = \frac{2R_S \omega}{R}\sin\theta \sin\phi\, \delta(r - R), \quad \nabla^2 \bar{h}_{0y} = \frac{2R_S \omega}{R}\sin\theta \cos\phi\, \delta(r - R) \tag{9.33}$$

These equations are integrated in the same manner as Eq. (9.22). For $\bar{h}_{00}$ we get

$$\frac{d\bar{h}_{00}}{dr} = \begin{cases} -2R_S/r^2, \; r \geq R \\ 0, \quad\quad\quad r < R \end{cases}.$$

(9.34)

Integrating once more and demanding that $\bar{h}_{00}$ is continuous at the shell give

$$\bar{h}_{00} = \begin{cases} 2R_S/r, \; r \geq R \\ 2R_S/R, \; r \leq R \end{cases}.$$

(9.35)

The second Eq. (9.33) and the requirement that there shall be Minkowski metric infinitely far from the shell gives $\bar{h}_{ii} = 0$.

Because of the spherical symmetry the two last equations in (9.33) will have the same solution, so it is sufficient to solve one of then, say the last one. Due to the form of the right-hand side we assume that $\bar{h}_{0y}$ has the form

$$\bar{h}_{0y} = f(r) \sin \theta \cos \phi.$$

(9.36)

This gives

$$\begin{aligned}
\nabla^2 \bar{h}_{0y} &= \left[\frac{1}{r^2} \frac{\partial}{\partial r} r^2 \frac{\partial}{\partial r} + \frac{1}{r^2 \sin^2 \theta} \frac{\partial^2}{\partial \phi^2} + \frac{1}{r^2 \sin^2 \theta} \frac{\partial}{\partial \theta} \sin \theta \frac{\partial}{\partial \theta} \right] f(r) \sin \theta \cos \phi \\
&= \left(f'' + \frac{2}{r} f' - \frac{2}{r^2} f \right) \sin \theta \cos \phi
\end{aligned}$$

(9.37)

where $f' = df/dr$. Inserting this into the last Eq. (9.33) leads to

$$f'' + \frac{2}{r} f' - \frac{2}{r^2} f = -\frac{2R_S \omega}{R} \delta(r - R).$$

(9.38)

This equation may be written as

$$\left[\frac{1}{r^2} (r^2 f)' \right]' = -\frac{2R_S \omega}{R} \delta(r - R).$$

(9.39)

Using that $\delta(r - T) = -\delta(R - r)$ we have

$$\left[\frac{1}{r^2} (r^2 f)' \right]' = \frac{2R_S \omega}{R} \delta(R - r).$$

(9.40)

Integration gives

$$\frac{1}{r^2} (r^2 f)' = \begin{cases} 0, \quad\quad r > R \\ \frac{2R_S \omega}{R}, \; r < R \end{cases}.$$

(9.41)

Hence,

$$(r^2 f)' = \begin{cases} 0, & r > R \\ \frac{2R_S\omega}{R}r^2, & r < R \end{cases}.$$ (9.42)

Integrating once more gives

$$r^2 f = \begin{cases} K_{\text{out}}, & r > R \\ \frac{2R_S\omega}{3R}r^3 + K_{\text{in}}, & r < R \end{cases}.$$ (9.43)

Hence,

$$f = \begin{cases} K_{\text{out}}/r^2, & r > R \\ \frac{2R_S\omega}{3R}r + \frac{K_{\text{in}}}{r^2}, & r < R \end{cases}.$$ (9.44)

The requirement that the metric is well defined at $r = 0$ implies that $K_{\text{in}} = 0$. Furthermore the requirement that the metric is continuous at the shell then implies that $K_{\text{out}} = 2R_S R^2\omega/3$. Hence the solution of Eq. (9.33) outside and inside the shell is

$$f(r) = \begin{cases} \frac{2R_S\omega}{3}\left(\frac{R}{r}\right)^2, & r > R \\ \frac{2R_S\omega}{3}\frac{r}{R}, & r < R \end{cases}.$$ (9.45)

Inserting the expressions for $\bar{h}_{\mu\nu}$ from Eqs. (9.35), (9.36) and (9.45) into Eq. (9.17) gives the line element

$$ds^2 = \begin{cases} -\left(1 - \frac{R_S}{R}\right)c^2 dt^2 + \left(1 + \frac{R_S}{R}\right)\left(dr^2 + r^2 d\Omega^2\right) - \frac{4R_S\omega}{3R}r^2 \sin\theta d\phi dt, & r < R \\ -\left(1 - \frac{R_S}{r}\right)c^2 dt^2 + \left(1 + \frac{R_S}{r}\right)\left(dr^2 + r^2 d\Omega^2\right) - \frac{4R_S R^2\omega}{3r} \sin\theta d\phi dt, & r > R \end{cases}.$$ (9.46)

The angular momentum of the shell is

$$S = (1/3)R_S R^2\omega.$$ (9.47)

Hence the line-element (9.46) can be written

$$ds^2 = \begin{cases} -\left(1 - \frac{R_S}{R}\right)c^2 dt^2 + \left(1 + \frac{R_S}{R}\right)\left(dr^2 + r^2 d\Omega^2\right) - \frac{4S}{R^3}r^2 \sin\theta d\phi dt, & r < R \\ -\left(1 - \frac{R_S}{r}\right)c^2 dt^2 + \left(1 + \frac{R_S}{r}\right)\left(dr^2 + r^2 d\Omega^2\right) - \frac{4S}{r} \sin\theta d\phi dt, & r > R \end{cases}.$$ (9.48)

Let us see how the shell modifies space inside it.

9.3 Inertial Dragging

We shall first find a formula for the inertial dragging effect inside the shell. Consider a free observer moving in the equator plane, $\theta = \pi/2$, in the space described by the line-element (9.46) for $r < R$. The Lagrange function of the observer is

$$
L = -\frac{1}{2}\left(1 - \frac{R_S}{R}\right)c^2\dot{t}^2 + \frac{1}{2}\left(1 + \frac{R_S}{R}\right)\dot{r}^2 + \frac{1}{2}\left(1 + \frac{R_S}{R}\right)r^2\dot{\phi}^2 - \frac{2R_S\omega}{3R}r^2\dot{\phi}\dot{t}.
$$

(9.49)

The momentum p_ϕ conjugates to the cyclic coordinate ϕ,

$$
p_\phi = \frac{\partial L}{\partial \dot{\phi}} = \left(1 + \frac{R_S}{R}\right)r^2\dot{\phi} - \frac{2R_S\omega}{3R}r^2\dot{t}.
$$

(9.50)

An observer with $p_\phi = 0$ is called a zero-angular-momentum-observer, ZAMO. The coordinate angular velocity of the ZAMO is

$$
\Omega_{IN} = \frac{d\phi}{dt} = \frac{\dot{\phi}}{\dot{t}} = \frac{2}{3}\frac{R_S}{R + R_S}\omega.
$$

(9.51)

The metric is time independent. This means that the physical distance between arbitrary coordinate points is independent of the time. Hence the coordinates are co-moving with a stiff reference frame. At the origin the $d\phi dt$—term of the line-element (9.46) vanishes, and close to the origin the line element approaches the linearized line-element of the Schwarzschild spacetime. This means that as seen by a non-rotating observer at the origin the reference frame of the coordinate system is non-rotating.

Imagine that the observer at the origin throws out a particle. It is a free particle with zero angular momentum, a ZAMO. This particle will however not move along a radial line. It has a constant coordinate angular velocity given by Eq. (9.51) and moves along a spiral path. The particle is dragged in the same direction at the shell rotates. This phenomenon has several names. It was originally called the Lense–Thirring effect because it was first described in a published article by the Austrian Physicist Lense and Thirring in 1918. Later it has been called *inertial dragging*.

The reason for the latter name is that a free particle with vanishing angular momentum represents a local inertial frame. Hence inside the shell *inertial frames are dragged around by the rotation of the shell*—inertial dragging.

There is a similar effect outside the shell. Calculating the angular velocity of a ZAMO outside the shell in the same manner as above leads to

$$\Omega_{\text{IN}} = \frac{2}{3} \frac{R_S}{r + R_S} \left(\frac{R}{r}\right)^2 \omega. \tag{9.52}$$

The dragging angular velocity outside the shell decreases from the value (9.51) at the shell to zero infinitely far from the shell.

Note that the expressions (9.51) and (9.52) are valid only in the linear field approximation, i.e. for $R \gg R_S$. Corresponding expressions for arbitrarily strong gravitational fields will be found in the next chapter.

9.4 Gravitoelectromagnetism

In the context of gravitoelectromagnetism the Newtonian gravitational potential

$$\phi = -\frac{GM}{r} = -\frac{R_S}{2r} \tag{9.53}$$

of a rotating body with mass M is called the "gravito-electric" potential. One also introduces a "gravitomagnetic" vector potential $\vec{A}$ with components (including the velocity of light in this section)

$$A_i = \frac{G}{c} \varepsilon_{ijk} \frac{S^j x^k}{r^3}, \tag{9.54}$$

where ε_{ijk} is the antisymmetric Levi-Civita symbol, and S^j is the j-component of the angular momentum of the rotating body. In 3-vector notation this takes the form

$$\vec{A} = \frac{G}{c} \frac{\vec{S} \times \vec{r}}{r^3}. \tag{9.55}$$

The angular momentum of the rotating mass distribution is related to the energy–momentum tensor by

$$S^i = 2 \int \varepsilon^i_{jk} x^j j^k \mathrm{d}^3 r, \tag{9.56}$$

where

$$j^k = T^{0k}/c \tag{9.57}$$

is the mass current density. Then the line element of the linear field approximation can be written

$$\mathrm{d}s^2 = -\left(1 + \frac{2\phi}{c^2}\right) c^2 \mathrm{d}t^2 + \left(1 - \frac{2\phi}{c^2}\right) \delta_{ij} \mathrm{d}x^i \mathrm{d}x^j - \frac{4}{c} A_i \mathrm{d}x^i \mathrm{d}t. \tag{9.58}$$

In terms of the gravitoelectromagnetic potentials the Lorenz gauge condition (9.10) takes the form

$$\frac{1}{c}\frac{\partial \phi}{\partial t} + \frac{1}{2}\nabla \cdot \vec{A} = 0. \tag{9.59}$$

The "gravito-electric" field is

$$\vec{E}_G = -\nabla \phi - \frac{1}{2c}\frac{\partial \vec{A}}{\partial t}, \tag{9.60}$$

and the gravitomagnetic field is

$$\vec{B}_G = \frac{1}{2}\nabla \times \vec{A}. \tag{9.61}$$

It follows from the field Eqs. (9.13) together with Eqs. (9.59)–(9.61) that the gravito-electric and -magnetic fields satisfy equations of similar form as Maxwell's equations for electromagnetic fields,

$$\nabla \cdot E_G = -4\pi G\rho, \tag{9.62}$$

$$\nabla \cdot B_G = 0, \tag{9.63}$$

$$\nabla \times E_G = -\frac{\partial \vec{B}_G}{\partial t}, \tag{9.64}$$

$$\nabla \times B_G = \frac{1}{c^2}\frac{\partial \vec{E}_G}{\partial t} + \frac{4\pi G}{c^2}\vec{j}. \tag{9.65}$$

The gravitoelectromagnetic analog of the Lorentz force is

$$\vec{F} = m\left(\vec{E}_G + 4\vec{v} \times \vec{B}_G\right). \tag{9.66}$$

In the linear field approximation the gravitomagnetic field can be written as a dipole field,

$$\vec{B}_G = -\frac{G}{2cr^3}\left[\vec{S} - 3\frac{\left(\vec{S} \cdot \vec{r}\right)\vec{r}}{r^2}\right]. \tag{9.67}$$

If a body with an angular momentum $\vec{L}$ is in a gravitomagnetic field $\vec{B}_G$, the field will cause a torque

$$\vec{\tau} = \frac{1}{2c}\vec{L} \times \vec{B}_G. \tag{9.68}$$

The torque is equal to the time derivative of the angular momentum

$$\vec{\tau} = \frac{d\vec{L}}{dt}. \tag{9.69}$$

The rate of change of angular momentum is

$$\frac{d\vec{L}}{dt} = \vec{\Omega} \times \vec{L}, \tag{9.70}$$

where $\vec{\Omega}$ is the angular velocity of the inertial dragging. Equations (9.63)–(9.65) give

$$\vec{\Omega} = -\frac{1}{2c}\vec{B}_G. \tag{9.71}$$

Outside a body, for example the Earth, producing the gravitomagnetic field (9.67), the dragging angular velocity is

$$\vec{\Omega} = \frac{G}{4c^2 r^3}\left[\vec{S} - 3\frac{\left(\vec{S} \cdot \vec{r}\right)\vec{r}}{r^2}\right]. \tag{9.72}$$

At the equatorial plane $\vec{S}$ is orthogonal to $\vec{r}$ so $\vec{S} \cdot \vec{r} = 0$ and the equation reduces to

$$\vec{\Omega} = \frac{G}{4c^2 r^3}\vec{S}. \tag{9.73}$$

In the case of a rotating body, for example a gyroscope, this dragging field represents the rate of angular precession of the rotation axis (which was observed in the Gravity probe B experiment), and in the case of orbital motion it represents the perihelion precession of the orbit (for example the orbit of the planet Mercury), or the precession of the orbital plane (which was observed for the LAGEOS satellites).

9.5 Gravitational Waves

We will here consider gravitational waves in the weak field approximation of Einstein's equations using the Maxwell like equations for the gravitoelectromagnetic fields.

Inserting Eq. (9.60) into Eq. (9.62) gives

$$-\nabla^2 \phi - \frac{1}{2c}\frac{\partial}{\partial t}\left(\nabla \cdot \vec{A}\right) = -4\pi G\rho. \tag{9.74}$$

From the Lorenz gauge condition we have

$$\nabla \cdot \vec{A} = -\frac{2}{c}\frac{\partial \phi}{\partial t}. \tag{9.75}$$

Hence

$$\left(-\frac{1}{c^2}\frac{\partial^2}{\partial t^2} + \nabla^2\right)\phi = \Box\phi = 4\pi G\rho. \tag{9.76}$$

Furthermore, inserting Eqs. (9.60) and (9.61) into Eq. (9.65) gives

$$\frac{1}{2}\nabla \times \left(\nabla \times \vec{A}\right) = -\frac{1}{c}\nabla\frac{\partial \phi}{\partial t} - \frac{1}{2c^2}\frac{\partial^2 \vec{A}}{\partial t^2} + \frac{4\pi G}{c^2}\vec{j}. \tag{9.77}$$

Using the identity

$$\nabla \times \left(\nabla \times \vec{A}\right) = \nabla\left(\nabla \cdot \vec{A}\right) - \nabla^2 \vec{A} \tag{9.78}$$

we get

$$\nabla\left(\frac{1}{c}\frac{\partial \phi}{\partial t} + \frac{1}{2}\nabla \cdot \vec{A}\right) = \frac{1}{2}\left(-\frac{1}{c^2}\frac{\partial^2}{\partial t^2} + \nabla^2\right)\vec{A} + \frac{4\pi G}{c^2}\vec{j} = \frac{1}{2}\Box\vec{A} + \frac{4\pi G}{c^2}\vec{j}. \tag{9.79}$$

Due to the Lorenz gauge condition (9.59) the left-hand side vanishes. Hence we obtain

$$\Box\vec{A} = -\frac{8\pi G}{c^2}\vec{j}. \tag{9.80}$$

In empty space Eqs. (9.76) and (9.80) reduce to

$$\Box\phi = 0, \quad \Box\vec{A} = 0. \tag{9.81}$$

These are equations of gravitational waves moving through empty space with the velocity of light.

9.5.1 What Sort of Gravitational Waves Is Predicted by Einstein's Theory?

We shall now investigate the nature of these waves. In order to simplify the treatment we shall consider plane gravitational waves. There are three types of such gravitational waves, transverse (T), shear (S) and longitudinal (L). We shall here investigate what type Einstein's theory predicts the existence of and follow a procedure given by Eddington in an article published in 1922 [1].

We consider plane gravitational waves moving in the x-direction with velocity v. Hence $h_{\mu\nu}$ are periodic functions of $x + vt$. Differentiation with respect to $x + vt$ will be denoted by ', and the coordinates by $(x, y, z, t) = (x_1, x_2, x_3, x_4)$. Then the only non-vanishing derivatives are

$$\frac{\partial^2 g_{\mu\nu}}{\partial x_1^2} = h''_{\mu\nu}, \quad \frac{\partial^2 g_{\mu\nu}}{\partial x_1 \partial x_4} = v h''_{\mu\nu}, \quad \frac{\partial^2 g_{\mu\nu}}{\partial x_4^2} = v^2 h''_{\mu\nu}. \tag{9.82}$$

The covariant components of the linearized Riemann tensor are

$$R_{\mu\nu\alpha\beta} = \frac{1}{2}\left(\frac{\partial^2 g_{\mu\nu}}{\partial x_\alpha \partial x_\beta} + \frac{\partial^2 g_{\alpha\beta}}{\partial x_\mu \partial x_\nu} - \frac{\partial^2 g_{\mu\alpha}}{\partial x_\beta \partial x_\nu} - \frac{\partial^2 g_{\nu\beta}}{\partial x_\mu \partial x_\alpha} \right). \tag{9.83}$$

For a spacetime with plane gravitational waves the Riemann tensor has 21 independent components, six of them vanishing. The 15 non-vanishing components were given by Eddington as follows for i and j equal to 1, 2 and 3,

$$R_{i4j4} = \frac{1}{2}v^2 h''_{ij}, \quad R_{1ij4} = -R_{1ji4} = -\frac{1}{2}v h''_{ij}, \quad R_{1i1j} = \frac{1}{2}h''_{ij},$$
$$R_{14i4} = -v R_{1i14} = \frac{v}{2}(v h''_{1i} - h''_{i4}), \quad R_{1414} = \frac{1}{2}v^2 h''_{11} - v h''_{14} + \frac{1}{2}h''_{44} \tag{9.84}$$

The corresponding non-vanishing components of the Einstein tensor are

$$G_{11} = \frac{v}{2}(v h''_{11} - 2h''_{14}) + \frac{1}{2}(h''_{44} - h''_{22} - h''_{33}), \quad G_{kk} = \frac{1}{2}(v^2 - 1)h''_{kk}, \quad k = 2, 3$$
$$G_{44} = \frac{1}{2}(h''_{44} - 2v h''_{14}) - \frac{1}{2}(h''_{11} + h''_{22} + h''_{33}), \quad G_{1k} = v G_{k4} = \frac{v}{2}(v h''_{1k} - h''_{k4}),$$
$$G_{23} = \frac{1}{2}(v^2 - 1)h''_{23}, \quad G_{14} = -\frac{v}{2}(h''_{22} + h''_{33}), \tag{9.85}$$

Einstein's field equations for empty space are $G_{\mu\nu} = 0$, which is a set of second-order differential equations for $h_{\mu\nu}$. Since $h_{\mu\nu}$ are periodic functions, the second derivatives of the functions are equal to minus the functions themselves. This leads to field equations

$$v(v h_{11} - 2h_{14}) + h_{44} - h_{22} - h_{33} = 0, \; h_{kk} = 0, k = 2, 3$$

$$h_{44} - 2vh_{14} - h_{11} - h_{22} - h_{33} = 0,$$
$$vh_{1k} - h_{k4} = 0, h_{23} = 0, v(h_{22} + h_{33}) = 0. \tag{9.86}$$

These equations can be reduced to the following seven conditions,

$$h_{22} + h_{33} = 0, \quad (1 - v^2)(h_{22}, h_{33}, h_{23}) = 0,$$
$$h_{24} = vh_{12}, \quad h_{34} = vh_{13}, \quad v^2 h_{11} - 2vh_{14} + h_{44} = 0. \tag{9.87}$$

Eddington pointed out that for T-waves h_{22}, h_{33}, h_{23} cannot all vanish. Hence for these waves Eq. (9.87) implies $v = \pm 1$, meaning that according to the general theory of relativity transverse gravitational waves move with a coordinate independent velocity equal to the velocity of light in empty space. For S- and L-waves h_{22}, h_{33}, h_{23} are zero, and there is no coordinate independent equation determining v. The value of v found from the three last relationships in (9.87) depends upon the metric components and is hence coordinate dependent. Furthermore from (9.84) it is seen that for S- and L-waves the relationships (9.87) imply that the Riemann curvature tensor vanishes so that spacetime is flat and the periodic changes of the metric components are coordinate artifacts.

The conclusion is that the general theory of relativity predicts the existence of transverse gravitational waves travelling with the velocity of light.

9.5.2   Polarization of the Gravitational Waves

The gravitational waves may be represented by the metric functions

$$\bar{h}_{\mu\nu} = A_{\mu\nu} \cos(k_\alpha x^\alpha), \tag{9.88}$$

where $A_{\mu\nu}$ are the components of a symmetric tensor of rank 2, and k_α are the components of a constant wave vector. Inserting (9.88) into the field Eqs. (9.13) leads to

$$k_\alpha k^\alpha = 0. \tag{9.89}$$

Hence k_α is a null-vector, meaning again that the gravitational waves move with the velocity of light. An observer with four-velocity U^μ would observe the wave to have an angular frequency

$$\omega = -k_\alpha U^\alpha. \tag{9.90}$$

In the co-moving frame of the observer, where $U^\mu = (1, 0, 0, 0)$, the so-called transverse traceless gauge condition takes the form

$$h_{\mu 0} = 0, \quad h^j_{k,j} = 0, \quad h_{ii} = 0. \tag{9.91}$$

The first of these equations tells that only the spatial components of the metric perturbations are nonzero. The second says that the spatial components are divergence free, and the third says that they are trace free. It should also be noted that since $h = h^\mu_\mu = 0$ there is no difference between $\bar{h}_{\mu\nu}$ and $h_{\mu\nu}$ in this gauge.

We now choose the orientation of the coordinate system such that the gravitational wave travels along the z-axis. The metric perturbation then takes the form

$$h_{\mu\nu} = \begin{bmatrix} 0 & 0 & 0 & 0 \\ 0 & h_{xx} & h_{xy} & 0 \\ 0 & h_{xy} & -h_{xx} & 0 \\ 0 & 0 & 0 & 0 \end{bmatrix}. \tag{9.92}$$

There are only two free metric functions. This corresponds to the fact that there are only two different polarizations of gravitational waves according to Einstein's theory.

9.6   The Effect of Gravitational Waves upon Matter

We shall investigate physical effects of gravitational waves upon systems they pass. A gravitational wave is a "curvature wave". We shall therefore study the physical effects of gravitational waves by utilizing the Eq. (6.94) of geodesic deviation,

$$\frac{d^2 s^i}{dt^2} = -R^i_{0j0} s^j. \tag{9.93}$$

Here s^i are the components of the separation vector between two nearby geodesics. From the expression (9.5) for the components of the Riemann curvature tensor in the linear field approximation we get in the transverse traceless gauge,

$$R^i_{0j0} = (1/2) h^i_{j,00}. \tag{9.94}$$

Note that the mixed components of the metric perturbations are not equal to the Kronecker symbols, so the derivatives do not in general vanish. Equation (9.93) now takes the form

$$\frac{d^2 s^i}{dt^2} = -\frac{1}{2} h^i_{j,00} s^j. \tag{9.95}$$

Inserting the components of the metric perturbation from Eq. (9.87) we obtain

$$\frac{d^2 s^x}{dt^2} = \frac{1}{2} s^x h^x_{x,tt} + \frac{1}{2} s^y h^x_{y,tt}, \quad \frac{d^2 s^y}{dt^2} = \frac{1}{2} s^x h^y_{x,tt} - \frac{1}{2} s^y h^x_{x,tt}, \quad \frac{d^2 s^z}{dt^2} = 0. \quad (9.96)$$

These equations show that only the s^x and s^y components of the separation vector between two nearby free particles will be disturbed by a gravitational wave travelling in the z-direction.

A gravitational antenna is a system able to shape form, and according to Eq. (9.96) it is only disturbed in directions perpendicular to the wave propagation. Let us use the above equations to describe what happens to a ring in the xy-plane of free test particles at rest as a gravitational wave passes in the z-direction. To lowest order we can then neglect the terms with s^y at the right-hand side of the two first Eqs. (9.96) so that

$$\frac{d^2 s^x}{dt^2} = -\frac{1}{2} h^x_{y,tt} s^x, \quad \frac{d^2 s^y}{dt^2} = -\frac{1}{2} h^x_{x,tt} s^y. \quad (9.97)$$

Suppose a wave propagating in the z-direction with

$$h_{xx} = h \sin[\omega(t - z)], \quad h_{xy} = 0 \quad (9.98)$$

hits the particles. Let us consider two particles that are hit by the wave. One is at the origin, and the other has initially the position $x = \varepsilon \cos\theta, y = \varepsilon \sin\theta, z = 0$. The initial separation vector has the components $s^x(0) = \varepsilon \cos\theta, s^y(0) = \varepsilon \sin\theta$. Then Eqs. (9.97) and (9.98) give

$$\frac{d^2 s^x}{dt^2} = -\frac{1}{2} \varepsilon h \omega^2 \cos\theta \sin\omega t, \quad \frac{d^2 s^y}{dt^2} = \frac{1}{2} \varepsilon h \omega^2 \sin\theta \sin\omega t. \quad (9.99)$$

These equations have the solutions

$$s^x = \varepsilon \cos\omega\left(1 + \frac{h}{2}\sin\omega t\right), \quad s^y = \varepsilon \sin\omega\left(1 - \frac{h}{2}\sin\omega t\right), \quad (9.100)$$

satisfying the initial conditions. It follows from these expressions that

$$\frac{(s^x)^2}{a^2} + \frac{(s^y)^2}{b^2} = 1, \quad a = \varepsilon\left(1 + \frac{h}{2}\sin\omega t\right), \quad b = \varepsilon\left(1 - \frac{h}{2}\sin\omega t\right). \quad (9.101)$$

This shows that an originally circular ring is deformed to an oscillating elliptical shape by the gravitational wave, varying between being stretched in the y-direction and compressed in the x-direction and then stretched in the x-direction and compressed in the y-direction. This is called the $+$ polarization and is shown in Fig. 9.1.

We now consider the case when the wave has

Fig. 9.1 Deformation of a ring of free particles caused by a gravitational wave with + polarization

$$h_{xx} = 0, \quad h_{xy} = h \sin[\omega(t - z)]. \tag{9.102}$$

In this case Eqs. (9.97) and (9.98) give

$$\frac{d^2 s^x}{dt^2} = -\frac{1}{2}\varepsilon h\omega^2 \sin\theta \sin\omega t, \quad \frac{d^2 s^y}{dt^2} = -\frac{1}{2}\varepsilon h\omega^2 \cos\theta \sin\omega t. \tag{9.103}$$

These equations have the solutions

$$s^x = \varepsilon\cos\theta + \frac{1}{2}\varepsilon h \sin\theta \sin\omega t, \quad s^y = \varepsilon\sin\theta + \frac{1}{2}\varepsilon h \cos\theta \sin\omega t. \tag{9.104}$$

Rotating the coordinate axes through an angle $\pi/4$ we obtain new components of the separation vector

$$s^{x'} = \frac{1}{\sqrt{2}}(s^x - s^y), \quad s^{y'} = \frac{1}{\sqrt{2}}(s^x + s^y). \tag{9.105}$$

Inserting the expressions (9.104) gives

$$s^{x'} = \varepsilon\cos\left(\theta + \frac{\pi}{4}\right)\left(1 - \frac{h}{2}\sin\omega t\right), \quad s^{y'} = \varepsilon\cos\left(\theta + \frac{\pi}{4}\right)\left(1 + \frac{h}{2}\sin\omega t\right). \tag{9.106}$$

It follows from these expressions that

$$\frac{\left(s^{x'}\right)^2}{a^2} + \frac{\left(s^{y'}\right)^2}{b^2} = 1, \quad a = \varepsilon\left(1 - \frac{h}{2}\sin\omega t\right), \quad b = \varepsilon\left(1 + \frac{h}{2}\sin\omega t\right). \tag{9.107}$$

In this case the gravitational wave produces elliptical deformations rotated an angle $\pi/4$ relative to those of Fig. 9.1, as shown in Fig. 9.2. This is called the x polarization.

Since h_{xx} and h_{xy} are independent, Figs. 9.1 and 9.2 show the existence of two different states of polarization that are oriented at an angle $\pi/4$ relative to each other.

Fig. 9.2 Deformation of a ring of free particles caused by a gravitational wave with x polarization

9.7 The LIGO-Detection of Gravitational Waves

As noted in Sect. 9.6 the effect of a gravitational wave, GW, upon two freely falling nearby particle is to change the distance, L, between them. If the induced change of distance is ΔL, the fractional change in length $h = \Delta L / L$ is called the GW-strain. This is a dimensional measure of the amplitude of the gravitational wave.

LIGO consists of two detectors, one at the West coast of USA and one at the East coast. They are separated by about 3000 km. Hence light uses about 10 ms to move the distance between them. This means that signals of a certain type separated in time by less than 10 ms can be a sign that a gravitational wave has passed through the detectors.

The detectors are Michelson interferometers as illustrated in Fig. 9.3. The interferometer is constructed so that in normal modus the light that has been reflected from the upper mirror and the right hand mirror is in opposite phase and there is destructive interference when it meets after the reflections. Hence no light arrives at the detector in normal modus.

The two mirrors of the Fabry-Pérot cavities are hanged up so that they are free to move along the 4 km long L's. When a gravitational wave, say from a system of black holes with some tens of solar masses a billion light years away that spiral towards each other and collide, passes the detector, the distances between the two mirrors of the Fabry–Pérot cavities will get a length change with opposite signs, one is shortened, and the other gets longer. Hence the phases of the reflected light change, and a light signal arrives at the detector. In this situation typically the GW-strain is 10^{-21}. Hence with $L = 4 \, \text{km}$ one has to be able to detect a change of length of the order 10^{-18} m.

The shape of the signal has been calculated and is as shown in Fig. 9.4.

The signal has three phases. The first phase is the *inspiral* when the objects are outside the innermost stable orbit. In this phase the frequency and the amplitude increase. It is said to *chirp*. The second phase is the *merger* when they are inside and collide, and the third phase is the *ringdown* when the object formed by the collision, vibrates and falls to rest. The shape of the signal in the inspiral phase can be calculated analytically using post-Newtonian theory, the merger using numerical calculations and the ringdown using perturbation theory.

The first detection of gravitational waves by LIGO came 14 September 2015 [2]. The most recent was detected by LIGO and Virgo. The detected signals are shown in Fig. 9.5.

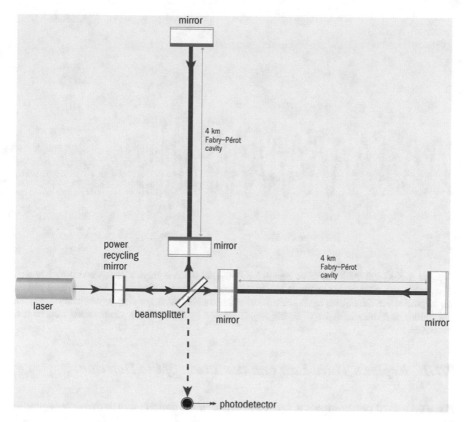

Fig. 9.3 LIGO gravitational wave detector

Order of magnitude estimates for the amplitudes of the waves emitted at different phases are given in [3], and a simple, but quantitative analysis has recently been given by Mathur et al. [4]. The main points when it comes to extracting information of the physical properties of the source of gravitational waves from the signals in an interferometric gravitational wave detector such as LIGO when a gravitational wave is detected, will be presented below.

The LiGO–Virgo-team detected a signal of this type the 14 August 2017. Figure 9.5 is from this report. An analysis of the signal that lasted for about 0.2 s showed that they had detected a gravitational wave emitted by a system of two black holes spiralling towards each other and merging at a distance of 1,8 billion light years. The frequency increased from 35 to 250 Hz meaning that the number of times the black holes went around each other per second increased from 17 to 125 during less than the fifth of a second. The masses of the black holes were detected to be $31 M_\odot$ and $25 M_\odot$, where $1 M_\odot$ is the mass of the Sun. The mass of the final black hole was $54 M_\odot$. Hence an energy corresponding to $3 M_\odot$ was emitted in the form of gravitational waves. The relative velocity of the black holes increased from 0.32c to 0.57c in this brief time interval.

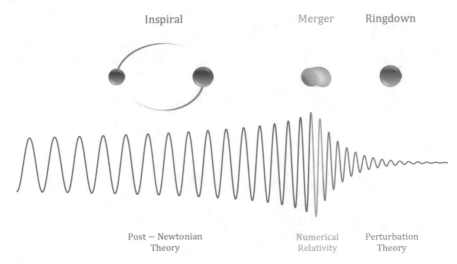

Inspiral   Merger   Ringdown

Post − Newtonian   Numerical   Perturbation
Theory   Relativity   Theory

Fig. 9.4 LIGO-gravitational wave signal. LIGO-signal when a gravitational wave from a system of two compact objects, for example two black holes, passes the detector. There are three phases, the *inspiral* when the objects are outside the innermost stable orbit, the *merger* when they are inside and collide and the *ringdown* when the object formed by the collision vibrates and falls to rest. (Nobel.org)

9.7.1 Kepler's Third Law and the Strain of the Detector

For a system of two compact objects with masses M_1 and M_2 one defines the so-called *chirp mass*,

$$M_{\mathrm{ch}} = \left(\frac{(M_1 M_2)^3}{M} \right)^{1/5} \tag{9.108}$$

where $M = M_1 + M_2$ is the total mass of the system, and the reduced mass is

$$\mu = \frac{M_1 M_2}{M}. \tag{9.109}$$

One also defines the dimensionless mass ratio for the system,

$$\eta = \frac{\mu}{M} = \frac{M_1 M_2}{M^2}. \tag{9.110}$$

The expressions for the chirp mass and the total mass can be solved for the masses of the compact objects, giving

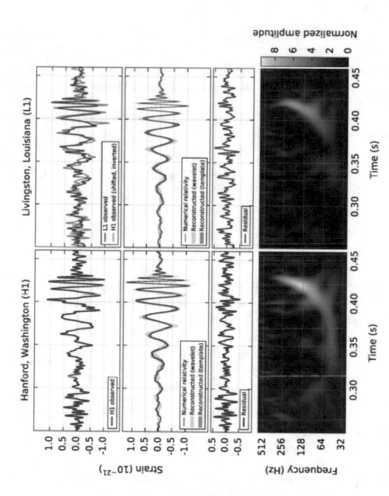

Fig. 9.5 LIGO gravitational wave discovery signal. Discovery signals registered at Hanford, Livingston and Virgo of a gravitational wave from colliding black holes. Reprinted from [2] by The Author(s) licensed under CC BY 3.0

$$M_1 = \frac{M}{2} + \sqrt{\left(\frac{M}{2}\right)^2 - M^{1/3}M_{\mathrm{ch}}^{5/3}}, \quad M_2 = \frac{M}{2} - \sqrt{\left(\frac{M}{2}\right)^2 - M^{1/3}M_{\mathrm{ch}}^{5/3}},$$

$$(9.111)$$

which requires $M_{\mathrm{ch}} < 2^{-6/5}M$.

In the inspiral phase there will be a maximal orbital angular frequency, Ω, given by Kepler's third law

$$2\Omega^2 R^3 = c^2 R_S. \tag{9.112}$$

when two objects with equal masses are about to merger, where $R = R_1 + R_2$ is the sum of the radial coordinates of the compact objects, and $R_S = 2GM/c^2$ is the Schwarzschild radii of the total mass.

At a distance r from the source of two inspiralling compact objects, the strain due to a gravitational wave is

$$h = \frac{4R_S R^2 \Omega^2}{c^2 r}. \tag{9.113}$$

It follows from the last two equations that

$$h = 2\frac{R_S^2}{rR}. \tag{9.114}$$

Let us use this formula to make an estimate of the expected magnitude of the strain for a gravitational wave detector on the Earth. The maximal value of h is obtained just before the objects merger, when $R \approx 2R_S$. Then

$$h_{\mathrm{max}} \approx R_S/r. \tag{9.115}$$

The Schwarzschild radius of the Sun is $R_{S\odot} = 3\,\mathrm{km} = 3 \cdot 10^{-13}l \cdot y$. Hence for a system of for example black holes with total mass equal to 50 solar masses one billion light years from the Earth the strain is $h = 1.5 \cdot 10^{-20}$. For an interferometer with 4 km long arms this means that the distance between two mirrors at the ends of the arms will change by $\Delta L = hL = 6 \cdot 10^{-17}$ m, which is less than the magnitude of an atomic nucleus.

Combining Eq. (9.114) with Eq. (9.110) and introducing the Schwarzschild radii R_{S1} and R_{S2} of the two objects, the strain can be expressed as

$$h = \frac{2}{\eta} \frac{R_{S1} R_{S2}}{rR}. \tag{9.116}$$

Sometimes the strain is expressed by the total mass and the chirp mass. Introducing the Schwarzschild radius of the total mass, $R_S = 2GM$, and the chirp mass, $R_{\mathrm{Sch}} = 2GM_{\mathrm{ch}}/c^2$, and using Eq. (9.108) the strain can be expressed as

$$h = \frac{2}{\eta} \frac{R_S^{1/3} R_{Sch}^{5/3}}{rR}. \tag{9.117}$$

Using Eq. (9.112) the strain as measured at a distance r far away from the colliding objects can then be expressed as

$$h = \Omega^{2/3} R_{Sch}^{5/3} \frac{2^{4/3}}{c^{2/3} \eta r}. \tag{9.118}$$

9.7.2 Newtonian Description of a Binary System

We use a coordinate system co-moving with the mass centre of the binary system, $M_1 R_1 = M_2 R_2$. Using this together with $R = R_1 + R_2$ we get

$$R_1 = \frac{M_2}{M} R, \quad R_2 = \frac{M_1}{M} R. \tag{9.119}$$

Furthermore we introduce an angular coordinate, $\theta = \theta_1 = \theta_2 - \pi$. The kinetic energy of the system is

$$E_K = \frac{1}{2} M_1 v_1^2 + \frac{1}{2} M_2 v_2^2. \tag{9.120}$$

where the square of the velocities of the compact objects are

$$v_1^2 = \dot{R}_1^2 + R_1^2 \dot{\theta}_1^2 = \left(\frac{M_2}{M}\right)^2 v^2,$$

$$v_2^2 = \dot{R}_2^2 + R_2^2 \dot{\theta}_2^2 = \left(\frac{M_1}{M}\right)^2 v^2,$$

$$v^2 = \dot{R}^2 + R^2 \dot{\theta}^2. \tag{9.121}$$

Inserting these expressions into Eq. (9.120) and using Eq. (9.109) gives

$$E_K = \frac{1}{2} \mu v^2. \tag{9.122}$$

The objects move along elliptical paths. In the Newtonian theory of binary systems it is usual to write the equation of the elliptic orbit as

$$p = R(1 + e \cos \theta), \tag{9.123}$$

where p is called the semi-latus rectum, and e is the eccentricity of the ellipse, i.e. the ratio of the semi-minor and semi-major half axis of the ellipse, $e = b/a$. The specific angular momentum, $L = R^2\dot{\theta}$, is a constant of motion. The relation $L^2/p = GM$ will be needed below. Differentiation of Eq. (9.123) with respect to time gives

$$\dot{R} = \frac{R^2 e \sin\theta}{R(1 + e\cos\theta)} = \frac{Le}{p}\sin\theta, \quad R\dot{\theta} = \frac{R^2\dot{\theta}}{R} = \frac{L}{p}(1 + e\cos\theta). \tag{9.124}$$

Using these expressions and that $p = a(1 - e^2)$ one finds that the kinetic energy is

$$E_K = \frac{1}{2}\mu GM\left(\frac{2}{R} - \frac{1}{a}\right) = \frac{GM_1 M_2}{R} - \frac{GM_1 M_2}{2a}. \tag{9.125}$$

The potential energy of the system is

$$E_P = -\frac{GM_1 M_2}{R}. \tag{9.126}$$

It follows that the mechanical energy of the system, $E = E_K + E_P$, has the value

$$E = -\frac{GM_1 M_2}{2a}, \tag{9.127}$$

which is constant. Approximating the distance between the objects by the semi-major half axis of the ellipse, $R \approx a$ we can use Eq. (9.112) to express the mechanical energy in terms of the angular frequency,

$$E = -\frac{GM_1 M_2}{R_S^{1/3}}\left(\frac{\Omega}{2}\right)^{2/3}. \tag{9.128}$$

9.7.3 Gravitational Radiation Emission

According to the general theory of relativity the binary system will emit gravitational waves and thereby loose mechanical energy. The emitted power is given by Eq. (9.121) in Ref. [5] for a system of two compact objects with equal masses. In the general case allowing different masses this expression is generalized to

$$P = \frac{32}{5}Gc\mu^2 R^4\Omega^6. \tag{9.129}$$

Using Eq. (9.108) this may be written as

$$P = \frac{32}{5} \frac{cG^4 M_1^2 M_2^2 M}{R^5}. \tag{9.130}$$

The emission of gravitational waves causes the compact object to move towards each other along a spiral path. The orbital energy may be approximated by

$$E = -\frac{GM_1 M_2}{2R}. \tag{9.131}$$

The rate of change of the orbital energy,

$$\dot{E} = \frac{GM_1 M_2}{2} \frac{\dot{R}}{R^2}, \tag{9.132}$$

is equal to the power given in Eq. (9.130). Hence

$$R^3 \dot{R} = -\frac{8}{5} c R_{S1} R_{S2} R_S. \tag{9.133}$$

where $R_S = R_{S1} + R_{S2}$. The distance between the objects at a point of time t is

$$R = \left(R_0^4 - \frac{32}{5} R_{S1} R_{S2} R_S ct \right)^{1/4}, \tag{9.134}$$

where $R_0 = R(0)$. This shows that it takes a finite time

$$t_{GW} = \frac{5 R_0^4}{32 c R_{S1} R_{S2} R_S} \tag{9.135}$$

before two objects moving around an elliptical orbit with initial distance R_0, collides with each other. If a detection signal lasts for a time t_D until the sign that a merger has happened appears, then the objects had an initial distance

$$R_0 = \left(\frac{32}{5} R_{S1} R_{S2} R_S ct_D \right)^{1/4} \tag{9.136}$$

when the initial part of the signal was emitted.

9.7.4 The Chirp

I order to extract information from the measured data, and it will be useful to express the emitted effect in terms of the frequency f of the observed radiation. As shown in Eq. (9.110) of Ref. [5] the frequency of the radiation is twice the orbital frequency. Hence $\Omega = \pi f$. Inserting this into Eq. (9.112) gives

$$R = \left(\frac{c^2 R_S}{2\pi^2 f^2}\right)^{1/3}. \tag{9.137}$$

Hence Eqs. (9.129) and (9.131) take the form

$$P = \frac{32}{5}\pi^{10/3} cG\mu^2\left(\frac{c^2 R_S}{2}\right)^{4/3} f^{10/3}, \tag{9.138}$$

$$E = -\frac{GM_1 M_2}{2}\left(\frac{2\pi^2}{c^2 R_S}\right)^{1/3} f^{2/3}. \tag{9.139}$$

Differentiation gives

$$\dot{E} = -\pi^{2/3}\frac{GM_1 M_2}{3}\left(\frac{2}{c^2 R_S}\right)^{1/3} f^{-1/3}\dot{f}. \tag{9.140}$$

Equating (9.138) and (9.140) leads to

$$f^{-11/3}\dot{f} = \frac{96\,\pi^{8/3}}{5c^5}(GM_{ch})^{5/3}, \tag{9.141}$$

where M_{ch} is the chirp mass defined in Eq. (9.108). Integration gives

$$f = f_0\left(1 - \frac{64}{5}2^{1/3}\pi^{8/3}\left(\frac{R_{Sch}}{c}\right)^{5/3} f_0^{-8/3} t\right)^{-3/8}. \tag{9.142}$$

where $f_0 = f(0)$. Solving this equation with respect to the Schwarzschild radius of the chirp mass leads to

$$R_{Sch} = c\left(\frac{5\Delta_f}{64\cdot 2^{1/3}\pi^{8/3}\Delta t}\right)^{3/5}, \quad \Delta_f \equiv \frac{1}{f_1^{8/3}} - \frac{1}{f_2^{8/3}}, \quad \Delta t = t_2 - t_1, \tag{9.143}$$

which may be written

$$M_{ch} = 7.8\cdot 10^2\, Hz\left(\frac{\Delta_f}{\Delta t}\right)^{3/5} M_\odot. \tag{9.144}$$

The objects begin to coalesce when their separation is equal to the sum of their Schwarzschild radii, $R = 2R_S$. Inserting this into Eq. (9.112) gives

$$R_S = \frac{c}{4\pi f_c}, \tag{9.145}$$

where f_c is the frequency of the emitted gravitational radiation when the objects begin to coalesce. This may be written

$$M = \frac{2.0 \cdot 10^4 \, Hz}{f_c} M_\odot, \qquad (9.146)$$

where $M_\odot$ is the mass of the Sun.

Let us apply Eqs. (9.144) and (9.146) to the signals shown in Fig. 9.5. We shall only illustrate the method and therefore do not include the uncertainties. The figure shows that the initial frequency is $f_1 = 30$ Hz at the point of time $t_1 = 0.46$ s and the frequency is $f_2 \approx 150$ Hz at $t_2 = 0.50$ s. This gives $\Delta_f = 1.1 \cdot 10^{-4}$, $\Delta t = 0.04$ s with a rather large uncertainty. Hence $M_{ch} \approx 23 M_\odot$. The measured peak frequency was around $f_p = 200$ Hz. This gives from Eq. (9.146) a total mass $M = 100 M_\odot$. More accurate relativistic calculations give a somewhat lower mass. Also there is a rather large uncertainty in the value of the maximal frequency, which may be somewhat higher than the measured peak frequency due to the very rapid merger. Assuming that $f_c \approx 350$ Hz, we get $M \approx 57 M_\odot$. Inserting these values for the total mass and the chirp mass into Eq. (9.111) gives for the masses of the two black holes, $M_1 \approx 38 M_\odot$, $M_2 \approx 19 M_\odot$. The masses determined by the more accurate calculations of the LIGO–VIRGO-teams were $M_{ch} \approx 24 M_\odot$, $M \approx 56 M_\odot$, $M_1 \approx 31 M_\odot$, $M_2 \approx 25 M_\odot$. The Schwarzschild radii of the black holes are $R_{S1} = 91$ km and $R_{S2} = 74$ km. Inserting these radii and the duration of the signal before merger, $t_D \approx 0.07$ s, into Eq. (9.136) gives the distance between the black holes when they emitted the first detected gravitational wave, $R_0 = 630$ km.

On 17 August 2017 LIGO and Virgo detected a gravitational wave from colliding neutron stars, and 1.7 s later ESA's INTEGRAL telescope and NASA's Fermi gamma ray space telescope detected a short gamma ray burst from the same source. The next days and weeks the afterglow of the burst was observed by a large number of telescopes. This was the first time electromagnetic signals were observed from the source of directly detected gravitational waves. It opened a new window for observing the universe and gave a great stimulus to the new research area called *multi-messenger astronomy*.

References

1. Eddington, A.S.: The propagation of gravitational waves. Proc. Roy. Soc. Lond. 268–281 (1922)
2. Abbott, P.B., et al.: Observation of gravitational waves from a binary black hole merger. PRL **116**, 061102 (2016). https://doi.org/10.1103/PhysRevLett.116.061102
3. Satyaprakash, B.S., Schutz, B.F.: Physics, astrophysics and cosmology with gravitational waves. Living Rev. Relativity **12**, 2 (2009)
4. Mathur, H., Brown, K., Lowenstein, A.: An analysis of the LIGO discovery based on introductory physics. Am. J. Phys. **85**(9) (2017)
5. Grøn, Ø., Hervik, S.: Einstein's General Theory of Relativity. Springer (2007)

Chapter 10
Black Holes

Abstract Spacetime outside black holes with and without rotation—i.e. the Kerr and Schwarzschild spacetimes—is studied. By considering the motion of free particles in the Kerr spacetime we find an exact expression for the angular velocity of the inertial dragging. Hawking radiation from a non-rotating black hole is also studied.

10.1 "Surface Gravity": Acceleration of Gravity at the Horizon of a Black Hole

The quantity which is called surface gravity is equal to the acceleration of gravity at the horizon of a black hole. The acceleration of gravity is equal to the acceleration of a freely falling particle instantaneously at rest as observed by a static observer in the coordinate system. However, it has become usual to express the acceleration of gravity measured by an observer in terms of the acceleration scalar *of the observer*. Note that the 4-acceleration, and hence, the acceleration scalar of a free particle vanishes, so this cannot be used in the same way.

It is tempting to define the acceleration of gravity mathematically as equal to the acceleration scalar a of the observer, since this is an invariant quantity representing the acceleration of the observer relatively to a freely falling particle, as measured by the standard measuring rods and clocks of the observer. But we saw in Eq. (8.36) that the standard clocks do not proceed at the horizon of a black hole. Therefore the acceleration scalar of the observer diverges there. For this reason it has become usual to define the acceleration of gravity as

$$g = a\frac{\mathrm{d}\tau}{\mathrm{d}t} = \frac{a}{u^t},\tag{10.1}$$

where u^t is the time component of the 4-velocity of the observer. This quantity has a finite value at the horizon of a black hole.

In this chapter we use units so that $c = 1$. Surface gravity is denoted by κ and is defined by

© Springer Nature Switzerland AG 2020
Ø. Grøn, *Introduction to Einstein's Theory of Relativity*,
Undergraduate Texts in Physics, https://doi.org/10.1007/978-3-030-43862-3_10

$$\kappa = \lim_{r \to r_+} \frac{a}{u^t}, \quad a = \sqrt{a_\mu a^\mu}, \tag{10.2}$$

where r_+ is the horizon radius, $r_+ = R_S$ for the Schwarzschild spacetime, and u^t is the time component of the 4-velocity.

The 4-velocity of an observer permanently at rest in the Schwarzschild spacetime is

$$\vec{u} = u^t \vec{e}_t = \frac{dt}{d\tau} \vec{e}_t = \frac{1}{\sqrt{-g_{tt}}} \vec{e}_t = \frac{\vec{e}_t}{\sqrt{1 - \frac{R_S}{r}}}. \tag{10.3}$$

The only component of the 4-acceleration different from zero is a_r. The 4-acceleration of the observer is

$$\vec{a} = \nabla_{\vec{u}} \vec{u} = u^\mu_{;\nu} u^\nu \vec{e}_\mu = (u^\mu_{,\nu} + \Gamma^\mu_{\alpha\nu} u^\alpha) u^\nu \vec{e}_\mu. \tag{10.4}$$

The covariant, radial component of the 4-acceleration is

$$a_r = (u_{r,\nu} + \Gamma_{r\alpha\nu} u^\alpha) u^\nu = \underbrace{u_{r,\nu} u^\nu}_{=0} + \Gamma_{rtt}(u^t)^2 = \frac{\Gamma_{rtt}}{1 - \frac{R_S}{r}}. \tag{10.5}$$

The Christoffel symbol is

$$\Gamma_{rtt} = -\frac{1}{2} \frac{\partial g_{tt}}{\partial r} = -\frac{R_S}{2r^2}. \tag{10.6}$$

Inserting this into Eq. (10.5) gives

$$a_r = \frac{R_S/2r^2}{1 - \frac{R_S}{r}}. \tag{10.7}$$

The contravariant component of the 4-acceleration is

$$a^r = g^{rr} a_r = \frac{a_r}{g_{rr}} = (1 - \frac{R_S}{r}) a_r = \frac{R_S}{2r^2}. \tag{10.8}$$

The positive sign of a^r means that an observer permanently at rest in the curvature coordinate system accelerates outwards relative to an inertial (freely falling) reference frame which has vanishing 4-acceleration.

The acceleration scalar is

$$a = \sqrt{a_r a^r} = \frac{R_S/2r^2}{\sqrt{1 - \frac{R_S}{r}}}. \tag{10.9}$$

This represents the acceleration as measured with standard instruments at the position of the particle. The acceleration of gravity as defined in Eq. (10.1) is

$$g = \frac{R_S}{2r^2}. \tag{10.10}$$

The surface gravity is equal to the acceleration of gravity at the horizon of a black hole,

$$\kappa = \lim_{r \to R_S} g = \frac{1}{2R_S}. \tag{10.11}$$

Inserting the velocity of light we get

$$\kappa = \frac{c^2}{2R_S}. \tag{10.12}$$

On the horizon of a black hole with one solar mass $\kappa_\odot = 2 \times 10^{13} \text{m/s}^2$.

10.2 Hawking Radiation: Radiation from a Black Hole

The radiation from a black hole has a thermal spectrum. Following Hawking we shall write down an expression for the temperature of a Schwarzschild black hole of mass M. The Planck spectrum has an intensity maximum at a wavelength given by Wien's displacement law,

$$\Lambda = \frac{N\hbar c}{kT}. \tag{10.13}$$

where k is Boltmann's constant and $N = 0.2014$. For radiation emitted from a black hole Hawking derived the following expression for the wavelength at a maximum intensity,

$$\Lambda = 4\pi N R_S = \frac{8\pi NGM}{c^2}. \tag{10.14}$$

Substituting for Λ from Wien's displacement law leads to

$$T = \frac{\hbar c^3}{8\pi GkM} = \frac{\hbar c}{2\pi k}\kappa. \tag{10.15}$$

Inserting values for $\hbar$, c and k gives

$$T \approx \frac{2 \times 10^{-4}\text{m}}{R_S} K. \tag{10.16}$$

For a black hole with one solar mass, we have $T_\odot \approx 10^{-7}$. When the mass decreases because it radiates, the temperature *increases*. So a black hole has a negative heat capacity. The energy loss of a black hole because of radiation is given by the Stefan–Boltzmann law,

$$-\frac{\mathrm{d}M}{\mathrm{d}t} = \sigma T^4 \frac{A}{c^2}, \tag{10.17}$$

where A is the area of the horizon

$$A = 4\pi R_S^2 = \frac{16\pi G^2 M^2}{c^4}, \tag{10.18}$$

In Eq. (10.17)

$$\sigma = \frac{\pi^2 k^4}{60\hbar^3 c^2} \tag{10.19}$$

is Stefan–Boltzmann's constant. Combining these equations we get

$$\frac{\mathrm{d}M}{\mathrm{d}t} = -\frac{K_H}{M^2}, \ K_H = \frac{\hbar c^4}{15360\pi G^2} = 2.2 \cdot 10^{15} \, \text{kg}^3/\text{s}. \tag{10.20}$$

Hence

$$\int_{M_0}^{M} M^2 \mathrm{d}M = -K_H t, \tag{10.21}$$

giving

$$M(t) = \left(M_0^3 - 3K_H t\right)^{1/3}. \tag{10.22}$$

The time taken for a black hole to evaporate is

$$t_{ev} = \frac{M_0^3}{3K_H} = \left(\frac{M_0}{M_\odot}\right)^3 t_{ev\odot}, \tag{10.23}$$

where $t_{ev\odot} = 2.1 \cdot 10^{67}$ years is the evaporation time of the Sun. The evaporation time of a black hole with one Planck mass is

$$t_{evP} = 5120\pi \sqrt{\frac{\hbar G}{c^5}} = 5120\pi \, t_P. \tag{10.24}$$

where $t_P = 5.4 \cdot 10^{-44}$ s is the Planck time. Hence $t_{evP} = 8.67 \cdot 10^{-40}$ s. A primordial black hole made at the Big Bang and exploding at the present time when the age of the universe is $t_0 = 13.7 \cdot 10^9$ years, is

$$M_0 = (3K_H t_0)^{1/3},\tag{10.25}$$

giving $M_0 \approx 10^{11}$kg which is around the mass of Mount Everest. It is called a "mini-black hole".

10.3 Rotating Black Holes: The Kerr Metric

The solution of Einstein's field equations describing spacetime outside a rotating mass distribution was found by Roy Kerr in 1963 and is therefore called the Kerr spacetime.

A time-independent, time-orthogonal metric is known as a *static* metric. A time-independent metric is known as a *stationary* metric. A stationary metric allows rotation.

Consider a stationary metric which describes an axially symmetric space

$$
\begin{aligned}
ds^2 &= -e^{2\nu}dt^2 + e^{2\mu}dr^2 + e^{2\psi}(d\phi - \omega dt)^2 + e^{2\lambda}d\theta^2 \\
&= -\left(e^{2\nu} - e^{2\psi}\omega^2\right)dt^2 + e^{2\mu}dr^2 + e^{2\psi}d\phi^2 + e^{2\lambda}d\theta^2 - 2e^{2\psi}\omega\, d\phi\, dt
\end{aligned} \tag{10.26}
$$

where ν, μ, ψ, λ and ω are functions of r and θ. The covariant components of the metric tensor are

$$g_{tt} = e^{2\psi}\omega^2 - e^{2\nu}, \quad g_{rr} = e^{2\mu}, \quad g_{\phi\phi} = e^{2\psi}, \quad g_{\theta\theta} = e^{2\lambda}, \quad g_{t\phi} = -e^{2\psi}\omega. \tag{10.27}$$

Hence,

$$\omega = -\frac{g_{t\phi}}{g_{\phi\phi}}, \quad g_{tt} = g_{\phi\phi}\omega^2 - e^{2\nu}. \tag{10.28}$$

By solving the vacuum field equations for this line element, Kerr found the solution:

$$
\begin{aligned}
&e^{2\nu} = \frac{\rho^2 \Delta}{\Sigma^2}, \quad e^{2\mu} = \frac{\rho^2}{\Delta}, \quad e^{2\psi} = \frac{\Sigma^2}{\rho^2}\sin^2\theta, \quad e^{2\lambda} = \rho^2, \\
&\omega = \frac{R_S ar}{\Sigma^2}, \quad \text{where} \quad \rho^2 = r^2 + a^2\cos^2\theta, \\
&\Delta = r^2 + a^2 - R_S r, \\
&\Sigma^2 = (r^2 + a^2)^2 - a^2\Delta\sin^2\theta.
\end{aligned} \tag{10.29}
$$

This is the Kerr solution expressed in *Boyer–Lindquist coordinates*. The function ω is an angular velocity. As mentioned above the Kerr solution is the metric for spacetime outside a rotating mass distribution. The constant a represents spin per mass unit for the mass distribution and R_S its Schwarzschild radius. With the units we use, a has dimension length.

The line element has the form

$$ds^2 = -\left(1 - \frac{R_S r}{\rho^2}\right)dt^2 + \frac{\rho^2}{\Delta}dr^2 - \frac{2R_S ar}{\rho^2}\sin^2\theta\, dt\, d\phi + \rho^2 d\theta^2$$
$$+ \left(r^2 + a^2 + \frac{R_S a^2 r}{\rho^2}\sin^2\theta\right)\sin^2\theta\, d\phi^2. \tag{10.30}$$

Light emitted from the surface, $r = r_0$, where $g_{tt}(r_0) = 0$, is infinitely redshifted further out. Observed from the outside time does not proceed at this surface. The radial coordinate r_0 is given by

$$\rho^2 = R_S r_0, \tag{10.31}$$

or

$$r_0^2 + a^2\cos^2\theta = R_S r_0, \tag{10.32}$$

giving

$$r_0 = R_S/2 \pm \sqrt{(R_S/2)^2 - a^2\cos^2\theta}. \tag{10.33}$$

This is the equation of two surfaces with the property that light emitted from the surfaces is infinitely redshifted. In the equatorial plane, $\theta = \pi/2$, the radius of the surface with infinitely redshifted light has a coordinate distance from the origin equal to the Schwarzschild radius of the mass distribution,

$$r_0 = R_S. \tag{10.34}$$

10.3.1   Zero-Angular Momentum Observers

The Lagrange function of a free particle in the equator plane, $\theta = \pi/2$, is

$$L = -\frac{1}{2}(e^{2\nu} - \omega^2 e^{2\psi})\dot{t}^2 + \frac{1}{2}e^{2\mu}\dot{r}^2 + \frac{1}{2}e^{2\psi}\dot{\phi}^2 + \frac{1}{2}e^{2\lambda}\dot{\theta}^2 - \omega e^{2\psi}\dot{t}\dot{\phi}. \tag{10.35}$$

Here $\dot{\theta} = 0$. The constant momentum p_ϕ of the cyclic coordinates ϕ is

$$p_\phi \equiv \frac{\partial L}{\partial \dot\phi} = e^{2\psi}(\dot\phi - \omega \dot t), \quad \dot t = \frac{dt}{d\tau}, \quad \dot\phi = \frac{d\phi}{d\tau}. \tag{10.36}$$

The angular velocity of the particle relative to the coordinate system is

$$\Omega = \frac{d\phi}{dt} = \frac{\dot\phi}{\dot t}. \tag{10.37}$$

Hence

$$p_\phi = e^{2\psi} \dot t (\Omega - \omega). \tag{10.38}$$

Thus a zero-angular momentum observer, a ZAMO, has an angular velocity in the Boyer–Lindquist coordinate system, $\Omega = \omega$. This gives the physical interpretation of ω. It is the angular velocity of the ZAMOs in the Kerr spacetime relative to an observer at rest infinitely far from the mass distribution. Hence ω *is the angular velocity of the inertial dragging in the Kerr spacetime.* It follows from Eqs. (10.28) and (10.29) that

$$\omega = -\frac{g_{t\phi}}{g_{\phi\phi}} = \frac{R_S a}{r^3 + a^2(r + R_S)}. \tag{10.39}$$

Note that $\omega \to 0$ when $r \to \infty$. The Kerr spacetime approaches the Minkowski spacetime for large r. Calculating the inertial dragging angular velocity at the surface of the Earth one finds $\omega_{\text{Earth}} = 10^{-14}$ rad/s. At the infinite redshift surface with radius (10.34) the dragging angular velocity is

$$\omega(r_0) = \frac{a}{R_S^2 + 2a^2}. \tag{10.40}$$

10.3.2 Does the Kerr Spacetime Have a Horizon?

Definition 10.3.1 (*Horizon*) A horizon is a one-way surface through which anything can enter, but nothing can exit, not even light. Hence the region inside the horizon is invisible for an outside observer.

Consider a particle in an orbit with constant r and θ. Its 4-velocity is

$$\vec u = \frac{d\vec x}{d\tau} = \frac{dt}{d\tau}\frac{d\vec x}{dt} = (-g_{tt} - 2g_{t\phi}\Omega - g_{\phi\phi}\Omega^2)^{-\frac{1}{2}}(1, \Omega). \tag{10.41}$$

The existence of a stationary orbit requires

$$g_{\phi\phi}\Omega^2 + 2g_{t\phi}\Omega + g_{tt} < 0. \tag{10.42}$$

This implies that

$$\Omega_{\min} < \Omega < \Omega_{\max}, \tag{10.43}$$

where

$$\Omega_{\min} = \omega - \sqrt{\omega^2 - \frac{g_{tt}}{g_{\phi\phi}}} \text{ and } \Omega_{\max} = \omega + \sqrt{\omega^2 - \frac{g_{tt}}{g_{\phi\phi}}}. \tag{10.44}$$

Outside the surface with infinite redshift $g_{tt} < 0$. That is Ω can be negative, zero and positive. Inside the surface $r = r_0$ with infinite redshift $g_{tt} > 0$. Here $\Omega_{\min} > 0$ and static particles, $\Omega = 0$, cannot exist. This is due to the inertial dragging effect. The surface $r = r_0$ is therefore known as "the static border". The interval of Ω, where stationary orbits are allowed, is reduced to zero when $\Omega_{\min} = \Omega_{\max}$, that is

$$g_{tt} = \omega^2 g_{\phi\phi}, \tag{10.45}$$

which is the equation of the horizon. For the Kerr metric we have

$$g_{tt} = \omega^2 g_{\phi\phi} - e^{2\nu}. \tag{10.46}$$

Comparing with the second of Eq. (10.28), we see that the equation of the horizon becomes

$$e^{2\nu} = 0, \tag{10.47}$$

which requires $\Delta = 0$ or

$$r^2 - R_S r + a^2 = 0. \tag{10.48}$$

The exterior solution is

$$r_+ = R_S/2 + \sqrt{(R_S/2)^2 - a^2}, \tag{10.49}$$

and this is the equation for a spherical surface. Note that the radius of the horizon is smaller than the Schwarzschild radius of the rotating mass. The dragging angular velocity at the horizon is

$$\omega(r_+) = \frac{a}{R_S r_+}. \tag{10.50}$$

The static border and horizon of a Kerr black hole are shown in Fig. 10.1.

Going from a region outside the static border and inwards we have the following situation. In the outside region it is possible for an observer to stay at rest if a suitable non-gravitational force acts upon the observer. The surface $r = r_0$ has two roles;

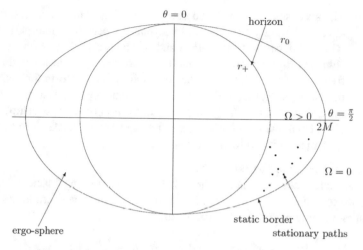

Fig. 10.1 Static border and horizon of a Kerr black

light emitted from sources at rest on it are infinitely redshifted, and inside the surface the inertial dragging is so strong that it is impossible to stay at rest even by trying to do so by means of a motor. But it is possible to move along a path with constant r, i.e. there exist stationary orbits. Also it is possible to come out of this region. But inside the horizon, everything fall inwards, and it is not possible neither to move with constant r nor to move outwards.

Exercises

10.1. *A spaceship falling into a black hole*

(a) In this problem we will consider a spaceship (A) falling radially from the Earth (neglecting the gravitational field of the earth) into a Schwarzschild black hole with mass the mass of the Sun, $M = M_{\text{Sun}}$ at the position of the Sun, 150 million km from the Earth.

What is the Schwarzschild radius of the black hole?

Find the equations of motion of the spaceship in curvature coordinates r and t, using the proper time τ as time parameter.

Solve the equations of motion with the initial condition that the space ship falls from rest at the Earth at proper time $\tau = 0$.

When (in terms of τ) does the spaceship reach the Schwarzschild radius? And the singularity?

(b)   The spaceship (A) has radio contact with a stationary space station (B) at $r_B = 1$ light years. The radio signals are sent with intervals ΔT and with frequency ω from both A and B. The receivers at A and B receive signals with frequency ω_A and ω_B, respectively. Find ω_A and ω_B as a function of the position of the spaceship. Hint: Perform the calculation in two steps. At first find the change in frequency between two stationary inertial systems in the points r_A (the position of the spaceship) and r_B. Then calculate the change in frequency due to a transfer into an inertial system with the velocity of the spaceship.

10.2.   *Kinematics in the Kerr spacetime*

A Kerr black hole is an electrically neutral, rotating black hole. When spacetime outside a Kerr black hole is described in Boyer–Lindquist coordinates, the line element is the following,

$$ds^2 = -e^{2\nu}dt^2 + e^{2\mu}dr^2 + e^{2\lambda}d\theta^2 + e^{2\psi}(d\phi - \omega dt)^2,$$

where

$$e^{2\nu} = \frac{\rho^2 \Delta}{\Sigma^2}, e^{2\mu} = \frac{\rho^2}{\Delta}, e^{2\lambda} = \rho^2,$$

$$e^{2\psi} = \left(\frac{\Sigma^2}{\rho^2}\right)\sin^2\theta, \omega = -\frac{g_{t\phi}}{g_{\phi\phi}} = \frac{2Mar}{\Sigma^2},$$

$$\rho^2 = r^2 + a^2\cos^2\theta, \ \Delta = r^2 + a^2 - 2Mr, \ \Sigma^2 = (r^2 + a^2)^2 - a^2\Delta\sin^2\theta.$$

Here M is the mass of the hole and a its spin per unit mass.

(a)   Consider light moving in negative and positive direction of ϕ in the equatorial plane, $\theta = \pi/2$. What is the coordinate velocity $c_\phi = d\phi/dt$ of light?

We now want to investigate the Sagnac effect in the Kerr space. An emitter–receiver is attached to a point in the BL coordinate system. Light signals with the frequency ν are sent by means of mirrors in both directions along the circle $r = r_0, \theta = \pi/2$. Find the travel time difference of light travelling in opposite directions, when the signals reach the receiver.

(b)   A *ZAMO* is an observer with vanishing angular momentum. In the following a *ZAMO* in the Kerr spacetime will be describing particles with fixed r- and θ-coordinates. Introduce an orthonormal basis $(\vec{e}_{\hat{t}'}, \vec{e}_{\hat{r}'}, \vec{e}_{\hat{\theta}'}, \vec{e}_{\hat{\phi}'})$, where $\vec{e}_{\hat{t}'}$ is the 4-velocity of a *ZAMO*. The dual basis 1-forms are

$$\underline{\omega}^{\hat{t}'} = e^\nu \underline{\omega}^t, \quad \underline{\omega}^{\hat{r}'} = e^\mu \underline{\omega}^r,$$

$$\underline{\omega}^{\hat{\theta}'} = e^\lambda \underline{\omega}^\theta, \quad \underline{\omega}^{\hat{\phi}'} = e^\psi(\underline{\omega}^\phi - \omega \underline{\omega}^t). \tag{10.51}$$

Show that

$$\vec{e}_{\hat{t}'} = e^{-\nu}(\vec{e}_t + \omega\vec{e}_\phi), \quad \vec{e}_{\hat{r}'} = e^{-\mu}\vec{e}_r,$$
$$\vec{e}_{\hat{\theta}'} = e^{-\lambda}\vec{e}_\theta, \quad \vec{e}_{\hat{\phi}'} = e^{-\psi}\vec{e}_\phi, \tag{10.52}$$

where $(\vec{e}_t, \vec{e}_r, \vec{e}_\theta, \vec{e}_\phi)$ are the coordinate basis vectors in the BL coordinate system.

Given a particle with 4-velocity components

$$U^\mu = (\dot{t}, \ \dot{\phi}) = \dot{t}(1, \ \Omega), \quad \Omega = \frac{d\phi}{dt} \tag{10.53}$$

in the Boyer–Lindquist coordinate system.

Show that the physical velocity of the particle, measured by a *ZAMO*, is

$$v^{\hat{\phi}'} = e^{\psi-\nu}(\Omega - \omega). \tag{10.54}$$

What is the velocity $v_0^{\hat{\phi}'}$ of a fixed coordinate point measured by a *ZAMO*?

(c) Introduce an orthonormal basis field given by the expressions

$$\vec{e}_{\hat{0}} = (-g_{00})^{-1/2}\vec{e}_0, \quad \vec{e}_{\hat{i}} = (\gamma_{ii})^{-1/2}[\vec{e}_i - (g_{i0}/g_{00})\vec{e}_0], \tag{10.55}$$

where

$$\gamma_{ii} = g_{ii} - g_{i0}^2/g_{00}.$$

Show that

$$\vec{e}_{\hat{t}} = \hat{\gamma}e^{-\nu}\vec{e}_t, \quad \vec{e}_{\hat{r}} = e^{-\mu}\vec{e}_r,$$
$$\vec{e}_{\hat{\theta}} = e^{-\lambda}\vec{e}_r, \quad \vec{e}_{\hat{\phi}} = \hat{\gamma}^{-1}e^{-\psi}\vec{e}_\phi + \hat{\gamma}e^{-\nu}v_0^{\hat{\phi}'}\vec{e}_t, \tag{10.56}$$

where $\hat{\gamma} = (1 - (v_0^{\hat{\phi}'})^2)^{-1/2}$. Find the dual basis 1-forms.

10.3. *A gravitomagnetic clock effect*

This problem is concerned with the difference of proper time shown by two clocks moving freely in opposite directions in the equatorial plane of the Kerr spacetime outside a rotating body. The clocks move along a path with $r = $ constant and $\theta = \pi/2$.

(a) Show that in this case the radial geodesic equation reduces to

$$\Gamma_{tt}^r dt^2 + 2\Gamma_{\phi t}^r d\phi dt + \Gamma_{\phi\phi}^r d\phi^2 = 0. \tag{10.57}$$

Calculate the Christoffel symbols and show that the equation takes the form

$$\left(\frac{dt}{d\phi}\right)^2 - 2a\frac{dt}{d\phi} + a^2 - \frac{r^3}{M} = 0, \qquad (10.58)$$

where M is the mass of the rotating body and a its angular momentum per unit mass, $a = J/M$.

(b) Show that the time difference for one closed orbit in ($\phi \to \phi + 2\pi$) the direct and the retrograde direction is $t_+ - t_- \approx 4\pi a = 4\pi J/M$, or in S.I. units,

$$t_+ - t_- = 4\pi a = 4\pi J/mc^2. \qquad (10.59)$$

Estimate this time difference for clocks in satellites moving in the equatorial plane of the Earth. (The mass of the Earth is $m = 6 \cdot 10^{26}$ kg and its angular momentum $J = 10^{34}$ kg m^2/s.)

Chapter 11
Sources of Gravitational Fields

Abstract In this chapter we shall first find a general expression of the acceleration of gravity due to a mass distribution. Then we shall deduce the solution of Einstein's field equations inside an incompressible star—the internal Schwarzschild solution. Furthermore we shall present Israel's formalism for describing singular mass shells in the general theory of relativity, and apply this first to a shell consisting of dust particles, and then to find a source of the conformally flat, spherically symmetric Levi-Civita—Bertotti—Robinson metric, and finally to the Kerr spacetime. Lastly we shall introduce a river model of space.

11.1   The Pressure Contribution to the Gravitational Mass of a Static, Spherically Symmetric System

According to the Definition (10.1) of the acceleration of gravity (including a minus sign denoting inwards acceleration, here),

$$g = -\frac{a}{u^t}, \quad a = \sqrt{a_\mu a^\mu}. \tag{11.1}$$

where a is the acceleration scalar and u^t the time component of the 4-velocity of an observer permanently at rest in the coordinate system. We have the line element,

$$ds^2 = -e^{2\alpha(r)}dt^2 + e^{2\beta(r)}dr^2 + r^2 d\Omega^2. \tag{11.2}$$

Hence

$$g_{tt} = -e^{2\alpha}, \quad g_{rr} = e^{2\beta}, \tag{11.3}$$

which gives for the acceleration of gravity,

$$g = -e^{\alpha-\beta}\alpha'. \tag{11.4}$$

© Springer Nature Switzerland AG 2020
Ø. Grøn, *Introduction to Einstein's Theory of Relativity*,
Undergraduate Texts in Physics, https://doi.org/10.1007/978-3-030-43862-3_11

From the expressions (8.15) for the components of the Einstein tensor $E_{\hat{t}\hat{t}}$, $E_{\hat{r}\hat{r}}$, $E_{\hat{\theta}\hat{\theta}}$, $E_{\hat{\phi}\hat{\phi}}$, it follows that

$$E_{\hat{t}}^{\hat{t}} - E_{\hat{r}}^{\hat{r}} - E_{\hat{\theta}}^{\hat{\theta}} - E_{\hat{\phi}}^{\hat{\phi}} = -2e^{-2\beta}\left(\frac{2\alpha'}{r} + \alpha'' + \alpha'^2 - \alpha'\beta'\right). \tag{11.5}$$

We also have

$$(r^2 e^{\alpha-\beta}\alpha')' = r^2 e^{\alpha-\beta}\left(\frac{2\alpha'}{r} + \alpha'' + \alpha'^2 - \alpha'\beta'\right), \tag{11.6}$$

which gives

$$g = \frac{1}{2r^2}\int (E_{\hat{t}}^{\hat{t}} - E_{\hat{r}}^{\hat{r}} - E_{\hat{\theta}}^{\hat{\theta}} - E_{\hat{\phi}}^{\hat{\phi}})r^2 e^{\alpha+\beta}dr. \tag{11.7}$$

By applying Einstein's field equations

$$E_{\hat{\nu}}^{\hat{\mu}} = 8\pi G T_{\hat{\nu}}^{\hat{\mu}} \tag{11.8}$$

we get

$$g = \frac{4\pi G}{r^2}\int (T_{\hat{t}}^{\hat{t}} - T_{\hat{r}}^{\hat{r}} - T_{\hat{\theta}}^{\hat{\theta}} - T_{\hat{\phi}}^{\hat{\phi}})r^2 e^{\alpha+\beta}dr. \tag{11.9}$$

This is the Tolman–Whittaker expression for gravitational acceleration. The corresponding Newtonian expression is

$$g_N = -\frac{4\pi G}{r^2}\int \rho r^2 dr. \tag{11.10}$$

The relativistic gravitational mass–density is therefore defined as

$$\rho_G = -T_{\hat{t}}^{\hat{t}} + T_{\hat{r}}^{\hat{r}} + T_{\hat{\theta}}^{\hat{\theta}} + T_{\hat{\phi}}^{\hat{\phi}}. \tag{11.11}$$

For an isotropic fluid with

$$T_{\hat{t}}^{\hat{t}} = -\rho, \quad T_{\hat{r}}^{\hat{r}} = T_{\hat{\theta}}^{\hat{\theta}} = T_{\hat{\phi}}^{\hat{\phi}} = p \tag{11.12}$$

we get $\rho_G = \rho + 3p$ (with $c = 1$), which becomes

$$\rho_G = \rho + \frac{3p}{c^2}. \tag{11.13}$$

It follows that in relativity, pressure has a gravitational effect. Greater pressure gives increasing gravitational attraction. Strain ($p < 0$) decreases the gravitational attraction. In the Newtonian limit, $c \to \infty$, pressure has no gravitational effect.

Inserting the equation of state (7.17) of LIVE into Eq. (11.13) gives $\rho_G = -2\rho < 0$. Hence *the assumption that it is not possible to measure velocity relative to the vacuum energy, and that the vacuum energy can be described as a perfect fluid, together with the general theory of relativity, imply that the vacuum energy produces repulsive gravitation.*

11.2   The Tolman–Oppenheimer–Volkoff Equation

We are going to solve Einstein's field equations inside a spherically symmetric, static mass distribution with density $\rho(r)$. Then the line-element can be given the form (11.2). Using Eq. (8.15) for the components of the Einstein curvature tensor the field equation for $E_{\hat{t}\hat{t}}$ can be written

$$\frac{1}{r^2}\frac{d}{dr}[r(1 - e^{-2\beta})] = 8\pi G\rho. \tag{11.14}$$

Integration gives

$$r(1 - e^{-2\beta}) = 2G\int_0^r 4\pi\rho r^2 dr. \tag{11.15}$$

The mass inside a spherical surface with coordinate radius r is

$$m(r) = \int_0^r 4\pi\rho r^2 dr. \tag{11.16}$$

Hence, Eq. (11.15) can be written as

$$e^{-2\beta} = 1 - \frac{2Gm(r)}{r} = \frac{1}{g_{rr}}. \tag{11.17}$$

From the field equation for $E_{\hat{r}\hat{r}}$, we have

$$\frac{2}{r}\frac{d\alpha}{dr}e^{-2\beta} - \frac{1}{r^2}(1 - e^{-2\beta}) = 8\pi Gp. \tag{11.18}$$

This leads to

$$\frac{2}{r}\frac{d\alpha}{dr}\left(1 - \frac{2Gm(r)}{r}\right) - \frac{2Gm(r)}{r^3} = 8\pi Gp, \tag{11.19}$$

or

$$\frac{d\alpha}{dr} = G\frac{m(r) + 4\pi r^3 p(r)}{r(r - 2Gm(r))}. \tag{11.20}$$

The relativistic equation of hydrostatic equilibrium is

$$T^{\hat{r}\hat{v}}_{;v} = 0. \tag{11.21}$$

Written out

$$T^{\hat{r}\hat{v}}_{,\hat{v}} + \Gamma^{v}_{\hat{\alpha}\hat{v}}T^{\hat{r}\hat{\alpha}} + \Gamma^{\hat{r}}_{\hat{\alpha}\hat{v}}T^{\hat{\alpha}\hat{v}} = 0. \tag{11.22}$$

The first term is

$$T^{\hat{r}\hat{v}}_{,\hat{v}} = T^{\hat{r}\hat{r}}_{,\hat{r}} = p_{,\hat{r}} = \vec{e}_{\hat{r}}(p). \tag{11.23}$$

Since $\vec{e}_r \cdot \vec{e}_r = g_{rr}$ we have $|\vec{e}_r| = \sqrt{g_{rr}}$. Hence, $\vec{e}_{\hat{r}} = \left(1/\sqrt{g_{rr}}\right)\vec{e}_r$. This gives

$$T^{\hat{r}\hat{v}}_{,\hat{v}} = \vec{e}_{\hat{r}}(p) = \frac{1}{\sqrt{g_{rr}}}\vec{e}_r(p) = e^{-\beta}\frac{dp}{dr}. \tag{11.24}$$

We here use ordinary derivatives instead of partial derivatives since p only depends upon r. The second term is

$$\Gamma^{\hat{v}}_{\hat{\alpha}\hat{v}}T^{\hat{r}\hat{\alpha}} = \Gamma^{\hat{v}}_{\hat{r}\hat{v}}T^{\hat{r}\hat{r}} = \Gamma^{\hat{v}}_{\hat{r}\hat{v}}p = \Gamma^{\hat{t}}_{\hat{r}\hat{t}}p + \Gamma^{\hat{i}}_{\hat{r}\hat{i}}p. \tag{11.25}$$

The third term is

$$\Gamma^{\hat{r}}_{\hat{\alpha}\hat{v}}T^{\hat{\alpha}\hat{v}} = \Gamma^{\hat{r}}_{\hat{v}\hat{v}}T^{\hat{v}\hat{v}} = \Gamma^{\hat{r}}_{\hat{t}\hat{t}}\rho + \Gamma^{\hat{r}}_{\hat{i}\hat{i}}p. \tag{11.26}$$

In orthonormal basis, we have

$$\Omega_{\hat{v}\hat{\mu}} = -\Omega_{\hat{\mu}\hat{v}} \Rightarrow \Gamma_{\hat{v}\hat{\mu}\hat{\alpha}} = \Gamma_{\hat{\mu}\hat{v}\hat{\alpha}} \quad \text{and} \quad \Gamma^{\hat{i}}_{\hat{r}\hat{i}} = \Gamma_{\hat{i}\hat{r}\hat{i}} = -\Gamma_{\hat{r}\hat{i}\hat{i}} = -\Gamma^{\hat{r}}_{\hat{i}\hat{i}}. \tag{11.27}$$

Hence the last terms in the expressions for $\Gamma^{\hat{v}}_{\hat{\alpha}\hat{v}}T^{\hat{r}\hat{\alpha}}$ and $\Gamma^{\hat{r}}_{\hat{\alpha}\hat{v}}T^{\hat{\alpha}\hat{v}}$ cancel each other, and the hydrostatic equation then takes the form

$$e^{-\beta}\frac{dp}{dr} + \Gamma^{\hat{t}}_{\hat{r}\hat{t}}p + \Gamma^{\hat{r}}_{\hat{t}\hat{t}}\rho = 0. \tag{11.28}$$

Furthermore

$$\Gamma^{\hat{t}}_{\hat{r}\hat{t}} = -\Gamma_{\hat{t}\hat{r}\hat{t}} = \Gamma_{\hat{r}\hat{t}\hat{t}} = \Gamma^{\hat{r}}_{\hat{t}\hat{t}}, \tag{11.29}$$

and

$$\Gamma^{\hat{r}}_{\hat{t}\hat{t}} = e^{-\beta}\frac{d\alpha}{dr}. \tag{11.30}$$

The hydrostatic equation then takes the form

$$\frac{dp}{dr} + (p+\rho)\frac{d\alpha}{dr} = 0. \tag{11.31}$$

Inserting Eq. (11.20) into Eq. (11.31) gives

$$\frac{dp}{dr} = -G(\rho+p)\frac{m(r)+4\pi r^3 p(r)}{r(r-2Gm(r))}. \tag{11.32}$$

This is *the Tolman–Oppenheimer–Volkov (TOV) equation* which can be used to construct relativistic star models. It can be written in the form

$$\frac{dp}{dr} = -\frac{G\rho m(r)}{r^2}\frac{\left(1+\frac{p(r)}{\rho}\right)\left(1+\frac{4\pi r^3 p(r)}{m(r)}\right)}{1-\frac{2Gm(r)}{r}}. \tag{11.33}$$

In Newtonian theory, only the first factor appears. The three last terms are relativistic corrections. They show that pressure, $p > 0$, makes relativistic gravity stronger than Newtonian gravity. This is significant for the possibility of collapse to black holes.

11.3   An Exact Solution for Incompressible Stars—Schwarzschild's Interior Solution

The metric component $g_{tt} = -e^{2\alpha(r)}$ can now be calculated from Eq. (11.31) in the form

$$d\alpha = -\frac{dp}{\rho+p} \tag{11.34}$$

We now consider an incompressible star, say a neutron star, with $\rho =$ constant and radius R. Integration of eq. (11.34) with vanishing pressure at the surface then gives

$$\int_{\alpha(R)}^{\alpha(r)} d\alpha = -\int_0^p \frac{dp}{\rho+p} \Rightarrow \alpha(r) = \alpha(R) - \ln\left(1+\frac{p}{\rho}\right). \tag{11.35}$$

Requiring continuity of the metric at $r = R$, and using the exterior Schwarzschild metric, we get

$$g_{tt}(r) = e^{2\alpha(r)} = \frac{g_{tt}(R)}{\left(1 + \frac{p}{\rho}\right)^2} = -\frac{1 - \frac{R_S}{R}}{\left(1 + \frac{p}{\rho}\right)^2}. \tag{11.36}$$

For an incompressible star, the mass inside a radius r is

$$m(r) = \frac{4\pi}{3} \rho r^3. \tag{11.37}$$

We then get

$$e^{-2\beta} = 1 - \frac{2Gm(r)}{r} = 1 - \frac{8\pi G\rho}{3} r^2. \tag{11.38}$$

Defining a constant a with dimension length by

$$a^2 = \frac{3}{8\pi G\rho}, \tag{11.39}$$

this may be written as

$$e^{-2\beta} = 1 - \frac{r^2}{a^2}. \tag{11.40}$$

It may be noted that

$$R_S = 2Gm(R) = \frac{8\pi G\rho}{3} R^3 = \frac{R^3}{a^2}. \tag{11.41}$$

Hence

$$a^2 = \frac{R^3}{R_S} \quad \text{and} \quad \frac{R^2}{a^2} = \frac{R_S}{R}. \tag{11.42}$$

The TOV-equation may now be written

$$\frac{dp}{dr} = -\frac{1}{2a^2\rho}(\rho + 3p)(\rho + p)\frac{r}{1 - \frac{r^2}{a^2}}. \tag{11.43}$$

So

$$\int_0^p \frac{dp}{(\rho + 3p)(\rho + p)} = -\frac{1}{2a^2\rho} \int_R^r \frac{r}{1 - \frac{r^2}{a^2}} dr. \tag{11.44}$$

which leads to

$$\frac{\rho + p}{\rho + 3p} = \sqrt{\frac{a^2 - R^2}{a^2 - r^2}}, \tag{11.45}$$

giving the pressure distribution

$$p(r) = \frac{\sqrt{a^2 - r^2} - \sqrt{a^2 - R^2}}{3\sqrt{a^2 - R^2} - \sqrt{a^2 - r^2}}\rho = \frac{\sqrt{1 - \frac{R_s}{R^3}r^2} - \sqrt{1 - \frac{R_s}{R}}}{3\sqrt{1 - \frac{R_s}{R}} - \sqrt{1 - \frac{R_s}{R^3}r^2}}\rho, \quad r < R. \tag{11.46}$$

This gives

$$1 + \frac{p}{\rho} = \frac{2\sqrt{1 - \frac{R_s}{R}}}{3\sqrt{1 - \frac{R_s}{R}} - \sqrt{1 - \frac{R_s}{R^3}r^2}}. \tag{11.47}$$

Combining this with Eq. (11.36) we get

$$g_{tt} = -\frac{1}{4}\left(3\sqrt{1 - \frac{R_s}{R}} - \sqrt{1 - \frac{R_s}{R^3}r^2}\right)^2. \tag{11.48}$$

Hence the line-element of the spacetime inside an incompressible star is

$$ds^2 = -\frac{1}{4}\left(3\sqrt{1 - \frac{R_s}{R}} - \sqrt{1 - \frac{R_s}{R^3}r^2}\right)^2 dt^2 + \frac{dr^2}{1 - \frac{R_s}{R^3}r^2} + r^2 d\Omega^2. \tag{11.49}$$

The spacetime described by this line-element is called the *internal Schwarzschild spacetime*.

In Newtonian gravity, there is no limit to how large a star can be if there exists a sufficiently effective mechanism for generating a pressure gradient which may resist gravity. We shall see that it is not so in relativistic gravity.

In order to satisfy the condition of hydrostatic equilibrium, the pressure must be positive. This must be valid at every distance from the centre of the star, also at the centre, $p(0) > 0$. It follows from the expression for the pressure distribution that

$$p(0) = \frac{1 - \sqrt{1 - \frac{R_s}{R}}}{3\sqrt{1 - \frac{R_s}{R}} - 1} > 0, \tag{11.50}$$

which requires that $R_S < (8/9)R$. Including here the velocity of light, this leads to the condition

$$M < \frac{c^2}{G\sqrt{3\pi G\rho}} = \frac{2\sqrt{2}}{3}\frac{c^2}{G}a, \quad a = c\sqrt{\frac{3}{8\pi G\rho}}. \tag{11.51}$$

A star with a larger mass will collapse to a black hole independent of the mechanism that generates a pressure gradient in the star.

Let us consider an incompressible neutron star as an illustrating example. A typical density is then $\rho_n = 5 \cdot 10^{17}$ kg/m^3. This gives $a_n = 18$ km and $M < 22 \cdot 10^{30}$ kg $= 11\,M_\odot$.

11.4   The Israel Formalism for Describing Singular Mass Shells in the General Theory of Relativity

We shall here follow the descriptions of this formalism as presented by Israel [1], Misner, Thorne Wheeler [2], and Lightman, Press, Price and Teukolsky [3].

Let S_{ij} be the components of the energy–momentum tensor of a singular shell, and $T_{\alpha\beta}$ the components of the energy–momentum tensor of the medium filling the region at each side of the shell. Here the indices i, j mark the coordinates in the shell and α, β the coordinates of the spacetime outside the shell. Israel has shown that the equation of continuity can be written

$$u^j S^m_{j|m} = -\left[T_{\alpha\beta}u^\alpha n^\beta\right], \tag{11.52}$$

where u^j is the j-component of the 4-velocity of a point on the shell, n^β the $\beta-$component of its normal vector, the vertical line denotes covariant derivative with respect to the intrinsic metric on the shell, and the bracket denotes the difference between the values of the quantity inside the bracket at the two sides of the shell. The energy–momentum tensor of the shell has the form

$$S_{ij} = \sigma u_i u_j + t_{ij}, \tag{11.53}$$

where σ is the mass per unit area of the shell, and $t_{i\,j}$ are the components of the stress tensor of the shell.

The components of the extrinsic curvature tensor of the shell are ([2] p. 513)

$$K_{ij} = -n_{i;j} = -n_{i,j} + n_\alpha\Gamma^\alpha_{ij}, \tag{11.54}$$

where Γ^α_{ij} are the Christoffel symbols of the spacetime. The equation of motion of the shell is ([3] exercise 21.8)

$$[K_{ij}] = \kappa \left(S_{ij} - \frac{1}{2} S^{(3)} g_{ij} \right), \quad S = S_k^k = g^{ij} S_{ij}. \tag{11.55}$$

Contraction gives

$$[K] = -(\kappa/2)S. \tag{11.56}$$

Hence, solving Eq. (11.55) with respect to S_j^i gives the energy–momentum tensor of the shell in terms of the discontinuity of the extrinsic curvature tensor of the shell, as evaluated inside and outside the shell

$$\kappa S_j^i = [K_j^i] - K \delta_j^i. \tag{11.57}$$

Example 11.1 (*Thin dust shell described by the Israel formalism*) We shall describe a spherical shell consisting of dust particles in empty space. The shell will collapse, and we shall first investigate whether the rest mass of the shell is constant and then find the equation of motion of the shell.

In empty space, the energy–momentum tensor at the right-hand side of equation of continuity (11.52) vanishes. Then it takes the form,

$$u^j S_{j|m}^m = 0. \tag{11.58}$$

Also the stress tensor of the shell vanishes. Then the energy–momentum tensor of the shell as given in Eq. (11.53) has the form

$$S_{ij} = \sigma u_i u_j. \tag{11.59}$$

Hence,

$$u^j S_{j|m}^m = u^i \left(\sigma u_j u^m \right)_{|m} - u^j \sigma_{|m} u_j u^m \mid u^j \sigma u_{j|m} u^m \mid \sigma u^i u_j u_{|m}^m. \tag{11.60}$$

We now use the 4-velocity identity which here takes the form $u^j u_j = -1$, and that the 4-acceleration is

$$a_j = u^m u_{j|m}. \tag{11.61}$$

Equation (11.60) then takes the form

$$u^j S_{j|m}^m = -u^m \sigma_{|m} + \sigma u^j a_j - \sigma u_{|j}^j. \tag{11.62}$$

Here

$$u^m \sigma_{|m} = \frac{d\sigma}{d\tau} \tag{11.63}$$

is the derivative of the mass density of the shell with respect to its proper time. Furthermore, since the 4-acceleration is orthogonal to the 4-velocity, $u^j a_j = 0$. Thus we get

$$u^j S^m_{j|m} = -\frac{d\sigma}{d\tau} - \sigma u^j u_{|j}.$$ (11.64)

The continuity equation of the shell then takes the form

$$\frac{d\sigma}{d\tau} = -\sigma u^j u_{|j}.$$ (11.65)

Writing this out we have

$$\frac{d\sigma}{d\tau} = -\sigma \left({}^{(3)}g \right)^{-1/2} \left[\left({}^{(3)}g \right)^{1/2} u^j \right]_{,j},$$ (11.66)

where ${}^{(3)}g$ is the determinant of the three-dimensional metric on the shell.

The motion of the shell is given by $r = R(\tau)$ where $4\pi R^2(\tau)$ is the area of the shell at the point of time τ. Let the intrinsic geometry of the shell be described in terms of the coordinates (τ, θ, ϕ). Then the line-element of the intrinsic geometry of the shell may be written

$$ds^2 = -d\tau^2 + R^2(\tau)d\Omega^2.$$ (11.67)

Since

$${}^{(3)}g = -R^4(\tau) \sin^2 \theta,$$ (11.68)

and the only non-vanishing component of the 4-velocity in the intrinsic coordinate system is

$$u^\tau = \frac{d\tau}{d\tau} = 1,$$ (11.69)

the continuity equation of the shell reduces to

$$\frac{1}{\sigma} \frac{d\sigma}{d\tau} = -\frac{1}{R^2} \frac{dR^2}{d\tau} = -\frac{2}{R} \frac{dR}{d\tau}.$$ (11.70)

Integration gives

$$\sigma R^2 = \text{constant}.$$ (11.71)

Hence, the rest mass of the shell

$$M_S = 4\pi\sigma R^2 \tag{11.72}$$

is constant.

We shall now find the equation of motion of the shell. For a shell consisting of dust particles, Eq. (11.55) reduces to

$$\left[K_{ij}\right] = \kappa\sigma\left(u_i u_j + \frac{1}{2}\,^{(3)}g_{ij}\right). \tag{11.73}$$

The equation of motion will be found from the $\theta\theta-$ component of this equation. Since $u_\theta = 0$ and $^{(3)}g_{\theta\theta} = R^2$ this gives

$$[K_{\theta\theta}] = (\kappa/2)\sigma R^2 = R_{SS}/2, \tag{11.74}$$

where R_{SS} is the Schwarzschild radius of the shell. As applied to a spherical surface with the line-element (11.67), Eq. (11.54) gives

$$K_{\theta\theta} = n_\alpha\Gamma^\alpha_{\theta\theta} = n^r\Gamma_{r\theta\theta} = -Rn^r. \tag{11.75}$$

Hence, Eq. (11.74) takes the form

$$n^{r+} - n^{r-} = -\frac{R_{SS}}{2R}. \tag{11.76}$$

We now utilize that

$$\vec{n}\cdot\vec{n} = 1, \quad \vec{u}\cdot\vec{u} = -1, \quad \vec{u}\cdot\vec{n} = 0. \tag{11.77}$$

In the exterior region with Schwarzschild metric, these equations take the form

$$-\left(1 - \frac{R_S}{r}\right)(n^{t+})^2 + \frac{(n^{r+})^2}{1 - \frac{R_S}{r}} = 1, \tag{11.78}$$

$$-\left(1 - \frac{R_S}{r}\right)(u^t)^2 + \frac{(u^r)^2}{1 - \frac{R_S}{r}} = -1, \tag{11.79}$$

$$-\left(1 - \frac{R_S}{r}\right)u^t n^{t+} + \frac{u^r n^{r+}}{1 - \frac{R_S}{r}} = 0, \tag{11.80}$$

where R_S is the constant Schwarzschild radius of the shell. Equations (11.78–11.80) give

$$n^{t+} = \frac{u^r}{1 - R_S/R}, \quad n^{r+} = \left(1 + (u^r)^2 - \frac{R_S}{R}\right)^{1/2}. \tag{11.81}$$

The radial component of the 4-velocity of a particle on the shell is

$$u^r = \frac{dR}{d\tau} \equiv \dot{R}.$$

(11.82)

Hence,

$$n^{r+} = \left(1 + \dot{R}^2 - \frac{R_S}{R}\right)^{1/2}.$$

(11.83)

Inside the shell $R_S = 0$, giving

$$n^{r-} = \left(1 + \dot{R}^2\right)^{1/2}.$$

(11.84)

Inserting the expressions (11.83) and (11.84) into Eq. (11.76) gives the equation of motion of the shell

$$\left(1 + \dot{R}^2 - \frac{R_S}{R}\right)^{1/2} - \left(1 + \dot{R}^2\right)^{1/2} = \frac{R_{SS}}{2R}.$$

(11.85)

Solved with respect to $M = R_S/2$ (using units so that Newton's gravitational constant and the velocity of light are equal to one), we get

$$M = M_S\left(1 + \dot{R}^2\right)^{1/2} - M_S^2/2R.$$

(11.86)

Let T be the time measured on a standard clock at rest just inside the shell. Then

$$dT = \left[1 - (dR/dT)^2\right]^{-1/2} d\tau.$$

(11.87)

This gives

$$1 + \dot{R}^2 = \left[1 - (dR/dT)^2\right]^{-1}.$$

(11.88)

Hence Eq. (11.86) takes the form

$$M = \frac{M_S}{\sqrt{1 - (dR/dT)^2}} - \frac{M_S^2}{2R}.$$

(11.89)

This equation expresses energy conservation. The mass M of the external Schwarzschild spacetime represents the total energy of the shell. The quantity $M_S\left[1 - (dR/dT)^2\right]^{-1/2}$ is the relativistic energy of the shell as measures by an observer at rest in the coordinate system at the position of the shell, i.e. the sum of rest mass energy and kinetic energy, not including any gravitational potential energy.

The term $M_S^2/2R$ is the Newtonian expression for the gravitational self energy—or potential energy—of a spherical shell.

Differentiation of Eq. (11.86) with constant M gives

$$\ddot{R} = -\left(1 + \dot{R}^2\right)^{1/2} M_S/2R^2.$$

(11.90)

The acceleration of a shell instantaneously at rest is

$$\ddot{R} = -M_S/2R^2.$$

(11.91)

It accelerates as if half of its mass is inside the shell and half is outside.

11.5 The Levi-Civita—Bertotti—Robinson Solution of Einstein's Field Equations

We shall here derive and give a physical interpretation of a conformally flat, static, spherically symmetric solution of Einstein's field equation which was found independently by Levi-Civita in 1917 ([4]), and by B. Bertotti and I. Robinson in 1959 ([5] and [6]), and which is therefore called the Levi-Civita—Bertotti—Robinson solution of Einstein's field equations.

A spacetime is said to be *conformally flat* if it admits a coordinate system such that the line-element can be written as a function times the line element of the Minkowski spacetime. In the static spherically symmetric case the line element of a conformally flat spacetime can be written

$$ds^2 = e^{2\alpha(r)}\left(-c^2dt^2 + dr^2 + r^2d\Omega^2\right)$$

(11.92)

Einstein's field equations take the form

$$G_t^t = e^{-2\alpha}\left(2\alpha'' + \alpha'^2 + \frac{4}{r}\alpha'\right) = -e^{-4\alpha}\frac{GQ^2}{4\pi\varepsilon_0 c^4 r^4},$$

(11.93)

$$G_r^r = e^{-2\alpha}\left(3\alpha'^2 + \frac{4}{r}\alpha'\right) = -e^{-4\alpha}\frac{GQ^2}{4\pi\varepsilon_0 c^4 r^4},$$

(11.94)

$$G_\theta^\theta = G_\phi^\phi = e^{-2\alpha}\left(2\alpha'' + \alpha'^2 + \frac{2}{r}\alpha'\right) = e^{-4\alpha}\frac{GQ^2}{4\pi\varepsilon_0 c^4 r^4}.$$

(11.95)

Here $Q = Q(r)$ is the charge inside a spherical shell with radius r. Subtracting Eq. (11.93) from (11.95) we obtain

$$e^{2\alpha}\alpha' = -\frac{GQ^2}{4\pi\varepsilon_0 c^4 r^3}.$$

(11.96)

Inserting this into Eq. (11.94) leads to two different solutions, either

$$Q(r) = 0, \tag{11.97}$$

or

$$Q(r) = \pm c^2 \sqrt{\frac{4\pi\varepsilon_0}{G}} r e^\alpha. \tag{11.98}$$

The first solution gives $\alpha(r) = $ constant, which can be chosen to be zero with a suitable choice of coordinates. Hence this solution represents the Minkowski spacetime.

Inserting Eq. (11.98) into (11.96) gives

$$\alpha' = -\frac{1}{r}. \tag{11.99}$$

Integration gives

$$e^\alpha = \frac{R_Q}{r}, \tag{11.100}$$

where R_Q is a constant of integration with dimension length. Inserting this into Eq. (11.64) leads to

$$Q(r) = \pm c^2 \sqrt{\frac{4\pi\varepsilon_0}{G}} R_Q. \tag{11.101}$$

This means that the charge inside a spherical surface with radius r is constant. Hence the charge density vanishes in the considered region. Thus the assumption of a conformally flat, static and spherically symmetric spacetime implies that there is no charge in this region, although there is a radial electric field there. So there must be a charge inside the region described by the solution (11.100). The line element of the spacetime outside the charge distribution has the form

$$ds^2 = \frac{R_Q^2}{r^2}\left(-c^2 dt^2 + dr^2 + r^2 d\Omega^2\right). \tag{11.102}$$

Since the metric is static the coordinate clocks go at the same rate everywhere, equal to the rate of standard clocks at $r = R_Q$. There is continuous metric at $r = R_Q$ with Minkowski spacetime inside this radius. Hence the charge is situated on a spherical shell with radius $r = R_Q$.

The line-element (11.102) shows that in this spacetime standard clocks at rest slow down for increasing r. Hence, the gravitational field points in the direction of increasing r. This means that the mass and charge distribution inside the region with this geometry causes repulsive gravitation.

The Kretschmann curvature scalar of this spacetime is

$$R^{\mu\nu\alpha\beta} R_{\mu\nu\alpha\beta} = \frac{8}{R_Q^4}, \tag{11.103}$$

which is constant. Hence there is no physical singularity in this spacetime.

Let us introduce a new radial coordinate equal to the physical radial distance,

$$d\hat{r} = \frac{R_Q}{r} dr. \tag{11.104}$$

Integration with the boundary condition that $r = R_Q$ corresponds to $\hat{r} = R_Q$ gives

$$\hat{r} = R_Q \left(1 + \ln \frac{r}{R_Q} \right). \tag{11.105}$$

The inverse transformation is

$$r = R_Q e^{(\hat{r} - R_Q)/R_Q}. \tag{11.106}$$

With the radial coordinate $\hat{r}$ the line-element takes the form

$$ds^2 = -e^{-2(\hat{r} - R_Q)/R_Q} c^2 dt^2 + d\hat{r}^2 + R_Q^2 d\Omega^2. \tag{11.107}$$

The geometry of space has a very strange property. The physical area of a spherical surface with radius $\hat{r}$ is independent of the radius and equal to the area of a surface with radius $\hat{r} = R_Q$.

We shall now find the physical properties of source of this spacetime [7]. These properties will presumably provide an explanation for the phenomenon of repulsive gravity outside the charged shell.

11.6 The Source of the Levi-Civita—Bertotti—Robinson Spacetime

We shall now find the energy–momentum tensor of a static spherical shell as given by Eq. (11.57) with Minkowski spacetime inside the shell and a curved spacetime outside. In order to provide a rather general result, we shall first consider a line-element,

$$ds^2 = -e^\alpha c^2 dt^2 + e^\beta dr^2 + e^\gamma d\Omega^2, \tag{11.108}$$

describing spacetime outside the shell, where α, β and γ are functions of r, and then specialize to the Levi-Civita—Bertotti—Robinson metric.

The unit normal vector to a spherical surface about the origin is

$$\vec{n} = e^{-\beta/2}\vec{e}_r. \tag{11.109}$$

The covariant components of the extrinsic curvature tensor are given by Eq. (11.54). Inserting the expression (5.65) for the Christoffel symbols with the line element (11.108) gives

$$K_{ij} = -\frac{1}{2}n^r g_{ij,r} = -\frac{1}{2}e^{-\beta/2}g_{ij,r}. \tag{11.110}$$

The mixed components are

$$K_t^t = -\frac{1}{2}e^{-\beta/2}\alpha_{,r}, \quad K_\theta^\theta = K_\phi^\phi = -\frac{1}{2}e^{-\beta/2}\gamma_{,r}. \tag{11.111}$$

Inserting these expressions into Eq. (11.57) gives the mixed components of the energy–momentum tensor of the shell

$$\kappa S_t^t = -2[K_\theta^\theta] = [e^{-\beta/2}\gamma_{,r}],$$
$$\kappa S_\theta^\theta = \kappa S_\phi^\phi = -[K_t^t] - [K_\theta^\theta] = \frac{1}{2}[e^{-\beta/2}(\alpha_{,r} + \gamma_{,r})]. \tag{11.112}$$

We shall now specialize to the Levi-Civita—Bertotti—Robinson spacetime with the line element (11.102) outside the shell. Then $\alpha_+ = \beta_+ = 2\ln(R_Q/r)$ and $\gamma_+ = 2\ln R_Q$ outside the shell, and $\alpha_- = \beta_- = 0$, $\gamma_- = 2\ln r$ inside it. This gives

$$\kappa S_t^t = \kappa S_\theta^\theta = \kappa S_\phi^\phi = -\frac{2c^2}{R_Q}. \tag{11.113}$$

Surprisingly, the mass density and strain of the shell are independent of its radius. It follows from Eq. (11.113) that the components of the energy–momentum tensor of the singular shell may be written as

$$S_j^i = -\sigma \delta_j^i, \quad \sigma = \frac{2c^2}{\kappa R_Q}, \tag{11.114}$$

where σ is the mass density of the shell. Hence, the shell has a mass

$$M = 4\pi \sigma R_Q^2 = R_Q c^2 / G. \tag{11.115}$$

Note that the radius of the shell is only half its Schwarzschild radius.

Equation (11.114) shows that the shell is a domain wall. Furthermore the Levi-Civita—Bertotti—Robinson spacetime is a solution of Einstein's field equations with a radial electric field. Hence the domain wall is charged. Using Eqs. (11.101) and (11.115) can be written $Q = \sqrt{4\pi \varepsilon_0 G} M$, which shows that the charge of the domain wall is proportional to its mass.

From Eq. (11.113) we can understand why there is repulsive gravity in the Levi-Civita—Bertotti—Robinson spacetime. This equation shows that there is a strain in the charged domain wall equal to minus its energy density. Hence, according to Eq. (11.11) is has negative gravitational mass, and according to the Tolman—Whittaker formula (11.9) the strain of the domain wall causes repulsive gravity in the space outside the charged domain wall.

11.7 A Source of the Kerr–Newman Spacetime

The most general black holes can have three properties: mass, charge and angular momentum. The spacetime outside a black hole with all three properties is a generalization of the Kerr spacetime with an electric field. It is called the Kerr–Newman spacetime. The line element of this spacetime in Boyer–Lindquist coordinates may be written

$$ds^2 = -c^2 dt^2 + \rho^2 \left(\frac{dr^2}{\Delta} + d\theta^2 \right) + \left(r^2 + a^2 \right) \sin^2 \theta d\phi^2 + \frac{r_{\text{eff}}^2}{\rho^2} \left(c dt - a \sin^2 \theta d\phi \right)^2,$$

$$r_{\text{eff}}^2 = R_S r - 2 R_Q^2, \quad \rho^2 = r^2 + a^2 \cos^2 \theta, \quad \Delta = r^2 + a^2 - r_{\text{eff}}^2, \qquad (11.116)$$

where R_S is the Schwarzschild radius of the mass distribution, R_Q the radius given in Eq. (11.101) corresponding to its charge Q, and $a = J/Mc$ a length representing its angular per unit mass.

We shall need both the covariant and the contravariant components of the metric tensor,

$$g_{\mu\nu} = \begin{bmatrix} -\left(1 - \frac{r_{\text{eff}}^2}{\rho^2}\right) & 0 & 0 & -\frac{a r_{\text{eff}}^2 \sin^2 \theta}{\rho^2} \\ 0 & \frac{\rho^2}{\Delta} & 0 & 0 \\ 0 & 0 & \rho^2 & 0 \\ -\frac{a r_{\text{eff}}^2 \sin^2 \theta}{\rho^2} & 0 & 0 & \left(r^2 + a^2 + \frac{a^2 r_{\text{eff}}^2 \sin^2 \theta}{\rho^2} \right) \sin^2 \theta \end{bmatrix}, \qquad (11.117)$$

$$g^{\mu\nu} = \begin{bmatrix} -\frac{\Sigma^2}{\rho^2 \Delta} & 0 & 0 & -\frac{a r_{\text{eff}}^2}{\rho^2 \Delta} \\ 0 & \frac{\Delta}{\rho^2} & 0 & 0 \\ 0 & 0 & \frac{1}{\rho^2} & 0 \\ -\frac{a r_{\text{eff}}^2}{\rho^2 \Delta} & 0 & 0 & \frac{\Delta - a^2 \sin^2 \theta}{\rho^2 \Delta \sin^2 \theta} \end{bmatrix}, \qquad (11.118)$$

$$\Sigma^2 = \left(r^2 + a^2 \right)^2 - a^2 \Delta \sin^2 \theta$$

The unit normal vector of the shell is $\vec{n} = n^r \vec{e}_r$. Hence, $\vec{n} \cdot \vec{n} = 1$ gives $g_{rr}(n^r)^2 = 1$. With the metric (11.117) this leads to

$$\vec{n} = \frac{\sqrt{\Delta}}{\rho} \vec{e}_r. \tag{11.119}$$

From Eq. (11.110) we now find the components of the external curvature tensor of a surface with $r = $ constant,

$$K_{\theta\theta} = n^r \Gamma_{r\theta\theta} = -\frac{1}{2} n^r \frac{\partial g_{\theta\theta}}{\partial r} = -\frac{r\sqrt{\Delta}}{\rho},$$

$$K_{\phi\phi} = -\frac{\sqrt{\Delta}}{2\rho} \frac{\partial g_{\phi\phi}}{\partial r} = -\left[2r + \left(R_S - \frac{2r_{eff}^2 r}{\rho^2} \right) \frac{a^2}{2\rho^2} \sin^2 \theta \right] \frac{\sqrt{\Delta}}{\rho} \sin^2 \theta,$$

$$K_{tt} = -\frac{\sqrt{\Delta}}{2\rho} \frac{\partial g_{tt}}{\partial r} = -\left(R_S - \frac{2r_{eff}^2 r}{\rho^2} \right) \frac{\sqrt{\Delta}}{2\rho^3},$$

$$K_{\phi t} = -\frac{\sqrt{\Delta}}{2\rho} \frac{\partial g_{\phi t}}{\partial r} = -\left(R_S - \frac{2r_{eff}^2 r}{\rho^2} \right) \frac{a\sqrt{\Delta}}{2\rho^3} \sin^2 \theta. \tag{11.120}$$

Lopez [8] has considered a shell with radius

$$r_0 = \frac{2R_Q^2}{R_S} = \frac{Q^2}{4\pi \varepsilon_0 M c^2}, \tag{11.121}$$

so that $r_{eff} = 0$. In the case that Q is equal to the elementary charge, and M is the electron mass, r_0 is equal to *the classical electron radius*, 2.8×10^{-15} m. At the shell the line-element (11.82) reduces to

$$ds^2 = -c^2 dt^2 + \frac{\rho_0^2}{r_1^2} dr^2 + \rho_0^2 d\theta^2 + r_1^2 \sin^2 \theta d\phi^2,$$

$$\rho_0^2 = r_0^2 + a^2 \cos^2 \theta, \quad r_1^2 = r_0^2 + a^2. \tag{11.122}$$

This line-element describes Minkowski spacetime at the shell in oblate spherical coordinates. One can introduce ordinary spherical coordinates (R, Θ) by the transformation

$$R \cos \Theta = r \cos \theta, \quad R \sin \Theta = \sqrt{r^2 + a^2} \sin \theta. \tag{11.123}$$

The ordinary coordinates are given explicitly in terms of the oblate ones by

$$R = r\sqrt{1 + \frac{a^2}{r^2} \sin^2 \theta}, \quad \tan \Theta = \sqrt{1 + \frac{a^2}{r^2}} \tan \theta. \tag{11.124}$$

The inverse transformation is

$$r = R\sqrt{\frac{1 + \frac{a^2}{R^2}\cos^2\theta}{1 + \frac{a^2}{R^2}}}, \qquad \sin\theta = \frac{\sin\Theta}{\sqrt{1 + \frac{a^2}{R^2}}}. \tag{11.125}$$

Taking the differential and inserting them into the line element (11.88) gives

$$ds^2 = -c^2 dT^2 + dR^2 + R^2(d\Theta^2 + \sin^2\Theta d\phi^2), \tag{11.126}$$

which describes Minkowski spacetime in ordinary spherical coordinates.

We now assume that there is Minkowski spacetime inside the shell, so that the metric is continuous at the shell. However the derivatives are not continuous. At the shell the expressions (11.120) for the components of the external curvature tensor then reduce to

$$K_{\theta\theta} = -\frac{r_0 r_1}{\rho_0}, \quad K_{\phi\phi} = -\left(r_0 + \frac{R_S}{2}\frac{a^2}{\rho_0^2}\sin^2\theta\right)\frac{r_1}{\rho_0}\sin^2\theta,$$

$$K_{tt} = -\frac{R_S}{2}\frac{r_1}{\rho_0^3}, \quad K_{\phi t} = \frac{R_S}{2}\frac{a r_1}{\rho_0^3}\sin^2\theta, \tag{11.127}$$

and the metric components (11.117) and (11.118) are

$$g_{\mu\nu} = \text{diag}\left(-1, \frac{\rho_0^2}{r_1^2}, \rho_0^2, r_1^2\sin^2\theta\right), \quad g^{\mu\nu} = \text{diag}\left(-1, \frac{r_1^2}{\rho_0^2}, \frac{1}{\rho_0^2}, \frac{1}{r_1^2\sin^2\theta}\right). \tag{11.128}$$

Hence,

$$K_\theta^{\theta+} = -\frac{r_0 r_1}{\rho_0^3}, \quad K_\phi^{\phi+} = -\left(r_0 + \frac{R_S}{2}\frac{a^2}{\rho_0^2}\sin^2\theta\right)\frac{1}{\rho_0 r_1},$$

$$K_t^{t+} = \frac{R_S}{2}\frac{r_1}{\rho_0^3}, \quad K_t^{\phi+} = \frac{R_S}{2}\frac{a}{r_1\rho_0^3} \tag{11.129}$$

outside the shell. In the oblate spherical coordinates inside the shell $a \neq 0$, but there $M = 0$. Thus the discontinuities of the mixed components of the exterior curvature tensor at the shell are

$$[K_\theta^\theta] = 0, \quad [K_\phi^\phi] = -\frac{R_S}{2}\frac{a^2}{r_1\rho_0^3}\sin^2\theta,$$

$$[K_t^t] = \frac{R_S}{2}\frac{r_1}{\rho_0^3}, \quad [K_t^\phi] = \frac{R_S}{2}\frac{a}{r_1\rho_0^3}. \tag{11.130}$$

The components of the energy–momentum tensor are now calculated from Eq. (11.57) with the result

$$\kappa S_t^t = -\left(\left[K_\theta^\theta\right] + \left[K_\phi^\phi\right]\right) = \frac{R_S a^2}{2r_1 \rho_0^3} \sin^2 \theta, \tag{11.131}$$

$$\kappa S_\theta^\theta = -\left(\left[K_t^t\right] + \left[K_\phi^\phi\right]\right) = -\frac{R_S}{2r_1 \rho_0}, \quad \kappa S_\phi^\phi = -\left(\left[K_t^t\right] + \left[K_\theta^\theta\right]\right) = -\frac{R_S r_1}{2\rho_0^3}, \tag{11.132}$$

$$\kappa S_t^\phi = \left[K_t^\phi\right] = \frac{R_S a}{2r_1 \rho_0^3}, \tag{11.133}$$

$$\kappa S_\phi^t = \kappa g^{tt} S_{t\phi} = \kappa g^{tt} S_{\phi t} = \kappa g^{tt} g_{\phi\phi} S_t^\phi = -\frac{R_S r_1 a}{2\rho_0^3} \sin^2 \theta. \tag{11.134}$$

The energy–momentum tensor of the shell may be written

$$2\kappa S_j^i = -\frac{R_S}{r_1 \rho_0}\left(u^i u_j + \delta_j^i\right), \quad u^i = \frac{1}{r_1 \rho_0}\left(r_1^2, 0, 0, a\right). \tag{11.135}$$

This is Lopez's source of the Kerr–Newman spacetime. The first term represents dust with negative rest mass density, and the second term a domain wall. The four velocity of the dust particles show that they move in the $\phi-$ direction. Thus the source may be described as a domain wall with "bubbles" of negative rest energy within it rotating around in the wall.

11.8   Physical Interpretation of the Components of the Energy–Momentum Tensor by Means of the Eigenvalues of the Tensor

Let a system be described by an energy–momentum tensor with mixed components S_j^i. The eigenvalues, λ, of the tensor are defined by the matrix equation

$$\left|S_j^i - \lambda \delta_j^i\right| = 0. \tag{11.136}$$

In an n-dimensional space (11.136) is an n-degree equation for λ so that S_j^i has n eigenvalues. Furthermore S_j^i has n eigenvectors, $\vec{u}^{(k)}$, given by

$$S_j^i u^{(k)j} = \lambda_{(k)} u^{(k)i}. \tag{11.137}$$

Let $\vec{u}$ and $\vec{v}$ be two different eigenvectors of S_j^i associated with two different eigenvalues λ_1 and λ_2. Then

$$S^i_j u^i = \lambda_1 u^i, \quad S^i_j v^i = \lambda_1 v^i. \tag{11.138}$$

giving

$$S_{ij} u^i v^j - S_{ij} u^j v^i = (\lambda_2 - \lambda_1) u_i v^i. \tag{11.139}$$

Interchanging the summation indices in the last term and using that the tensor S is symmetrical, $S_{ij} = S_{ji}$, we get

$$(\lambda_2 - \lambda_1)(\vec{u} \cdot \vec{v}) = 0. \tag{11.140}$$

Since in general $\lambda_2 \neq \lambda_1$ this requires

$$\vec{u} \cdot \vec{v} = 0. \tag{11.141}$$

Hence, the eigenvectors of S are orthogonal.

For the physical problems considered, Eq. (11.137) will give one time like eigenvector, and the other are space like. We now require that all of the eigenvectors shall be unit vectors. The eigenvectors of S will form an orthonormal basis. The time like eigenvector is the 4-velocity of a point of the system described by the tensor S. This means that the eigenvectors form an orthonormal basis co-moving with the system.

Let this four-velocity be $\vec{u}$. Then the eigenvectors of S have the following physical interpretation,

$$\lambda_{(t)} = \rho_0, \quad \lambda_{(k)} = -p_k, \tag{11.142}$$

where ρ_0 is the proper energy density measured by an observer at rest with the physical system, and p_k is the stress in the k-direction, i.e. the stress towards a surface with normal vector in the k-direction. The components of the energy–momentum tensor S may in general be written as

$$S_{ij} = \rho_0 u_i u_j + \sum_k p_k v_i^{(k)} v_j^{(k)}, \tag{11.143}$$

where $\vec{v}^{(k)}$ is a space like eigenvector of S.

Example 11.2 (*Lopez's source of the Kerr–Newman metric*) The energy–momentum tensor of Lopez's source of the Kerr–Newman tensor has components given in Eqs. (11.132–11.134). Equation (11.136) here takes the form

$$\begin{vmatrix} S^t_t - \lambda & 0 & S^\phi_t \\ 0 & S^\theta_\theta - \lambda & 0 \\ S^t_\phi & 0 & S^\phi_\phi - \lambda \end{vmatrix} = 0. \tag{11.144}$$

Thus

$$\left(S_t^t - \lambda\right)\left(S_\theta^\theta - \lambda\right)\left(S_\phi^\phi - \lambda\right) - S_t^\phi S_\phi^t \left(S_\theta^\theta - \lambda\right) = 0. \tag{11.145}$$

One solution of this equation is

$$\lambda_\theta = S_\theta^\theta. \tag{11.146}$$

The other two solutions are found from

$$\lambda^2 - \left(S_t^t + S_\phi^\phi\right)\lambda + S_t^t S_\phi^\phi - S_t^\phi S_\phi^t = 0, \tag{11.147}$$

giving

$$\lambda_t = \frac{1}{2}\left(S_t^t + S_\phi^\phi\right) - \sqrt{\frac{1}{4}\left(S_t^t - S_\phi^\phi\right)^2 + S_t^\phi S_\phi^t},$$

$$\lambda_\phi = \frac{1}{2}\left(S_t^t + S_\phi^\phi\right) + \sqrt{\frac{1}{4}\left(S_t^t - S_\phi^\phi\right)^2 + S_t^\phi S_\phi^t}. \tag{11.148}$$

Inserting the components (11.133–11.134) of the energy–momentum tensor of Lopez's source of the Kerr–Newman spacetime we get

$$\lambda_t = 0, \quad \lambda_\theta = \lambda_\phi = -\frac{R_S}{2r_1\rho_0}. \tag{11.149}$$

The eigenvectors are now found from Eq. (11.137). The time like eigenvector, i.e. the 4 velocity field of the source is given by

$$S_t^t u^t + S_\phi^t u^\phi = 0, \quad S_t^\phi u^t + S_\phi^\phi u^\phi = 0. \tag{11.150}$$

The 4-velocity identity gives

$$g_{tt}\left(u^t\right)^2 + g_{\phi\phi}\left(u^\phi\right)^2 + 2g_{\phi t}u^\phi u^t = -1. \tag{11.151}$$

Since $g_{\phi t} = 0$ at the shell this reduces to

$$g_{tt}\left(u^t\right)^2 + g_{\phi\phi}\left(u^\phi\right)^2 = -1. \tag{11.152}$$

Equations (11.150) and (11.152) give

$$u^t = \left[-g_{tt} - g_{\phi\phi}\left(\frac{S_t^t}{S_\phi^t}\right)^2\right]^{-1/2}, \quad u^\phi = \left[-g_{tt}\left(\frac{S_\phi^t}{S_t^t}\right)^2 - g_{\phi\phi}\right]^{-1/2}. \tag{11.153}$$

Inserting the values of the metric tensor at the shell,

$$g_{tt} = -1, \quad g_{\phi\phi} = r_1^2 \sin^2 \theta, \tag{11.154}$$

and

$$\left(\frac{S_t^t}{S_\phi^t}\right)^2 = \frac{a^2}{r_1^4} \tag{11.155}$$

gives the eigenvector

$$\vec{u} = \frac{1}{r_1 \rho_0}\left(r_1^2 \vec{e}_t + a\vec{e}_\phi\right). \tag{11.156}$$

This is the 4-velocity of particles following the shell. According to the interpretation at the end of the previous section, this means that the bubbles with negative rest energy within the shell rotate rigidly with an angular velocity

$$\Omega = \frac{u^\phi}{u^t} = \frac{a}{r_1^2}c = \frac{a}{r_0^2 + a^2}c, \tag{11.157}$$

where we have used the last of Eq. (11.122) and inserted the velocity of light.

11.9 The River of Space

When teaching the theory of relativity one sometimes gets the question: What is space? It is here understood that the question is concerned with ordinary three-space and not the four-dimensional spacetime. The first part of the answer is to make clear that space is a theory dependent concept. The second is to try to explain what we mean by "space" according to the general theory of relativity.

One definition is to say that space is a set of simultaneous events. Even if this is an essential part of what we mean by space, this definition is not sufficient to give us a picture of space which makes us understand, for example, why light cannot be emitted from the horizon of a black hole, and what is meant by an "expanding space". We must demand from the properties of space that it provides an interpretation of the expression the "expanding space" according to the general theory of relativity. Furthermore, the definition should capture the phenomenon of inertial dragging.

We shall here define the concept of "physical space" as a continuum of freely moving reference particles with specified initial conditions. In a homogeneous and isotropic universe, the only motion of the reference particles is that due to a universal change of the distances between the particles. This may be described by a single scale factor and defines the *Hubble flow* which obeys the Hubble–Lemaître law. These particles make up what we call the "river of space" ([10]).

We shall here describe the river of space in the Schwarzschild–de Sitter space-time with the line element (11.172). One may think of the cosmological constant as representing the constant density of LIVE (see 7.2.1) causing repulsive gravity.

Let

$$f(r) = 1 - \frac{R_S}{r} - \frac{r^2}{R_H^2}, \quad R_S = \frac{2GM}{c^2}, \quad R_H = \sqrt{\frac{3}{\Lambda}}. \tag{11.158}$$

The value of the cosmological constant representing the density of LIVE in our universe is $\Lambda \sim 10^{52} \, m^{-2}$. Hence $R_H \sim 1,7 \cdot 10^{26} m \sim 1.7 \cdot 10^{10} l.y.$ The horizons of this spacetime are given by $f(r) = 0$, giving

$$r_1 = \frac{2}{\sqrt{3}} R_H \sin \theta, \quad r_2 = R_H \cos \theta - \frac{R_H}{\sqrt{3}} \sin \theta, \quad \sin(3\theta) = \frac{3\sqrt{3}}{2} \frac{R_S}{R_H}. \tag{11.159}$$

We shall assume that $0 < \theta < \pi/6$.

Let us first calculate the 4-acceleration of a particle at rest in the coordinate system. Hence we consider a particle with $dr = d\theta = d\phi = 0$ in the line element (11.172). Then the only non-vanishing component of the particle's 4-velocity is $u^t = dt/d\tau = f(r)^{-1/2}$, where τ is the proper time of the particle. Since u^t is independent of the time, the components of the particle's 4-acceleration are given by $u^\alpha = \Gamma^\alpha_{tt}(u^t)^2$, where the only non-vanishing Christoffel symbol is Γ^r_{tt}. Hence the only nonzero component of the 4-acceleration is

$$a^r = \frac{2r^3 - R_S R_H^2}{2R_H^2 r^2}. \tag{11.160}$$

Accordingly, the 4-acceleration vanishes on a 3-surface with radius

$$r_0 = \left(\frac{R_S R_H^2}{2} \right)^{1/3}. \tag{11.161}$$

Inserting, for example, the Schwarzschild radius of the Sun and the density of dark energy in the universe we get $r_0 \approx 200 l.y..$ It is clear that $R_H \gg r_0$ for all localized systems we are aware of in the local part of the universe. A particle at rest at $r = r_0$ is a free particle. Note further that $a^r < 0$ for $r < r_0$ and $a^r > 0$ for $r > r_0$.

In this spacetime we define the *river of space* by a continuum of free particles that are at rest at $r = r_0$. Hence the river of space flows inwards towards the central mass for $r < r_0$ and outwards for $r > r_0$. The reason for the outwards flow outside $r = r_0$ is the repulsive gravity of the LIVE.

Let us now calculate the velocity of the river of space in the Schwarzschild–de Sitter spacetime. It consists of falling particles with a Lagrangian

$$L = -\frac{1}{2}f(r)c^2\dot{t}^2 + \frac{1}{2}\frac{\dot{r}^2}{f(r)}, \tag{11.162}$$

where the dots represent differentiation with respect to the proper time of a particle. Since L does not depend upon the time, the conjugate momentum to the time coordinate is a constant of motion,

$$p_t = \frac{\partial L}{\partial \dot{t}} = -f(r)c^2\dot{t} = \text{constant}. \tag{11.163}$$

From the four-velocity identity, we obtain

$$-\frac{p_t^2}{f(r)} + \frac{\dot{r}^2}{f(r)} = -c^2. \tag{11.164}$$

Inserting the boundary condition $\dot{r}(r_0) = 0$ gives

$$p_t = -c\sqrt{f(r_0)}, \quad f(r_0) = 1 - 3\left(\frac{R_S}{2R_H}\right)^{2/3}. \tag{11.165}$$

Hence

$$\dot{r} = \pm c\sqrt{f(r_0) - f(r)}, \tag{11.166}$$

with $+$ for $r > r_0$ and $-$ for $r < r_0$. Inserting the expression (11.165) into Eq. (11.163) gives

$$\dot{t} = \frac{\sqrt{f(r_0)}}{f(r)}. \tag{11.167}$$

The coordinate velocity of the river of space is then

$$\frac{dr}{dt} = \frac{\dot{r}}{\dot{t}} = \pm f(r)\sqrt{1 - \frac{f(r)}{f(r_0)}}. \tag{11.168}$$

We now introduce an orthonormal basis vector field associated with observers at rest in the coordinate system. The relationship between the coordinate basis vectors and those of the orthonormal basis vector field is

$$e_{\hat{t}} = \frac{1}{\sqrt{f(r)}}e_t, \quad e_{\hat{r}} = \sqrt{f(r)}e_r. \tag{11.169}$$

Hence, the velocity of the river of space as measured by an observer at rest in the static reference frame is

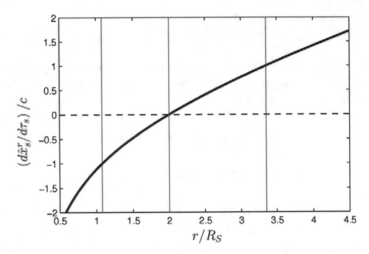

Fig. 11.1 River of space in the Schwarzschild–de Sitter spacetime. The centre vertical line marks the equilibrium radius, where the river of space is at rest, and the left and right lines mark the horizons where the river of space flows with the velocity of light (From Braeck and Grøn [10])

$$\frac{d\hat{r}}{d\hat{t}} = \frac{1}{f(r)}\frac{dr}{dt} = \pm\sqrt{1 - \frac{f(r)}{f(r_0)}} = \pm\sqrt{\frac{r^2 + 2r_0^3/r - 3r_0^2}{R_H^2 - 3r_0^2}}. \tag{11.170}$$

Inserting the horizon radii (11.159) gives $d\hat{r}/d\hat{t} = \pm c$ with minus at the Schwarzschild horizon and plus at the de Sitter horizon. Hence, the river of space flows at the velocity of light into a black hole around the origin of the coordinate system and there is an outwards flow with the velocity of light at the outer horizon. The function (11.170) is plotted in Fig. 11.1.

The river of space flows with superluminal velocity inside the Schwarzschild horizon and outside the de Sitter horizon.

Note that for $r \gg r_0$ the expression (11.170) reduces to

$$\frac{d\hat{r}}{d\hat{t}} \approx H_0 r, \quad H_0 = \frac{c}{\sqrt{R_H^2 - 3r_0^2}} \approx \frac{c}{R_H} = c\sqrt{\frac{\Lambda}{3}}. \tag{11.171}$$

Hence, far outside the equilibrium radius, the velocity of the river of space is proportional to the distance. This is the Hubble–Lemaître expansion law which will be discussed in the next chapter. Thus, the flow of the river approaches the Hubble flow far from the mass distribution.

Exercises

11.1 The Schwarzschild–de Sitter metric

ShowSchwarzschild–de Sitter metric that the solution of Einstein's field equations with a cosmological constant in a static, spherically symmetric space is

$$ds^2 = -\left(1 - \frac{R_S}{r} + \frac{r^2}{R_H^2}\right)c^2dt^2 + \frac{dr^2}{1 - \frac{R_S}{r} + \frac{r^2}{R_H^2}} + r^2d\Omega^2, \qquad (11.172)$$

where $R_S = 2GM/c^2$ is the Schwarzschild radius of the central mass, and $R_H = \sqrt{3/\Lambda}$ is the de-Sitter horizon radius which is the horizon radius in the case that there is no central mass, $R_S = 0$.

11.2 A spherical domain wall described by the Israel formalism

Consider a static, spherically symmetric domain wall in empty space with mass density σ and radius R. Show that the mass M of the Schwarzschild spacetime outside the wall is

$$M = (1 - 2\pi\sigma R)4\pi\sigma R^2. \qquad (11.173)$$

References

1. Israel, W.: Singular hypersurfaces and thin shells in general relativity. Nuovo Cimento **44B**, 1–14 (1966)
2. Misner C.W., Thorne, K.S., Wheeler, J.A.: Gravitation. Princeton University Press (1973)
3. Lightman, A.P., Press, W.H., Price, R.H., Teukolsky, S.A.: Problem Book in Relativity and Gravitation. Princeton University Press (1975)
4. An English translation of the 1917-article is: T. Levi-Civita, The physical reality of some normal spaces of Bianchi. Gen. Rel. Grav. **43**, 2307 (2011)
5. Bertotti, B.: Uniform electromagnetic field in the theory of general relativity. Phys. Rev. **116**, 1331–1333 (1959)
6. Robinson, I.: A solution of the Maxwell-Einstein equations. Bull. Acad. Pol. Sci. Ser. Sci. Math. Astr. Phys. **7**, 351 (1959)
7. Grøn, Ø., Johannesen, S.: A solution of the Einstein-Maxwell equations describing conformally flat spacetime outside a charged domain wall. Eur. Phys. J. Plus. **126**, 89 (2011)
8. Lopez, C.A.: Extended model of the electron in general relativity. Phys. Rev. **D30**, 313–316 (1984)
9. Grøn, Ø.: New derivation of Lopez's source of the Kerr-Newman field. Phys. Rev. **D32**, 1588–1589 (1985)
10. Braeck, S., Grøn, Ø.: A river model of space. Eur. Phys. J. Plus **128**, 24 (2013)

Chapter 12
Cosmology

Abstract The Lemaître–Friedmann–Robertson–Walker (LFRW) universe models are deduced as solutions of Einstein's field equations, and the Hubble–Lemaître expansion law is found as a general property of these models. It is shown that the cosmic redshift due to the expansion of the space contains both the kinematic Doppler effect due to the velocity of the emitter relative to the observer and the gravitational shift of wavelength for light moving vertically in a gravitational field. Several observational properties of a flat LFRW universe model with dust and Lorentz invariant vacuum energy (LIVE) are deduced. Also anisotropic and inhomogeneous universe models are considered. Finally some inflationary universe models are discussed, and their predictions for some observational properties are confronted with observed data.

12.1 Co-moving Coordinate System

In this chapter we will first consider expanding homogeneous and isotropic models of the universe. They are called *Lemaître-Friedmann–Robertson–Walker universe models*. We introduce an expanding frame of reference with the galactic clusters as reference particles. Then we introduce a "co-moving coordinate system" in this reference frame with spatial coordinates χ, ϑ, ϕ. Time measured on standard clocks carried by the galactic clusters is used as coordinate time.

The line element can then be written in the form

$$ds^2 = -c^2 dt^2 + a(t)^2 [R_0^2 d\chi^2 + r(\chi)^2 d\Omega^2], \tag{12.1}$$

Here χ is a dimensionless radial coordinate, and R_0 represents the present value of the curvature radius of the 3-space. The coordinate time t is shown on standard clocks co-moving with the reference particles of the expanding system, $d\chi = d\Omega = 0$ and $ds^2 = -c^2 d\tau^2 = -c^2 dt^2$. The function $a(t)$ is called the *scale factor*, and t is called *cosmic time*. The *age of the universe* is the present value of the cosmic time, t_0. The scale factor tells how the reference particles move radially. It is normalized to have the value 1 now, $a(t_0) = 1$. Hence the scale factor represents the ratio of the

© Springer Nature Switzerland AG 2020

Ø. Grøn, *Introduction to Einstein's Theory of Relativity*,
Undergraduate Texts in Physics, https://doi.org/10.1007/978-3-030-43862-3_12

distance between the reference particles at an arbitrary point of time and their present distance.

We shall now investigate whether the reference particles are freely moving, i.e. whether they obey the geodesic equation. Since a reference particle is permanently at rest in the coordinate system, and the derivative of the coordinate time with respect to its proper time is $dt/d\tau = 1$, its 4-velocity has components

$$u^\mu = \frac{dx^\mu}{d\tau} = \frac{dx^\mu}{dt} = (c, 0, 0, 0). \tag{12.2}$$

This applies at an arbitrary time, so $\frac{du^\mu}{dt} = 0$. Hence, the geodesic equation

$$\frac{du^\mu}{dt} + \Gamma^\mu_{\alpha\beta} u^\alpha u^\beta = 0 \tag{12.3}$$

reduces to

$$\Gamma^\mu_{tt} = 0. \tag{12.4}$$

Since

$$\Gamma^\mu_{tt} = \frac{1}{2} g^{\mu\nu} (\overbrace{g_{\nu t,t}}^{0} + \overbrace{g_{t\nu,t}}^{0} + \overbrace{g_{tt,\nu}}^{0}) = 0, \tag{12.5}$$

The geodesic equation is fulfilled for the reference particles. Hence they are freely falling.

12.2   Curvature Isotropy—The Robertson–Walker Metric

We introduce an orthonormal form basis,

$$\underline{\omega}^{\hat{t}} = \underline{dt}, \quad \underline{\omega}^{\hat{\chi}} = R_0 a(t)\underline{d\chi}, \quad \underline{\omega}^{\hat{\theta}} = a(t)r(\chi)\underline{d\theta},$$
$$\underline{\omega}^{\hat{\phi}} = a(t)r(\chi) \sin\theta \underline{d\phi}. \tag{12.6}$$

Then we use Cartan's 1. and 2. structure equations,

$$\underline{d\omega}^{\hat{\mu}} = -\underline{\Omega}^{\hat{\mu}}_{\hat{\nu}} \wedge \underline{\omega}^{\hat{\nu}}, \quad R^{\hat{\mu}}_{\hat{\nu}} = \underline{d\Omega}^{\hat{\mu}}_{\hat{\nu}} + \underline{\Omega}^{\hat{\mu}}_{\hat{\lambda}} \wedge \underline{\Omega}^{\hat{\lambda}}_{\hat{\nu}} \tag{12.7}$$

to find the connection forms and the curvature forms. Calculations give (notation: $\dot{} = \frac{d}{dt}$, $' = \frac{d}{d\chi}$)

$$\underline{R}^{\hat{i}}_{\;\hat{j}} = \frac{\ddot{a}}{a}\underline{\omega}^{\hat{i}} \wedge \underline{\omega}^{\hat{j}}, \quad \underline{\omega}^{\hat{i}} = \underline{\omega}^{\hat{\chi}}, \underline{\omega}^{\hat{\theta}}, \underline{\omega}^{\hat{\phi}},$$

$$\underline{R}^{\hat{\chi}}_{\;\hat{j}} = \left(\frac{\ddot{a}^2}{a^2} - \frac{r''}{ra^2}\right)\underline{\omega}^{\hat{\chi}} \wedge \underline{\omega}^{\hat{j}}, \quad \underline{\omega}^{\hat{j}} = \underline{\omega}^{\hat{\theta}}, \quad \underline{\omega}^{\hat{\phi}}, \tag{12.8}$$

$$\underline{R}^{\hat{\theta}}_{\;\hat{\phi}} = \left(\frac{\ddot{a}^2}{a^2} + \frac{1}{r^2 a^2} - \frac{r'^2}{r^2 a^2}\right)\underline{\omega}^{\hat{\theta}} \wedge \underline{\omega}^{\hat{\phi}}.$$

The present curvature of 3-space ($dt = 0$) can be found by putting $a = 1$. That is

$$_3\underline{R}^{\hat{\chi}}_{\;\hat{j}} = -\frac{r''}{r}\underline{\omega}^{\hat{\chi}} \wedge \underline{\omega}^{\hat{j}},$$

$$_3\underline{R}^{\hat{\theta}}_{\;\hat{\phi}} = \left(\frac{1}{r^2} - \frac{r'^2}{r^2}\right)\underline{\omega}^{\hat{\theta}} \wedge \underline{\omega}^{\hat{\phi}}. \tag{12.9}$$

The 3-space is assumed to be isotropic and homogeneous. This demands

$$-\frac{r''}{r} = \frac{1 - r'^2}{r^2} = \frac{k}{R_0^2}, \tag{12.10}$$

where R_0 is the present value of the curvature radius of the 3-space. Here k is a dimensionless constant which has the value 1 for a positively curved (spherical) space, 0 for Euclidean (flat) space, and -1 for negatively curved (hyperbolic) space.

$$r'' + \frac{k}{R_0^2}r = 0 \quad \text{and} \quad r' = \sqrt{1 - k(r/R_0)^2}. \tag{12.11}$$

The solutions with the boundary conditions $r(0) = 0$, $r'(0) = 1$ are

$$\begin{aligned} r &= R_0 \sinh \chi, \quad dr = \sqrt{R_0^2 + r^2}d\chi \quad (k = -1),\\ r &= R_0\chi, \quad dr = R_0 d\chi \quad (k = 0),\\ r &= R_0 \sin \chi, \quad dr = \sqrt{R_0^2 - r^2}d\chi \quad (k = 1). \end{aligned} \tag{12.12}$$

In all three cases one may write

$$dr = R_0\sqrt{1 - k(r/R_0)^2}d\chi. \tag{12.13}$$

We now substitute

$$R_0^2 d\chi^2 = \frac{dr^2}{1 - k(r/R_0)^2} \tag{12.14}$$

into the line element (12.1). In standard coordinates it then takes the form

$$ds^2 = -c^2 dt^2 + R_0^2 a^2(t)\left(d\chi^2 + S_k^2(\chi)d\Omega^2\right). \tag{12.15}$$

where

$$
S_k(\chi) = \begin{cases} \sin \chi & , \ k = 1 \\ \chi & , \ k = 0 \\ \sinh \chi, & k = -1 \end{cases}.
\tag{12.16}
$$

With the radial coordinate r the line-element takes the form

$$
ds^2 = -c^2 dt^2 + a^2(t) \left(\frac{dr^2}{1 - k(r/R_0)^2} + r^2 d\Omega^2 \right).
\tag{12.17}
$$

This is called the *Robertson–Walker line element*.

The 3-space has constant curvature and is spherical for $k = 1$, Euclidean for $k = 0$ and hyperbolic for $k = -1$. Universe models with $k = 1$ are known as "closed", and models with $k = -1$ are known as "open". Models with $k = 0$ are called "flat" even though also these models have curved spacetime.

12.3   Cosmic Kinematics and Dynamics

12.3.1   The Hubble–Lemaître Law

The physical distance at a point of time t, to a particle with an instantaneous coordinate distance χ from an observer at the origin, is

$$
l = \sqrt{g_{\chi\chi}} \, \chi = R_0 a(t) \chi.
\tag{12.18}
$$

The value of χ determines which reference, particle' (galactic cluster) we are observing, and $a(t)$ how it is moving. The velocity of the particle relative to the observer at the origin is

$$
v = R_0(\dot{a}\chi + a\dot{\chi}) = R_0 \left(\frac{\dot{a}}{a} a\chi + a\dot{\chi} \right).
\tag{12.19}
$$

The *Hubble parameter* is defined as

$$
H = \frac{\dot{a}}{a}.
\tag{12.20}
$$

The present value of the Hubble parameter, $H_0 = H(t_0)$ is called the *Hubble constant*. A universe expanding with constant velocity has an age equal to the inverse value of the Hubble constant. This is called the *Hubble age* of the universe,

$$t_H = \frac{1}{H_0}.$$ (12.21)

We shall call $l_H = c\,t_H$ the *Hubble length*.

A large number of observations have been performed in order to determine the age of the universe and the value of the Hubble constant. Favoured values are $t_0 = 13.8 \times 10^9$ years, and $H_0 = 21.5$ km/s per million light years corresponding to a Hubble age $t_H = 13.85 \times 10^9$ years. It may be noted that the astronomers use the length unit parsec which is equal to 3.26 light years. So they give the Hubble constant in units km/s per Megaparsec, $H_0 = 70.1$ km/s per Mpc.

Equation (12.19) may be written as

$$v = H\,l + aR_0\dot\chi = v_H + v_P.$$ (12.22)

Here

$$v_H = H\,l$$ (12.23)

is the velocity of the *Hubble flow*, which represent the expansion of the universe. It says that *the velocity of the Hubble flow is proportional to the distance from the observer*. This is the *Hubble–Lemaître law*. The general relativistic interpretation of this law is that *space expands*.

Furthermore

$$v_P = aR_0\dot\chi$$ (12.24)

is called the *peculiar velocity* of the considered particle. It is a velocity peculiar to the considered particle due to a local gravitational field at the position of the particle. In other words: v_H is the velocity of space, and v_P is a velocity through space. The velocity of space is permitted to be larger than the velocity of light, which is the case farther away from the observer than c/H, and the velocity though space is always smaller than the velocity of light.

12.3.2 Cosmological Redshift of Light

We consider light emitted from a position with standard radial coordinate χ_e and received by the observer at $\chi = 0$. Let Δt_e be the period of the light as measured in the emitter position at the emission time, and Δt_0 the period as measured in the receiver position at the receiving time (Fig. 12.1).

Light follows curves with $ds^2 = 0$, with $d\vartheta = d\phi = 0$, and we have

$$c\,dt = -a(t)R_0 d\chi.$$ (12.25)

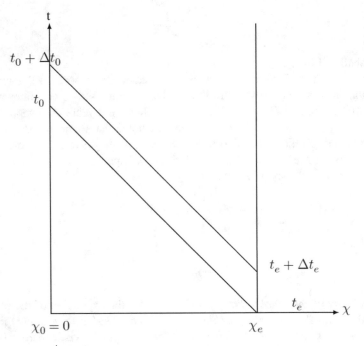

Fig. 12.1 Cosmological redshift

Integration from emitter event to receiver event gives

$$\frac{c}{R_0} \int_{t_e}^{t_0} \frac{dt}{a(t)} = -\int_{\chi_e}^{0} d\chi = \chi_e,$$

$$\frac{c}{R_0} \int_{t_e+\Delta t_e}^{t_0+\Delta t_0} \frac{dt}{a(t)} = -\int_{\chi_e}^{0} d\chi = \chi_e,$$

(12.26)

which gives

$$\int_{t_e+\Delta t_e}^{t_0+\Delta t_0} \frac{dt}{a} - \int_{t_e}^{t_0} \frac{dt}{a} = 0$$

(12.27)

or

$$\int_{t_0}^{t_0+\Delta t_0} \frac{dt}{a} - \int_{t_e}^{t_e+\Delta t_e} \frac{dt}{a} = 0.$$

(12.28)

During the integration time interval from t_e to $t_e + \Delta t_e$ the expansion factor $a(t)$ can be considered a constant with value $a(t_e)$ and during the integration time interval from t_0 to $t_0 + \Delta t_0$ with value $a(t_0)$, giving

$$\frac{\Delta t_e}{a(t_e)} = \frac{\Delta t_0}{a(t_0)}. \tag{12.29}$$

Δt_0 and Δt_e are periods of the light at the receiving and emitting time. Since the wavelength of the light is $\lambda = c\Delta t$ we have

$$\frac{\lambda_0}{a(t_0)} = \frac{\lambda_e}{a(t_e)}. \tag{12.30}$$

This can be interpreted as a "stretching" of the electromagnetic waves due to the expansion of space [1]. The *cosmological redshift* is denoted by z and is given by

$$z = \frac{\lambda_0 - \lambda_e}{\lambda_e} = \frac{a(t_0)}{a(t_e)} - 1. \tag{12.31}$$

The scale factor is usually normalized to have a present value equal to one, $a(t_0) = 1$, so that

$$1 + z = \frac{1}{a}. \tag{12.32}$$

12.3.3 Cosmic Fluids

The energy–momentum tensor for a perfect fluid with mass density ρ as measured by a co-moving observer in the fluid, and pressure p (no viscosity and no thermal conductivity), is

$$T_{\mu\nu} = (\rho + p/c^2)u_\mu u_\nu + pg_{\mu\nu}. \tag{12.33}$$

In an orthonormal basis

$$T_{\hat{\mu}\hat{\nu}} = (\rho + p/c^2)u_{\hat{\mu}} u_{\hat{\nu}} + p\eta_{\hat{\mu}\hat{\nu}}, \tag{12.34}$$

where $\eta_{\hat{\mu}\hat{\nu}}$ is the Minkowski metric. We consider three types of cosmic fluid.

1. Dust: $p = 0$,

$$T_{\hat{\mu}\hat{\nu}} = \rho u_{\hat{\mu}} u_{\hat{\nu}}. \tag{12.35}$$

2. Radiation: $p = \frac{1}{3}\rho c^2$,

$$T_{\hat{\mu}\hat{\nu}} = \frac{4}{3}\rho u_{\hat{\mu}}u_{\hat{\nu}} + p\eta_{\hat{\mu}\hat{\nu}}$$

$$= \frac{\rho}{3}(4u_{\hat{\mu}}u_{\hat{\nu}} + \eta_{\hat{\mu}\hat{\nu}}). \tag{12.36}$$

The trace vanishes for radiation,

$$T = T_{\hat{\mu}}^{\hat{\mu}} = \frac{\rho}{3}\left(4u^{\hat{\mu}}u_{\hat{\mu}} + \delta_{\mu}^{\mu}\right) = 0. \tag{12.37}$$

3. Lorentz invariant vacuum energy, LIVE, $p_L = -\rho_L c^2$,

$$T_{\hat{\mu}\hat{\nu}} = -\rho\eta_{\hat{\mu}\hat{\nu}}. \tag{12.38}$$

The density of LIVE can be related to the cosmological constant by

$$\Lambda = \kappa\,\rho_L. \tag{12.39}$$

One has also introduced a more general type of vacuum energy given by the equation of state

$$p_\phi = w\rho_\phi c^2, \tag{12.40}$$

where ϕ represents a scalar field, and w is a factor which is often assumed to be constant. In a homogeneous universe the pressure and density are given by

$$p_\phi = \frac{1}{2}\dot{\phi}^2 - V(\phi), \quad \rho_\phi = \frac{1}{2}\dot{\phi}^2 + V(\phi), \tag{12.41}$$

where $V(\phi)$ is the potential for the scalar field. Then we have

$$w = \frac{\frac{1}{2}\dot{\phi}^2 - V(\phi)}{\frac{1}{2}\dot{\phi}^2 + V(\phi)}. \tag{12.42}$$

The special case $\dot{\phi} = 0$ gives LIVE with $w = -1$. The more general vacuum is called "quintessence".

12.3.4 Isotropic and Homogeneous Universe Models

We will discuss isotropic and homogenous universe models with perfect fluid and a non-vanishing cosmological constant Λ. Calculating the components of the Einstein tensor from the line-element (12.17) we find in an orthonormal basis

$$E_{\hat{t}\hat{t}} = \frac{3\dot{a}^2}{a^2} + \frac{3kc^2}{R_0^2 a^2}, \quad E_{\hat{m}\hat{m}} = -\frac{2\ddot{a}}{a} - \frac{\dot{a}^2}{a^2} - \frac{kc^2}{R_0^2 a^2}. \tag{12.43}$$

Hence, the trace of the Einstein tensor is

$$E_{\hat{\mu}}^{\hat{\mu}} = -\frac{6}{a^2}\left(a\ddot{a} + \dot{a}^2 + kc^2/R_0^2\right). \tag{12.44}$$

The components of the energy–momentum tensor of a perfect fluid in a co-moving orthonormal basis are

$$T_{\hat{t}\hat{t}} = \rho, \quad T_{\hat{m}\hat{m}} = p. \tag{12.45}$$

Hence the $\widehat{t}\widehat{t}$ component of Einstein's field equations is

$$3\frac{\dot{a}^2 + kc^2/R_0^2}{a^2} = \kappa\rho + \Lambda \tag{12.46}$$

where κ is Einstein's gravitational constant as given in Eq. (7.37). The $\hat{m}\hat{m}$ component of Einstein's field equations is

$$-2\frac{\ddot{a}}{a} - \frac{\dot{a}^2}{a^2} - \frac{kc^2}{R_0^2 a^2} = \kappa p c^2 - \Lambda, \tag{12.47}$$

where ρ is the energy density and p is the pressure. These equations are called the *Friedmann–Lemaître equations*.

There are three cases of empty universe models, i.e. models with $\rho = 0$.

1. Empty, flat universe model with vanishing cosmological constant: $\rho = p = k = \Lambda = 0$. Then Eq. (12.44) gives $\ddot{a} = 0$. Integrating with the normalization $a(t_0) = 1$ gives $a(t) = 1$. The line element then takes the form

$$ds^2 = -c^2 dt^2 + dr^2 + r^2 d\Omega^2 \tag{12.48}$$

This represents the Minkowski spacetime in spherical coordinates.

2. Empty universe model with curved 3-space and vanishing cosmological constant: $\rho = p = \Lambda = 0, \quad k \neq 0$. Then Eq. (12.44) gives $\dot{a}^2 + kc^2/R_0^2 = 0$. This requires $k = -1$, i.e. hyperbolic 3-space. For an expanding universe model we then get $\dot{a} = 1$. Integrating with the normalization $a(t_0) = 1$ gives $a(t) = t/t_0$. The line-element then takes the form

$$ds^2 = -c^2 dt^2 + \left(\frac{t}{t_0}\right)^2\left(\frac{dr^2}{1 + (r/R_0)^2} + r^2 d\Omega^2\right). \tag{12.49}$$

The universe model represented by this line-element is called *the Milne universe model*.

In this universe model the Hubble parameter (12.20) is

$$H = \frac{1}{t}. \tag{12.50}$$

Hence the age of this universe model is related to the Hubble constant by

$$t_0 = \frac{1}{H_0} = t_{\mathrm{H}}. \tag{12.51}$$

Applying the coordinate transformation

$$ct = \sqrt{c^2 T^2 - R^2}, \quad r = \frac{ct_0 R}{\sqrt{c^2 T^2 - R^2}} \tag{12.52}$$

transforms the line-element to the form (12.48) which represents the Minkowski spacetime. *When there exists a coordinate transformation between two line-elements they represent the same spacetime in two coordinate systems, which may be co-moving with different reference frames.* The coordinate R is co-moving with a static reference frame, SR. The coordinate r is co-moving with another reference frame, RF. We can find the motion of the reference particles of RF relative to those of SR as follows.

Solving the last of the two transformation equations with respect to R gives

$$R = \frac{rcT}{\sqrt{c^2 t_0^2 + r^2}} \tag{12.53}$$

The reference particles of RF have $r = $ constant. Hence for these particles R increases linearly with T. This means that the frame in which r is co-moving is expanding with a constant expansion velocity. Hence the Milne universe is nothing but the Minkowski spacetime as described in an expanding reference frame.

3. Expanding, flat, empty universe model with positive cosmological constant, $\rho = p = k = 0$, $\Lambda > 0$ For this universe model Eq. (12.46) reduces to

$$H = \frac{\dot{a}}{a} = \sqrt{\frac{\Lambda}{3}}. \tag{12.54}$$

Hence the Hubble parameter is constant. Integration with $a(t_0) = 1$ gives

$$a(t) = e^{H(t-t_0)}. \tag{12.55}$$

The line-element takes the form

$$ds^2 = -c^2 dt^2 + e^{2H(t-t_0)}(dr^2 + r^2 d\Omega^2). \tag{12.56}$$

The spacetime represented by this line-element is called *the De Sitter spacetime*.

It was represented by De Sitter in 1917 as a static and spherically symmetric solution of Einsteins's equations with a cosmological constant for empty space. Five years later it was shown that the reference particles of the static frame were not freely moving, and that when transforming the solution to a coordinate system co-moving with freely moving reference particles, one obtained the line-element above. Also Georges Lemaître showed in 1933 that the cosmological constant could be interpreted to represent the constant energy density of Lorentz Invariant Vacuum Energy, LIVE.

Equations (12.46) and (12.47) then give

$$\ddot{a} = -\frac{\kappa}{6}a(\rho + 3p/c^2) + \frac{\Lambda}{3}. \tag{12.57}$$

Inserting the *gravitational mass–density* ρ_G from Eq. (11.13) this equation takes the form

$$\ddot{a} = -\frac{\kappa}{6}a\rho_G + \frac{\Lambda}{3}. \tag{12.58}$$

Inserting $p = w\rho c^2$ into Eq. (11.13) gives

$$\rho_G = (1 + 3w)\rho, \tag{12.59}$$

which is negative for $w < -1/3$, i.e. according to Eq. (12.42), for $\dot{\phi}^2 < V(\phi)$. Special cases:

- matter in the form of dust: $w = 0$, $\rho_G = \rho_m$,
- radiation: $w = \frac{1}{3}$, $\rho_G = 2\rho_R$,
- LIVE: $w = -1$, $\rho_G = -2\rho_L$.

In a universe dominated by LIVE the acceleration of the cosmic expansion is

$$\ddot{a} = \frac{\kappa}{6}a\rho_L > 0, \tag{12.60}$$

that is *accelerated expansion*. This means that LIVE acts upon itself with repulsive gravitation.

The field equations can be combined into

$$H^2 \equiv \left(\frac{\dot{a}}{a}\right)^2 = \frac{\kappa}{3}\rho_m + \frac{\Lambda}{3} - \frac{kc^2}{R_0^2 a^2}, \tag{12.61}$$

where ρ_m is the density of matter in the form of dust,

$$\Lambda = \kappa \rho_{\mathrm{L}}, \tag{12.62}$$

and ρ_{L} is the constant density of LIVE. Then we may write

$$H^2 = \frac{\kappa}{3}\rho - \frac{kc^2}{R_0^2 a^2}, \tag{12.63}$$

where $\rho = \rho_{\mathrm{m}} + \rho_{\mathrm{L}}$. The critical density ρ_{cr} is the density in a universe with Euclidean space-like geometry, $k = 0$, which gives

$$\kappa \rho_{\mathrm{cr}} = 3H^2. \tag{12.64}$$

The present value of the Hubble parameter, i.e. the Hubble constant, is

$$H_0 = \sqrt{(\kappa/3)\rho_{\mathrm{cr}0}} \tag{12.65}$$

We introduce the density parameters

$$\Omega_{\mathrm{m}} = \frac{\rho_{\mathrm{m}}}{\rho_{\mathrm{cr}}}, \quad \Omega_{\Lambda} = \frac{\rho_{\mathrm{L}}}{\rho_{\mathrm{cr}}}. \tag{12.66}$$

Note that these quantities are only well defined for expanding universe models with $H \neq 0$. Furthermore we introduce a dimensionless parameter which represents the curvature of 3-space

$$\Omega_{\mathrm{k}} = -\frac{kc^2}{R_0^2 a^2 H^2}. \tag{12.67}$$

with present value

$$\Omega_{\mathrm{k}0} = -\frac{kc^2}{R_0^2 H_0^2}. \tag{12.68}$$

Note that $\Omega_k < 0$ means positive spatial curvature, and $\Omega_k > 0$ means negative spatial curvature. Equation (12.63) can now be written as

$$\Omega_{\mathrm{m}} + \Omega_{\mathrm{L}} + \Omega_{\mathrm{k}} = 1. \tag{12.69}$$

This equation shows that an empty, expanding universe has $\Omega_k = 1$; i.e. an empty, expanding universe has negative spatial curvature. It may be noted that since the present value of the scale parameter is normalized to $a(t_0) = 1$ the curvature parameter may be written

$$\Omega_{\mathrm{k}} = \frac{\Omega_{\mathrm{k}0} H_0^2}{a^2 H^2} = \frac{1 - \Omega_0}{a^2} \frac{H_0^2}{H^2}, \quad \Omega_0 = \Omega_{\mathrm{m}0} + \Omega_{\mathrm{L}0}. \tag{12.70}$$

Hence a sufficiently large mass density $\Omega_0 > 1$ gives spherical spatial geometry, $\Omega_0 = 1$ gives Euclidean spatial geometry, and $\Omega_0 < 1$ gives hyperbolic spatial geometry. Inserting $\Omega_{k0} = 1 - \Omega_0$ into Eq. (12.68) and solving the resulting equation with respect to R_0 gives

$$R_0 = \frac{l_H}{\sqrt{|1 - \Omega_0|}}. \tag{12.71}$$

The Hubble length in our universe is 13.8 billion light years. Observations of temperature fluctuations in the CMB-radiation show that $\Omega_0 \approx 1$. Hence the curvature radius of the 3-space is extremely large, much larger than the Hubble length of the universe.

12.3.5 Cosmic Redshift

We consider a homogeneous and isotropic universe with standard coordinates so that the line-element has the form (12.1). An observer is at $\chi = 0$ and an emitter at $\chi = \chi_e$. A light signal is emitted at a point of time t_e and received by the observer at the present time t_0. The scale factor is normalized to have the value $a(t_0) = 1$ at the present time. The emitter is relatively close to the observer in the sense that the expansion velocity of the emitter relatively to the observer is much smaller than the velocity of light, $v_H \ll c$. Due to the Hubble–Lemaître law that the expansion velocity is proportional to the distance, this condition takes the form

$$v_H = H\,l = R_0 H(t_e) a(t_e) \chi_e \ll c, \tag{12.72}$$

The equation of motion of a light signal is given in Eq. (12.25). Integrating between the emitter and the observer gives

$$c(t_0 - t_e) = R_0 \int_0^{\chi_e} a(t) \mathrm{d}\chi. \tag{12.73}$$

To second order in $c(t_0 - t_e)$ this gives

$$R_0 \chi_e = \int_{t_e}^{t_0} \frac{c\,\mathrm{d}t}{a(t)} = c(t_0 - t_e) + \frac{1}{2} H_0 c^2 (t_0 - t_e)^2, \tag{12.74}$$

where we have used that $a(t_0) = 1$ and $\dot{a}(t_0) = H_0$. In the condition (12.72) it is sufficient to keep the first term. Hence the condition takes the form $H(t_e) a(t_e)(t_0 - t_e) \ll 1$. In this inequality we can approximate the values of the Hubble parameter and the scale factor at the emission point of time by their values at the time that the light signal is received by the observer. Hence we arrive at the condition $H_0(t_0 - t_e) \ll 1$. In the following we shall include only terms up to 2 order in $H_0(t_0 - t_e)$.

The cosmic redshift is given in Eq. (12.32). Introducing the *deceleration parameter*,

$$q = -\frac{a\,\ddot{a}}{\dot{a}^2}, \tag{12.75}$$

we have

$$\dot{z} = -\frac{\dot{a}}{a^2} = -\frac{H}{a}, \quad \ddot{z} = 2\frac{\dot{a}^2}{a^3} - \frac{\ddot{a}}{a^2} = 2\frac{H^2}{a} + \frac{H^2 q}{a}, \tag{12.76}$$

giving $\dot{z}(t_0) = -H_0$, $\ddot{z}(t_0) = H_0^2(2 + q_0)$. Hence a Taylor expansion of z about $t = t_0$ to 2 order in $H_0(t_0 - t_e)$ gives

$$z(t_e) = H_0(t_0 - t_e) + \left(1 + \frac{1}{2}q_0\right)H_0^2(t_0 - t_e)^2. \tag{12.77}$$

We now introduce an observer G instantaneously at the position of the emitter at the point of time t_e, but permanently at rest relative to the observer O at the origin. The velocity of the emitter relative to this observer is

$$V_e = \dot{a}(t_e)\,\chi_e. \tag{12.78}$$

Due to the Doppler effect G measures a redshift

$$z_D = [(1 + V_e/c)/(1 - V_e/c)]^{1/2} - 1 \tag{12.79}$$

of the emitted signal. Since $|V_e| \ll c$ we can use the two first terms of a Mc Laurin expansion, giving

$$z_D \approx \frac{V_e}{c} + \frac{1}{2}\left(\frac{V_e}{c}\right)^2. \tag{12.80}$$

A Taylor expansion of $\dot{a}$ about t_0 gives to 2 order in $H_0(t_0 - t_e)$,

$$\dot{a}(t_e) \approx \dot{a}(t_0) + \ddot{a}(t_0)(t_0 - t_e) = H_0 + q_0 H_0^2(t_0 - t_e). \tag{12.81}$$

Inserting the expressions (12.74) and (12.81) in Eq. (12.78) we get

$$V_e \approx H_0 c(t_0 - t_e) + \left(\frac{1}{2} + q_0\right)H_0^2 c^2(t_0 - t_e)^2. \tag{12.82}$$

Substituting this into Eq. (12.80) gives

$$z_D = H_0(t_0 - t_e) + (1 + q_0)H_0^2(t_0 - t_e)^2. \tag{12.83}$$

Inserting an observer in the homogeneous LFRW universe model breaks the homogeneous symmetry. Hence the gravitational potential depends upon the distance from the observer. From Eq. (12.57) with a vanishing cosmological constant we see that in the theory of relativity it is natural to generalize the Newtonian gravitational potential of Eq. (1.29) and define a relativistic gravitational potential ϕ_G by

$$\nabla^2 \phi_G = 4\pi G(\rho + 3p/c^2) \tag{12.84}$$

Using Eq. (12.57) we then have

$$\nabla^2 \phi_G = 3qH^2. \tag{12.85}$$

The potential at the present time is given by

$$\nabla^2 \phi_G = 3q_0 H_0^2. \tag{12.86}$$

Integrating this equation with the conditions $(d\phi_G/dr)_{r=0} = \phi_G(0) = 0$ gives

$$\phi_G = \frac{1}{2} q_0 H_0^2 r^2. \tag{12.87}$$

Note that there is accelerated expansion for $q_0 < 0$ and decelerated expansion for $q_0 > 0$. Hence the gravitational potential decreases in the outwards direction when there is accelerated expansion and increases for decelerated expansion.

The distance from the emitter to the observer at a time close to the present time, when the scale factor is equal to one, is

$$r = c(t_0 - t_e). \tag{12.88}$$

Hence the gravitational potential at the emitter is

$$\phi_G = \frac{1}{2} q_0 H_0^2 c^2 (t_0 - t_e)^2. \tag{12.89}$$

We now consider the Newtonian limit where the contribution of the pressure to the acceleration of gravity is negligible and the relativistic gravitational potential reduces to the Newtonian gravitational potential, $\phi_G \to \phi$, and we can neglect the expansion of the region between the emitter and the observer. Then the gravitational shift of the wavelength of light is given by Eqs. (5.108) and (5.111),

$$z_G = \frac{\lambda_0}{\lambda_e} - 1 = \sqrt{\frac{(g_{tt})_0}{(g_{tt})_e}} - 1. \tag{12.90}$$

According to Eq. (9.20) the *tt*-component of the metric tensor is then

$$g_{tt} = -\left(1 + \frac{2\phi}{c^2}\right),$$ (12.91)

where $2|\phi|/c^2 \ll 1$. We then get

$$z_G = \sqrt{\frac{1 + 2\phi_0/c^2}{1 + 2\phi_e/c^2}} - 1.$$ (12.92)

Noting that $\phi_0 = 0$ and making a series expansion to 1 order in ϕ/c^2 we have

$$z_G = -\phi_e/c^2 \approx -\phi_G/c^2.$$ (12.93)

Inserting the potential (12.89) gives

$$z_G = -\frac{1}{2}q_0 H_0^2 (t_0 - t_e)^2.$$ (12.94)

It follows from Eqs. (12.78), (12.84) and (12.95) that in the limit of small cosmic redshifts,

$$z = z_D + z_G.$$ (12.95)

In this limit the cosmic redshift due to the expansion of the universe is equal to the sum of the kinematic redshift due to the Doppler effect and the gravitational shift of wavelength due to the gravitational field between the emitter and the observer. The Doppler effect is a redshift since the emitter moves away from the observer, but the gravitational effect is a blueshift in a universe with retarded expansion due to attractive gravitation, since the light moves downwards in this gravitational field. In a universe with accelerated expansion due to the repulsive gravity of the vacuum energy, light moves upwards in the gravitational field between the emitter and the observer, and the gravitational effect is a redshift.

Hence *the cosmic redshift due to the expansion of the space contains both the kinematic Doppler effect due to the velocity of the emitter relative to the observer and the gravitational shift of wavelength for light moving vertically in a gravitational field.*

12.3.6 Energy–Momentum Conservation

From the 2 Bianchi identity and Einstein's field equations, it follows that the energy–momentum density tensor is covariantly divergence free. The time component expresses the equation of continuity and takes the form

$$[(\rho + p/c^2)u^{\hat{i}}u^{\hat{v}}]_{;\hat{v}} + (p\eta^{\hat{i}\hat{v}})_{;\hat{v}} = 0. \tag{12.96}$$

Since $u^{\hat{i}} = 1$, $u^{\hat{m}} = 0$ and $\eta^{\hat{i}\hat{i}} = -1$, $\eta^{\hat{i}\hat{m}} = 0$, we get

$$(\rho + p/c^2)^{\cdot} + (\rho + p/c^2)u^{\hat{v}}_{;\hat{v}} - \dot{p} = 0 \tag{12.97}$$

or

$$\dot{\rho} + (\rho + p/c^2)(u^{\hat{v}}_{,\hat{v}} + u^{\hat{a}}\Gamma^{\hat{v}}_{\hat{a}\hat{v}}) = 0. \tag{12.98}$$

Here $u^{\hat{v}}_{,\hat{v}} = 0$ and $\Gamma^{\hat{i}}_{\hat{t}\hat{t}} = 0$. Calculating $\Gamma^{\hat{m}}_{\hat{t}\hat{m}}$ from Cartan's 1. structure equation in the form

$$\underline{d}\,\underline{\omega}^{\hat{\mu}} = -\Gamma^{\hat{\mu}}_{\hat{\alpha}\hat{\beta}}\underline{\omega}^{\hat{\alpha}} \wedge \underline{\omega}^{\hat{\beta}} \tag{12.99}$$

with the basis forms (12.6) we get

$$\Gamma^{\hat{m}}_{\hat{t}\hat{m}} = \Gamma^{\hat{r}}_{\hat{t}\hat{r}} + \Gamma^{\hat{\theta}}_{\hat{t}\hat{\theta}} + \Gamma^{\hat{\phi}}_{\hat{t}\hat{\phi}} = 3\frac{\dot{a}}{a}. \tag{12.100}$$

Inserting this into Eq. (12.98) gives

$$(\rho c^2 a^3)^{\cdot} + p(a^3)^{\cdot} - 0. \tag{12.101}$$

Let $V = a^3$ be a co-moving volume in the universe and $U = \rho c^2 V$ the energy in the co-moving volume. Then we may write

$$dU + pdV = 0. \tag{12.102}$$

This is the first law of thermodynamics for an adiabatic expansion. It follows that *the universe expands adiabatically*. The adiabatic equation can be written as

$$\frac{\dot{\rho}}{\rho + p/c^2} = -3\frac{\dot{a}}{a}. \tag{12.103}$$

With $p = w\rho c^2$ we get

$$\frac{d\rho}{\rho} = -3(1 + w)\frac{da}{a}. \tag{12.104}$$

Assuming that $w = $ constant this equation can be integrated to give

$$\ln \frac{\rho}{\rho_0} = \ln \left(\frac{a}{a_0} \right)^{-3(1+w)}. \tag{12.105}$$

It follows that

$$\rho = \rho_0 \left(\frac{a}{a_0} \right)^{-3(1+w)} \quad \text{or} \quad \rho a^{3(1+w)} = \text{constant} \tag{12.106}$$

We have three particularly important special cases:

- Matter in the form of dust: $w = 0$ gives

$$\rho_m a^3 = \text{constant.} \tag{12.107}$$

Thus, the mass in a co-moving volume is constant.
- Radiation: $w = 1/3$ gives

$$\rho_{\text{rad}} a^4 = \text{constant.} \tag{12.108}$$

Thus, the radiation energy density decreases faster than the density of the dust when the universe is expanding. The energy in a co-moving volume is decreasing because of the thermodynamic work on the surface of a co-moving volume. In a remote past, the density of radiation was greater than the density of dust.
- LIVE: $w = -1$ gives $\rho_L = $ constant. The vacuum energy in a co-moving volume is increasing proportionally to a^3. In spite of this energy is conserved locally because of the negative work performed at a co-moving, expanding surface. This work transfers energy from the region outside the surface to the region inside it, maintaining the constant value of the energy density of LIVE in the expanding universe. Since we can choose the surface to have an arbitrarily large radius this amounts to extracting energy from an infinitely remote region to the region in a finite distance from the observer at the origin. Hence there is a difficulty with global energy conservation in cosmology.

The Friedmann–Lemaître Eq. (12.63) can be written

$$\frac{H^2}{H_0^2} = \frac{\Omega_{\text{rad}0}}{a^4} + \frac{\Omega_{m0}}{a^3} + \frac{\Omega_{k0}}{a^2} + \Omega_{L0}. \tag{12.109}$$

where $\Omega_{\text{rad}0}$ and Ω_{m0} are the present values of the radiation density and the density of dust, respectively, and Ω_0 is given in Eq. (12.70). Since $H = (1/a)(da/dt)$, the scale factor can be found as a function of time by integrating the equation

$$\frac{da}{dt} = H_0 \sqrt{\frac{\Omega_{\text{rad}0}}{a^2} + \frac{\Omega_{m0}}{a} + \Omega_{k0} + \Omega_{L0} a^2}. \tag{12.110}$$

By means of Eq. (12.32) this equation can be expressed in terms of the redshift of the source. In this way we obtain an equation for the Hubble parameter at the time of emission of light emitted from a source with redshift z and received at the present time,

$$\frac{H^2}{H_0^2} = \Omega_{rad0}(1+z)^4 + \Omega_{m0}(1+z)^3 + \Omega_{k0}(1+z)^2 + \Omega_{L0}. \tag{12.111}$$

Similarly this equation can be integrated to find the redshift, and hence the scale factor as a function of time by utilizing that

$$H = -\frac{1}{1+z}\frac{dz}{dt}. \tag{12.112}$$

In combination with Eq. (12.111) this gives an expression for age of the universe at the time a source with observed redshift z emits the observed light, i.e. the *emission point of time*,

$$t_E(z) = t_H \int_z^{\infty} \frac{dz}{(1+z)\sqrt{\Omega_{rad0}(1+z)^4 + \Omega_{m0}(1+z)^3 + \Omega_{k0}(1+z)^2 + \Omega_{L0}}}. \tag{12.113}$$

where the Hubble age t_H is defined in Eq. (12.21). This is called the *age-redshift relationship*. The age of the universe is found by letting the redshift of the source be zero, $t_0 = t(0)$. Taking the integral from 0 to z gives the time taken for the radiation to move from the source to the observer and arrive at the present time. This is called the *lookback time*.

Equation (12.113) is in general an elliptic integral, but it can be integrated in terms of elementary functions in the case that the cosmological constant vanishes, so that $\Omega_\Lambda = 0$. This gives the age-redshift relationship for a curved, dust-dominated universe.

$$t_E(z) = t_H \left[\frac{\sqrt{1+\Omega_{m0}z}}{(1-\Omega_{m0})(1+z)} - \frac{\Omega_{m0}}{(1-\Omega_{m0})^{3/2}} \text{arc sinh} \sqrt{\frac{1-\Omega_{m0}}{\Omega_{m0}(1+z)}} \right]. \tag{12.114}$$

The age of this universe is

$$t_0 = t_E(0) = \frac{t_H}{1-\Omega_{m0}} \left(1 - \frac{\Omega_{m0}}{\sqrt{1-\Omega_{m0}}} \text{arc sinh} \sqrt{\frac{1-\Omega_{m0}}{\Omega_{m0}}} \right). \tag{12.115}$$

12.4 Some LFRW Cosmological Models

12.4.1 Radiation-Dominated Universe Model

The energy–momentum tensor for radiation is trace free. According to the Einstein's field equations the Einstein tensor must then be trace free. Using Eq. (12.44) we then have

$$a\ddot{a} + \dot{a}^2 + kc^2/R_0^2 = 0,$$
$$(a\dot{a} + ktc^2/R_0^2)^{\cdot} = 0. \tag{12.116}$$

Integration gives

$$a\dot{a} + ktc^2/R_0^2 = B. \tag{12.117}$$

Integrating once more gives

$$\frac{1}{2}a^2 + \frac{1}{2}\frac{kc^2}{R_0^2}t^2 = Bt + C. \tag{12.118}$$

The initial condition $a(0) = 0$ gives $C = 0$. Hence

$$a = \sqrt{2Bt - kt^2c^2/R_0^2}. \tag{12.119}$$

For $k = 0$ we have

$$a = \sqrt{2Bt}, \quad \dot{a} = \sqrt{\frac{B}{2t}}. \tag{12.120}$$

The expansion velocity reaches infinity at $t = 0$, $\lim_{t \to 0} \dot{a} = \infty$ (Fig. 12.2). According to the Stefan–Boltzmann law we have

$$\rho_R = \sigma T^4. \tag{12.121}$$

Combining this with Eq. (12.108) we have

$$T^4 a^4 = \text{constant}, \tag{12.122}$$

or

$$T = \frac{T_0}{a}, \tag{12.123}$$

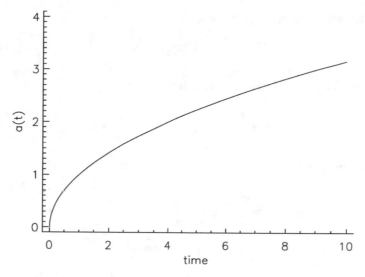

Fig. 12.2 Expansion of a radiation dominated universe. In a radiation-dominated universe the expansion velocity reaches infinity at $t = 0$

where T_0 is the temperature of the radiation at the present time. Equations (12.120) and (12.123) give

$$T = T_0 \sqrt{\frac{t_0}{t}}. \tag{12.124}$$

12.4.2 Dust-Dominated Universe Model

From the first of the Friedmann–Lemaître equations we have

$$\dot{a}^2 + \frac{kc^2}{R_0^2} = \frac{\kappa}{3} \rho a^2. \tag{12.125}$$

We now introduce a so-called *conformal time* η by

$$\frac{dt}{d\eta} = a(\eta) \Rightarrow \frac{d}{dt} = \frac{1}{a} \frac{d}{d\eta}. \tag{12.126}$$

Hence,

$$\dot{a} = \frac{da}{dt} = \frac{1}{a} \frac{da}{d\eta}. \tag{12.127}$$

We also introduce

$$A \equiv \frac{\kappa}{3}\rho_{m0}. \tag{12.128}$$

The first Friedmann–Lemaître equation then gives

$$a\dot{a}^2 + \frac{kc^2}{R_0^2}a = \frac{\kappa}{3}\rho_m a^3 = \frac{\kappa}{3}\rho_{m0} = A. \tag{12.129}$$

Using η we get

$$\frac{1}{a}\left(\frac{da}{d\eta}\right)^2 = \frac{c^2}{R_0^2}(A - ka),$$

$$\tag{12.130}$$

$$\frac{da}{d\eta} = \frac{ca}{R_0}\sqrt{\frac{A}{a} - k} = \frac{c}{R_0}\sqrt{aA}\sqrt{1 - \frac{ka}{A}},$$

where we have chosen the positive root. We now introduce u by

$$a = Au^2, \quad u = \sqrt{\frac{a}{A}}, \tag{12.131}$$

and get

$$\frac{da}{d\eta} = 2Au\frac{du}{d\eta}, \tag{12.132}$$

which together with Eq. (12.131) gives

$$\frac{du}{\sqrt{1 - ku^2}} = \frac{c}{2R_0}d\eta. \tag{12.133}$$

This equation will first be integrated for $k = -1$. Then

$$\int \frac{du}{\sqrt{1 + u^2}} = \frac{c\eta}{2R_0} + K, \tag{12.134}$$

or

$$\operatorname{arsinh}(u) = \frac{c\eta}{2R_0} + K. \tag{12.135}$$

The condition $u(0) = 0$ gives $K = 0$. Hence

$$\frac{1}{A}a = \sinh^2\frac{c\eta}{2R_0} = \frac{1}{2}\left(\cosh\frac{c\eta}{R_0} - 1\right) \tag{12.136}$$

or

$$a = \frac{A}{2}\left(\cosh \frac{c\eta}{R_0} - 1\right). \tag{12.137}$$

From Eqs. (12.128), (12.65) and (12.66) we have

$$A = \frac{\kappa}{3}\rho_{m0} = H_0^2 \frac{\rho_{m0}}{\rho_{cr0}} = H_0^2 \Omega_{m0}. \tag{12.138}$$

From Eqs. (12.67) and (12.69) we get

$$k = \frac{R_0^2 H_0^2}{c^2}(\Omega_{m0} - 1). \tag{12.139}$$

Hence, the scale factor of the negatively curved, dust-dominated universe model is

$$a(\eta) = \frac{1}{2}\frac{\Omega_{m0}}{1 - \Omega_{m0}}\left(\cosh \frac{c\eta}{R_0} - 1\right). \tag{12.140}$$

Inserting this into Eq. (12.126) and integrating with $t(0) = \eta(0)$ leads to

$$t(\eta) = \frac{\Omega_{m0}}{2H_0(1 - \Omega_{m0})^{3/2}}\left(\sinh \frac{c\eta}{R_0} - \frac{c\eta}{R_0}\right). \tag{12.141}$$

Integrating Eq. (12.118) for $k = 0$ leads to an Einstein–de Sitter universe

$$a(t) = \left(\frac{t}{t_0}\right)^{\frac{2}{3}}. \tag{12.142}$$

Finally integrating Eq. (12.125) for $k = 1$ gives

$$a(\eta) = \frac{1}{2}\frac{\Omega_{m0}}{\Omega_{m0} - 1}\left(1 - \cos \frac{c\eta}{R_0}\right), \tag{12.143}$$

$$t(\eta) = \frac{\Omega_{m0}}{2H_0(\Omega_{m0} - 1)^{3/2}}\left(\frac{c\eta}{R_0} - \sin \frac{c\eta}{R_0}\right). \tag{12.144}$$

This is a parametric representation of a cycloid (Fig. 12.3).
In the Einstein–de Sitter model the Hubble parameter is

$$H = \frac{2}{3}\frac{1}{t}. \tag{12.145}$$

The age of this universe model is

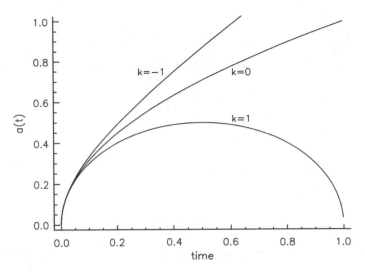

Fig. 12.3 Expansion of matter dominated universe models. For $k = 1$ the density is larger than the critical density, and the universe is closed. For $k = 0$ we have $\rho = \rho_{cr}$ and the expansion velocity of the universe will approach zero as $t \to \infty$. For $k = -1$ we have $\rho < \rho_{cr}$. The universe is then open and will continue expanding forever

$$t_0 = (2/3)\, t_{\mathrm{H}}. \tag{12.146}$$

With a Hubble age $t_{\mathrm{H}} = 13.85$ billion years the age of the flat, dust-dominated universe becomes $t_0 = 9.2$ billion years, in conflict with the age $t_0 = 13.8$ years based upon a large set of different observations (Fig. 12.4).

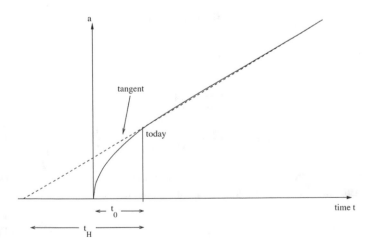

Fig. 12.4 Hubble age. Here t_H is the Hubble age, i.e. the age of the universe if the expansion had been constant. But in the dust-dominated universe the expansion rate was faster closer to the Big Bang, so the age is lower

Example 12.4.1 (*Lookback Time for Flat Dust-dominated Universe*) In the case of a flat universe $\Omega_0 = 1$. With $\Omega_{L0} = 0$ Eq. (12.113) with modified limits corresponding to the lookback time reduces to

$$t_{LB} = t_H \int_0^z \frac{dz}{(1+z)^{\frac{5}{2}}} = \frac{2}{3} t_H \left[1 - \frac{1}{(1+z)^{\frac{3}{2}}} \right]. \tag{12.147}$$

Using Eq. (12.146) for the age of a flat, dust-dominated universe the expression for the lookback time can be written

$$t_{LB} = t_0 \left[1 - \frac{1}{(1+z)^{3/2}} \right]. \tag{12.148}$$

Hence, the redshift of an object with lookback time t_{LB} is

$$z = \frac{1}{\left(1 - \frac{t_{LB}}{t_0} \right)^{2/3}} - 1. \tag{12.149}$$

12.4.3 Transition from Radiation-Dominated to Matter-Dominated Universe

We consider the early universe filled with radiation and matter, but where vacuum energy can be neglected. The universe is assumed to be flat. Then Friedmann's 1. equation takes the form

$$\dot{a}^2 = \frac{\kappa}{3}(\rho_m + \rho_r)a^2. \tag{12.150}$$

For matter,

$$\rho_m a^3 = \rho_{m0}. \tag{12.151}$$

For radiation,

$$\rho_r a^4 = \rho_{r0}. \tag{12.152}$$

Hence

$$a^2 \dot{a}^2 = \frac{\kappa}{3}(\rho_{m0} a + \rho_{r0}). \tag{12.153}$$

The present values of the critical density and the density parameters are

$$\kappa \rho_{cr0} = 3H_0^2, \quad \Omega_{m0} = \frac{\rho_{m0}}{\rho_{cr0}}, \quad \Omega_{r0} = \frac{\rho_{r0}}{\rho_{cr0}}, \tag{12.154}$$

giving

$$a\dot{a} = H_0(\Omega_{m0}a + \Omega_{r0})^{1/2}. \tag{12.155}$$

Integration with $a(0) = 0$ leads to

$$H_0 t = \frac{4}{3}\frac{\Omega_{r0}^{3/2}}{\Omega_{m0}^2} + \frac{2}{3}\frac{(\Omega_{m0}a - 2\Omega_{r0})(\Omega_{m0}a + \Omega_{r0})^{1/2}}{\Omega_{m0}^2}. \tag{12.156}$$

From Eqs. (12.155) and (12.156) it follows that at the transition time t_{eq} when $\rho_m = \rho_r$, the scale factor has the value

$$a_{eq} = \frac{\rho_{r0}}{\rho_{m0}} = \frac{\Omega_{r0}}{\Omega_{m0}}. \tag{12.157}$$

Inserting this into Eq. (12.156) gives

$$t_{eq} = \frac{2}{3}\left(2 - \sqrt{2}\right)\frac{\Omega_{r0}^{3/2}}{\Omega_{m0}^2}t_H. \tag{12.158}$$

The microwave background radiation has a temperature 2.73 K corresponding to a density parameter $\Omega_{r0} = 8.4 \times 10^{-5}$. In a flat universe without vacuum energy $\Omega_{m0} = 1 - \Omega_{r0}$. Inserting $t_H = 13.85 \times 10^9$ years then leads to $t_{eq} = 47 \times 10^3$ years.

12.4.4 The de Sitter Universe Models

These are universe models without radiation and matter, containing only LIVE. There are three models depending upon the spatial curvature. For these models the Friedmann Eq. (12.46) takes the form

$$\dot{a}^2 - \frac{\Lambda}{3}a^2 = -\frac{kc^2}{R_0^2}. \tag{12.159}$$

This equation has the following solution

$$a(t) = \begin{cases} \frac{R_0}{c}\sqrt{\frac{3}{\Lambda}}\cosh\left(\sqrt{\frac{\Lambda}{3}}t\right), & k = 1 \\ e^{\sqrt{\Lambda/3}\,t}, & k = 0 \\ \frac{R_0}{c}\sqrt{\frac{3}{\Lambda}}\sinh\left(\sqrt{\frac{\Lambda}{3}}t\right), & k = -1 \end{cases} \tag{12.160}$$

Fig. 12.5 The scale factor of
the De Sitter universe models

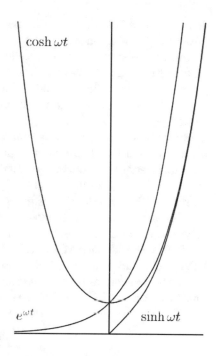

The scale factor as a function of time is shown for these universe models in
Fig. 12.5. Note that the flat De Sitter universe is infinitely old. It had no Big Bang.

12.4.5   The Friedmann–Lemaître Model

The dynamics of galaxies and clusters of galaxies has made it clear that far stronger
gravitational fields are needed to explain the observed motions than those produced
by visible matter [2]. At the same time it has become clear that the density of this
dark matter is only about 30% of the critical density, although it is a prediction by the
usual versions of the inflationary universe models that the total density of all that is
contained in the universe ought to be equal to the critical density [3]. Also the recent
observations of the temperature fluctuations of the cosmic microwave radiation have
shown that space is either flat or very close to flat [4–6]. The energy which fills up
the universe to the critical density must be evenly distributed in order not to affect
the dynamics of the galaxies and the clusters.

Furthermore, in 1998 observations of supernovae of type-I a with high cosmic
redshifts indicated that the expansion of the universe is accelerating [7, 8]. This was
explained as a result of repulsive gravitation due to some sort of vacuum energy.
Thereby the missing energy needed to make space flat was identified as vacuum
energy. Hence, it seems that we live in a flat universe with vacuum energy having a

density around 70% of the critical density and with matter having a density around 30% of the critical density.

Until the discovery of the accelerated expansion of the universe the standard model of the universe was assumed to be the Einstein–de Sitter model, which is a flat universe model dominated by cold matter. Now it seems that we must replace this model with a new "standard model" containing both dark matter and vacuum energy [9].

Recently several types of vacuum energy or so-called quintessence energy have been discussed [10, 11]. However, the most simple type of vacuum energy is the Lorentz invariant vacuum energy, LIVE, which has constant energy density during the expansion of the universe [12, 13]. This type of energy can be mathematically represented by including a cosmological constant in Einstein's gravitational field equations. The flat universe model with cold dark matter and this type of vacuum energy is the Friedmann–Lemaître model. This universe model is usually denoted the Lambda-Cold-Dark-Matter (Λ CDM) model, but a better name would be the LIVE-Cold-Dark-Matter (LCDM) model.

The field equations for the flat Friedmann–Lemaître universe model are found by putting $k = p = 0$ in Eq. (12.62). This gives

$$2\frac{\ddot{a}}{a} + \frac{\dot{a}^2}{a^2} = \Lambda, \quad \Lambda = \kappa\rho_L \tag{12.161}$$

Integration leads to

$$a\dot{a}^2 = \frac{\Lambda}{3}a^3 + K, \tag{12.162}$$

where K is a constant of integration. Since the amount of matter in a volume co-moving with the cosmic expansion is constant, $\rho_m a^3 = \rho_{m0} a_0^3$, where the index 0 refers to values at the present time. Normalizing the expansion factor so that $a_0 = 1$ and comparing Eqs. (12.46) and (12.162) then give $K = (\kappa/3)\rho_{m0}$. Introducing a new variable x by $a^3 = x^2$ and integrating once more with the initial condition $a(0) = 0$, we obtain

$$a^3 = \frac{3K}{\Lambda}\sinh^2\left(\frac{t}{t_L}\right), \tag{12.163}$$

where

$$t_L = \frac{2}{\sqrt{3\Lambda}} = \frac{2}{3H_0\sqrt{\Omega_{L0}}} = \frac{2}{3\sqrt{\Omega_{L0}}}t_H, \tag{12.164}$$

where L denotes LIVE. Since there is, at the present time (July 2019), a rather large disagreement between the value of H_0 as determined from supernova observations and from observations of temperature fluctuations in the cosmic microwave background radiation, we shall here determine the parameters of the universe model from the observed values of the age of the universe and the value of the density parameters of LIVE. From a large number of different types of observations we have $t_0 = (13.8 \pm 0.02) \times 10^9$ years and $\Omega_{L0} = 0.694 \pm 0.007$ [14].

Since the present universe model has flat space, the total density is equal to the critical density, i.e. $\Omega_m + \Omega_L = 1$. Equation (12.162) with the normalization $a(t_0) = 1$ gives $3H_0^2 = 3K + \Lambda$. Equation (12.46) with $k = 0$ gives $\kappa\rho_0 = 3H_0^2 - \Lambda$. Hence $K = \kappa\rho_0/3$ and

$$\frac{3K}{\Lambda} = \frac{\kappa\rho_0}{\Lambda} = \frac{\rho_0}{\rho_L} = \frac{\Omega_{m0}}{\Omega_{L0}}. \tag{12.165}$$

In terms of the values of the relative densities at the present time the expression for the scale factor then takes the form

$$a = A^{1/3} \sinh^{2/3}\left(\frac{t}{t_L}\right), \quad A = \frac{\Omega_{m0}}{\Omega_{L0}} = \frac{1 - \Omega_{L0}}{\Omega_{L0}}. \tag{12.166}$$

Inserting $\Omega_{L0} = 0.694$ gives $A = 0.44$. Using the identity $\sinh(x/2) = \sqrt{(\cosh x - 1)/2}$ this expression may be written as

$$a^3 = \frac{A}{2}\left[\cosh\left(\frac{2t}{t_L}\right) - 1\right]. \tag{12.167}$$

The age t_0 of the universe is found from $a(t_0) = 1$, which, by use of the formula $\operatorname{arctanh} x = \operatorname{arsinh}(x/\sqrt{1 - x^2})$, leads to the expression

$$t_0 = t_L \operatorname{arsinh}\sqrt{\frac{\Omega_{L0}}{\Omega_{m0}}} = t_L \operatorname{artanh}\sqrt{\Omega_{L0}}. \tag{12.168}$$

We use this to determine t_L, writing

$$t_L = \frac{t_0}{\operatorname{artanh}\sqrt{\Omega_{L0}}}. \tag{12.169}$$

Inserting $t_0 = 13.8 \times 10^9$ years and $\Omega_{L0} = 0.694$ gives $t_L = 11.5 \times 10^9$ years. Substituting the values $A = 0.44$ and $t_L = 11.5 \times 10^9$ years into Eq. (12.166) gives

$$a = 0.76 \sinh^{2/3}\left(1.2\frac{t}{t_0}\right). \tag{12.170}$$

This function is plotted in Fig. 12.6.

A consistency requirement on the Hubble constant H_0 follows from Eqs. (12.164) and (12.158). If the general theory of relativity is correct, and if it is correct to describe the universe as homogeneous and isotropic on a large scale, and dominated by dust and LIVE, then the Hubble constant must obey the equation

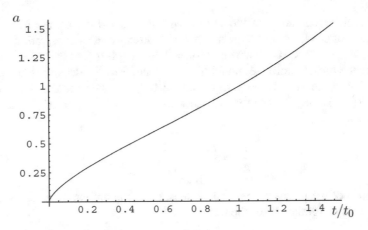

Fig. 12.6 The scale factor of the flat ΛCDM universe model

$$H_0 = \frac{2}{3} \frac{\text{artanh}\sqrt{\Omega_{L0}}}{\sqrt{\Omega_{L0}}} \frac{1}{t_0}. \tag{12.171}$$

Inserting $t_0 = 13.8 \times 10^9$ years and $\Omega_{L0} = 0.694$ gives $H_0 = 68.3$ km/s per Mpc. This is the predicted value of the Hubble constant. The corresponding Hubble age is $t_H = (977/H_0) \times 10^9$ years, giving $t_H = 14.3 \times 10^9$ years.

The ratio of the age of the universe and its Hubble age, $t_0/t_H = H_0 t_0$, is plotted in Fig. 12.7. The age of the universe increases with increasing density of vacuum energy. In the limit that the density of the vacuum approaches the critical density, there is no dark matter, and the universe model approaches the de Sitter model with

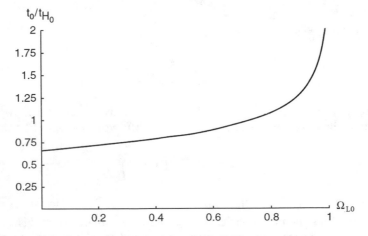

Fig. 12.7 The ratio of age and Hubble age of the flat ΛCDM universe model

exponential expansion and no Big Bang. This model behaves in the same way as the steady-state cosmological model and is infinitely old.

Invoking the phenomenon of perfect inertial dragging it has been argued that the validity of the principle of relativity for rotational motion requires that the age of the universe must be equal to its Hubble age, $H_0 t_0 = 1$ [15]. According to Eq. (12.171) this requires that

$$\tanh\left(\frac{3}{2}\sqrt{\Omega_{L0}}\right) = \sqrt{\Omega_{L0}}. \tag{12.172}$$

The positive, real solution of this equation is $\Omega_{L0} = 0.737$.

Using Eqs. (12.166) and (12.168) we obtain a nice form of the age-redshift relationship, giving the emission point of time of an object with redshift z,

$$t_E = \frac{\text{arsinh}\left(\sqrt{\frac{\Omega_{L0}}{\Omega_{m0}}}\frac{1}{(1+z)^{3/2}}\right)}{\text{arsinh}\left(\sqrt{\frac{\Omega_{L0}}{\Omega_{m0}}}\right)} t_0. \tag{12.173}$$

Inserting $\Omega_{L0} = 0.7$ and $\Omega_{m0} = 0.3$ with $t_0 = 13.8 \cdot 10^9$ years gives

$$t_E = 12.5 \times 10^9 \, \text{arsinh}\left(\frac{1.53}{(1+z)^{3/2}}\right) \text{ years.} \tag{12.174}$$

This is plotted in Fig. 12.8.

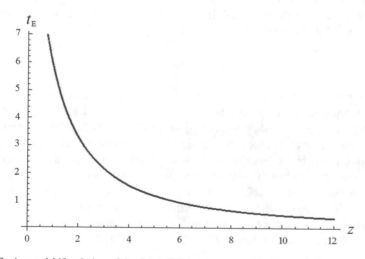

Fig. 12.8 Age-redshift relation of the flat ΛCDM universe model. The emission time is given in billion years

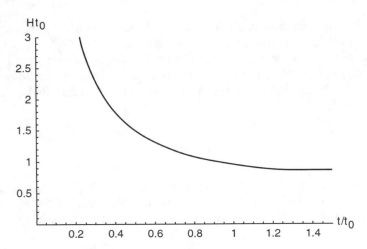

Fig. 12.9 The Hubble parameter of the flat ΛCDM universe model

The time which the radiation has taken in moving from the source to the observer and being observed at the present time, i.e. the lookback time, is $t_{LB} = t_0 - t_E$. The distance which the radiation has travelled is t_{LB} light years.

The Hubble parameter as a function of time is

$$H = \frac{2}{3\,t_L}\coth\left(\frac{t}{t_L}\right). \tag{12.175}$$

Inserting $t_0 = 1.2\,t_L$ we get $H\,t_0 = 0.8\coth(1.2\,t/t_0)$, which is plotted in Fig. 12.9.

It may be noted that the Hubble parameter is given as a function of the redshift in Eq. (12.111) with $\Omega_{RAD0} = 0$ and $\Omega_0 = 1$. The graph in Fig. 12.8 shows that the Hubble parameter decreases all the time and approaches a constant value $H_\infty = 2/3\,t_L$ in the infinite future. The Hubble age is

$$t_H = (3/2)t_L\sqrt{\Omega_{L0}}. \tag{12.176}$$

Inserting numerical values gives $t_H = 14.4 \cdot 10^9$ years. In this universe model the age of the universe is nearly as large as the Hubble age, while in the Einstein–de Sitter model the corresponding age is $t_{0ED} = (2/3)t_H = 9.5 \times 10^9$ years. The reason for this difference is that in the Einstein–de Sitter model the expansion is decelerated all the time, while in the Friedmann–Lemaître model the repulsive gravitation due to the vacuum energy has made the expansion accelerate lately (see below). Hence, for a given value of the Hubble parameter the previous velocity was larger in the Einstein–de Sitter model than in the Friedmann–Lemaître model.

A dimensionless quantity representing the rate of change of the cosmic expansion velocity is the *deceleration parameter*, which is defined as

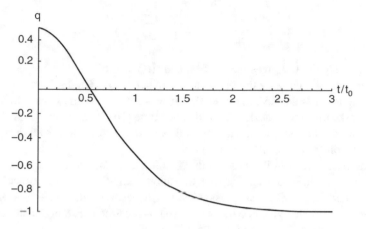

Fig. 12.10 The deceleration parameter of the flat ΛCDM universe model

$$q = -\ddot{a}/aH^2. \tag{12.177}$$

For the present universe model the deceleration parameter as a function of time is

$$q = \frac{1}{2}\left[1 - 3\tanh^2(t/t_{\rm L})\right], \tag{12.178}$$

which is shown graphically in Fig. 12.10.

From Eqs. (12.169) and (12.178) the deceleration parameter at the present point of time may be written as

$$q_0 = -\frac{1}{2}(3\Omega_{\rm L0} - 1). \tag{12.179}$$

With $\Omega_{\rm L0} = 0.7$ we get $q_0 = -0.55$.

The inflection point of time t_1 when deceleration turned into acceleration is given by $q = 0$. This leads to

$$t_1 = t_\Lambda \operatorname{artanh}\left(1/\sqrt{3}\right) \tag{12.180}$$

or is expressed in terms of the age of the universe

$$t_1 = \frac{\operatorname{artanh}\left(1/\sqrt{3}\right)}{\operatorname{artanh}\sqrt{\Omega_{\Lambda 0}}} t_0. \tag{12.181}$$

The corresponding cosmic redshift is

$$z(t_1) = \frac{1}{a(t_1)} - 1 = \left(\frac{2\Omega_{L0}}{1 - \Omega_{L0}}\right)^{1/3} - 1. \qquad (12.182)$$

Inserting $\Omega_{L0} = 0.7$ gives $t_1 = 0.54t_0$ and $z(t_1) = 0.67$.

The results of analyzing the observations of supernova SN 1997 at $z = 1.7$, corresponding to an emission time $t_E = 0.30t_0 = 4.5 \times 10^9$ years, have provided evidence that the universe was decelerated at that time [16]. Turner and Riess [17] have recently argued that the other supernova data favour a transition from deceleration to acceleration for a redshift around $z = 0.5$.

It may be noted from Eq. (12.57) that in a flat universe with dust and LIVE and vanishing cosmological constant, the transition from deceleration to acceleration happens when the gravitational mass density vanishes, i.e. when $\rho_m + \rho_L + 3p_L = 0$. Since $p_L = -\rho_L$, this gives $\rho_m(t_1) = 2\rho_L(t_1)$. Hence the constant density of LIVE is half the density of the dust at this transition.

Note that the expansion velocity given by Hubble's law, $v = H\,l$, always decreases as seen from Fig. 11.8. This is the velocity away from the Earth of the cosmic fluid at a fixed physical distance l from the Earth $\dot{a}$. The quantity $\dot{a}$, on the other hand, is the velocity of a fixed fluid particle co-moving with the expansion of the universe. If such a particle accelerates, the expansion of the universe is said to accelerate. While $\dot{H}$ tells how fast the expansion velocity changes at a fixed distance from the Earth, the quantity $\ddot{a}$ represents the acceleration of a free particle co-moving with the expanding universe. The connection between these two quantities is $\ddot{a} = a(\dot{H} + H^2)$.

The ratio of the inflection point of time and the age of the universe, as given in Eq. (12.168), is depicted graphically as a function of the present relative density of vacuum energy in Fig. 12.11. The turnover point of time happens earlier the greater the vacuum density is. The change from deceleration to acceleration would happen at the present time if $\Omega_{L0} = 1/3$.

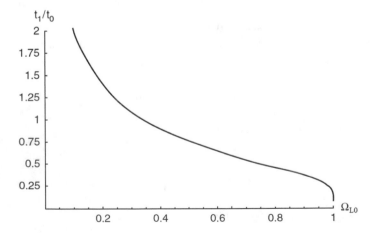

Fig. 12.11 Point of time for deceleration–acceleration turnover

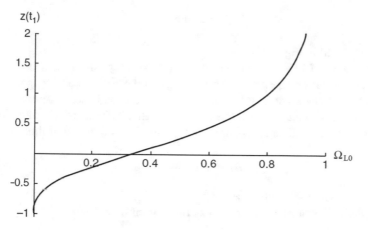

Fig. 12.12 Cosmic redshift at the deceleration–acceleration turnover

The redshift of the inflection point given in Eq. (12.181) as a function of vacuum energy density is plotted in Fig. 12.12 Note that the redshift of future points of time is negative, since then $u > u_0$. If $\Omega_{L0} < 1/3$ the transition to acceleration will happen in the future.

The critical density is

$$\rho_{cr} = \rho_L \tanh^{-2}(t/t_L).$$ (12.183)

This is plotted in Fig. 12.13. The critical density decreases with time. Equation (12.182) shows that the density parameter of LIVE is

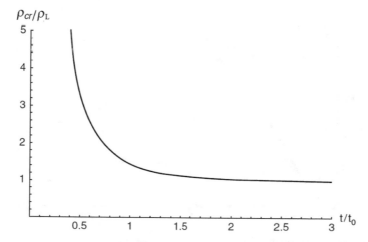

Fig. 12.13 Critical density as function of time

$$\Omega_L = \tanh^2(t/t_L), \tag{12.184}$$

which is plotted in Fig. 12.14.

The density of LIVE approaches the critical density. Since the density of LIVE is constant, this is better expressed by saying that the critical density approaches the density of the vacuum energy. Furthermore, since the total energy density is equal to the critical density all the time, this also means that the density of matter decreases faster than the critical density. The density of matter as function of time is

$$\rho_m = \rho_\Lambda \sinh^{-2}(t/t_L), \tag{12.185}$$

which is shown graphically in Fig. 12.15.

The density parameter of the dust as function of time is

$$\Omega_m = \cosh^{-2}(t/t_L), \tag{12.186}$$

which is shown in Fig. 12.16. Adding the density parameters of the expressions (12.183) and (12.185) we get the total density parameter $\Omega_{TOT} = \Omega_m + \Omega_\Lambda = 1$.

The universe became vacuum dominated at a point of time t_2 when $\rho_L(t_2) = \rho_m(t_2)$. From Eq. (12.184) it follows that this point of time is given by $\sinh(t_2/t_L) = 1$. Using Eq. (12.142) we get

$$t_2 = \frac{\text{arsinh}(1)}{\text{artanh}(\sqrt{\Omega_{L0}})} t_0. \tag{12.187}$$

It follows that the corresponding redshift is

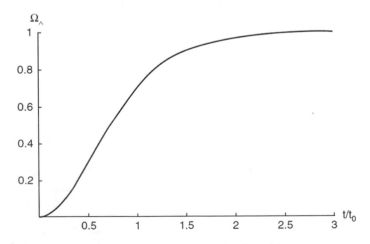

Fig. 12.14 The density parameter of vacuum energy as function of time

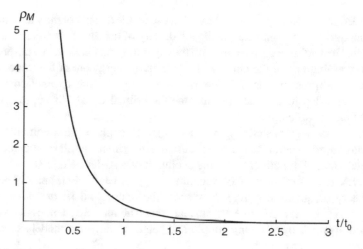

Fig. 12.15 The density of matter as function of time

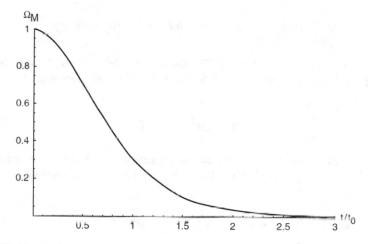

Fig. 12.16 The density parameter of matter as a function of time

$$z(t_2) = A^{-1/3} - 1. \tag{12.188}$$

Inserting $\Omega_{L0} = 0.7$ gives $t_2 = 0.73\,t_0$ and $z(t_2) = 0.32$. Hence, the transition to accelerated expansion happens before the universe becomes vacuum dominated.

As mentioned above, many different observations indicate that we live in a universe with critical density, where cold matter contributes with about 30% of the density and vacuum energy with about 70%. Such a universe is well described by the Friedmann–Lemaître universe model which has been presented above.

However, this model is not quite without problems in explaining the observed properties of the universe. In particular there is now much research directed at solving the so-called *coincidence problem*. As we have seen, the density of the vacuum energy is constant during the expansion, while the density of the matter decreases inversely proportional to a volume co-moving with the expanding matter. Yet, one observes that the density of matter and the density of the vacuum energy are of the same order of magnitude at the present time. This seems to be a strange and unexplained coincidence in the model. Also just at the present time the critical density is approaching the density of the vacuum energy.

There is now a great activity in order to try to explain these coincidences by introducing more general forms of vacuum energy called quintessence and with a density determined dynamically by the evolution of a scalar field [18].

However, the simplest type of vacuum energy is LIVE. One may hope that a future theory of quantum gravity may settle the matter and let us understand the vacuum energy. In the meantime we can learn much about the dynamics of a vacuum-dominated universe by studying simple and beautiful universe models such as the Friedmann–Lemaître model.

12.4.6   Flat Universe with Dust and Phantom Energy

We shall here consider a flat LFRW universe model with dust and phantom energy [18]. Differentiating Eq. (12.63) with $k = 0$ and using Eq. (12.103) we get

$$2\dot{H} + 3H^2 = -\kappa p/c^2. \tag{12.189}$$

The pressure is due to the phantom energy which can be described as due to a scalar field ϕ with a potential $V(\phi)$. The density and pressure of the phantom energy are

$$\kappa\rho_\phi = -\dot{\phi}^2/2 + V(\phi), \quad \kappa p_\phi/c^2 = -\dot{\phi}^2/2 - V(\phi). \tag{12.190}$$

Equation (12.189) then takes the form

$$2\dot{H} + 3H^2 = \frac{1}{2}\dot{\phi}^2 + V(\phi). \tag{12.191}$$

The evolution equation for the phantom energy field is

$$\ddot{\phi} + 3H\dot{\phi} = V'(\phi), \tag{12.192}$$

where the prime denotes differentiation with respect to ϕ.

It is now assumed that the time derivative of the phantom field is proportional to the Hubble parameter,

$$\dot{\phi} = -\alpha H = -\alpha \frac{\dot{a}}{a}. \tag{12.193}$$

where α is a positive constant. Integration gives

$$\phi = \phi_0 - \alpha \ln a. \tag{12.194}$$

Hence the phantom field decreases during the expansion of the universe.

We shall now deduce a differential equation for the relationship between the Hubble parameter and

the scalar field itself (not its derivative). We use that

$$\frac{d}{dt} = \dot{\phi} \frac{d}{d\phi}. \tag{12.195}$$

Differentiating Eq. (12.191) and substituting for V' from Eq. (12.192) we then get

$$\left(2H'' - 3H\right)\dot{\phi}^2 + 2\left(H' - \dot{\phi}\right)\ddot{\phi} + 6HH'\dot{\phi} = 0. \tag{12.196}$$

Using the relationship (12.193) in this equation gives

$$HH'' + H'^2 + \left(\alpha - \frac{3}{\alpha}\right)HH' - \frac{3}{2}H^2 = 0. \tag{12.197}$$

Introducing $y = H^2$ this equation can be written as

$$y'' + \left(\alpha - \frac{3}{\alpha}\right)y' - 3y = 0. \tag{12.198}$$

This is a 2 order linear differential equation with constant coefficients. The general solution is

$$H^2(\phi) = Ae^{3\phi/\alpha} + Be^{-\alpha\phi}. \tag{12.199}$$

It follows from Eqs. (12.191), (12.193) and (12.199) that the potential of the phantom field is

$$V(\phi) = \left(3 + \frac{\alpha^2}{2}\right)Be^{-\alpha\phi} - \frac{\alpha^2}{2}Ae^{3\phi/\alpha}. \tag{12.200}$$

Using Eqs. (12.190), (12.193) and (12.200) we find the energy density of the phantom field

$$\kappa\rho_\phi = 3Be^{-\alpha\phi} - \alpha^2 Ae^{3\phi/\alpha}. \tag{12.201}$$

Similarly the pressure is

$$\kappa p_\phi = -(3 + \alpha^2) B e^{-\alpha\phi}. \tag{12.202}$$

The mass density of the dust is

$$\kappa \rho_m = \kappa(\rho_{cr} - \rho_\phi) = 3H^2 - \kappa\rho_\phi = (3 - \alpha^2) A e^{3\phi/\alpha}. \tag{12.203}$$

Hence the constant A represents the dust. Note that there is no future singularity in this model.

Inserting the expression (12.199) for H into Eq. (12.193) gives a differential equation for the time evolution of the phantom field, but this equation has no analytical solution. The equations simplify if we assume that there is no dust, only the phantom field. Then $A = 0$, and the equation of state parameter of the phantom field is

$$w = \frac{p_\phi}{\rho_\phi} = -1 - \frac{\alpha^2}{3}, \tag{12.204}$$

which is less than minus one. This is characteristic of phantom energy. In the present case the potential of the phantom field is

$$V(\phi) = \left(3 + \frac{\alpha^2}{2}\right) B e^{-\alpha\phi}, \tag{12.205}$$

and the Hubble parameter is

$$H(\phi) = \sqrt{B}\, e^{-(\alpha/2)\phi}. \tag{12.206}$$

Letting $H_0 = H(0)$ we have $B = H_0^2$. Inserting Eq. (12.206) into Eq. (12.193) gives

$$e^{(\alpha/2)\phi}\dot{\phi} = -\alpha H_0. \tag{12.207}$$

Integration with $\phi(0) = \phi_0$ gives

$$\phi(t) = \ln\left[e^{(\alpha/2)\phi_0} - \frac{\alpha^2}{2}H_0 t\right]^{2/\alpha}. \tag{12.208}$$

The field vanishes at a point of time

$$t_0 = \frac{2}{\alpha^2}\left(e^{\frac{\alpha}{2}t_0} - 1\right)t_H, \tag{12.209}$$

where $t_H = 1/H_0$. After this point of time the field is negative and diverges at

$$t_1 = \frac{2}{\alpha^2} e^{(\alpha/2)\phi_0} t_H.$$

(12.210)

Inserting Eq. (12.208) into (12.206) gives

$$H = \frac{\dot{a}}{a} = \frac{H_0}{e^{(\alpha/2)\phi_0} - (\alpha^2/2)H_0 t}.$$

(12.211)

Hence $H(0) = e^{-(\alpha/2)\phi_0} H_0$ which is finite, $H(t_0) = H_0$ and $H(t_1) = \infty$, which is called the Big Rip. Integration with $a(t_0) = 1$ gives

$$a(t) = \left[e^{(\alpha/2)\phi_0} - \frac{\alpha^2}{2} H_0 t \right]^{-2/\alpha^2}.$$

(12.212)

Hence at the Big Rip $a(t_1) = \omega$. It should be noted that the presence of dust removes the Big Rip.

12.5 Flat Anisotropic Universe Models

We shall consider anisotropic world models with flat 3-space of Bianchi type-I following Ref. [19]. The line-element has the form

$$ds^2 = -c^2 dt^2 + a_1^2 (dx^1)^2 + a_2^2 (dx^2)^2 + a_3^2 (dx^3)^2.$$

(12.213)

The directional Hubble parameters are defined by

$$H_i = \frac{\dot{a}_i}{a_i}.$$

(12.214)

The mean scale factor is defined as

$$a = (a_1 a_2 a_3)^{1/3}.$$

(12.215)

A "volume scale factor",

$$V = a^3 = a_1 a_2 a_3$$

(12.216)

will also be useful. The average Hubble parameter is

$$H = \frac{1}{3}(H_1 + H_2 + H_3).$$

(12.217)

These definitions give

$$H = \frac{1}{3}(\ln V)^{\cdot} = (\ln a)^{\cdot}. \tag{12.218}$$

The *anisotropy parameter* is defined by

$$A = \frac{1}{3}\sum_{i=1}^{3}\left(\frac{H_i - H}{H}\right)^2. \tag{12.219}$$

Einstein's field equations with a cosmological constant for this class of universe models filled by a perfect fluid may be written

$$(\ln V)^{\cdot} + H_1^2 + H_2^2 + H_3^2 = \frac{\kappa}{2}\left(\rho + \frac{3p}{c^2}\right) + \Lambda, \tag{12.220}$$

$$\frac{1}{V}(V\,H_i)^{\cdot} = \frac{\kappa}{2}\left(\rho - \frac{p}{c^2}\right) + \Lambda. \tag{12.221}$$

We shall consider some special cases.

1. *Anisotropic empty universe with vanishing cosmological constant*

In this case Eqs. (12.220) and (12.221) reduces to

$$(\ln V)^{\cdot} + H_1^2 + H_2^2 + H_3^2 = 0, \tag{12.222}$$

$$\frac{1}{V}(V\,H_i)^{\cdot} = 0. \tag{12.223}$$

The last equation gives

$$H_i = \frac{\alpha_i}{V}, \tag{12.224}$$

This gives

$$3H = \sum_{i=1}^{3} H_i = \frac{\displaystyle\sum_{i=1}^{3}\alpha_i}{V}. \tag{12.225}$$

Combining this with Eq. (12.218) gives

$$\dot{V} = \sum_{i=1}^{3}\alpha_i. \tag{12.226}$$

Integration with $V(0) = 0$ gives

$$V = \sum_{i=1}^{3} \alpha_i \, t. \tag{12.227}$$

Inserting this into Eq. (12.224) gives

$$H_i = \frac{\dot{a}_i}{a_i} = \frac{\alpha_i}{\sum_{i=1}^{3} \alpha_i} \frac{1}{t} = \frac{p_i}{t}, \quad p_i = \frac{\alpha_i}{\sum_{i=1}^{3} \alpha_i}, \quad \sum_{i=1}^{3} p_i = 1. \tag{12.228}$$

Integration gives

$$a_i = a_{i0} t^{p_i}. \tag{12.229}$$

Inserting this into Eq. (12.222) gives

$$\sum_{i=1}^{3} p_i^2 = 1. \tag{12.230}$$

The directional and average Hubble parameters are

$$H_i = \frac{p_i}{t}, \quad H = \frac{1}{3t}. \tag{12.231}$$

Inserting this into Eq. (12.219) gives

$$A = \frac{1}{3} \sum_{i=1}^{3} (3p_i - 1)^2 = 2. \tag{12.232}$$

The universe described in this subsection is called the *Kasner universe*. It represents the Minkowski universe as described from a reference frame with anisotropic expansion. Due to the conditions (12.228) and (12.230) it is not possible to have equal values for the three constants p_i. Hence there is no isotropic special case of the Kasner universe. In particular, the Milne universe is not a special case since the Kasner universe is spatially flat, while the Milne universe has curved 3-space. It may be noted that the Kasner universe has a constant anisotropy parameter equal to 2.

2. *Anisotropic universe dominated by LIVE*

In this case Eqs. (12.220) and (12.221) take the form

$$(\ln V)^{\cdot\cdot} + H_1^2 + H_2^2 + H_3^2 = \Lambda, \tag{12.233}$$

$$\frac{1}{V}(V H_i)^{\cdot} = \Lambda, \tag{12.234}$$

where the cosmological constant represents the constant density of LIVE. It follows from Eqs. (12.217), (12.218) and (12.234) that

$$\dot{H} + 3H^2 = \Lambda. \tag{12.235}$$

The solution of this equation with the boundary condition $H(0) = H_0$ is

$$H = H_0 \coth(3H_{\mathrm{L}}t), \quad H_{\mathrm{L}} = \sqrt{\Lambda/3}. \tag{12.236}$$

Inserting this into Eq. (12.218) gives

$$(\ln V)^{\cdot} = 3H_0 \coth(3H_{\mathrm{L}}t). \tag{12.237}$$

The general solution of this equation is

$$V = V_0 \sinh^{H_0/H_{\mathrm{L}}}(3H_{\mathrm{L}}t). \tag{12.238}$$

In order to have a simple illustrating example we choose $H_0 = H_{\mathrm{L}}$ giving

$$V = V_0 \sinh(3H_{\mathrm{L}}t). \tag{12.239}$$

Equation (12.234) can be written

$$\dot{H}_i + \frac{\dot{V}}{V}H_i = 3H_{\mathrm{L}}^2. \tag{12.240}$$

Inserting (12.239) gives

$$\dot{H}_i + 3H_{\mathrm{L}} \coth(3H_{\mathrm{L}}t)\,H_i = 3H_{\mathrm{L}}^2. \tag{12.241}$$

The general solution of this equation is

$$H_i = \frac{\dot{a}_i}{a_i} = \frac{C_i}{\sinh(3H_{\mathrm{L}}t)} + H_{\mathrm{L}} \coth(3H_{\mathrm{L}}t), \tag{12.242}$$

where C_i are integration constants. Integration with the initial condition $a_i(0) = 0$ gives

$$a_i = 2^{1/3} \sinh^{p_i}\left(\frac{3}{2}H_{\mathrm{L}}t\right) \cosh^{\frac{2}{3}-p_i}\left(\frac{3}{2}H_{\mathrm{L}}t\right), \tag{12.243}$$

where $p_i = \frac{1}{3}\left(1 + \frac{C_i}{H_{\mathrm{L}}}\right)$. Inserting (12.236) and (12.242) into Eq. (12.219) and using that $C_i = (3p_i - 1)H_{\mathrm{L}}$ together with the conditions $\sum_{i=1}^{3} p_i = 1$ and $\sum_{i=1}^{3} p_i^2 = 1$, give the anisotropy parameter

Fig. 12.17 Evolution of
anisotropy in a
LIVE-dominated
Bianchi-type I universe.
Initially the anisotropy
parameter A has the maximal
value of the empty Kasner
universe, but it decreases
rapidly towards zero

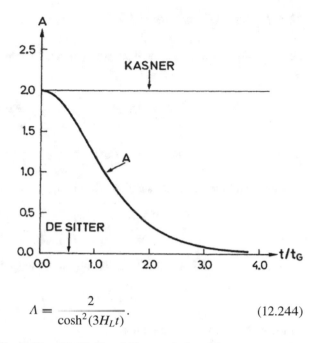

$$A = \frac{2}{\cosh^2(3H_L t)}. \tag{12.244}$$

This is shown graphically in Fig. 12.17. For $3H_L t \gg 1$ the anisotropy parameter approaches zero exponentially. Hence the *LIVE causes the universe model to
isotropize*. At the beginning, $3H_L t \rightarrow 0$, and the expression (12.244) gives the value
$A = 2$ of the Kasner universe.

12.6   Inhomogeneous Universe Models

We shall here consider the Lemaître–Tolman–Bondi (LTB) class of universe models following [20–23]. These universe models are inhomogeneous and spherically
symmetric. The line-element has the form

$$ds^2 = -c^2 dt^2 + \frac{F'^2(r, t)}{1 - k(r)} dr^2 + F^2(r, t) d\Omega^2, \tag{12.245}$$

where $k(r)$ is an arbitrary function of r which represents the curvature of 3-space.
We use the notation $' = d/dr$ and $\cdot = d/dt$.

12.6.1 Dust-Dominated Model

The Einstein equations for the dust-dominated LTB-universe models can be written as

$$H_\perp^2 + 2H_\parallel H_\perp + \frac{k}{F^2} + \frac{k'}{FF'} = \kappa\rho_m, \tag{12.246}$$

$$(1 - 3q)2H_\perp^2 - 2H_\parallel H_\perp + 2\frac{k}{F^2} - \frac{k'}{FF'} = -\kappa\rho_m. \tag{12.247}$$

where the Hubble parameters and the deceleration parameter are

$$H_\perp = \frac{\dot{F}}{F}, \quad H_\parallel = \frac{\dot{F}'}{F'}, \quad q = -\frac{F\ddot{F}}{\dot{F}^2}. \tag{12.248}$$

Adding Eqs. (12.246) and (12.247) and using the Definition (12.248) gives

$$2F\ddot{F} + \dot{F}^2 = -k, \tag{12.249}$$

Integration of this equation leads to

$$H_\perp^2 = \frac{\alpha}{F^3} - \frac{k}{F^2}, \tag{12.250}$$

where α is a function of r. The functions α and k are not determined by the field equations. They must be specified as boundary conditions. Differentiating Eq. (12.250) with respect to r and inserting the result into Eq. (12.247) gives

$$\kappa\rho_m = \frac{\alpha'}{F^2 F'}. \tag{12.251}$$

Substituting Eqs. (12.249) and (12.250) into the expression (12.248) for the deceleration parameter yields

$$q = \frac{\alpha}{2(\alpha - kF)}. \tag{12.252}$$

Let $F(r, \ t = 0) = F_0(r)$. We now introduce a conformal time η by $k^{1/2}dt = -(cF/R_0)\,d\eta$, where R_0 is the present value of the curvature radius of the 3-space. The solution of Eqs. (12.246) and (12.247) with $k < 0$ can then be written

$$F = -\frac{\alpha}{2k}\left(\cosh\frac{c\eta}{R_0} - 1\right) + F_0\left(\cosh\frac{c\eta}{R_0} + \sqrt{1 - \frac{\alpha}{kF_0}}\,\sinh\frac{c\eta}{R_0}\right), \tag{12.253}$$

$$\sqrt{-k}\,t = -\frac{\alpha}{2k}\left(\sinh\frac{c\eta}{R_0} - \frac{c\eta}{R_0}\right) + F_0\left[\sinh\frac{c\eta}{R_0} + \sqrt{1 - \frac{\alpha}{kF_0}}\left(\cosh\frac{c\eta}{R_0} - 1\right)\right]$$

$$(12.254)$$

The dust-dominated solution given in Eqs. (12.140) and (12.141) for a homogeneous dust-dominated LFRW universe with negative spatial curvature is found by choosing $F_0 = 0$ and $\alpha = H_0^2\Omega_{m0}r^3$, $k = -H_0^2(1 - \Omega_{m0})r^2$. Note that in this case $F = r\,a(\eta)$.

12.6.2 Inhomogeneous Universe Model with Dust and LIVE

We shall here consider an inhomogeneous generalization of the LCDM-universe which was presented in Sect. 12.4.5. In this case Eq. (12.246) is generalized to

$$H_\perp^2 + 2H_\parallel H_\perp + \frac{k}{F^2} + \frac{k'}{FF'} = \kappa(\rho_m + \rho_L), \qquad (12.255)$$

and the acceleration equation is

$$2\frac{\ddot{F}}{F} + H_\perp^2 + \frac{k}{F^2} = \kappa\rho_L. \qquad (12.256)$$

Here $\rho_L = \rho_{L0}$ is the constant density of LIVE. Equation (12.250) now takes the form

$$H_\perp^2 = \frac{\alpha}{F^3} - \frac{k}{F^2} + \frac{\kappa}{3}\rho_L. \qquad (12.257)$$

Equation (12.251) is still valid. Combining Eqs. (12.255) and (12.256) we get

$$2\frac{\ddot{F}}{F} + \frac{\ddot{F}'}{F'} = \frac{\kappa}{2}(2\rho_L - \rho_m). \qquad (12.258)$$

The present values of the density parameters are defined by the equations

$$\alpha = H_{\perp 0}^2 F_0^2 \Omega_m, \qquad (12.259)$$

$$k = H_{\perp 0}^2 F_0^2(\Omega_m + \Omega_L - 1), \qquad (12.260)$$

where $F_0 = F_0(r) = F(r, t_0)$, $H_{\perp 0} = H_{\perp 0}(r) = H(r, t_0)$ and $\Omega_{L0} = \Omega_{L0}(r) = \kappa\rho_L/3H_{\perp 0}^2(r)$. With these definitions Eq. (12.257) of the Hubble parameter $H_\perp$ takes the form

$$\frac{\dot{F}}{F} = H_\perp(r, \ t) = H_0 \left[\Omega_{m0}\left(\frac{F_0}{F}\right)^3 + \Omega_{L0} + (1 - \Omega_{m0} - \Omega_{L0})\left(\frac{F_0}{F}\right)^2 \right]^{1/2}.$$

(12.261)

It follows that the scale factor F is given implicitly as a function of time by

$$H_0 t = \int \frac{dF}{\left[\Omega_{m0}\left(\frac{F_0}{F}\right)^3 + \Omega_{L0} + (1 - \Omega_{m0} - \Omega_{L0})\left(\frac{F_0}{F}\right)^2 \right]^{1/2}}.$$

(12.262)

We now consider an inhomogeneous universe corresponding to the LCDM-universe where the density parameters of dust and LIVE obey the condition $\Omega_{m0} + \Omega_{L0} = 1$. Then Eq. (12.262) reduces to

$$H_0 t = \int \frac{dF}{\left[\Omega_{m0}\left(\frac{F_0}{F}\right)^3 + \Omega_{L0} \right]^{1/2}}.$$

(12.263)

Integrating this equation with the initial condition $F(r, \ 0) = 0$ gives

$$H_0 t = \frac{2}{3\sqrt{\Omega_{L0}}} \text{arsinh}\left[\sqrt{\frac{\Omega_{L0}}{\Omega_{m0}}}\left(\frac{F}{F_0}\right)^{\frac{3}{2}} \right].$$

(12.264)

Hence

$$F = F_0 \left(\frac{\Omega_{m0}}{\Omega_{L0}}\right)^{1/3} \sinh^{\frac{2}{3}}\left(\frac{3}{2}\sqrt{\Omega_{L0}} \, H_0 t\right).$$

(12.265)

In the present universe model $\Omega_{m0}, \ \Omega_{L0}$ and H_0 are functions of r, while in the LCDM-universe they are constants.

12.7 The Horizon and Flatness Problems

12.7.1 The Horizon Problem

The cosmic microwave background (CMB) radiation from two points A and B in opposite directions has the same temperature. This means that it has been radiated by sources of the same temperature at these points. Thus, the universe must have been in thermic equilibrium at the decoupling time, $t_d = 3 \times 10^5$ years. This implies that points A and B, "at opposite sides of the universe" as seen by an observer, had been in causal contact already at that time. That is, a light signal must have had time to move from A to B during the time from $t = 0$ to $t = 3 \times 10^5$ years. The points A and B must have been within each other's horizons at the decoupling.

Consider a photon moving radially in space described by the Robertson–Walker metric (12.17) with $k = 0$. Light follows a null-geodesic curve; i.e. the curve is defined by $ds^2 = 0$. We get (using units so that $c = 1$),

$$dr = \frac{dt}{a(t)}. \tag{12.266}$$

The *particle horizon* (also called the *cosmological horizon*) is the maximum distance from which light from particles could have travelled to the observer during the time which the universe has existed. It represents the boundary between the observable and the unobservable regions of the universe. Its distance at the present epoch defines the size of the observable part of the universe. The coordinate distance from an observer at the origin of the coordinate system to the particle horizon at the time t is

$$\Delta r = \int_0^t \frac{dt}{a(t)}, \tag{12.267}$$

The physical radius of the particle horizon is

$$l_{PH} = a(t)\Delta r = a(t) \int_0^t \frac{dt}{a(t)}. \tag{12.268}$$

To find a quantitative expression for the "horizon problem", we may consider a model with critical mass–density (Euclidean space-like geometry). Using $p = w\rho$ and $k = 0$, Eq. (12.63) takes the form

$$\frac{\dot{a}}{a} = \sqrt{\frac{\kappa\rho_0}{3}} a^{-\frac{3}{2}(1+w)} = H_0 a^{-\frac{3}{2}(1+w)} \tag{12.269}$$

Integration with the condition $a(t_0) = 1$ gives

$$a = \left(\frac{t}{t_0}\right)^{\frac{2}{3(1+w)}}, \quad t_0 = \frac{2}{3(1+w)}\frac{1}{H_0} = \frac{2}{3(1+w)} t_H. \tag{12.270}$$

where $t_H = 1/H_0$ is the Hubble age of the universe. Inserting this into the expression (12.268) and integrating gives

$$l_{PH} = \frac{3w+3}{3w+1} t. \tag{12.271}$$

Hence, the radius of the particle horizon increases proportionally to the cosmic time. Equation (12.270) also shows that the age of a flat radiation-dominated universe is only half of the Hubble age.

Let us call the volume inside the horizon the "horizon volume" and denote it by V_{PH}. From Eq. (12.271) it follows that $V_{PH} \propto t^3$. At the decoupling time, the horizon volume may therefore be written as

$$(V_{PH})_d = \left(\frac{t_d}{t_0}\right)^3 V_0, \tag{12.272}$$

where V_0 is the size of the present horizon volume. Events within this volume are causally connected, and a volume of this size may be in thermal equilibrium at the decoupling time.

Let $(V_0)_d$ be the size, at the decoupling, of the part of the universe that corresponds to the present horizon volume, i.e. the observable universe. For our Euclidean universe, Eq. (12.271) holds, giving

$$(V_0)_d = \frac{a^3(t_d)}{a^3(t_0)} V_0 = \left(\frac{t_d}{t_0}\right)^{\frac{2}{w+1}} V_0. \tag{12.273}$$

From Eqs. (12.272) and (12.273), we get

$$\frac{(V_0)_d}{(V_{PH})_d} = \left(\frac{t_d}{t_0}\right)^{-\frac{3w+1}{w+1}}. \tag{12.274}$$

Using that $t_d = 10^{-4} t_0$ and inserting $w = 0$ for dust, we find

$$\frac{(V_0)_d}{(V_{PH})_d} = 10^4. \tag{12.275}$$

Thus, there was room for 10^4 causally connected areas at the decoupling time within the region which represents our observable universe. Points at opposite sides of our observable universe were therefore not causally connected at the decoupling, according to the Friedmann models of the universe. These models therefore cannot explain that the temperature of the radiation from such points is the same.

12.7.2   The Flatness Problem

According to Eqs. (12.67) and (12.68), the total mass parameter $\Omega = \rho/\rho_{cr}$ is given by

$$\Omega - 1 = \frac{kc^2}{R_0^2 a^2 H^2}.$$ (12.276)

Using the expansion factor (12.269) for a universe near critical mass–density, we get

$$\frac{\Omega - 1}{\Omega_0 - 1} = \left(\frac{t}{t_0}\right)^{2\left(\frac{3w+1}{3w+3}\right)}.$$ (12.277)

For a radiation-dominated universe, we get

$$\frac{\Omega - 1}{\Omega_0 - 1} = \frac{t}{t_0}.$$ (12.278)

Measurements indicate that $\Omega_0 - 1$ is of the order of magnitude 1. The age of the universe is about $t_0 = 10^{17}$ s. When we stipulate initial conditions for the universe, it is natural to consider the Planck time, $t_P = 10^{-43}$ s, since this is the limit of the validity of general relativity. At earlier time, quantum effects will be important, and one cannot give a reliable description without using quantum gravitation. The stipulated initial condition for the mass parameter then becomes that $\Omega - 1$ is of order 10^{-60} at the Planck time. Such an extreme fine tuning of the initial value of the universe's mass–density cannot be explained within standard Big Bang cosmology. Since a universe with critical density has Euclidean spatial geometry and is called flat, this fine tuning requirement is called the "flatness problem".

12.8 Inflationary Universe Models

12.8.1 Spontaneous Symmetry Breaking and the Higgs Mechanism

The particles responsible for the electroweak force are the $W^\pm$ and Z^0 bosons. They are massive, causing the weak force to only have short-distance effects. This was originally a problem for the quantum field theory describing this force, since it made it difficult to create a renormalizable theory. This was solved by Higgs and Kibble in 1964 by introducing the so-called Higgs mechanism.

The main idea is that the massive bosons $W^\pm$ and Z^0 are originally massless, but are given a mass by interacting with a *Higgs field* ϕ. The effect causes the mass of the particles to be proportional to the value of the Higgs field in vacuum. It is therefore necessary that the Higgs field has a value different from zero in the vacuum (the *vacuum expectation value* must be nonzero).

Let us see how the Higgs field can get a nonzero vacuum expectation value. The important thing for our purpose is that the potential of the Higgs field may be temperature dependent. Let us assume that the potential of the Higgs field is described by the function

$$V(\phi) = \frac{1}{2}\mu^2\phi^2 + \frac{1}{4}\lambda\phi^4, \tag{12.279}$$

where the sign of μ^2 depends on whether the temperature is above or below a critical temperature T_{cr}. This sign has an important consequence on the shape of the potential V. The potential is shown in Fig. 12.18 for two different temperatures. For $T > T_{cr}$, $\mu^2 > 0$, and the shape is like in Fig. 12.18a, and there is a stable minimum for $\phi = 0$. However, for $T < T_{cr}$, $\mu^2 < 0$, the shape is like in Fig. 12.18b. In this case the potential has stable minima for $\phi = \pm\phi_0 = \pm\frac{|\mu|}{\sqrt{\lambda}}$ and an unstable maximum at $\phi = 0$.

The "real" vacuum state of the system is at a stable minimum of the potential. For $T > T_{cr}$, the minimum is in the "symmetric" state $\phi = 0$. On the other hand, for $T < T_{cr}$ this state is unstable. It is therefore called a "false vacuum". The system will move into one of the stable minima at $\phi = \pm\phi_0$. When the system is in one of these states, it is no longer symmetric under the change of sign of ϕ. Such a symmetry, which is not reflected in the vacuum state, is called *spontaneously broken*. From Fig. 12.16b we see that the energy of the false vacuum is larger than for the real vacuum.

The central idea, from which the "inflationary cosmology" originated, was to take into consideration the consequences of the unified quantum field theories, the gauge theories, at the construction of relativistic models for the early universe. According to the Friedmann models, the temperature was extremely high in the early history of the universe. If one considers Higgs fields associated with GUT models (grand

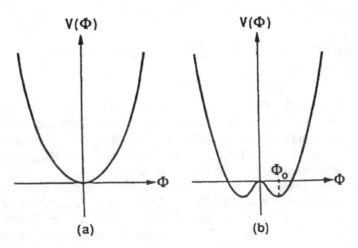

Fig. 12.18 Inflationary potentials. The shape of the potential depends on the sign of μ^2. **a** Higher temperature than the critical, with $\mu^2 > 0$. **b** Lower temperature than the critical, with $\mu^2 < 0$

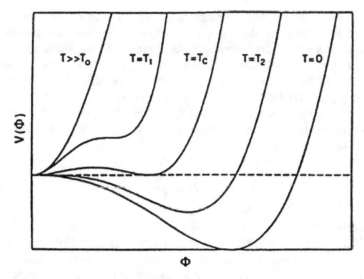

Fig. 12.19 Temperature dependence of Higgs potential

unified theories), one finds a critical temperature T_{cr} corresponding to the energy $kT_{cr} = 10^{14}$GeV, where k is the Boltzmann's constant. Before the universe was about $t_1 = 10^{-35}$s old, the temperature was larger than this. Thus, the Higgs field was in the symmetric ground state. According to most of the inflation models, the universe was dominated by radiation at this time.

When the temperature decreases, the Higgs potential changes. This could happen as shown in Fig. 12.19. Here, there is a potential barrier at the critical temperature, which means that there cannot be a classical phase transition. The transition to the stable minimum must happen by quantum tunnelling. This is called a first-order phase transition. Alan Guth's original inflation model [23] was based on a first-order phase transition.

12.8.2 Guth's Inflationary Model [24]

According to most of the inflationary models, the universe was dominated by radiation during the time before 10^{-35}s. The universe was then expanding so fast that there was no causal contact between the different parts of the universe that became our observable universe. Probably, the universe was rather inhomogeneous, with considerable space-like variations in temperature. There were also regions of false vacuum, with energy densities characteristic of the GUT energy scale, which also controlled the critical temperature. While the energy density of the radiation decreased quickly, as a^{-4}, the energy density of vacuum was constant. At the time $t = 10^{-35}$s, the energy density of the radiation became less than that of the vacuum.

At the same time the potential started to change such that the false vacuum went from being stable to unstable. Thus, there was a first-order phase transition to a real vacuum. Because of the inhomogeneity of the universe's initial condition, this happened with different speeds at differing places. The potential barrier slowed down the process, which happened by tunnelling, and the universe was at several places considerably under cooled. Then there appeared "bubbles" dominated by the energy of the false vacuum. These regions acted upon themselves with repulsive gravity.

By integrating the equation of motion for the expansion factor in such a vacuum-dominated bubble, one gets

$$a = e^{Ht}, \quad H = \sqrt{\frac{8\pi G \rho_{cr}}{3}}. \tag{12.280}$$

By inserting the GUT value above, we get $H = 6.6 \times 10^{34} s^{-1}$, i.e. $H^{-1} = 1.5 \times 10^{-35} s$. With reference to field theoretical works by Sidney Coleman and others, Guth reasoned that a realistic duration of the nucleation process happening during the phase transition is $10^{-33} s$. During this time, the expansion factor increases by a factor of 10^{28}. This vacuum-dominated epoch is called the *inflationary era*.

Let us look closer at what happened with the energy of the universe in the course of its development, according to the inflationary models. To understand this we first have to consider what happened at the end of the inflationary era. When the Higgs field reached the minimum corresponding to the real vacuum, it started to oscillate. According to the quantum description of the oscillating field, the energy of the false vacuum was converted to radiation and particles. In this way the equation of state for the energy dominating the development of the expansion factor changed from $p = -\rho$, characteristic of LIVE, to $p = (1/3)\rho$, characteristic of radiation.

The energy density and the temperature of the radiation then increased enormously. Before and after this short period around the time $t = 10^{-33} s$ the radiation energy increased adiabatically, such that $\rho a^4 =$ constant. According to Stefan–Boltzmann law of radiation, $\rho \propto T^4$. Therefore, $aT =$ constant during adiabatic expansion. This means that during the inflationary era, while the scale factor increased exponentially, the energy density and temperature of radiation decreased exponentially. At the end of the inflationary era, the radiation was reheated so that it returned to approximately the energy it had when the inflationary era started.

It may be interesting to note that the Newtonian theory of gravitation does not allow an inflationary era, since stress has no gravitational effect according to it.

12.8.3 The Inflationary Models' Answers to the Problems of the Friedmann Models

The horizon problem will be investigated here in the light of the inflationary universe models. The problem was that there was room for about 10,000 causally connected regions inside the region defined by our presently observable universe at the decoupling time when the universe became transparent to the background radiation. Let us calculate the horizon radius l_h and the radius a of the region presently within the particle horizon, $l_{PH} = 15 \times 10^9 \ ly = 1.5 \times 10^{26} cm$, at the time $t_1 = 10^{-35}$s when the inflation started. From Eq. (12.120) for the radiation-dominated period before the inflationary era, one gets

$$l_{PH} = 2t_1 = 6 \times 10^{-25} cm. \tag{12.281}$$

The radius, at time t_1, of the region corresponding to our observable universe is found by using $a \propto e^{Ht}$ during the inflation era from $t_1 = 10^{-35}$s to $t_2 = 10^{-33}$s, $a \propto t^{1/2}$ in the radiation-dominated period from t_2 to $t_3 = 10^{11}$s and $a \propto t^{2/3}$ in the matter-dominated period from t_3 until now, $t_0 = 10^{17}$s. This gives

$$a_1 = \frac{e^{Ht_1}}{e^{Ht_2}}\left(\frac{t_2}{t_3}\right)^{1/2}\left(\frac{t_3}{t_0}\right)^{2/3} l_{PH}(t_0) = 1.5 \times 10^{-28} cm. \tag{12.282}$$

Hence at the beginning of the inflationary era the horizon radius, l_{PH}, was larger than the radius a_1 of the region corresponding to our observable universe. The whole of this region was then causally connected, and thermic equilibrium was established. This equilibrium was preserved, and there was thermal equilibrium at the decoupling time about 400,000 years later. This explains the observed isotropy of the cosmic background radiation and solves the horizon problem.

We will now consider the flatness problem. This problem was the necessity, in the Friedmann models, of fine tuning the initial density in order to obtain the closeness of the observed mass–density to the critical density. Again, the inflationary models give another result, making fine tuning unnecessary.

Inserting the scale factor (12.280) into Eq. (12.276), we get

$$\Omega - 1 = \frac{k}{H^2}e^{-2Ht}, \tag{12.283}$$

where H is constant and given in Eq. (12.280). The ratio of $\Omega - 1$ at the end of the inflationary era to the beginning of the inflationary era becomes

$$\frac{\Omega_2 - 1}{\Omega_1 - 1} = e^{-2H(t_2-t_1)} = 10^{-56}. \tag{12.284}$$

Contrary to the Friedmann models, where the mass–density moves *away* from the critical density as time increases, the density approaches the critical density exponentially during the inflationary era. Within a large range of initial conditions, this means that according to the inflation models the universe should still have a density very close to the critical density.

12.8.4 Dynamics of the Inflationary Era

During the inflationary era the evolution of the universe is assumed to be dominated by a scalar field ϕ which is called the *inflaton field*. The first Friedmann equation is

$$H^2 = \frac{\kappa}{3}\rho = \frac{\kappa}{3}\left(\frac{1}{2}\dot{\phi}^2 + V\right), \qquad (12.285)$$

where ρ is the energy and $V = V(\phi)$ the potential of the inflaton field. The continuity equation is

$$\dot{\rho} + 3H\rho = -3Hp. \qquad (12.286)$$

It follows from these equations that

$$\dot{\rho} = -\sqrt{3\kappa\rho}(\rho + p). \qquad (12.287)$$

The equation for the evolution of the inflaton field which generates the dark energy causing repulsive gravity during the inflationary era is

$$\ddot{\phi} + 3H\dot{\phi} = -V', \qquad (12.288)$$

where $V' = dV/d\phi$.

It follows from the second Friedmann equation that the acceleration of the cosmic expansion is given by

$$\frac{\ddot{a}}{a} = -\frac{\kappa}{6}(\rho + 3p). \qquad (12.289)$$

The inflaton field is often described as a perfect fluid with density and pressure

$$\rho = \frac{1}{2}\dot{\phi}^2 + V, \quad p = \frac{1}{2}\dot{\phi}^2 - V. \qquad (12.290)$$

Hence, the fluid obeys the equation of state

$$p = w\rho, \quad w = \frac{(1/2)\dot{\phi}^2 - V}{(1/2)\dot{\phi}^2 + V}. \qquad (12.291)$$

It interpolates between a Lorentz invariant vacuum energy (LIVE) with $w = -1$ for a constant inflaton field and a Zel'dovich fluid with $w = 1$ for a flat potential with $V = 0$. Solved with respect to $\dot{\phi}^2$ the second of these equations gives

$$\dot{\phi}^2 = \frac{1+w}{1-w}2V. \qquad (12.292)$$

This equation shows that $V > 0$ for $|w| < 1$.

The acceleration Eq. (12.289) of the scale factor then takes the form

$$\frac{\ddot{a}}{a} = -\frac{\kappa}{3}(\dot{\phi}^2 - V).$$

$$(12.293)$$

Differentiating Eq. (2.285) and inserting Eq. (2.289) gives

$$\dot{H} = -(\kappa/2)\dot{\phi}^2,$$

$$(12.294)$$

or

$$\dot{\phi} = -(2/\kappa)H',$$

$$(12.295)$$

where $H' = \mathrm{d}H/\mathrm{d}\phi = \dot{H}/\dot{\phi}$. Equation (2.294) shows that H is a decreasing function of time. It follows from Eqs. (2.285) and (2.295) that

$$\kappa^2 V = 3\kappa H^2 - 2H'^2.$$

$$(12.296)$$

Equation (2.294) shows that the Hubble parameter is constant and there is exponential expansion for a constant inflaton field. This represents the case where the inflaton field behaves like LIVE with a constant density, which may be represented by a cosmological constant. Equation (2.294) implies that the Hubble parameter is a decreasing function of time for a variable scalar field.

During most of the inflationary era, i.e. except during the transient phases at the beginning and the end of the era, the scalar field changes very slowly so that $\ddot{\phi} \ll H\dot{\phi}$. If the potential V is not too small, the condition $\dot{\phi} \ll V$ may also be satisfied. Then $w \approx -1$ which means that the inflaton field behaves like LIVE with approximately constant energy density, and with exponential expansion of the space during most of the inflationary era.

In the so-called *slow-roll* approximation we shall assume that $\ddot{\phi} \ll H\dot{\phi}$, but not in general that $\dot{\phi}^2 \ll V$. Then Eq. (12.288) reduces to

$$\dot{\phi} \approx -\frac{V'}{3H}.$$

$$(12.297)$$

Equations (12.291) and (12.294) give

$$\dot{H} = -\kappa V \frac{1+w}{1-w}.$$

$$(12.298)$$

It follows from Eqs. (12.285) and (12.294) that

$$\kappa V = \dot{H} + 3H^2.$$

$$(12.299)$$

Hence

$$\dot{H} = -(3/2)(1+w)H^2. \tag{12.300}$$

Integration of this equation for constant $w \neq -1$ gives

$$a = a_1 \left(\frac{t}{t_1}\right)^{\frac{2}{3(1+w)}}. \tag{12.301}$$

Hence, power law expansion corresponds to a constant equation of state parameter $w \neq -1$ during the inflationary era, and exponential expansion to $w = -1$. Inserting the first of Eq. (2.291) into Eq. (12.287) gives

$$\dot{\rho} = -\sqrt{3\kappa\rho}(1+w)\rho. \tag{12.302}$$

Integrating this equation for $w \neq -1$ with $\rho(0) = \rho_0$ gives

$$\rho = \frac{\rho_0}{\left[1 + (1/2)(1+w)\sqrt{3\kappa\rho_0}\,t\right]^2}. \tag{12.303}$$

Hence for $\sqrt{\rho_0}\,t \gg M_P$ the energy density of an inflaton field with constant equation of state parameter, $w \neq -1$, decreases approximately inversely proportionally to the square of time.

We define the *slow-roll parameters* ε, η by

$$\varepsilon \equiv \frac{1}{2\kappa}\left(\frac{V'}{V}\right)^2, \quad \eta \equiv \frac{1}{\kappa}\frac{V''}{V}. \tag{12.304}$$

The absolute values of the slow-roll parameters are much less than one during the slow-roll period. This means that during a slow-roll period the graph of $V(\phi)$ is very flat and has small curvature.

Alternatively one defines "Hubble slow-roll parameters", ε_H, η_H in terms of the Hubble parameter and its derivatives with respect to the inflaton field

$$\varepsilon_H = \frac{2}{\kappa}\left(\frac{H'}{H}\right)^2, \quad \eta_H = \frac{2}{\kappa}\frac{H''}{H}. \tag{12.305}$$

Inserting the first of these expressions into Eq. (12.296) we get for the inflaton potential

$$\kappa V = (3 - \varepsilon_H)H^2. \tag{12.306}$$

It follows from Eq. (12.295) that during the slow-roll era differentiation with respect to time and with respect to the inflaton field are related by

$$\frac{d}{dt} = -\frac{2}{\kappa}H'\frac{d}{d\phi}. \tag{12.307}$$

Using this in the Definition (12.305) we get simple expressions for ε_H and η_H

$$\varepsilon_H \equiv -\frac{\dot{H}}{H^2}, \eta_H = -\frac{1}{2}\frac{\ddot{H}}{\dot{H}H}. \tag{12.308}$$

Using that $H = \dot{a}/a$ the first equation takes the form

$$\varepsilon_H = 1 - \frac{a\ddot{a}}{\dot{a}^2} = 1 + q, \tag{12.309}$$

where q is the deceleration parameter defined in Eq. (12.75). A requirement for inflation is that there is accelerated expansion, $\ddot{a} > 0$. Hence a necessary condition for inflation is that $\varepsilon_H < 1$.

It follows from Eq. (12.294) that

$$\frac{\ddot{\phi}}{\dot{\phi}} = \frac{1}{2}\frac{\ddot{H}}{\dot{H}}, \tag{12.310}$$

giving

$$\eta_H = -\frac{\ddot{\phi}}{H\dot{\phi}}. \tag{12.311}$$

This equation may be written

$$[(1/2)\dot{\phi}^2]^{\cdot} = -\eta_H H \dot{\phi}^2. \tag{12.312}$$

Hence the sign of the parameter η_H decides whether the kinetic energy of the inflaton field increases, $\eta_H < 0$, or decreases, $\eta_H > 0$. The kinetic energy is constant for $\eta_H = 0$.

To lowest order

$$\varepsilon_H = \varepsilon, \quad \eta_H = \eta - \varepsilon. \tag{12.313}$$

Equations (2.57), (12.63) and (2.309) give

$$w = -1 + (2/3)\varepsilon_H. \tag{12.314}$$

It follows that a universe with $\varepsilon_H = 0$ is dominated by LIVE with equation of state parameter $w = -1$ and a constant energy density.

From Eqs. (12.285), (2.294) and (2.308) we get

$$\varepsilon_H = 3\frac{\dot{\phi}^2/2}{\dot{\phi}^2/2 + V}. \qquad (12.315)$$

Hence the parameter ε_H represents the ratio of the kinetic energy and the total energy of the inflaton field. This is exact. It does not require the slow-roll approximation.

The ratio of the final value a_f of the scale factor during the inflationary era and the initial value $a(N)$ is

$$\frac{a_f}{a(N)} = e^N, \qquad (12.316)$$

where N is called *the number of e-folds* of the slow-roll era. Hence

$$N = \ln(a_f/a). \qquad (12.317)$$

Note that $N = 0$ at the *end* of inflation, so that N counts the number of e-folds until inflation ends and increases as we go backward in time. It follows from Eq. (12.317) that

$$\dot{N} = -H, \qquad (12.318)$$

or

$$\frac{d}{dN} = -\frac{1}{H}\frac{d}{dt}. \qquad (12.319)$$

If $\dot{H} \ll H^2$ Eq. (12.299) can be approximated by

$$\kappa V = 3H^2. \qquad (12.320)$$

Hence

$$\frac{V'}{V} = 2\frac{H'}{H}, \quad V'' = (6/\kappa)(H'^2 + HH'') \qquad (12.321)$$

Using this together with Eqs. (2.319) and $\dot{N} = N'\dot{\phi}$, we have

$$dN = -\frac{H}{\dot{\phi}}d\phi = \frac{\kappa}{2}\frac{H}{H'}d\phi = \kappa\frac{V}{V'}d\phi. \qquad (12.322)$$

This equation can be used to relate derivative with respect to N and derivative with respect to ϕ as

$$\frac{d}{dN} = \sqrt{\frac{2\varepsilon}{\kappa}}\frac{d}{d\phi}, \qquad (12.323)$$

which may be written

$$\varepsilon = \frac{\kappa}{2}\left(\frac{d\phi}{dN}\right)^2,$$

(12.324)

showing that $\varepsilon > 0$. From Eqs. (12.288) and (12.319) we have

$$\varepsilon_H = \frac{H'(N)}{H}.$$

(12.325)

Hence $H'(N) > 0$. From the Definition (12.304) and Eq. (12.323) we get

$$\varepsilon = \frac{1}{2}\frac{V'(N)}{V}.$$

(12.326)

Integration of Eq. (12.322) gives

$$N \approx \kappa \int_{\phi_f}^{\phi} \frac{V}{V'}d\phi = \int_{\phi_f}^{\phi}\sqrt{\frac{\kappa}{2\varepsilon}}d\phi < \sqrt{\frac{\kappa}{2\varepsilon_{min}}}(\phi - \phi_f),$$

(12.327)

where ε_{min} is the minimum value of ε. Note that if $V' > 0$ we must have $\phi_f < \phi$ in order that $N > 0$, and if $V' < 0$ we must have $\phi_f > \phi$. Equation (12.327) implies a bound on the change of the value of the scalar field during the inflationary era,

$$\Delta\phi > N\sqrt{\frac{2\varepsilon_{min}}{\kappa}} = N M_P\sqrt{2\varepsilon_{min}}.$$

(12.328)

This is called the *Lyth bound*.

There exists a third type of slow-roll parameters. They are defined by

$$\varepsilon_1 = \varepsilon_H, \varepsilon_{n+1} = -\frac{d\ln|\varepsilon_n|}{dN}.$$

(12.329)

Using Eq. (12.319) we have

$$\varepsilon_{n+1} = \frac{1}{H}\frac{\dot{\varepsilon}_n}{\varepsilon_n}.$$

(12.330)

12.8.5   Testing Observable Consequences of the Inflationary Era

It has turned out that observations of the polarization of the cosmic microwave background radiation (CMB) can be used to test predictions made by different inflationary models [25].

The polarization of the CMB may at every point be described by an amplitude of the oscillation and a direction. The field of polarization is decomposed in two modes: the E-mode and the B-mode. The E-mode is curl free like the electrical field of a charged particle. The B-mode is divergence free like the magnetic field of a current (Fig. 12.20).

One has classified the polarization in three types:

Scalar perturbation: Energy density fluctuations in the plasma (resulting in hotter and colder regions) cause velocity distributions that are out of phase with the acoustic density mode. The fluid velocity from hot to colder regions causes blueshift of the photons, resulting in E-mode polarization.

Vector perturbation: Vorticity in the plasma causes Doppler shifts resulting in the quadrupole lobes in the figure. However, vorticity would be damped by inflation and is expected to be negligible.

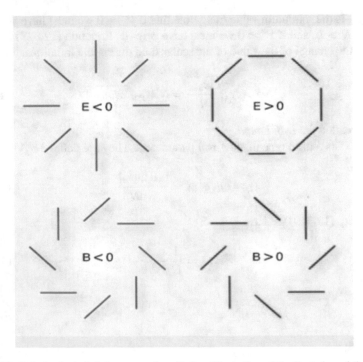

Fig. 12.20 Polarization of electromagnetic radiation. Illustration of the E-mode and the B-mode of polarization of electromagnetic radiation

Tensor perturbation: Gravity waves stretch and squeeze space in orthogonal directions. This also stretches the wavelength of radiation, therefore creating both E-mode and B-mode polarization in the radiation temperature field.

Gravity waves from inflation produce tensor perturbations.

The power spectra of scalar and tensor fluctuations are represented by

$$
P_S = A_S(k_*)\left(\frac{k}{k_*}\right)^{n_S - 1 + (1/2)\alpha_S \ln(k/k_*) + \cdots}, \quad P_T = A_T(k_*)\left(\frac{k}{k_*}\right)^{n_T + (1/2)\alpha_T \ln(k/k_*) + \cdots},
$$
$$
A_S = \frac{V}{24\pi^2 \varepsilon M_P^4}, \qquad\qquad A_T = \frac{2V}{3\pi^2 M_P^4}
$$

$$(12.331)$$

Here k is the wave number of the perturbation which is a measure of the average spatial extension for a perturbation with a given power. One often writes $k = \dot{a} = aH$, where a is the scale factor representing the ratio of the physical distance between reference particles in the universe relative to their present distance. The quantities A_S and A_T are amplitudes, and n_S and n_T are the *spectral indices* of the scalar and tensor fluctuations. The quantities $-\delta_{ns} \equiv n_S - 1$ and n_T are called the tilt of the power spectra of curvature perturbations and tensor modes, respectively, because they represent the deviation from the values $\delta_{ns} = n_t = 0$ that represent a scale invariant spectrum.

Furthermore α_S and α_T are factors representing the k-dependence of the spectral indices. They are called the running of the spectral indices and are defined by

$$
\alpha_S = \frac{dn_S}{d \ln k}, \quad \alpha_T = \frac{dn_T}{d \ln k}.
$$

$$(12.332)$$

If $n_S = 1$ the spectrum of the scalar fluctuations is said to be *scale invariant*. An invariant mass–density power spectrum is called a *Harrison-Zel'dovich spectrum*. One of the predictions of the inflationary universe models is that the cosmic mass distribution has a spectrum that is *nearly* scale invariant, but not exactly. The observations and analysis of the Planck team [26] have given the result $n_S = 0.968 \pm 0.006$. Hence we may use $n_S = 0.968$ as the preferred value of n_S, corresponding to $\delta_{ns} = 0.032$. Furthermore they have obtained $\alpha_S = -0.003 \pm 0.007$.

The *tensor-to-scalar ratio r* is defined by

$$
r \equiv \frac{P_T(k_*)}{P_S(k_*)} = \frac{A_T}{A_S},
$$

$$(12.333)$$

which is a positive quantity. From Eqs. (12.331) and (12.333) we have

$$
r = 16\varepsilon.
$$

$$(12.334)$$

The tensor-to-scalar ratio can be determined from observations of the B-mode of polarization of the CMB. In the measured wavelength region this B-mode pattern is partly due to radiation from galactic dust and partly to imprints on the CMB at the time 380,000 years after the Big Bang, when the universe became transparent for the CMB, from relic gravitational waves produced by quantum fluctuations in the inflationary era. At the present time (November 2019) the best restriction on the tensor-to-scalar ratio obtained from CMB-measurements is $r < 0.04$.

It follows from Eqs. (12.334) and (12.314) that the equation of state parameter during the slow-roll era is given in terms of the tensor-to-scalar ratio as

$$1 + w = r/24. \tag{12.335}$$

With $0 \leq r < 0.04$ this gives $-0.9983 \leq w \leq -1$ during the slow-roll era, which is equal to or very close to the equation of state of LIVE.

We shall now find how the spectral indices depend upon the slow-roll parameters. From Eq. (12.331) it follows that they are given by

$$\delta_{ns} = -\left[\frac{d \ln P_S(k)}{d \ln k}\right]_{k=aH}, \quad n_T = \left[\frac{d \ln P_T(k)}{d \ln k}\right]_{k=aH}. \tag{12.336}$$

The quantities inside the brackets are evaluated at the horizon crossing where $k = k_*$, and the wave number is equal to the scale factor times the Hubble parameter. It will be useful to write

$$\frac{d}{d \ln k} = \frac{d}{dN} \times \frac{dN}{d \ln k}. \tag{13.337}$$

Hence, using that $A_S \propto H^2/\varepsilon$, the scalar spectral indices may be written as

$$\delta_{ns} = \left(\frac{d \ln \varepsilon}{dN} - 2\frac{d \ln H}{dN}\right)\frac{dN}{d \ln k}, \quad n_T = 2\frac{d \ln H}{dN}\frac{dN}{d \ln k}. \tag{12.338}$$

Using Eqs. (12.318) and (12.308) we get in the slow-roll approximation

$$\frac{d \ln H}{dN} = -\frac{\dot{H}}{H^2} = \varepsilon_H. \tag{12.339}$$

From the condition that the spectral indices are calculated at the horizon crossing we have $k = aH$. Equation (12.317) gives $dN = -d \ln a$. Hence $d \ln k = d \ln a + d \ln H = -dN + d \ln H$. Since H is approximately constant during the slow-roll inflationary era, it follows that

$$\frac{dN}{d \ln k} \approx -1. \tag{12.340}$$

It follows from the Definition (12.304) that the derivative of the slow-roll parameter ε with respect to the scalar field is given by

$$\sqrt{\frac{2\varepsilon}{\kappa}}\,\varepsilon'(\phi) = 2\varepsilon(\eta - 2\varepsilon)\,. \tag{12.341}$$

Using Eq. (12.323) this may be written as

$$\frac{d\varepsilon}{dN} = 2\varepsilon(\eta - 2\varepsilon). \tag{12.342}$$

Inserting this together with Eqs. (12.313) and (12.339) into Eq. (12.338) leads to

$$\delta_{ns} = 2(3\varepsilon - \eta) \tag{12.343}$$

It follows from Eqs. (12.313) and (12.343) that the spectral tilt can be expressed in terms of the Hubble slow-roll parameters as

$$\delta_{ns} = 2(2\varepsilon_H - \eta_H). \tag{12.344}$$

Equations (12.338), (12.339), (12.340) and (12.313) give

$$n_T \approx -2\varepsilon \tag{12.345}$$

A consistency relation between r and n_T follows from Eqs. (12.334) and (12.245)

$$n_T = -\frac{r}{8}. \tag{12.346}$$

Example 12.8.1 Polynomial Inflation As a simple illustration we shall here use the formalism above to calculate the optical parameters δ_{ns}, n_T and r for the class of inflationary models called polynomial inflation.

The so-called *chaotic inflation* models are a class of polynomial models. The potential of the inflaton field in this type of inflationary models is

$$V = M^4 \hat{\phi}^p, \tag{12.347}$$

where $\hat{\phi} = \sqrt{\kappa}\phi$. M is the energy scale of the potential when the inflaton field has Planck mass, and it is assumed that p is constant and $\phi > 0$.

We shall now deduce expressions for the spectroscopic parameters of this model in terms of the number of e-folds and find the restrictions that the observational results $\delta_{ns} = 0.032$, $r < 0.04$ put on this class of inflationary models.

Differentiating the potential we get

$$\frac{V'}{V} = \frac{p}{\hat{\phi}}, \quad \frac{V''}{V} = \frac{p(p-1)}{\hat{\phi}^2}. \tag{12.348}$$

Hence, the slow-roll parameters for this model are

$$\varepsilon = \frac{1}{2\kappa}\left(\frac{p}{\phi}\right)^2, \quad \eta = \frac{1}{\kappa}\frac{p(p-1)}{\phi^2}, \tag{12.349}$$

Hence for this class of inflationary universe models we have

$$\eta = \frac{2(p-1)}{p}\varepsilon \quad . \tag{12.350}$$

Inserting the expressions (12.349) into Eqs. (12.343) and (12.334) we get

$$\delta_{ns} = \kappa\frac{p(p+2)}{\phi^2}, \quad r = \kappa\frac{8p^2}{\phi^2}. \tag{12.351}$$

This gives the δ_{ns}, $r-$ relation

$$r = \frac{8p}{p+2}\delta_{ns}, \tag{12.352}$$

which may be written

$$p = \frac{2r}{8\delta_{ns}-r}. \tag{12.353}$$

It follows from this equation that the observational results $\delta_{ns} = 0.032$, $r < 0.04$ give requirement $p < 0.37$.

From Eqs. (12.322) and (12.347) we have

$$dN = \frac{\kappa}{p}\phi\,d\phi. \tag{12.354}$$

Integrating through the slow-roll inflationary era, we get

$$N = \frac{\kappa}{2p}\left(\phi^2 - \phi_f^2\right). \tag{12.355}$$

where $\phi = \phi(N)$ is the value of the field strength when the slow-roll era with N e-folds begins. It is usual to define the end of the inflationary era by $\varepsilon\left(\phi_f\right) = 1$. From Eq. (12.349) we then get

$$\phi_f = \frac{p}{\sqrt{2\kappa}}. \tag{12.356}$$

Inserting this expression for ϕ_f into Eq. (12.356) gives

$$\phi = \sqrt{\frac{p}{2\kappa}(p+4N)}. \tag{12.357}$$

Inserting this into Eq. (12.349) shows that for an inflationary era in which the potential of the dark energy is a power of the scalar field, the slow-roll parameters are

$$\varepsilon = \frac{p}{p+4N}, \quad \eta = \frac{2(p-1)}{p+4N}. \tag{12.358}$$

From Eqs. (12.343), (12.345) and (12.334) we then get

$$\delta_{ns} = \frac{2(p+2)}{p+4N}, \quad n_T = -\frac{2p}{p+4N}, \quad r = \frac{16p}{p+4N}. \tag{12.359}$$

Hence

$$N = \frac{8 - r/2}{8\delta_{ns} - r}. \tag{12.360}$$

The observational results $\delta_{ns} = 0.032$, $r < 0.04$ then requires that for these models the number of e-folds is restricted to $N < 37$.

There is a consensus that in order to solve the horizon and flatness problems the number of e-folds of the universe during the inflationary era must be larger than 50. Hence polynomial inflation does not give a satisfactory solution of these problems.

Confrontations of observable consequences of the inflationary era and observational data have been thoroughly discussed for several inflationary universe models in [24].

12.9 The Significance of Inertial Dragging for the Relativity of Rotation

The first published paper on inertial dragging inside a rotating shell based on the general theory of relativity was published by *H*. Thirring in 1918. He calculated the angular velocity, Ω, of a Zero Angular Momentum Observer (ZAMO) inside a shell with Schwarzschild radius, R_S, and radius, r_0, rotating slowly with angular velocity, ω, in the weak field approximation, and found the inertial dragging angular velocity,

$$\Omega = \frac{8R_S}{3r_0}\omega. \tag{12.361}$$

Both the angular velocity of the shell and that of the ZAMO are defined with respect to a system that is non-rotating in the far away region with asymptotic Minkowski spacetime.

In 1966 Brill and Cohen [26] presented a calculation of the angular velocity of an inertial frame inside a rotating shell valid for arbitrarily strong gravitational fields, but still restricted to slow rotation. The calculation of Brill and Cohen gave the dragging angular velocity Ω inside a massive shell with radius r_0, Schwarzschild radius R_S and rotating with angular velocity ω,

$$\Omega = \frac{4R_S(2r_0 - R_S)}{(r_0 + R_S)(3r_0 - R_S)}\omega. \tag{12.362}$$

A detailed calculation of this expression is found in [26]. For weak fields, i.e. for $r_0 \gg R_S$, this expression reduces to that of Thirring. But if the shell has so great mass that its Schwarzschild radius is equal to the radius of the shell, $r_0 = R_S$, the expression above gives $\Omega = \omega$. Then there is *perfect dragging*. In this case the inertial properties of space inside the shell no longer depend on the properties of the ZAMO at infinity, but are completely determined by the shell itself. Brill and Cohen further wrote that a shell of matter with radius equal to its Schwarzschild radius together with the space inside it can be taken as an idealized cosmological model, and proceeded: "Our result shows that in such a model there cannot be a rotation of the local inertial frame in the centre relative to the large masses in the universe. In this sense our result explains why the "fixed stars" are indeed fixed in our inertial frame". This means that rotation is relative to the motion of the large scale cosmic masses, and hence that the principle of relativity is valid for rotational motion in a universe with perfect inertial dragging.

When we look outwards in space, we look backwards in time, because we see an object the way it was when it emitted the light that we receive. Remarkably, gravitational waves move at the velocity of light. Although it has a quantum mechanical explanation in the fact that both photons and gravitons are massless, it is a strange coincidence from a classical point of view, possibly indicating a deep connection between gravity and electromagnetism. It means that when we search for sources of gravitational effects that have propagated undisturbed from a changing source to an observer, neglecting tales of gravitational waves that can be contributions from the inside of the light cone, we must look at events along the past light cone.

12.9.1 The Cosmic Causal Mass in the Einstein-de Sitter Universe

We search for cosmic sources of inertial dragging here and now. Hence, we introduce the concept *causal mass*, i.e. the mass which produces gravitational effects here and now. When the causal mass at the point of time t_0 of an observer at $r = 0$ is calculated by performing an integral with a mass element formed as a spherical shell about the observer with coordinate radius and thickness r and dr, respectively, the mass of the element is calculated by inserting the value of the density at the emission time of the considered mass element on the past light cone.

If the causal mass inside the particle horizon of the universe is so great that its Schwarzschild radius is equal to or larger than the radius of the horizon, there will be perfect inertial dragging [26]. In this case the principle of relativity is valid for rotational motion in such a universe.

In order to give a simple illustration we first consider an example permitting analytical expressions in terms of elementary functions; i.e. we shall first consider the Einstein–de Sitter universe. This is a flat universe containing only dust. It has scale factor

$$a(t) = (t/t_0)^{2/3}, \tag{12.363}$$

where t_0 is the present age of the universe, and the scale factor has been normalized to $a(t_0) = 1$.

The Hubble age is $t_H = 1/H_0$, where H_0 is the present value of the Hubble parameter. Inserting the most recent value of the Hubble parameter gives $t_H = 13.9$ Gy. The age of this universe is

$$t_{0ED} = (2/3)t_H. \tag{12.364}$$

where ED means Einstein-de Sitter.

The physical radius of the particle horizon is (we are using units so that $c = 1$),

$$R_{PH}(t) = a(t) \int_0^1 \frac{1}{a(t)} dt. \tag{12.365}$$

Inserting the scale factor (12.363) gives the horizon radius of the Einstein–de Sitter universe,

$$R_{PHED}(t) = 3 t_0^{2/3} t^{1/3}. \tag{12.366}$$

The present horizon radius is

$$R_{PHED}(t_0) = 3t_{0ED} = 2t_H. \tag{12.367}$$

Hence, the present radius of the particle horizon is 27.8 Gly in an Einstein–de Sitter universe with the measured value of the Hubble parameter.

We shall now calculate the Schwarzschild radius of the causal mass inside the particle horizon. It is calculated by integrating along the past light cone; i.e. the density is evaluated at retarded points of time,

$$r_{SED} = 2GM = 8\pi G \int_0^{r_{0PH}} \left[\rho\sqrt{g}\right]_{t-r} dr = \int_0^{t_{0PH}} \left[\rho a^3\right]_{t-r} r^2 dr. \tag{12.368}$$

Here, g is the determinant of the spatial part of the metric, and r_{0PH} is the present radius of the particle horizon. Since the density of the dust is $\rho = \rho_0 a^{-3}$, we get

$$r_{SED} = (8\pi G/3)\rho_0 r_{0PH}^3. \tag{12.369}$$

Using Eq. (12.367) and given that the present density is equal to the present value of the critical density,

$$\rho_0 = \frac{3H_0^2}{8\pi G} = \frac{3}{8\pi G t_H^2}, \tag{12.370}$$

we get

$$r_{SED} = 8 t_H. \tag{12.371}$$

Inserting Planck observational data gives $r_{SED} = 108.8$ Gly. Hence for this universe model, $r_{SED} = 4R_{PHED}$, showing that the Schwarzschild radius of the causal mass inside the horizon is larger than the horizon radius. This indicates that the causal mass inside the horizon is so great that there is perfect inertial dragging in this universe, and hence that the principle of relativity is valid for rotational motion in this universe.

The lookback distance is the radius of a surface S around an observer equal to the velocity of light times the age of the universe, $r_{LED} = t_0$. Inserting this as the upper limit in the integral (12.215), we find that the Schwarzschild radius of the mass inside S is equal to the lookback distance, $r_{SHED} = r_{LED}$. This corresponds to the condition for perfect inertial dragging used in Sect. 12.7. However, from a causal point of view, the relevant surface is the particle horizon.

12.9.2   The Cosmic Causal Mass in the Flat ΛCDM Universe

We now consider the flat ΛCDM universe which has scale factor [9]

$$a(t) = A^{1/3} \sinh^{2/3}(t/t_\Lambda), \quad A = \frac{1 - \Omega_{\Lambda 0}}{\Omega_{\Lambda 0}}, \quad t_\Lambda = \frac{2}{\sqrt{3\Lambda}} = \frac{2t_H}{3\sqrt{\Lambda_0}}. \tag{12.372}$$

The present radius of the particle horizon in this universe is

$$r_{0PH} = r_{PH}(t_0) = \frac{2}{3} \frac{t_H}{\Omega_{\Lambda 0}^{1/6}(1 - \Omega_{\Lambda 0})^{1/3}} \int\limits_0^{\operatorname{arsinh}\sqrt{1/A}} \frac{dx}{\sinh^{2/3}(x)}. \tag{12.373}$$

Inserting the Planck values $\Omega_{\Lambda 0} = 0.68$ and $t_H = 13.9$ Gy and calculating the integral numerically gives $R_{0PH} = 45$ Gly.

The mass which acts causally at the present time t_0 upon the observer located at $r = 0$ in the ΛCDM universe is

$$M = 4\pi \int_0^{r_{0PH}} \left[\rho_{M0} + \rho\left(a(t_e(r))^3\right) \right] r^2 \, dr. \qquad (12.374)$$

Here ρ_{M0} and ρ_Λ denote the present density of the dust and the vacuum energy, respectively, and t_e denotes the emission time of a signal emitted at the coordinate distance r and received at $r = 0$ at the time t_0. Accordingly t_e is a function of the coordinate distance r the signal travels and is given implicitly by the relation

$$\frac{dr}{dt} = \frac{t}{a(t)} \qquad (12.375)$$

with the initial condition $r(-t_0) = 0$. The present Schwarzschild radius of this mass is $R_S = 2\,GM$. Solving these equations numerically and plotting the Schwarzschild radius of the causal mass inside the particle horizon and the present radius of the particle horizon as functions of $\Omega_{\Lambda0}$ we obtain the result shown in Fig. 12.21 which is taken from the article [27].

We see from this that the Schwarzschild radius of the causal mass inside the present particle horizon in a flat ΛCDM universe is larger than the present radius of the particle horizon. This means that there is perfect inertial dragging in the universe. Hence the motion of inertial frames in the universe is determined by the

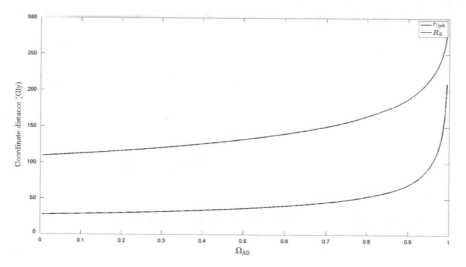

Fig. 12.21 Causal mass. The upper curve represents the Schwarzschild radius of the causal mass inside the present particle horizon in a flat ΛCDM universe, and the lower curve represents the present radius of the particle horizon

average motion of the causal mass in the universe. If this is a realistic model of our universe, we may conclude that rotational and accelerated motion is relative in our universe according to the general theory of relativity.

Exercises

12.1. *Gravitational collapse*

In this problem we shall find a solution to Einstein's field equations describing a spherically symmetric gravitational collapse. The solution shall describe the space-time both exterior and interior to the star. To connect the exterior and interior solutions, the metrics must be expressed in the same coordinate system. We will assume that the interior solution has the same form as a Friedmann solution. The Friedmann solutions are expressed in co moving coordinates, thus freely falling particles have constant spatial coordinates.

Let (ρ, τ) be the infalling coordinates, $i.$ co-moving coordinates with freely falling particles. τ is the proper time of a freely falling particle starting at infinity with zero velocity. These coordinates are connected to the curvature coordinates via the requirements

$$
\begin{aligned}
\rho &= r, \quad \text{for} \quad \tau = 0, \\
\tau &= t, \quad \text{for} \quad r = 0.
\end{aligned}
\tag{12.376}
$$

(a)   Show that the transformation between the in falling coordinates and the curvature coordinates is given by

$$
\tau = \frac{2}{3c\sqrt{R_S}}\left(\rho^{\frac{3}{2}} - r^{\frac{3}{2}}\right),
$$

$$
t = \tau - \frac{2}{c}(R_S r)^{\frac{1}{2}} + \frac{R_S}{c}\ln\left[\frac{\left(\frac{r}{R_S}\right)^{\frac{1}{2}} + 1}{\left(\frac{r}{R_S}\right)^{\frac{1}{2}} - 1}\right],
\tag{12.377}
$$

where R_S is the Schwarzschild radius of the star.

(b)   Show that the Schwarzschild metric in these coordinates takes the form

$$
ds^2 = -c^2 d\tau^2 + \left[1 - \frac{3}{2}(R_S)^{\frac{1}{2}}c\tau\rho^{-\frac{3}{2}}\right]^{-\frac{2}{3}} d\rho^2
$$

$$
+ \left[1 - \frac{3}{2}(R_S)^{\frac{1}{2}}c\tau\rho^{-\frac{3}{2}}\right]^{\frac{4}{3}} \rho^2\left(d\theta^2 + \sin^2\theta d\phi^2\right).
\tag{12.378}
$$

Assume the star has a position-dependent energy density $\rho(\tau)$, and that the pressure is zero. Assume further that the interior spacetime can be described with a Friedmann solution with Euclidean geometry ($k = 0$).

(c) Find the solution when the radius of the star is R_0 at $\tau = 0$.

12.2. The volume of a closed Robertson–Walker universe

Show that the volume of the region contained inside a radius $r = a\chi = a \arcsin r$ is

$$V = 2\pi a^3 \left(\chi - \frac{1}{2} \sin 2\chi \right). \tag{12.379}$$

Find the maximal volume. Find also an approximate expression for V when $\chi \ll R$.

12.3. Conformal time

Find the form of the line-element (12.1) if the cosmic time t is replaced by conformal time η defined by Eq. (12.126).

What is the equation of light moving radially when we use conformal time?

12.4. Lookback time and the age of the universe

The lookback time of an object is the time required for light to travel from an emitting object to the receiver. Hence, it is $t_L \equiv t_0 - t_e$, where t_0 is the point of time the object was observed and t_e is the point of time the light was emitted.

(a) Show that the lookback time is given by

$$t_L = \int_0^z \frac{dy}{(1 + y)H(y)}, \tag{12.380}$$

where z is the redshift of the object and the Hubble parameter $H(y)$ is given in Eq. (12.111).

(b) Show that the lookback time in the Milne universe model with $a(t) = (t/t_0)$, $k < 0$, is

$$t_L = \frac{1}{H_0} \frac{z}{1 + z}. \tag{12.381}$$

and find the age of this universe.

(c) Show that $t_L = t_0 \left[1 - (1 + z)^{-3/2} \right]$, where $t_0 = 2/(3H_0)$, in a flat, matter-dominated universe.

What is the age of this universe.?

(d) Find the lookback time of an object with redshift z and the age of the matter-dominated universe models with positive and negative spatial curvature.

(e) Find the age of a matter-dominated universe from the parametric solutions (12.140)–(12.144).

Are the resulting expressions in agreement with those found in d)?

(f) Find the lookback time–redshift relation and the age of a flat universe with dust and LIVE.

(g) Find the lookback time—redshift relation for a flat, LIVE-dominated universe.

12.5. *The LFRW universe models with a perfect fluid*

In this problem we will investigate FRW models with a perfect fluid. We will assume that the perfect fluid obeys the equation of state

$$p = w\rho, \tag{12.382}$$

where $-1 \le w \le 1$.

(a) Write down the Friedmann equations for a LFRW universe model with a w-law perfect fluid. Express the equations in terms of the scale factor a only.

(b) Assume that $a(0) = 0$. Show that when $-1/3 < w \le 1$, the closed model will recollapse. Explain why this does not happen in the flat and open models.

(c) Solve the Friedmann equation for a general $w \ne -1$ in the flat case. What is the Hubble parameter and the deceleration parameter? Also write down the time evolution for the matter density.

(d) Find the particle horizon distance in terms of H_0, w and z.

(e) Specialize the above to the dust-dominated, radiation-dominated and LIVE-dominated universe models.

(f) Find a general formula in terms of the density parameters for the present value of the deceleration parameter of a LFRW universe model.

12.6. *Age—density relation for a radiation-dominated universe*

Show that the age of a radiation-dominated universe model is given by

$$t_0 = \frac{t_H}{1 + \sqrt{\Omega_{rad0}}} \tag{12.383}$$

for all values of k.

12.7. Redshift–luminosity relation for matter-dominated universe: Mattig's formula

The *luminosity distance* of an object with redshift z is

$$d_L = \frac{1+z}{H_0\sqrt{|\Omega_{k0}|}} S_k\left[H_0\sqrt{|\Omega_{k0}|} \int_0^z \frac{dz}{H(z)} \right], \tag{12.384}$$

where the function S_k is defined in Eq. (12.17) and $H(z)$ is given in Eq. (12.111).

Show that the luminosity distance of an object with redshift z in a matter-dominated universe with relative density Ω_{m0} and Hubble constant H_0 is

$$d_L = \frac{2c}{H_0\Omega_{m0}^2}\left[\Omega_{m0}z + (\Omega_{m0} - 2)(\sqrt{1 + \Omega_{m0}z} - 1) \right]. \tag{12.385}$$

This is called Mattig's formula. Find the corresponding formula for the Einstein–de Sitter universe, with $\Omega_{m0} = 1$.

12.8. Newtonian approximation with vacuum energy

(a) Show that Einstein' linearized field equation for a static spacetime containing dust with density ρ and vacuum energy with density ρ_Λ takes the form of a modified Poisson equation

$$\nabla^2\phi = 4\pi G(\rho - 2\rho_\Lambda). \tag{12.386}$$

(b) Assume there is a particle with mass m at the origin. Solve Eq. (12.386) in the space outside the particle, and find the acceleration of gravity at a function of the distance from the origin.

Find the radius of a spherical surface where the acceleration of gravity vanishes.
How large is the mass inside this surface compared to the mass of the particle at the origin?
Evaluate the importance of LIVE for gravitational phenomena in the solar system.

12.9. Universe models with constant deceleration parameter

(a) Show that the universe with constant deceleration parameter q has expansion factor

$$a = \left(\frac{t}{t_0}\right)^{\frac{1}{1+q}}, \quad q \neq -1, \text{ and } a \propto e^{Ht}, \quad q = -1. \tag{12.387}$$

This is the scale factor of a flat universe with a perfect fluid. Find the equation of state of the fluid.

12.10. Density parameters as functions of the redshift

Show that the relative densities of LIVE and matter as functions of a are

$$\Omega_L = \frac{\Omega_{L0}a^3}{\Omega_{L0}a^3 + (1 - \Omega_{L0} - \Omega_{m0})a + \Omega_{m0}},$$

$$\Omega_m = \frac{\Omega_{m0}}{\Omega_{L0}a^3 + (1 - \Omega_{L0} - \Omega_{m0})a + \Omega_{m0}}. \qquad (12.388)$$

What can you conclude from these expressions concerning the universe at early and late times?

12.11. FRW universe with radiation and matter

Show that the scale factor and the cosmic time as functions of conformal time of a universe with radiation and matter are [28].

$$k > 0 : \begin{cases} a = a_0[\alpha(1 - \cos \eta) + \beta \sin \eta] \\ t = a_0[\alpha(\eta - \sin \eta) + \beta(1 - \cos \eta)], \end{cases} \qquad (12.389)$$

$$k = 0 : \begin{cases} a = a_0[\frac{1}{2}\alpha\eta^2 + \beta\eta] \\ t = a_0[\frac{1}{6}\alpha\eta^3 + \frac{1}{2}\beta\eta^2], \end{cases} \qquad (12.390)$$

$$k < 0 : \begin{cases} a = a_0[\alpha(\cosh \eta - 1) + \beta \sinh \eta] \\ t = a_0[\alpha(\sinh \eta - \eta) + \beta(\cosh \eta - 1)], \end{cases} \qquad (12.391)$$

where $\alpha = a_0^2 H_0^2 \Omega_{m0}/2$ and $\beta = (a_0^2 H_0^2 \Omega_{\gamma0})^{1/2}$, and $\Omega_{\gamma0}$ and Ω_{m0} are the present density parameters of radiation and matter, and H_0 is the present value of the Hubble parameter.

12.12. Event horizons in de Sitter universe models

Find the coordinate distances to the event horizons of the de Sitter universe models with $k > 0$, $k = 0$ and $k < 0$ as function of time.

12.13. Flat universe model with radiation and LIVE

(a) Find the scale factor as a function of time for a flat universe with radiation and Lorentz invariant vacuum energy represented by a cosmological constant Λ, and with present density parameter of radiation Ω_{rad0}.

(b) Calculate the Hubble parameter, H, as a function of time, and show that the model approaches a de Sitter model in the far future. Find also the deceleration parameter, $q(t)$.

    (c)   When is the inflection point, t_1, for which the universe went from decel-
eration to acceleration? What is the corresponding redshift observed at
the time t_0?

12.14 De Sitter spacetime

Consider a De Sitter spacetime with coordinates (t, r) and line element

$$ds^2 = -c^2 dt^2 + e^{2Ht}(dr^2 + r^2 d\Omega^2), \tag{12.393}$$

where the Hubble parameter H is constant.

(a)   Find the redshift of light emitter from a coordinate r as measured at a point of
time t_0 by an observer at the origin. The Hubble parameter H is assumed to be
known.

(b)   What is the 4-acceleration of a reference particle at rest in the coordinate system?
What does your result tell about the reference frame in which these coordinates
are co-moving?

     Will an observer with constant radial coordinate r experience an acceleration of
gravity?

     Introducing coordinates (T, R) by the transformation

$$R = re^{Ht}, \quad T = t - \ln(1 - H^2 r^2 e^{2Ht}) \tag{12.394}$$

or

$$r = \frac{R}{e^{HT}\sqrt{1 - H^2R^2/c^2}}, \quad e^{Ht} = e^{HT}\sqrt{1 - H^2R^2/c^2}, \tag{12.395}$$

the line element takes the form (you need not show this)

$$ds^2 = -(c^2 - H^2R^2)dT^2 + \frac{dR^2}{1 - H^2R^2/c^2} + R^2 d\Omega^2. \tag{12.396}$$

(c)   Find the redshift of light emitted from a coordinate R as measured by an observer
at the origin. Why is your result different to the one in a)?

(d)   What is the 4-acceleration of a reference particle at rest in the coordinate system?
What does your result tell about the reference frame in which these coordinates
are co-moving? Will an observer with constant radial coordinate r experience
an acceleration of gravity?

(e)   How does a reference particle with $r = r_0 = $ constant move in the (T, R)-
coordinate system.

(f)   How is the redshift of light explained in the (T, R)-coordinate system? How
is it explained in the (t, r)-system?

12.15. The Milne Universe

(a) The Milne Universe has a line element

$$ds^2 = -c\,dt^2 + \left(\frac{t}{t_0}\right)^2 \left(\frac{dr^2}{1 + r^2/c^2 t_0^2} + r^2 d\Omega^2\right). \tag{12.397}$$

The t, r coordinates are co-moving in a reference frame E. Give a physical interpretation of the line-element.

(b) The Hubble parameter is $H = \dot{a}/a$ where $\dot{a}$ is the derivative of the scale factor with respect to time. Calculate the Hubble parameter of this universe model at an arbitrary point of time. Find the age of the universe in terms of the present value of the Hubble parameter.

(c) The so-called deceleration parameter is given by $q \equiv -a\ddot{a}/\dot{a}^2$.

What is the value of q for this universe model? What does this tell about the expansion?

(d) Introduce new coordinates T and R by the transformation

$$T = t\sqrt{1 + \left(\frac{r}{ct_0}\right)^2}, \quad R = \frac{r\,t}{t_0}. \tag{12.398}$$

Show that the inverse transformation is

$$ct = \sqrt{c^2 T^2 - R^2} \quad, \quad r = \frac{ct_0 R}{\sqrt{c^2 T^2 - R^2}}, \tag{12.399}$$

and that

$$R = \frac{rcT}{\sqrt{c^2 t_0^2 + r^2}}. \tag{12.400}$$

(e) Make a Minkowski diagram with reference to the cT, R-system, and draw the world lines of the reference particles of E, i.e. those with $r = $ constant and the simultaneity curves of E, i.e. those with $t = $ constant.

(f) Use the transformation Eqs. (3) to show that the differentials of the coordinates co-moving in the reference frame E are

$$dt = \frac{c^2 T\,dT - R\,dR}{c\sqrt{c^2 T^2 - R^2}}, \quad dr = \frac{c^3 t_0 T(T\,dR - R\,dT)}{\left(c^2 T^2 - R^2\right)^{3/2}}. \tag{12.401}$$

Use these differentials and the expressions for t and r in Eq. (3) to calculate the line element in the T, R coordinate system. What does your result tell about the spacetime described by the line element you just have found and the line element (1)?

(g) Calculate the cosmic redshift, z, of a star with $r = r_1 = $ constant in terms of r_1 and the point of time t_0 of the observation. Then calculate the redshift of a star with $R = $ constant. Explain the results you found.

(h) Einstein's field equations as applied to an isotropic and homogeneous universe model lead to the Friedmann equations

$$\frac{\dot{a}^2 + kc^2}{a^2} = \frac{8\pi G}{3}\rho \quad , \quad \frac{\ddot{a}}{a} = -\frac{4\pi G}{3}\left(\rho + 3\frac{p}{c^2}\right) \tag{12.402}$$

where a is the scale factor, ρ and p are the density and pressure of the cosmic fluid, respectively, and k is the spatial curvature index.

Apply these equations to the line-element (1). What does your result tell about this universe model?

12.16. Natural Inflation

The natural inflation model has potential

$$V(\phi) = V_0\left(1 + \cos\tilde{\phi}\right), \tag{12.403}$$

where $\tilde{\phi} = \phi/M$, and M is the spontaneous symmetry breaking scale. We shall here write Einstein's gravitational constant as $\kappa = 1/M_P^2$, where M_P is the Planck mass.

(a) Show that for this model the spectral parameters are

$$\delta_{ns} = b\frac{3 - \cos\tilde{\phi}}{1 + \cos\tilde{\phi}}, \quad n_T = -b\frac{1 - \cos\tilde{\phi}}{1 + \cos\tilde{\phi}}, \quad r = 8b\frac{1 - \cos\tilde{\phi}}{1 + \cos\tilde{\phi}}, \quad b = \left(\frac{M_P}{M}\right)^2 \tag{12.404}$$

Observations have given the results $\delta_{ns} = 0.032$ and $r < 0.04$.

(b) Show that

$$b = \delta_{ns} - \frac{r}{4}, \tag{12.405}$$

and use this to calculate the requirement from the observations upon the symmetry breaking scale. Is there any problem with the result?

(c)   Show that for this model the number of e-folds is

$$N = \frac{1}{b} \ln \frac{1 - \cos\left(\tilde{\phi}_f\right)}{1 - \cos\left(\tilde{\phi}\right)}. \tag{12.406}$$

Use this to express the spectral parameters as

$$\delta_{ns} = b \frac{(2+b)e^{bN} + 2}{(2+b)e^{bN} - 2}, \quad n_T = -\frac{2b}{(2+b)e^{bN} - 2}, \quad r = \frac{16b}{(2+b)e^{bN} - 2}, \tag{12.407}$$

In order to solve the horizon- and flatness problems the number of e-folds must be larger than 50. Insert $N = 50$, $r = 0.04$ and make a judgement of this model.

References

1. Grøn, Ø., Elgarøy, Ø.: Is space expanding in the Friedmann universe models. Am. J. Phys. **75**, 151–157 (2007)
2. McGaugh, S.: Constraints on the radial mass distribution of dark matter halos from rotation curves. In: Natarajan, P. (ed.) The Shapes of Galaxies and their Dark Halos, pp. 186–193. World Scientific (2001)
3. Linde, A.:. Inflation and string cosmology. Int. J. Mod. Phys. **A17SI**, 89–104 (2001)
4. de Bernadis, P., et al.: Multiple peaks in the angular power spectrum of the cosmic microwave background: significance and consequences for cosmology. Astrophys. J. **584**, 559–566 (2001)
5. Stompor, R., et al.: Cosmological implications of the MAXIMA-1 high resolution cosmic microwave background anisotropy measurement. Astrophys. J. **561**, L7–L10 (2001)
6. Pryke, C., et al.: Cosmological parameter extraction from the first season of observations with DASI. Astrophys. J. **568**, 46–51 (2001)
7. Riess, A.G., et al.: Observational evidence from supernovae for an accelerating universe and a cosmological constant. Astron. J. **116**, 1009–1038 (1998)
8. Perlmutter, S., et al.: Measurements of omega and lambda from 42 high-redshift supernovae. Astrophys. J. **517**, 565–586 (1999)
9. Grøn, Ø.: A new standard model of the universe. Eur. J. Phys. **23**, 135–144 (2002)
10. Zlatev, I., Wang, L., Steinhardt, P.J.: Quintessence, cosmic coincidence, and the cosmological constant. Phys. Rev. Lett. **82**, 896–899 (1999)
11. Carroll, S.M.: Quintessence and the rest of the world. Phys. Rev. Lett. **81**, 3067–3070 (1998)
12. Zeldovich, Y.: The cosmological constant and the theory of elementary particles. Sov. Phys. Usp. **11**, 381–393 (1968)
13. Grøn, Ø.: Repulsive gravitation and inflationary universe models. Am. J. Phys. **54**, 46–52 (1986)
14. Lahav. O.: Cosmological Parameters. Particles data group. Updated 2017. http://pdg.lbl.gov/2017/reviews/rpp2017-rev-cosmological-parameters.pdf. (2017)
15. Grøn, Ø., Jemterud, T.: An interesting consequence of the principle of relativity. Euro. Phys. J. Plus. **131**, 91 (2016). https://doi.org/10.1140/epjp/i2016-16091-9
16. Riess, A.G.: The farthest known supernova: support for an accelerating universe and a glimpse of the epoch of deceleration. Astrophys. J. **560**, 49–71 (2001)

17. Wen-Fui, W., Zeng-Wei, S., Bin, T.: Exact solution of phantom dark energy model. Chin. Phys. B **19**, 119801 (2010)
18. Grøn, Ø.: Expansion isotropization during the inflationary era. Phys. Rev. D **32**, 2522–2527 (1985)
19. Alnes, H., Amarzguioui, M., Grøn, Ø.: An inhomogeneous alternative to dark energy? Phys. Rev. **D73**, 083519 (2006)
20. Enqvist, K.: Lemaître-Tolman-Bondi model and accelerating expansion. Gen. Rel. Grav. **40**, 451–466 (2008)
21. McGaugh, S.:Constraints on the radial mass distribution of dark matter halos from rotation curves. In: P. Natarajan (ed.) The Shapes of Galaxies and their Dark Halos, pp. 186–193. World Scientific (2001)
22. Turner, M.S.: Dark Energy and the New Cosmology, astro-ph/0108103 (2001)
23. Guth, A.H.: The inflationary universe: a possible solution to the horizon and flatness problems. Phys. Rev. D **23**, 347–356 (1981)
24. Grøn, Ø.: Predictions of spectral parameters by several inflationary universe models in light of the Planck results. Universe **4**(2), 15 (163 pages) (2018)
25. Ade, P., et al.: A joint analysis of BICEP2/Keck array and Planck Data. Phys. Rev. Lett. **114**, 101301 (2015)
26. Brill, D.R., Cohen, J.M.: Rotating masses and their effect on inertial frames. Phys. Rev.**143**, 1011–1015 (1966)
27. Braeck, S., Grøn, Ø.G., Farup, I.: The cosmic causal mass. Universe. **3**, 38 (2017). https://doi.org/10.3390/universe3020038
28. Johannesen. S.: Smooth Manifolds and Fibre Bundles with Application to Theoretical Physics. CRC Press (2017)

Appendix
Kaluza–Klein Theory

In the general theory of relativity gravitation is a property of 4-dimensional space-time and of the motion of the reference frame. Kaluza–Klein theory, [1, 2], is a unified theory of gravitation and electromagnetism in which electromagnetism is the projection of certain elements of gravitation in 5-dimensional spacetime into our 4-dimensional spacetime. The fifth dimension is spatial. According to this theory there is no electromagnetism in the 5-dimensional world, only gravitation which is described by the general theory of relativity in 5-dimensional spacetime.

A.1   The Structure of the Kaluza–Klein Theory

The Kaluza–Klein theory is constructed in the following way. Let the line-element in a Riemannian 5-dimensional space be

$$ds^2 = g_{\mu\nu}dx^\mu dx^\nu, \quad \mu, \nu = 1, 2, 3, 4, 5 \tag{A.1}$$

with signature $(-, +, +, +, +)$.

We experience a 4-dimensional world. We are flat-landers in a 4-dimensional hyper-surface R_4—our spacetime—in the 5-dimensional universe. It has become usual to consider the fifth dimension as a cylinder with axis orthogonal to 4-dimensional spacetime. Hence, the intersection of the fifth cylinder dimension with our 4-dimensional spacetime is a circle. Also it is assumed that this circle has an extremely small radius, since we have no direct experience of this fifth dimension. This means that what we ordinarily think of as a point in 4-dimensional spacetime is a small circle according to the Kaluza–Klein theory.

Let $\vec{e}_5$ be a basis vector orthogonal to our 4-dimensional spacetime. Einstein and Bergmann (1938) have shown that if one requires that closed geodesic curves in the fifth dimension—around the cylinder—shall be continuous, then it is necessary that

$$g_{\mu\nu,5} = 0. \tag{A.2}$$

© Springer Nature Switzerland AG 2020
Ø. Grøn, *Introduction to Einstein's Theory of Relativity*,
Undergraduate Texts in Physics, https://doi.org/10.1007/978-3-030-43862-3

Furthermore g_{55} is a constant along the curve. One can then require that g_{55} is generally constant, and by a suitable choice of coordinate one can normalize g_{55} so that

$$g_{55} = 1. \tag{A.3}$$

The conditions (A.2) and (A.3) are called the *cylinder conditions*.

In the same way that the spatial metric in the line-element (4.5) represents a projection of the metric in 4-dimensional spacetime into the 3-space orthogonal to the time direction, the metric of 4-dimensional spacetime is obtained in the Kaluza–Klein theory by projecting the metric of 5-dimensional spacetime onto the 4-dimensional spacetime orthogonal to the a basis vector pointing along the fifth dimension. Hence the line element of 4-dimensional spacetime is,

$$dl^2 = \gamma_{ij} dx^i dx^j, \quad \gamma_{ij} = g_{ij} - \frac{g_{5i} g_{5j}}{g_{55}} = g_{ij} - g_{5i} g_{5j}, \tag{A.4}$$

where we have used Eq. (A.3).

It follows from Eqs. (4.6)–(4.9) that dl^2 is invariant against a coordinate transformation of the form

$$x^{i'} = x^{i'}(x^i), \quad x^{5'} = x^5 + f(x^i). \tag{A.5}$$

Here the first equation represents an arbitrary transformation in 4-dimensional spacetime, and the second a gauge-transformation in the fifth dimension: the choice of origin on a circle $x^1, x^2, x^3, x^4 = $ constant depends upon the position in 4-dimensional spacetime. Note that the cylinder conditions (A.2) and (A.3) are invariant against a transformation of the form (A.5). Furthermore the transformation of g_{5i} against (A.5) has the form

$$g_{5v} = \frac{\partial x^{\mu'}}{\partial x^5} \frac{\partial x^{v'}}{\partial x^i} g_{\mu' v'} = \frac{\partial x^{5'}}{\partial x^5} \frac{\partial x^{v'}}{\partial x^v} g_{5'v'} + \frac{\partial x^{5'}}{\partial x^5} \frac{\partial x^{5'}}{\partial x^v} g_{5'5'} = \frac{\partial x^{v'}}{\partial x^v} g_{5'v'} + \frac{\partial f}{\partial x^v}. \tag{A.6}$$

Hence one can introduce a one-form in the 5-dimensional spacetime

$$\underline{g}_5 = \underline{g}_5' + \underline{d} f. \tag{A.7}$$

The physical interpretation of this form from the 4-dimensional point of view is that it represents the electromagnetic vector potential form,

$$\underline{g}_5 = \underline{A}. \tag{A.8}$$

This leads to the gauge transformation of the electromagnetic vector potential,

$$\underline{A} = \underline{A}' + \underline{d}f. \tag{A.9}$$

From this equation and Poincare's lemma it follows that the electromagnetic field form

$$\underline{F} = \underline{d}\underline{A} \tag{A.10}$$

is invariant against the gauge transformation (A.9). Also this invariance is here seen as a consequence of the invariance of the metric 1-form g_5 against a transformation representing a free choice of the origin on the fifth cylinder dimension.

A.2   Calculation of the 5-dimensional Curvature Scalar

We shall now calculate the curvature scalar of the Riemann tensor representing the curvature of the 5-dimensional spacetime using Cartan's structure equations.

The line-element of 5-dimensional spacetime may be written as

$$ds^2 = \gamma_{ij} dx^i dx^j + \left(dx^5 + A_i dx^i\right)^2, \quad A_i = g_{5i}, \tag{A.11}$$

where γ_{ij} are the components of the metric tensor of our 4-dimensional spacetime. We introduce a form basis

$$\underline{\omega}^i = \underline{d}x^i, \quad \underline{\omega}^5 = \underline{d}x^5 + A_i \underline{d}x^i, \tag{A.12}$$

Exterior differentiation gives

$$\underline{d}\underline{\omega}^i = 0, \quad \underline{d}\underline{\omega}^5 = \frac{1}{2}\left(\frac{\partial A_i}{\partial x^j} - \frac{\partial A_j}{\partial x^i}\right)\underline{d}x^i \wedge \underline{d}x^j. \tag{A.13}$$

Hence the electromagnetic field form has components

$$F_{ij} = \frac{\partial A_i}{\partial x^j} - \frac{\partial A_j}{\partial x^i} = \frac{\partial g_{5i}}{\partial x^j} - \frac{\partial g_{5j}}{\partial x^i}. \tag{A.14}$$

From Cartan's 1. structure equation in the form (5.180) we have

$$\underline{d}\underline{\omega}^5 = -\underline{\Omega}^5_j \wedge \underline{\omega}^j. \tag{A.15}$$

According to Eq. (5.164) the components of the connection form are

$$\underline{\Omega}^5_j = \Gamma^5_{j\nu}\underline{\omega}^\nu. \tag{A.16}$$

It follows from Eqs. (A.3) and (A.13)–(A.16) that

$$\Gamma_{ij}^k =^4 \Gamma_{ij}^k - \frac{1}{2}\left(A_i F_j^k + A_j F_i^k\right), \quad \Gamma_{5ji} = \Gamma_{ji}^5 = -\frac{1}{2}F_{ij} = \frac{1}{2}F_{ji}, \quad \Gamma_{55}^\mu = 0.$$

$$(A.17)$$

Hence

$$\Gamma_{5ij} = \Gamma_{ij}^5 = \frac{1}{2}F_{ij}.$$

$$(A.18)$$

With the basis (A.12) the metric has the form

$$g_{\mu\nu} = \begin{pmatrix} \gamma_{ij} & 0 \\ 0 & 1 \end{pmatrix}.$$

$$(A.19)$$

Hence

$$\underline{d}g_{5j} = \underline{d}g_{j5} = \underline{d}g_{55} = 0.$$

$$(A.20)$$

It follows from Eq. (5.169) and this equation that

$$\underline{\Omega}_{i5} = -\underline{\Omega}_{5i},$$

$$(A.21)$$

giving

$$\Gamma_{i5j} = -\Gamma_{5ij} = -\frac{1}{2}F_{ij}.$$

$$(A.22)$$

Thus

$$\Gamma_{5j}^i = -\frac{1}{2}F_j^i.$$

$$(A.23)$$

It follows from Eq. (5.132) that

$$\Gamma_{j5}^i = \Gamma_{5j}^i + c_{5j}^i,$$

$$(A.24)$$

where the structure coefficients are defined by Eq. (3.40) which here gives

$$\left[\vec{e}_5, \vec{e}_i\right] = c_{5i}^\mu \vec{e}_\mu.$$

$$(A.25)$$

Here

$$\vec{e}_5 = \frac{\partial}{\partial x^5}, \quad \vec{e}_i = \frac{\partial}{\partial x^i} - A_i \frac{\partial}{\partial x^5}.$$

$$(A.26)$$

Since A_i is independent of x^5 we get

$$\left[\frac{\partial}{\partial x^5}, \frac{\partial}{\partial x^i} - A_i \frac{\partial}{\partial x^5} \right] = 0. \tag{A.27}$$

It follows that

$$c^\mu_{5i} = 0. \tag{A.28}$$

Equations (A.23) and (A.24) then give

$$\Gamma^i_{j5} = \Gamma^i_{5j} = -\frac{1}{2} F^i_j. \tag{A.29}$$

We have

$$\underline{\Omega}^i_j = \Gamma^i_{jk}\underline{\omega}^k + \Gamma^i_{j5}\underline{\omega}^5. \tag{A.30}$$

The curvature forms are given in Cartan's 2. structure equation (6.24). In the present case they give for the curvature of the 5-dimensional spacetime,

$$\begin{aligned}
\underline{R}^i_j &= d\underline{\Omega}^i_j + \underline{\Omega}^i_\mu \wedge \underline{\Omega}^\mu_j = d\underline{\Omega}^i_j + \underline{\Omega}^i_k \wedge \underline{\Omega}^k_j + \underline{\Omega}^i_5 \wedge \underline{\Omega}^5_j \\
&= d\left(\Gamma^i_{jk}\underline{\omega}^k\right) + \underline{\Omega}^i_k \wedge \underline{\Omega}^k_j + d\left(\Gamma^i_{j5}\underline{\omega}^5\right) + \underline{\Omega}^i_5 \wedge \underline{\Omega}^5_j \\
&= {}^4\underline{R}^i_j + \Gamma^i_{j5,k}\underline{\omega}^k \wedge \underline{\omega}^5 + \Gamma^i_{j5}d\underline{\omega}^5 + \underline{\Omega}^i_5 \wedge \underline{\Omega}^5_j
\end{aligned} \tag{A.31}$$

From Eqs. (A.15), (A.16) and (A.19) we have

$$d\underline{\omega}^5 = \frac{1}{2} F_{ij}\underline{\omega}^i \wedge \underline{\omega}^j. \tag{A.32}$$

and

$$\underline{\Omega}^i_5 \wedge \underline{\Omega}^5_j - \frac{1}{4} F^i_m F_{nj}\underline{\omega}^m \wedge \omega^n, \tag{A.33}$$

which leads to

$$\underline{R}^i_j = {}^4\underline{R}^i_j - \frac{1}{2} F^i_{j,k}\underline{\omega}^k \wedge \underline{\omega}^5 - \frac{1}{4} F^i_j F_{mn}\underline{\omega}^m \wedge \underline{\omega}^n + \frac{1}{4} F^i_m F_{nj}\underline{\omega}^m \wedge \underline{\omega}^n. \tag{A.34}$$

The components of the Riemann curvature tensor are given as the components of the curvature forms by

$$\underline{R}^i_j = \frac{1}{2} R^i_{j\mu\nu}\underline{\omega}^\mu \wedge \underline{\omega}^\nu. \tag{A.35}$$

We need the components, R^i_{jmn}, of these curvature forms for the 5-dimensional spacetime in 4-dimensional spacetime. Then the second term in the right-hand side of

Eq. (A.34) does not contribute. Also the components of the 2-forms are antisymmetric. This is automatically taken care of in the third term due to the antisymmetry of the electromagnetic field tensor, but in the fourth term we have to take the antisymmetric combination when we write down the form-components. This gives

$$R^i_{jmn} =^4 R^i_{jmn} - \frac{1}{2}F^i_j F_{mn} + \frac{1}{4}F^i_m F_{nj} - \frac{1}{4}F^i_n F_{mj}. \tag{A.36}$$

We are going to calculate the Ricci curvature scalar in the 5-dimensional spacetime,

$$R = R^j_j + R^5_5, \tag{A.37}$$

where the terms at the right-hand side are the mixed components of the Ricci curvature tensor. These are calculated as follows.

The components of the Riemann curvature tensor are

$$R_{jn} = R^\mu_{j\mu n} = R^i_{jin} + R^5_{j5n} = R^i_{jin} + R_{5j5n} = R^i_{jin} + R_{j5n5}. \tag{A.38}$$

Using Eq. (A.36) and (A.48) we get

$$R_{jn} =^4 R_{jn} - \frac{1}{2}F^i_j F_{in} + \frac{1}{4}F^i_i F_{nj} - \frac{1}{4}F^i_n F_{ij} - \frac{1}{4}F_{ji} F^i_n. \tag{A.39}$$

The electromagnetic field tensor is trace free, so $F^i_i = 0$. Hence the third term at the right-hand side vanishes. Also since the electromagnetic field tensor is antisymmetric, the two last terms cancel each other. Hence we get

$$R_{jn} =^4 R_{jn} - \frac{1}{2}F^i_j F_{in}. \tag{A.40}$$

Thus

$$R^j_j =^4 R^j_j - \frac{1}{2}F^{ij} F_{ij} =^4 R^j_j - \frac{1}{2}F_{ij} F^{ij}. \tag{A.41}$$

In order to calculate R^5_5 we need the curvature forms

$$\underline{R}^i_5 = d\underline{\Omega}^i_5 + \underline{\Omega}^i_\mu \wedge \underline{\Omega}^\mu_5 = d\underline{\Omega}^i_5 + \underline{\Omega}^i_j \wedge \underline{\Omega}^j_5. \tag{A.42}$$

It follows from Eqs. (5.164) and (A.29) that

$$\underline{\Omega}^i_5 = \Gamma^i_{5j}\underline{\omega}^j = -\frac{1}{2}F^i_j\underline{\omega}^j. \tag{A.43}$$

Hence

$$\underline{d\Omega}_5^i = -\frac{1}{2} F_{j,k}^i \underline{\omega}^k \wedge \underline{\omega}^j. \tag{A.44}$$

Using Eqs. (A.30) and (A.43) we get

$$\begin{aligned}
\underline{\Omega}_j^i \wedge \underline{\Omega}_5^j &= \left(\Gamma_{jm}^i \underline{\omega}^m + \Gamma_{j5}^i \underline{\omega}^5\right) \wedge \Gamma_{5n}^j \underline{\omega}^n \\
&= \Gamma_{jm}^i \Gamma_{5n}^j \underline{\omega}^m \wedge \underline{\omega}^n + \Gamma_{j5}^i \Gamma_{5n}^j \underline{\omega}^5 \wedge \underline{\omega}^n
\end{aligned} \tag{A.45}$$

In the first term at the right-hand side $\Gamma_{jm}^i \Gamma_{5n}^j$ is symmetric in m and n while the basis is antisymmetric. Hence this term vanishes. Using Eq. (A.29) we then obtain

$$\underline{\Omega}_j^i \wedge \underline{\Omega}_5^j = \frac{1}{4} F_j^i F_m^j \underline{\omega}^5 \wedge \underline{\omega}^m. \tag{A.46}$$

Inserting the expressions (A.44) and (A.46) into Eq. (A.42) gives

$$\underline{R}_5^i = -\frac{1}{2} F_{j,k}^i \underline{\omega}^k \wedge \underline{\omega}^j + \frac{1}{4} F_j^i F_m^j \underline{\omega}^5 \wedge \underline{\omega}^m. \tag{A.47}$$

This gives

$$R_{5jk}^i = \frac{1}{2}\left(F_{k,j}^i - F_{j,k}^i\right) \tag{A.48}$$

and

$$R_{55m}^i = \frac{1}{4} F_j^i F_m^j \tag{A.49}$$

or

$$R_{5m5}^i = -\frac{1}{4} F_j^i F_m^j. \tag{A.50}$$

It follows that

$$R_5^5 = R_{55} = R_{5i5}^i = -\frac{1}{4} F_j^i F_i^j = -\frac{1}{4} F_{ij} F^{ij}. \tag{A.51}$$

Inserting Eqs. (A.41) and (A.51) into Eq. (A.37) finally gives the Kaluza–Klein expression for the Ricci curvature scalar of the 5-dimensional spacetime

$$R = {}^4 R_j^j - \frac{1}{2} F_{ij} F^{ij} + \frac{1}{4} F_{ij} F^{ij} = {}^4 R^j - \frac{1}{4} F_{ij} F^{ij} \tag{A.52}$$

Comparing with Eq. (7.65) the last term is recognized as the Lagrangian of an electromagnetic field. The variational principle (7.58) with the Lagrangian equal to the Ricci curvature scalar of 5-dimensional spacetime gives Einstein's gravitational field equations in this 5-dimensional word. But as interpreted from our 4-dimensional perspective, the right-hand side of Eq. (A.52) shows that we have the Lagrangian leading to the Einstein's gravitational equations and Maxwell's source free equations. So the Einstein's theory in the 5-dimensional world is interpreted as a unified theory of gravity and electromagnetism in our 4-dimensional subspace of the 5-dimensional world.

A.3 Field Equations for Kaluza–Klein Theory with $g_{55} = 1$

The theory as developed so far has one serious problem. The Einstein equations in 5-dimensional spacetime, (marking here the curvature tensors of 5-dimensional spacetime by an index 5)

$$^5E =^5 R_{\mu\nu} - \frac{1}{2} {}^5Rg_{\mu\nu} = \kappa T_{\mu\nu}, \qquad (A.53)$$

represent in general 15 equations. Since there are 5 Bianchi identities in 5-dimensional spacetime, there are 10 independent field equations. In general the 5-dimensional metric tensor has 15 independent components. But when the theory is projected down to our 4-dimensional spacetime g_{55} appears as an unwanted scalar field, i.e. a scalar field which is foreign to Einstein's theory of 4-dimensional spacetime and gravitation. This is eliminated by the cylinder condition (A.3). With this condition there are 14 functions to be determined. Because of the transformation (A.5) which contains 5 arbitrary functions, there are only 9 independent metric functions. Hence the system is over determined. Thus, with the cylinder condition (A.3) it is necessary to reduce the number of independent equations by 1. This can be performed by adding a term in the equations in such a way that a sixth algebraic identity results. In this way there will only be 9 independent equations.

We shall therefore require that the curvature tensor at the left-hand side of the field equations in 5-dimensional spacetime is both divergence free, trace free and symmetric. The trace of the 5-dimensional Einstein tensor is

$$^5E = -\frac{3}{2} {}^5R. \qquad (A.54)$$

So in order to obtain a trace-free curvature tensor we must add a symmetric, divergence-free tensor with trace (A.54). The tensor with components

$$N_{\mu\nu} = \frac{3}{2} {}^5Rg_{5\mu}g_{5\nu} \qquad (A.55)$$

fulfil these conditions. Thus a symmetric, divergence-free and trace-free curvature tensor in 5-dimensional spacetime is

$$^5\overset{\leftrightarrow}{E}_{\mu\nu} =\,^5 R_{\mu\nu} - \frac{1}{2}\left(g_{\mu\nu} - 3g_{5\mu}g_{5\nu}\right)^5 R. \tag{A.56}$$

It was originally proposed by Leibowitz and Rosen [3] and may therefore be called the Leibowitz-Rosen tensor. Hence, the modified field equations have the form

$$^5 R_{\mu\nu} - \frac{1}{2}\left(g_{\mu\nu} - 3g_{5\mu}g_{5\nu}\right)^5 R = \kappa\,^5 T_{\mu\nu}, \tag{A.57}$$

where $^5 T_{\mu\nu}$ are the components of the energy–momentum tensor in 5-dimensional spacetime. Since the left-hand side of this equation is trace free, the right-hand side must be so, too. From an arbitrary energy–momentum tensor $T_{\mu\nu}$ we can always construct a corresponding trace-free tensor

$$^5 T_{\mu\nu} = T_{\mu\nu} - T g_{5\mu}g_{5\nu}. \tag{A.58}$$

Since the electromagnetic energy–momentum tensor is trace free $^5 T_{\mu\nu} = T_{\mu\nu}$ for electromagnetic fields.

A.4 The 5-dimensional Counterpart of Electric Charge

From the tensor $^5 T_{\mu\nu}$ we can form a quantity with components

$$J^i = -\sqrt{\frac{2\kappa}{\mu_0}}\, T_5^i \tag{A.59}$$

that transform as the components of a 4-vector under the transformation (A.5). Here κ is Einstein's gravitational constant given in Eq. (7.37), and μ_0 is the permeability of empty space. The conservation equation

$$^5 T^{\mu\nu}_{\;;\nu} = 0, \tag{A.60}$$

following from the 5-dimensional field equations, then implies that

$$J^i_{\;;i} = 0, \tag{A.61}$$

which is the electromagnetic equation of continuity

$$\nabla \cdot \vec{j} + \frac{\partial \rho}{\partial t} = 0. \tag{A.62}$$

Hence conservation of charge follows from a geometric identity, the Bianchi identity, in this theory.

Let us consider a gas of dust particles in 5-dimensional spacetime. Its trace-free energy–momentum tensor has the components

$$^5T_{\mu\nu} = \rho\left(u_\mu u_\nu + g_{5\mu} g_{5\nu}\right).$$ (A.63)

where u^μ are the components of the 5-velocity of the dust particles. For this gas the conservation Eq. (A.60) takes the form

$$(\rho u^\nu)_{;\nu} u^\mu + \rho a^\mu = 0,$$ (A.64)

where

$$a^\mu = u^\nu u^\mu_{;\nu}$$ (A.65)

are the components of the 5-acceleration of the dust particles. The 5-velocity is a unit vector. It follows that the 5-velocity and the 5-acceleration are orthogonal. Thus, multiplying Eq. (A.64) by u_μ gives

$$(\rho u^\nu)_{;\nu} = 0$$ (A.66)

Inserting this into Eq. (A.64) gives

$$a^\mu = 0,$$ (A.67)

which shows that the dust particles follow geodesic curves in 5-dimensional spacetime. Written out this equation takes the form

$$\frac{du^\mu}{ds} + \Gamma^\mu_{\alpha\beta} u^\mu u^\nu = 0.$$ (A.68)

The k-component of this equation is

$$\frac{du^k}{ds} + \Gamma^k_{ij} u^i u^j + 2\Gamma^k_{5i} u^5 u^i + \Gamma^k_{55}\left(u^5\right)^2 = 0.$$ (A.69)

Inserting the expressions of the 5-dimensional Christoffel symbols from Eq. (A.17) gives

$$\frac{du^k}{ds} + {}^4\Gamma^k_{ij} u^i u^j = \frac{1}{2}\left(A_i F^k_j + A_j F^k_i\right) u^i u^j + F^k_i u^5 u^i.$$ (A.70)

Since $u^i u^j$ is symmetric in i and j this can be written as

$$\frac{du^k}{ds} + {}^4\Gamma^k_{ij} u^i u^j = A_j F^k_i u^i u^j + F^k_i u^5 u^i.$$ (A.71)

The Lagrange function of a free particle in the 5-dimensional spacetime is obtained from the line-element (A.11) and is

$$L = \frac{1}{2}\gamma_{ij}\dot{x}^i\dot{x}^j + \frac{1}{2}(\dot{x}^5 + A_i\dot{x}^i)^2, \tag{A.72}$$

where the dot denotes differentiation with respect to the proper arc length along the world line of the particle. Since x^5 is a cyclic coordinate the momentum p_5 conjugate to x^5 is a constant of motion,

$$p_5 = m_0(u^5 + A_j u^j) = m_0 u_5. \tag{A.73}$$

where m_0 is the rest mass of the particle. From this equation we get

$$u^5 = p_5 - A_j u^j. \tag{A.74}$$

Inserting this into Eq. (A.71) gives

$$m_0\left(\frac{du^k}{ds} +^4 \Gamma_{ij}^k u^i u^j\right) = p_5 F_i^k u^i, \tag{A.75}$$

or

$$m_0^4 a^k = p_5 F_i^k u^i, \tag{A.76}$$

where $^4 a^k$ is the k-component of the particle's 4-acceleration. It was shown by Lei-bowitz and Rosen [1] that the physical components of the electromagnetic field tensor are

$$\hat{F}^{ij} = \frac{1}{c\sqrt{\varepsilon_0\kappa}}F^{ij}. \tag{A.77}$$

where ε_0 is the permittivity of empty space, and we have used S.I. units here. Hence, Eq. (A.76) takes the form

$$m_0^4 a^k = c\sqrt{\varepsilon_0\kappa}\, p_5 \hat{F}_i^k u^i. \tag{A.78}$$

Comparing with the equation of motion of a charged particle in an electromagnetic field, Eq. (5.307) we get

$$q = c\sqrt{\varepsilon_0\kappa}\, p_5. \tag{A.79}$$

Hence *the conserved covariant momentum of the particle in the fifth direction, i.e. around the cylindrical fifth dimension, is interpreted as the charge of the particle from our 4-dimensional point of view.* Inserting the expression (7.37) for Einstein's gravitational constant gives

$$q = \frac{1}{c}\sqrt{8\pi\varepsilon_0 G}\,p_5.$$ (A.80)

The Kaluza–Klein theory may provide an explanation of the fact that gravitational waves move with the velocity of light. According to this theory all of electromagnetism comes from projection of gravity in a 5-dimensional world into our 4-dimensional spacetime. According to this theory electromagnetic waves are the projection of certain gravitational waves that propagate isotropically. Hence the projected waves, both those projections that are interpreted as electromagnetic waves from our 4-dimensional point of view, and the gravitational waves in our 4-dimensional spacetime, move with the same velocity. This should be expressed by saying that *electromagnetic waves move with the velocity of gravity*, i.e. of gravitational waves.

In this connection there appears a problem which has not, so far, been solved. According to the general theory of relativity the source of a gravitational wave must change its quadrupole moment; i.e. it must change its shape. A particle cannot emit gravitational waves. But if the particle is electrically charged, it can emit electromagnetic waves. According to Larmor's formula it will emit electromagnetic radiation when it is accelerated.

In the 5-dimensional world a particle which we call charged is a neutral particle moving around the fifth spatial cylinder dimension. Hence in order that a charged particle in our world should emit electromagnetic radiation when it is accelerated, a neutral particle moving around the compact fifth dimension should emit those gravitational waves that are interpreted as electromagnetic waves in our world, when it is accelerated in a direction orthogonal to its circular motion in the fifth direction. Whether this is really the case is still an unsolved problem.

A.5 Quantization of Charge as a Consequence of Quantization of Momentum Along a Closed Path Around the Fifth Cylinder Dimension

Due to the closed character of the fifth cylinder dimension the momentum p_5 is periodic. As shown by Klein [2] this implies a quantization of p_5 according to

$$r_5 p_5 = n\hbar,$$ (A.81)

where r_5 is the radius of the cylindrical fifth dimension. Thus

$$q = \frac{1}{r_5 c}\sqrt{8\pi\varepsilon_0 G}\,n\hbar.$$ (A.82)

This means that in the Kaluza–Klein theory quantization of charge need not be postulated. It follows as a consequence of quantization of momentum around the closed fifth dimension. This is again a consequence of the de Broglie relationship associating a wave with the momentum. Then quantization follows from the requirement of constructive interference, demanding a whole number of wavelengths around the fifth dimension.

The quantum of charge is the elementary charge e. Inserting $q = e$ into Eq. (A.82), solving the resulting equation with respect to r_5 and putting $n = 1$ gives the smallest allowed radius of the cylinder dimension,

$$r_5 = \frac{\hbar\sqrt{8\pi \varepsilon_0 G}}{ce}. \tag{A.83}$$

The Planck length is

$$l_P = \sqrt{\frac{\hbar G}{c^3}}. \tag{A.84}$$

Inserting the values of the constants gives $l_P \approx 1.6 \times 10^{-35}$m. The fine-structure constant is a dimensionless number

$$\alpha = \frac{e^2}{4\pi \varepsilon_0 \hbar c} \approx \frac{1}{137}. \tag{A.85}$$

Thus, in terms of the Planck length the radius of the fifth cylinder dimension is

$$r_5 = \sqrt{\frac{2}{\alpha}} l_P, \tag{A.86}$$

giving $r_5 \approx 16.6 l_P \approx 2.6 \times 10^{-34}$m. This may represent a quantum of length in 5-dimensional spacetime.

A.6   Electric Field from Inertial Dragging in the Fifth Dimension

In Kaluza–Klein theory electric charge in the projection of momentum in the fifth dimension. It is natural then to wonder. What in the 5-dimensional world with gravity only is the electrical field of a charged particle the projection of?

In order to answer this question we shall consider a solution of the field equations for empty 5-dimensional space that was found some years ago [4],

$$ds^2 = -\left(1 + \frac{b}{r} - \frac{3a^2}{r^2}\right)c^2 dt^2 + \frac{dr^2}{1 + \frac{b}{r} + \frac{a^2}{r^2}} + r^2 d\Omega^2 - \frac{a}{r} dt dx^5 + \left(dx^5\right)^2. \tag{A.87}$$

where a and b are constants. Projecting this line-element into ordinary 4-dimensional spacetime we obtain

$$ds^2 = ds^2 = -\left(1 + \frac{b}{r} + \frac{a^2}{r^2}\right)c^2 dt^2 + \frac{dr^2}{1 + \frac{b}{r} + \frac{a^2}{r^2}} + r^2 d\Omega^2. \qquad (A.88)$$

With the identification $b = R_S$ and $a = R_q$ as given in Eq. (8.164), this represents the Reissner–Nordström solution of Einstein's field equations in 4-dimensional spacetime with an electric field.

It remains to identify which part of the gravitational field in 5-dimensional spacetime is interpreted as an electric field from our 4-dimensional point of view. Let us first identify the physical meaning of the constant a in the line-element (A.87) from the 5-dimensional point of view.

Consider a test particle with $p_5 = 0$ corresponding to a zero-angular momentum particle in the Kerr spacetime as was described in Sect. 10.3.1. According to Eq. (A.74) it has

$$u^5 = \dot{x}^5 = -A_j u^j. \qquad (A.89)$$

Assume the particle is at rest in the spatial 3-space of the 4-dimensional spacetime. Then the only non-vanishing component of the 4-velocity is $u^0 = c\dot{t}$. Hence

$$\dot{x}^5 = -A_0 \dot{t}. \qquad (A.90)$$

According to Eq. (A.11) we have

$$A_0 = g_{50} = g_{05}. \qquad (A.91)$$

With the line-element (A.87)

$$g_{05} = -\frac{a}{r}. \qquad (A.92)$$

Thus

$$A_0 = -\frac{b}{r}, \qquad (A.93)$$

and

$$\dot{x}^5 = \frac{a}{r}\dot{t}. \qquad (A.94)$$

This gives

$$\frac{dx^5}{dt} = \frac{\dot{x}^5}{\dot{t}} = \frac{a}{r}.$$ (A.95)

This shows that a particle with vanishing conserved momentum in the 5-direction will drift through the coordinate system in the 5-direction. This is inertial dragging. Thus from the 5-dimensional point of view the constant a represents inertial dragging due to the motion of the central body moving in the 5-direction. Hence *the central body moving around the fifth direction causes an inertial dragging field. The projection of this field into 4-dimensional spacetime is interpreted from our point of view as the electric field due to the charge of this central body.*

Let us conclude with a speculation. One may wonder whether it is possible to use of our knowledge of quantum electrodynamics within the frame of Kaluza–Klein theory to shed some light upon how we should quantize gravity. In what way should one quantize gravity in order that the projection of quantized gravity into our 4-dimensional spacetime shall give us the quantum electrodynamic theory? The inertial dragging field may play a role in this programme.

References

1. Kaluza T.: Zum Unitätsproblem in der Physik. Sitzungsber. Preuss. Akad. Wiss. Berlin. Math. Phys. 966–972 (1921)
2. Klein, Q.: Quantentheorie und fünfdimensionale Relativitätstheorie. A **37**(12), 895–906 (1926)
3. Leibowitz, E., Rosen, N.: Five-dimensional relativity theory. Gen. Rel. Grav. **4**, 449–474 (1973)
4. Grøn, Ø.: Classical Kaluza-Klein description of the hydrogen atom. Nuovo Cimento **B91**, 57–66 (1986)

Solutions to the Exercises

Solutions for Chapter 1

1.1 *A tidal force pendulum*

Two particles each with mass m are connected by a rigid rod of length $2\,l$. The system is free to oscillate in any vertical plane about its centre of mass. The mass of the rod is negligible relative to m. The pendulum is at a distance R from the centre of a spherical distribution of a spherical distribution of matter with mass M (see Fig. 1.10).

$$|\mathbf{l} \times (\mathbf{F}_1 - \mathbf{F}_2)| = I\ddot{\theta}$$

where $I = 2ml^2$ is the moment of inertia of the pendulum.

By Newton's law of gravitation

$$\mathbf{F}_1 = -GMm\frac{\mathbf{R}+\mathbf{l}}{|\mathbf{R}+\mathbf{l}|^3}, \quad \mathbf{F}_2 = -GMm\frac{\mathbf{R}-\mathbf{l}}{|\mathbf{R}-\mathbf{l}|^3}$$

Thus

$$GMm\frac{\mathbf{l} \times \mathbf{R}}{|\mathbf{R}-\mathbf{l}|^3} - GMm\frac{\mathbf{l} \times \mathbf{R}}{|\mathbf{R}+\mathbf{l}|^3} = 2ml^2\ddot{\theta}.$$

It is seen from Fig. 1.10 that

$$|\mathbf{l} \times \mathbf{R}| = lR\sin\theta.$$

It is now assumed that $l \ll R$. Then we have to first order in l/R.

$$\frac{1}{|\mathbf{R}-\mathbf{l}|^3} - \frac{1}{|\mathbf{R}+\mathbf{l}|^3} = \frac{6l}{R^4}\cos\theta.$$

© Springer Nature Switzerland AG 2020
Ø. Grøn, *Introduction to Einstein's Theory of Relativity*,
Undergraduate Texts in Physics, https://doi.org/10.1007/978-3-030-43862-3

The equation of motion of the pendulum now takes the form

$$2\ddot{\theta} + \frac{3GM}{R^3}\sin 2\theta = 0.$$

This is the equation of motion of a simple pendulum in the variable 2θ instead of as usual θ as a variable. The equation shows that the pendulum oscillates about a vertical equilibrium position. The reason for 2θ instead of θ is that the tidal force pendulum is invariant under a change $\theta \to \theta + \pi$ while the simple pendulum is invariant under a change $\theta \to \theta + 2\pi$.

Assuming small angular displacements leads to

$$\ddot{\theta} + \frac{3GM}{R^3}\theta = 0.$$

This is the equation of a harmonic oscillator with period

$$T = 2\pi \left(\frac{R^3}{3GM}\right)^{1/3}.$$

Note that the period of the tidal force pendulum is independent of its length. This means that tidal forces can be observed on a system of arbitrarily small size. Also, from the equation of motion it is seen that in a uniform field, where $\mathbf{F}_1 = \mathbf{F}_2$, the pendulum does not oscillate.

The acceleration of gravity at the position of the pendulum is $g = GM/R^2$, so that the period of the tidal pendulum may be written

$$T = 2\pi \left(\frac{R}{3g}\right)^{1/2}.$$

The mass of a spherical body with density ρ is $M = (4\pi/3)\rho R^3$, which gives for the period of the tidal pendulum at its surface

$$T = \left(\frac{\pi G}{\rho}\right)^{1/2}.$$

Hence, the period depends only upon the density of the body. For a pendulum at the surface of the Earth the period is about 50 min. The region in spacetime needed in order to measure the tidal force is not arbitrarily small.

1.2 Newtonian potential for a spherically symmetric body

(a) We can find the potential using Gauss' law

$$\int_{\partial V} \vec{F}(r)d\vec{S} = \int_V \nabla \cdot \vec{F}(r)dV,$$

where the integration on the left-hand side is over a spherical surface (which may be inside or outside the mass shell), and the integral on the right-hand side is over the volume enclosed by this surface. Using Newton's law of gravity as formulated locally, we have

$$\nabla \cdot \vec{F}(\vec{r}) = -m\nabla^2\phi(\vec{r}) = -4\pi m G\rho(\vec{r}),$$

where $\phi(r)$ is the gravitational potential, $\rho(\vec{r})$ is the mass density at a point with position vector $\vec{r}$, and m is the mass of a test particle at this position. Inserting this into Gauss' law gives

$$4\pi r^2 F(r) = -4\pi Gm \int \rho(\vec{r})dV$$

The value of the integral at the right-hand side depends upon whether the boundary surface of the integration volume is inside or outside the mass distribution. We get

$$\vec{F}(\vec{r}) = \begin{cases} -\frac{GMm}{r^3}\vec{r} & r \geq R \\ 0 & r < R \end{cases}.$$

Defining zero potential infinitely far from the mass distribution, the potential at a finite distance r from the centre is found by calculating the (negative) work performed to move a particle downwards from the zero level to a distance r from the centre,

$$\phi(\vec{r}) = -\frac{1}{m}\int_\infty^r \vec{F} \cdot d\vec{s} = \begin{cases} -\int_\infty^r \frac{GM}{r^2}dr = -\frac{GM}{r} & r \geq R \\ \int_\infty^R \frac{GM}{r^2}dr = -\frac{GM}{R} & r < R \end{cases}.$$

(b) To calculate the gravitational potential outside an inside a sphere of constant density we use the same procedure as in point (a). Outside the sphere the result becomes the same as in point (a). Inside the sphere we now get

$$\int_0^r \rho(\vec{r})dV = \frac{4\pi}{3}\rho r^3 = M\frac{r^3}{R^3}.$$

Hence,

$$4\pi r^2 F(r) = -4\pi \frac{GMmr^3}{R^3}.$$

The gravitational force outside and inside the sphere is then

$$\vec{F}(\vec{r}) = \begin{cases} -\frac{GMm}{r^3}\vec{r} & r \geq R \\ -\frac{GMm}{R^3}\vec{r} & r < R \end{cases}.$$

This leads to the gravitational potential

$$\phi(\vec{r}) = \begin{cases} -\frac{GM}{r} & r \geq R \\ -\frac{GM(3R^2-r^2)}{2R^3} & r < R \end{cases}.$$

1.3 *Frictionless motion in a tunnel through the Earth*

In order to find the equation of motion of a particle moving without friction in the tube under the action of gravity only, we insert the expression for the gravitational force inside a sphere from Exercise 1.2 into Newton's 2. law and get

$$m\ddot{r} = -\frac{GMmr}{R^3}.$$

This is the equation for harmonic oscillations with period

$$T = 2\pi \sqrt{\frac{R^3}{GM}}.$$

This is just $\sqrt{3}$ times the period of the tidal force pendulum at the surface of the Earth (see Exercise 1.1). The period is 1 h and 24 min.

(b) The figure illustrates the situation with a tube not passing through the centre of the Earth, but having a closest distance s from the centre.

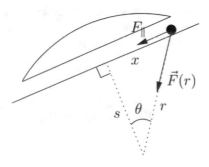

Here $x = r \sin \theta$ and the component of the gravitational force upon the direction of motion of the particle is

$$F_{\parallel} = F(r)\sin\theta = -\frac{GMmr}{R^3}\sin\theta = -\frac{GMm}{R^3}x.$$

Hence the equation of motion takes the form

$$m\ddot{x} = -\frac{GMm}{R^2}x.$$

This gives the same period as when the tube passes the centre of the Earth, showing that the period does not depend upon the direction of the tube. Using this procedure (and neglecting friction and the rotation of the Earth) it takes about 42 min to travel from an arbitrary point on the surface on the Earth to whatever other point on the Earth, and the traveller would be weightless during all of the travel.

(c) The equation of motion of a satellite moving in a circular path around the Earth at the surface of the Earth takes the form

$$\frac{GMm}{R^2} = m\frac{v^2}{R},$$

where v is the velocity of the satellite. Hence, the velocity is

$$v = \sqrt{\frac{GM}{R}}.$$

Since the magnitude of the velocity is constant, the time taken to move around the Earth is

$$T_C = \frac{2\pi R}{v} = 2\pi\sqrt{\frac{R^3}{GM}},$$

which is equal to the period of the frictionless motion through the tube.

1.4 The Earth-Moon System

(a) The Earth and the Moon have positions $\vec{r}_J$, $\vec{r}_M$ and masses M_J, M_M, respectively, as shown in the figure.

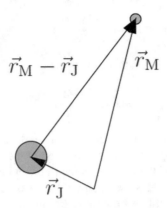

The equations of motion for the Earth and the Moon are

$$M_J \ddot{\vec{r}}_J = -M_M \ddot{\vec{r}}_M = \frac{GM_J M_M}{|\vec{r}_M - \vec{r}_J|^3}(\vec{r}_M - \vec{r}_J).$$

The position of the centre of mass is

$$\vec{r}_{CM} = \frac{M_J \vec{r}_J + M_M \vec{r}_M}{M_J + M_M}.$$

Since there are no external forces on the Earth-Moon system the mass centre is at rest. We may therefore choose a coordinate system co-moving with an inertial reference frame, in which the mass centre is at the origin, $M_J \vec{r}_J = -M_M \vec{r}_J$.

We shall show that there is a solution where the Earth and the Moon are moving in circular orbits around their common centre of mass. For the Earth such a motion may be represented by

$$x_J(t) = A \sin(\omega t), \quad y_J(t) = A \cos(\omega t).$$

where x_J and y_J are the components of the position vector $\vec{r}_J$, and A is the distance of the Earth from the mass centre. Hence the motion of the Moon is given by

$$x_M(t) = -\frac{M_J}{M_M} A \sin(\omega t), \quad y_M(t) = -\frac{M_J}{M_M} A \cos(\omega t).$$

The x-component of the equation of motion of the Earth can now be written

$$M_J \omega^2 A \sin(\omega t) = \frac{GM_J M_M}{|\vec{r}_M - \vec{r}_J|^3} A \left(1 + \frac{M_J}{M_M}\right) \sin(\omega t).$$

Giving

$$\omega^2 = \frac{GM_M}{|\vec{r}_M - \vec{r}_J|^3}\left(1 + \frac{M_M}{M_J}\right).$$

Furthermore

$$|\vec{r}_M - \vec{r}_J| = \sqrt{(x_M - x_J)^2 + (y_M - y_J)^2} = A\left(1 + \frac{M_J}{M_M}\right).$$

Hence the equation of motion is fulfilled for the circular path if

$$\omega^2 A^3 = \frac{GM_M}{(1 + M_J/M_M)^2}.$$

Inserting the values of the gravitational constant and the masses of the Earth and the Moon together with the period of the Moon's motion, $T = 27.3$ days, we get $\omega = 2\pi/T = 2.66 \times 10^{-6} s^{-1}$. Hence the radius of the circular motion of the Earth is $A = 4670$ km, and that of the Moon's motion $(M_J/M_M)A = 3.8 \times 10^5$ km.

(b) The gravitational potential of the Earth–Moon system is the sum of the potential of each of the bodies. Along the line connecting the Earth and the Moon we get

$$\phi(r) = -\frac{GM_J}{r} - \frac{GM_M}{R - r},$$

where r is the distance from the centre of the Earth, and R is the distance between the centres of the Earth and the Moon. The gravitational force upon a test particle with mass m is

$$F(r) = -\frac{GM_J m}{r^2} + \frac{GM_M m}{(R - r)^2}.$$

This force vanishes at two points given by

$$r = \frac{M_J \pm \sqrt{M_J M_M}}{M_J - M_M}R.$$

Inserting the Masses of the Earth and the Moon and the distance $R = 3.8 \times 10^5$ km gives $r_1 = 3.46 \times 10^5$ km and $r_2 = 4.32 \times 10^5$ km. The first one is between the Earth and the Moon, and the second one is beyond the Moon.

(c) The difference between the Moon's attraction upon a particle with mass m at a point on the surface of the Earth closest to the Moon and at the most remote point is

$$\Delta F = \frac{GM_M m}{(R - R_E)^2} - \frac{GM_M m}{(R + R_E)^2} \approx 4GM_M m \frac{R_E}{R^3},$$

where R_E is the radius of the Earth. This gives $\Delta F = 2.2 \times 10^{-6} N$.

1.5 *The Roche limit*

(a) A Moon with mass m moves around a planet with mass M. A stone with mass μ on the surface of the Moon is acted upon by the gravitational field of the planet and the Moon. Calculating with absolute values, the gravitational force acting upon the stone towards the planet is equal to the gravitational force due to the planet minus the gravitational force due to the Moon,

$$F = \frac{GM\mu}{(r-R)^2} - \frac{Gm\mu}{R^2},$$

where R is the radius of the Moon and r the distance between the centres of mass of the Moon and the planet. The centripetal acceleration of the stone is

$$a = \frac{F}{\mu} = \frac{GM}{(r-R)^2} - \frac{Gm}{R^2} = \frac{GM}{r^2}\left(1 - \frac{R}{r}\right)^{-2} - \frac{Gm}{R^2}.$$

Since $R \ll r$ we can use the series expansion

$$\left(1 - \frac{R}{r}\right)^{-2} \approx 1 + \frac{2R}{r},$$

giving

$$a \approx \frac{GM}{r^2} + \frac{2GMR}{r^3} - \frac{Gm}{R^2}.$$

The Moon has a centripetal acceleration

$$a_0 = \frac{GM}{r^2}.$$

In order that a stone at the point of the surface of the Moon closest to the Earth shall not be lifted up from the surface of the Moon, the centripetal acceleration of the stone cannot be larger than the centripetal acceleration of the Moon. So the acceleration of the stone must obey the condition $a \leq a_0$. Inserting the expressions above this gives

$$\frac{2MR}{r^3} \leq \frac{m}{R^2},$$

or

$$r \geq \left(\frac{2M}{m}\right)^{1/3} R.$$

The right-hand expression is called the *Roche limit*. If a Moon is inside the Roche limit of the Moon-planet system, matter will be drawn from the surface of the Moon towards the planet. Similarly, if a planet is inside the Roche limit of the planet-star system matter will be drawn from the planet onto the star. Or if a star is inside the Roche limit of a system consisting of the star and for example a super massive black hole at the centre of a galaxy, matter will be drawn from the star into the black hole. This is an essential part of the mechanism behind *quasars*.

(b) From the observations of what happened to the comet Shoemaker–Levy 9 when it passed Jupiter in 1992, it is reasonable to assume that the smallest distance of the comet from Jupiter was equal to the Roche limit of the comet-planet system. Then the radius of the comet nucleus is given by

$$R = \left(\frac{m}{2M}\right)^{1/3} s.$$

Inserting $m = 2.0 \times 10^{12}$ kg, $M = 1.9 \times 10^{27}$ kg and $s = 96,000$ km gives $R = 775$ m.

Solution for Chapter 2

2.1 Robb's Lorentz invariant spacetime interval formul

Consider Fig. 2.2. We shall find a formula for the invariant spacetime interval Δs_{AB} between the emission event A and the reflection event B of a light signal reflected by a mirror, in terms of the emission point of time t_A and the point of time t_C when the reflected light signal arrives back at the emission position.

Our point of departure is the usual Formula (2.41) for the invariant spacetime interval, the Einstein synchronization Formula (2.1) and Eq. (2.2). Hence we have

$$\Delta t_{AB} = \frac{1}{2}(t_A + t_C), \quad \Delta x_{AB} = \frac{c}{2}(t_C - t_A),$$

and

$$\Delta s_{AB}^2 = c^2 \Delta t_{AB}^2 - \Delta x_{AB}^2 = \frac{c^2}{4}(t_A + t_C)^2 - \frac{c^2}{4}(t_C - t_A)^2 = c^2 t_A t_C.$$

Hence we arrive at Robb's formula

$$\underline{\Delta s_{AB} = c\sqrt{t_A t_C}.}$$

2.2 The twin paradox

(a) In the reference frame of the Earth A travels with a velocity $v = 0.8c$. B remains at the Earth. Hence according to B the travel takes 8 light years/0.8 c = 10 years. Hence B sends 10 greetings.

Let us now describe the situation as seen from the reference frame of A. The twin A thinks of himself as at rest while the Earth and Alpha Centauri travels with the velocity $v = 0.8c$. Hence the distance between the Earth and Alpha Centauri becomes length contracted to $L_A = L\sqrt{1 - v^2/c^2} = 2.4\,l.y.$ The time taken for Alpha Centauri to arrive at A is therefore $2.41\,y/0.8\,c = 3$ years. The return trip takes the same time, so A sends 6 greetings.

(b) The figure shows the world lines of the twins and the signals.

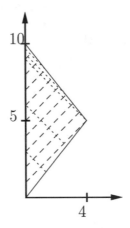

The figure shows that the twin A receives his first greeting at the moment he arrives at Alpha Centauri. He receives the 9 other signals from B at the return travel.

(c) The figure shows that B receives the first greeting from A after 3 years, the next after 6 years and the last year he receives the 3 signals A send during the return travel.

(d) Due to the Doppler effect a light signal sent by A with a frequency v_A is measured by B to have a frequency

$$v_B = \sqrt{\frac{1 - v/c}{1 + v/c}}\, v_A.$$

Hence, if the emitted frequency is 1/year, the received frequency is 1/3 years, and during the return travel the received frequency is 3 signals per year.

2.3 Faster than the speed of light?

(a) The situation is schematically illustrated on the Figure. The radiation source is moving from A towards B with a velocity v_0. The arrows at the bottom of the figure point towards an observer at the Earth. Here θ is the angle between the direction of motion of the light source and the direction of sight. We now consider light emit at A and B. The point A′ has the same distance from the Earth as B. We see from the figure that

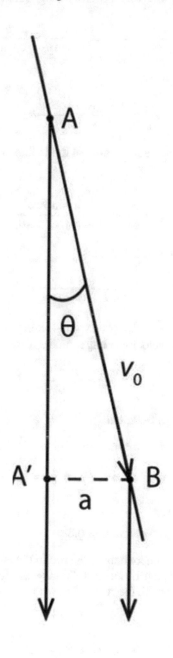

$$AB = \frac{a}{\sin\theta}, \quad AA' = \frac{a}{\tan\theta}.$$

The difference in arrival time on the Earth for light emitted from A and B is

$$\Delta t = \frac{AB}{v_0} - \frac{AA'}{c} = \frac{a}{v_0\sin\theta} - \frac{a}{c\tan\theta},$$

giving

$$\Delta t = \frac{1 - (v_0/c)\cos\theta}{v_0\sin\theta}a.$$

(b) The observed transverse velocity normal to the line of sight is

$$v = \frac{a}{\Delta t} = \frac{v_0\sin\theta}{1 - (v_0/c)\cos\theta}$$

Hence, the velocity of the light source is

$$v_0 = \frac{v}{\sin\theta + (v/c)\cos\theta}.$$

Inserting $v = 10c$ and $\theta = 10^0$ gives $v_0 = 0.998c$.
The observed transverse velocity is larger than c if

$$v_0 > \frac{c}{\sin\theta + \cos\theta}.$$

with $\theta = 10^0$ this requires $v_0 > 0.86c$.

2.4 *Time dilation and Lorentz contraction*

(a) Theformula for the time dilation is $\frac{\Delta t}{\Delta t_0} = \frac{1}{\gamma} = \sqrt{1 - \frac{v^2}{c^2}}$. Hence $\frac{1}{\gamma^2} = 1 - \frac{v^2}{c^2}$
  and $v = c\sqrt{1 - \frac{1}{\gamma^2}}$.

  In the present case $\frac{1}{\gamma} = 0.6$ giving $v = 0.8c$.

(b) Let L_x and L_y be the components of the rod in the x- and y-directions, respectively, as measured by an observer following the rod. As measured by this observer the angle θ_0 which the rod makes with the x-axis, is given by

$$\tan\theta_0 = \frac{L_y}{L_x}$$

As measured by an observer at rest on the x-axis the component in the y-direction is unchanged, and the component in the x-direction is L_x/γ. Hence this observer measures that the rod makes an angle θ given by

$$\tan \theta = \gamma \frac{L_y}{L_x}.$$

Thus

$$\frac{1}{\gamma} = \sqrt{1 - \frac{v^2}{c^2}} = \frac{\tan \theta_0}{\tan \theta},$$

giving

$$v = c\sqrt{1 - \left(\frac{\tan \theta_0}{\tan \theta}\right)^2}.$$

With $\theta_0 = \pi/4$ and $\theta = \pi/3$ we get $v = \sqrt{2/3}c \approx 0.82c$.

2.5 *Atmospheric mesons reaching the surface of the Earth.*

(a) The travel time of a muon is $T = \frac{L}{v} = 34 \times 10^{-6}$s, or 21.8 half lives. Thus the survival rate is $I/I_0 = 2^{-21.8} \approx 0.3 \times 10^{-6}$. Hence only 3 of ten million created muons in the upper atmosphere reaches the surface of the Earth.

(b) Due to the relativistic time dilation the actual half life of the travelling muons is $t = \gamma t_0 = \frac{t_0}{\sqrt{1-\frac{v^2}{c^2}}} \approx 5t_0 - 7.8 \times 10^{-6}$s. Then the survival rate is 0.049 meaning that 4900 of ten million muons reach the surface of the Earth.

(c) As observed by a co-moving observer with a muon its half life is $1.56 \cdot 10^{-6}$s. But as measured by this observer the distance is Lorenz contracted to $L = (1/\gamma)L_0 = 2$ km. Hence the survival rate is in agreement with the result of the Earth-bound observer.

2.6 *Relativistic Doppler effect*

Solving the equation for the redshift with respect to the velocity we get

$$v = \frac{(1+z)^2 - 1}{(1+z)^2 + 1}.$$

Since $|z| \ll 1$ in the present case, we can use the approximation $(1+z)^2 \approx 1 + 2z$. This gives to 1 order in z, $v \approx zc$. For the centre of the Andromeda galaxy $z = -0.0004$, giving $v = -120$ km/s. This means that the Andromeda galaxy moves towards the Milky Way galaxy with this velocity.

2.7 *The velocity of light in a moving medium*

The velocity of light in a medium with index of refraction, n, moving with velocity v is

$$u = \frac{u_0 + v}{1 + \frac{u_0 v}{c^2}} = \frac{\frac{c}{n} + v}{1 + \frac{1}{n}\frac{v}{c}} = \frac{c + nv}{nc + v}c.$$

This formula is exact. In most applications $v \ll c$ and one uses an approximate formula,

$$u = \frac{c}{n}\frac{1 + \frac{nv}{c}}{1 + \frac{v}{vc}} \approx \frac{c}{n}\left(1 + n\frac{v}{c}\right)\left(1 - \frac{1}{n}\frac{v}{c}\right) \approx \frac{c}{n} + v\left(1 - \frac{1}{n^2}\right).$$

2.8 *Cherenkov radiation*

The geometry of the light cone formed when a source of light passes through a medium having index of refraction n with a velocity v greater than that of light in the medium is shown in the Figure (From Wikipedia: https://en.wikipedia.org/wiki/Cherenkov_radiation#/media/File:Cherenkov.svg

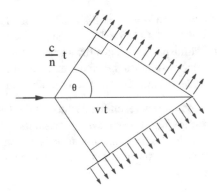

We see from the figure that the angle θ is given by $\cos\theta = \frac{c}{nv}$.

(a) In order that the electron shall emit Cherenkov radiation it must move through the glass with a velocity $v > c/n$. Hence the threshold kinetic energy in order that it shall emit Cherenkov radiation is

$$K = \left(\frac{1}{\sqrt{1 - v^2/c^2}} - 1\right)m_e c^2 = \left(\frac{n}{\sqrt{n^2 - 1}} - 1\right)m_e c^2.$$

Inserting the given data gives $\underline{K = 0.29 \text{ MeV}}$.

(b) The limiting velocity for increasing energy is $v \to c$. In this limit the angle θ approaches θ_{max} given by $\cos \theta_{max} = \frac{1}{n}$. From the figure is seen that the corresponding maximal half angle is given by $\sin \phi_{max} = \frac{1}{n}$, Inserting the refractive index for glass gives $\phi_{max} \approx 50°$.

2.9 Relativistic form of Newton's 2. law

Using the chain rule of differentiation we first differentiate $\gamma = \left(1 - v^2/c^2\right)^{-1/2}$,

$$\frac{d\gamma}{dt} = -\frac{1}{2}\left(1 - v^2/c^2\right)^{-1/2}\left(-2\frac{v}{c^2}\right)\frac{dv}{dt} = \gamma^3 \frac{v}{c^2}\frac{dv}{dt}.$$

The momentum of the particle is $p = \gamma m_0 v$. Differentiating we get

$$\frac{1}{m_0}\frac{dp}{dt} = v\frac{d\gamma}{dt} + \gamma\frac{dv}{dt} = \gamma^3\frac{v^2}{c^2}\frac{dv}{dt} + \gamma\frac{dv}{dt} = \left(\gamma^2\frac{v^2}{c^2} + 1\right)\gamma\frac{dv}{dt} = \gamma^3\frac{dv}{dt}.$$

Hence Newton's 2. law has the relativistic form

$$F = \gamma^3 m_0 \frac{dv}{dt}.$$

2.10 Lorentz transformation of electric and magnetic fields

We write out Eqs. (5.281) and (5.282) in component form

$$\frac{\partial B_x}{\partial x} + \frac{\partial B_y}{\partial y} + \frac{\partial B_z}{\partial z} = 0,$$

$$\frac{\partial E_z}{\partial y} - \frac{\partial E_y}{\partial z} = -\frac{\partial B_x}{\partial t}, \quad \frac{\partial E_x}{\partial z} - \frac{\partial E_z}{\partial x} = -\frac{\partial B_y}{\partial t}, \quad \frac{\partial E_y}{\partial x} - \frac{\partial E_x}{\partial y} = -\frac{\partial B_z}{\partial t}.$$

Inserting the transformations of the partial derivatives into the y-component of the second equation, we get

$$\frac{\partial E_x}{\partial z'} - \gamma\left(\frac{\partial E_z}{\partial x'} + \frac{v}{c^2}\frac{\partial E_z}{\partial t'}\right) = -\left(\frac{\partial B_y}{\partial t'} + v\frac{\partial B_y}{\partial x'}\right).$$

This equation shall be compared with the similar Maxwell equation in the marked coordinate system,

$$\frac{\partial E'_x}{\partial z'} - \frac{\partial E'_z}{\partial x'} = -\frac{\partial B'_y}{\partial t'}.$$

Hence, we rearrange the above equation as follows

$$\frac{\partial E'_x}{\partial z'} - \frac{\partial}{\partial x'}\gamma\left(E_z - cB_y\right) = -\frac{\partial}{\partial t'}\gamma\left(B_y - \frac{v}{c^2}E_z\right).$$

Requiring equivalence of these equations gives the transformation equations

$$E'_x = E_x, \; E'_z = \gamma\left(E_z - vB_y\right), \; B'_y = \gamma\left(B_y - \frac{v}{c^2}E_z\right).$$

Using all the Maxwell-equations in this way we end up with the transformation equations of the electric and magnetic field

$$
\begin{aligned}
E'_x &= E_x, & B'_x &= B_x, \\
E'_y &= \gamma\left(E_y + vB_z\right), & B'_y &= \gamma\left(B_y - \tfrac{v}{c^2}E_z\right) \\
E'_z &= \gamma\left(E_z - vB_y\right), & B'_z &= \gamma\left(B_z + \tfrac{v}{c^2}E_y\right)
\end{aligned}
$$

Solutions for Chapter 3

3.1 *Four-vectors*

(a) The scalar products of the vectors with themselves are

$$
\begin{aligned}
\vec{A}\cdot\vec{A} &= \left(4\vec{e}_t + 3\vec{e}_x + 2\vec{e}_y + \vec{e}_z\right)\cdot\left(4\vec{e}_t + 3\vec{e}_x + 2\vec{e}_y + \vec{e}_z\right) = -2, \\
\vec{B}\cdot\vec{B} &= \left(5\vec{e}_t + 4\vec{e}_x + 3\vec{e}_y\right)\cdot\left(5\vec{e}_t + 4\vec{e}_x + 3\vec{e}_y\right) = 0, \\
\vec{C}\cdot\vec{C} &= \left(\vec{e}_t + 2\vec{e}_x + 3\vec{e}_y + 4\vec{e}_z\right)\cdot\left(\vec{e}_t + 2\vec{e}_x + 3\vec{e}_y + 4\vec{e}_z\right) = 28
\end{aligned}
$$

(b) We now assume that $\vec{A}\cdot\vec{B} = 0$. If $\vec{A}$ is time-like, i.e. $\vec{A}\cdot\vec{A} < 0$, we may chose a basis where $\vec{A}$ points in the time direction, $\vec{A} = A^{t'}\vec{e}_{t'}$. Then $\vec{A}\cdot\vec{B} = -A^{t'}B^{t'}$. Hence, the condition $\vec{A}\cdot\vec{B} = 0$ requires that $B^{t'} = 0$, which means that $\vec{B}$ is space-like.

If $\vec{A}$ is light-like, then $\vec{A}\cdot\vec{A} = 0$. Assume now that $\vec{B}$ is time-like. Hence, we can chose a basis where $\vec{B} = B^{t'}\vec{e}_{t'}$. This gives $\vec{A}\cdot\vec{B} = -A^{t'}B^{t'}$. The vector component $A^{t'}$ cannot vanish since $\vec{A}$ is light-like. This means that $\vec{A}\cdot\vec{B} \neq 0$ in contradiction to the initial assumption. Thus $\vec{B}$ cannot be time-like.

If both $\vec{A}$ and $\vec{B}$ are light-like we may choose a basis where $\vec{A} = A(\vec{e}_{t'} + \vec{e}_{x'})$. Then $\vec{A}\cdot\vec{B} = A(-B^{t'} + B^{x'})$. In this case $\vec{A}\cdot\vec{B} = 0$ implies that $B^{x'} = B^{t'}$, giving $\vec{B}\cdot\vec{B} = \left(B^{y'}\right)^2 + \left(B^{z'}\right)^2$. Since $\vec{B}$ is light-like $\vec{B}\cdot\vec{B} = \left(B^{y'}\right)^2 + \left(B^{z'}\right)^2 = 0$ which requires $B^{y'} = B^{z'}$. Hence $\vec{B} = B^{t'}(\vec{e}_{t'} + \vec{e}_{x'})$ showing that the vectors $\vec{A}$ and $\vec{B}$ have the same direction; i.e. they are proportional to each other.

If $\vec{A}$ is space-like we may chose a basis in which only one of the spatial components of the vectors are is different from zero, say $\vec{A} = A^{x'}\vec{e}_{x'}$. This gives $\vec{A}\cdot\vec{B} = A^{x'}B^{x'}$, and hence $B^{x'} = 0$. There are, however, no further requirements on the other components of $\vec{B}$. This means that the vector $\vec{B}$ can be either time-like, light-like or space-like.

(c) A Lorentz transformation along the x-axis has the form, measuring distance in light seconds so that $c = 1$,

$$t = \frac{t' + vx'}{\sqrt{1 - v^2}}, \quad x = \frac{x' + vt'}{\sqrt{1 - v^2}}, \quad y = y', \quad z = z'$$

where v is the relative velocity of the reference frames. We now define the velocity parameter θ by $\tanh \theta = v$. Using the relationships

$$\cosh \theta = \frac{1}{\sqrt{1 - \tanh^2 \theta}}, \quad \sinh \theta = \frac{\tanh \theta}{\sqrt{1 - \tanh^2 \theta}}$$

The Lorentz transformation takes the form

$$t = t' \cosh \theta + x' \sinh \theta, \quad x = x' \cosh \theta + t' \sinh \theta, \quad y = y', \quad z = z'.$$

Hence the transformation matrix becomes

$$\Lambda^{\mu}_{\mu'} = \frac{\partial x^{\mu}}{\partial x^{\mu'}} = \begin{pmatrix} \cosh \theta & \sinh \theta & 0 & 0 \\ \sinh \theta & \cosh \theta & 0 & 0 \\ 0 & 0 & 1 & 0 \\ 0 & 0 & 0 & 1 \end{pmatrix}.$$

(d) The 4-velocity $\vec{u}$ is a vector in 4-dimensional spacetime. This vector may be described by expressing its components in a chosen basis. But the vector itself exist independently of any basis. This is expressed mathematically by

$$\vec{u} = u^{\mu} \vec{e}_{\mu} = u^{\mu'} \vec{e}_{\mu'},$$

where the components u^{μ} and $u^{\mu'}$ depend upon the basis, but $\vec{u}$ is independent of it. The component of the 4-velocity in an arbitrary basis are given by

$$\vec{u} = \gamma(1, \vec{v}),$$

where $\vec{v} = (dx^i / dt)\vec{e}_i$ is the ordinary velocity in 3-space and $\gamma = (1 - v^2)^{-1/2}$. The 3-velocity is not a vector in 4-dimensional spacetime. It depends upon the reference frame of the observer.

It follows from the 4-velocity identity, $\vec{u} \cdot \vec{u} = -1$ that $\vec{u}$ is a time-like vector. Also the 4-momentum $\vec{p} = m\vec{u}$ is time-like. In the limit $m \to 0$ the 4-momentum becomes light-like.

The energy of the particle with rest mass m and 3-velocity $\vec{v}$ is $E = \gamma m$. The velocity of the observer is $\vec{u}_{obs} = (1, 0)$. Hence

$$E = -\vec{u}_{obs} \cdot \vec{p}.$$

This formula is valid independent of the basis.

3.2 Tensor product

(a) We get

$$\underline{\alpha} \otimes \underline{\beta}(\vec{e}_0, \vec{e}_1) = \alpha_0 \beta_1 = 0, \quad \underline{\beta} \otimes \underline{\alpha}(\vec{e}_0, \vec{e}_1) = \beta_0 \alpha_1 = -1.$$

Hence $\underline{\alpha} \otimes \underline{\beta} \neq \underline{\beta} \otimes \underline{\alpha}$.

(b) Calculation of all the components of $\underline{\alpha} \otimes \underline{\beta}$ give

$$\alpha_\mu \beta_\nu = \begin{pmatrix} -1 & 0 & 1 & 0 \\ -1 & 0 & 1 & 0 \\ 0 & 0 & 0 & 0 \\ 0 & 0 & 0 & 0 \end{pmatrix}.$$

The symmetric part of $\underline{\alpha} \otimes \underline{\beta}$ has components

$$\alpha_{(\mu} \beta_{\nu)} = \frac{1}{2}(\alpha_\mu \beta_\nu + \alpha_\nu \beta_\mu) = \frac{1}{2} \begin{pmatrix} -2 & -1 & 1 & 0 \\ -1 & 0 & 1 & 0 \\ 1 & 1 & 0 & 0 \\ 0 & 0 & 0 & 0 \end{pmatrix},$$

and the antisymmetric part of $\underline{\alpha} \otimes \underline{\beta}$ has components

$$\alpha_{[\mu} \beta_{\nu]} = \frac{1}{2}(\alpha_\mu \beta_\nu - \alpha_\nu \beta_\mu) = \frac{1}{2} \begin{pmatrix} 0 & 1 & 1 & 0 \\ -1 & 0 & 1 & 0 \\ -1 & -1 & 0 & 0 \\ 0 & 0 & 0 & 0 \end{pmatrix}.$$

3.5 Symmetric and antisymmetric tensors

We have

$$T^{\alpha\beta} = \begin{pmatrix} 0 & 1 & 0 & 0 \\ 1 & -1 & 0 & 2 \\ 2 & 0 & 0 & 1 \\ 1 & 0 & -2 & 0 \end{pmatrix}.$$

1. The symmetric part of this tensor is

$$T^{(\alpha\beta)} = \frac{1}{2}\left(T^{\alpha\beta} + T^{\beta\alpha}\right) = \frac{1}{2}\begin{pmatrix} 0 & 2 & 2 & 1 \\ 2 & -2 & 0 & 1 \\ 2 & 0 & 0 & -1 \\ 1 & 2 & -1 & 0 \end{pmatrix}.$$

The antisymmetric part of the tensor is

$$T^{[\alpha\beta]} = \frac{1}{2}\left(T^{\alpha\beta} - T^{\beta\alpha}\right) = \frac{1}{2}\begin{pmatrix} 0 & 0 & -2 & -1 \\ 0 & 0 & 0 & 1 \\ 2 & 0 & 0 & 3 \\ 1 & -1 & -3 & 0 \end{pmatrix}.$$

2. The mixed components of the tensor are

$$T^{\alpha}_{\beta} = \eta_{\beta\mu}T^{\alpha\mu} = \begin{pmatrix} 0 & 1 & 0 & 0 \\ -1 & -1 & 0 & 2 \\ -2 & 0 & 0 & 1 \\ -1 & 0 & -2 & 0 \end{pmatrix}.$$

3. The covariant components are

$$T_{\alpha\beta} = \eta_{\alpha\mu}\eta_{\beta\nu}T^{\mu\nu} = \begin{pmatrix} 0 & -1 & 0 & 0 \\ -1 & -1 & 0 & 2 \\ -2 & 0 & 0 & 1 \\ -1 & 0 & -2 & 0 \end{pmatrix}.$$

(b) It is not possible to define a symmetric or antisymmetric part of a mixed tensor because it is not allowed to exchange a vector by a form in the argument.

3.4 *Contractions of tensors with different symmetries*

Since **A** is antisymmetric we may write

$$A^{\alpha\beta} = \frac{1}{2}\left(A^{\alpha\beta} - A^{\beta\alpha}\right),$$

and since **B** is symmetric we may write

$$B_{\alpha\beta} = \frac{1}{2}\left(B_{\alpha\beta} + B_{\beta\alpha}\right).$$

Hence, we get

$$A^{\alpha\beta} B_{\alpha\beta} = \frac{1}{4}\left(A^{\alpha\beta} - A^{\beta\alpha}\right)\left(B_{\alpha\beta} + B_{\beta\alpha}\right)$$

$$= \frac{1}{4}\left(A^{\alpha\beta} B_{\alpha\beta} + A^{\alpha\beta} B_{\beta\alpha} - A^{\beta\alpha} B_{\alpha\beta} - A^{\beta\alpha} B_{\beta\alpha}\right),$$

$$A^{\alpha\beta} C_{\alpha\beta} = \frac{1}{2}\left(A^{\alpha\beta} - A^{\beta\alpha}\right)C_{\alpha\beta} = \frac{1}{2}\left(A^{\alpha\beta} C_{\alpha\beta} - A^{\beta\alpha} C_{\alpha\beta}\right)$$

$$= \frac{1}{2}\left(A^{\alpha\beta} C_{\alpha\beta} - A^{\alpha\beta} C_{\beta\alpha}\right) = A^{\alpha\beta} \frac{1}{2}\left(C_{\alpha\beta} - C_{\beta\alpha}\right) = A^{\alpha\beta} C_{[\alpha\beta]},$$

$$B_{\alpha\beta} D^{\alpha\beta} = \frac{1}{2}\left(B_{\alpha\beta} D^{\alpha\beta} + B_{\beta\alpha} D^{\alpha\beta}\right) = B_{\alpha\beta} \frac{1}{2}\left(D^{\alpha\beta} + D^{\beta\alpha}\right) = B_{\alpha\beta} D^{(\alpha\beta)}.$$

3.5 Coordinate transformation in an Euclidean plane

(a) The given transformation can be written

$$\begin{pmatrix} x' \\ y' \end{pmatrix} = \begin{pmatrix} 2 & -1 \\ 1 & 1 \end{pmatrix}\begin{pmatrix} x \\ y \end{pmatrix}.$$

This means that we have a transformation matrix $M^{i'}_{j}$ given by $x^{i'} = M^{i'}_{i} x^i$ which is

$$M^{i'}_{i} = \begin{pmatrix} 2 & -1 \\ 1 & 1 \end{pmatrix}.$$

The inverse transformation matrix is given by $M^{i'}_{i} M^{i}_{j'} = \delta^{i'}_{j'}$ and is

$$M^{i}_{j'} = \frac{1}{3}\begin{pmatrix} 1 & -1 \\ 1 & 2 \end{pmatrix}.$$

(b) The transformation of the basis vectors is $\vec{e}_{i'} = M^{i}_{i'}\vec{e}_i$, giving

$$\vec{e}_{x'} = (1/3)(\vec{e}_x - \vec{e}_y), \quad \vec{e}_{y'} = (1/3)(\vec{e}_x + 2\vec{e}_y).$$

(c) The scalar products of the basis vectors in the $\{x', y'\}$-system are

$$\vec{e}_{x'} \cdot \vec{e}_{x'} = \frac{1}{9}(\vec{e}_x - \vec{e}_y) \cdot (\vec{e}_x - \vec{e}_y) = \frac{2}{9}, \quad \vec{e}_{y'} \cdot \vec{e}_{y'} = \frac{1}{9}(\vec{e}_x + 2\vec{e}_y) \cdot (\vec{e}_x + 2\vec{e}_y)$$

$$= \frac{5}{9} \quad \vec{e}_{x'} \cdot \vec{e}_{y'} = \frac{1}{9}(\vec{e}_x - \vec{e}_y) \cdot (\vec{e}_x + 2\vec{e}_y) = -\frac{1}{9} = \vec{e}_{y'} \cdot \vec{e}_{x'}$$

Hence, in this system the covariant components of the metric tensor are

$$g_{i'j'} = \frac{1}{9}\begin{pmatrix} 2 & -1 \\ -1 & 5 \end{pmatrix}.$$

(d) The line-element in the $\{x', y'\}$-system is

$$ds^2 = g_{i'j'}dx^{i'}dx^{j'} = (1/9)(2dx'^2 + 5dy'^2 - 2dx'dy').$$

(e) The vectors $\vec{\omega}^{i'}$ can be expressed in terms of the Cartesian basis vectors $\vec{e}_x$ and $\vec{e}_y$ by

$$\vec{\omega}^{i'} = M_i^{i'}\vec{e}_i = \begin{pmatrix} 2 & -1 \\ 1 & 1 \end{pmatrix}\begin{pmatrix} \vec{e}_x \\ \vec{e}_y \end{pmatrix},$$

giving

$$\vec{\omega}^{x'} - 2\vec{e}_x - \vec{e}_y, \quad \vec{\omega}^{y'} - \vec{e}_x + \vec{e}_y.$$

The scalar products of these vectors are

$$\vec{\omega}^{x'} \cdot \vec{\omega}^{x'} = (2\vec{e}_x - \vec{e}_y) \cdot (2\vec{e}_x - \vec{e}_y) = 5, \quad \vec{\omega}^{y'} \cdot \vec{\omega}^{y'} = (\vec{e}_x + \vec{e}_y) \cdot (\vec{e}_x + \vec{e}_y) = 2$$
$$\vec{\omega}^{y'} \cdot \vec{\omega}^{x'} = \vec{\omega}^{x'} \cdot \vec{\omega}^{y'} = (2\vec{e}_x - \vec{e}_y) \cdot (\vec{e}_x + \vec{e}_y) = 1$$

Hence, the contravariant components of the metric tensor in the $\{x', y'\}$-system are

$$g^{i'j'} = \begin{pmatrix} 5 & 1 \\ 1 & 2 \end{pmatrix}.$$

Solutions for Chapter 4

4.1 *Relativistic rotating disk*

We use units so that the velocity of light is $c = 1$.

(a) Let dl be the physical distance measured by standard measuring rods at rest in the reference frame between two points with coordinates (r, ϕ) and $(r + dr, \phi + d\phi)$, where (r, ϕ) are co-moving coordinates with a frame RF rotating with angular velocity ω. A radial distance $dl_r = dr$ is equal to that of the non-rotating inertial system IF of the axis. A physical distance, In the tangential direction the standard measuring rods will be Lorentz contracted, and therefore the measured distance will be larger in RF than in IF, $dl_\phi = (1 - r^2\omega^2)^{-1/2}rd\phi$.

Using the Pythagorean rule the square of the distance of an interval with both a radial and tangential component is

$$dl^2 = dl_r^2 + dl_\phi^2 = dr^2 + \frac{r^2}{1 - r^2\omega^2} d\phi^2.$$

Hence

$$f_1(r, \phi) = 1, \quad f_2(r, \phi) = \frac{r^2}{1 - r^2\omega^2}.$$

The distance in the radial direction between the axis and a point with coordinates R is $l_r = R$. The distance around a circle with this radius is

$$l_\phi = \frac{2\pi R}{\sqrt{1 - R^2\omega^2}}.$$

Since the observer finds that $l_\phi \neq 2\pi R$ he will conclude that the geometry is not Euclidean. On a positively curved surface (for example a part of a spherical surface) the periphery is less than $2\pi R$ and on a part of a negatively curves surface (saddle like) it is larger. Hence the observer concludes that the simultaneity surface with constant z is negatively curved in the rotating frame.

(b) In order to express the line-element

$$ds^2 = -dt^2 + dx^2 + dy^2$$

in terms of $\tilde{t}, \tilde{x}$ and $\tilde{y}$ we must first use Eqs. (4.54) to express t, x and y in terms of $\tilde{t}, \tilde{x}$ and $\tilde{y}$. This gives $t = \tilde{t}$ and

$$x = r\cos(\omega t + \phi) = r\cos\phi\cos(\omega t) - r\sin\phi\sin(\omega t) = \tilde{x}\cos(\omega t) - \tilde{y}\sin(\omega t),$$
$$y = r\sin(\omega t + \phi) = r\sin\phi\cos(\omega t) + r\cos\phi\sin(\omega t) = \tilde{y}\cos(\omega t) + \tilde{x}\sin(\omega t).$$

The differentials are $dt = d\tilde{t}$ and

$$dx = \cos(\omega\tilde{t})d\tilde{x} - \sin(\omega\tilde{t})d\tilde{y} + \left(-\omega\tilde{x}\sin(\omega\tilde{t}) - \omega\tilde{y}\cos(\omega\tilde{t})\right)d\tilde{t},$$
$$dy = \sin(\omega\tilde{t})d\tilde{x} - \cos(\omega\tilde{t})d\tilde{y} + \left(-\omega\tilde{y}\sin(\omega\tilde{t}) - \omega\tilde{x}\cos(\omega\tilde{t})\right)d\tilde{t}.$$

Squaring and inserting the result into the line-element above gives the form of the line-element in the rotating frame

$$ds^2 = -\left[1 - (\tilde{x}^2 + \tilde{y}^2)\omega^2\right]d\tilde{t}^2 + d\tilde{x}^2 + d\tilde{y}^2 - 2\tilde{y}\omega d\tilde{t}d\tilde{x} + 2\omega\tilde{x}d\tilde{t}d\tilde{y}.$$

Fig. 4.12 In a rotating frame
the path of a light signal
emitted from the axis is an
Archimedian spiral

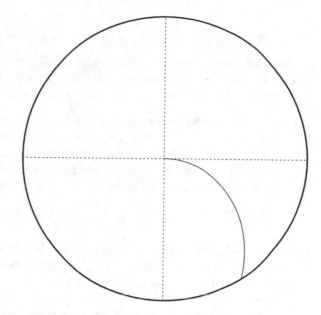

(c) A light signal is emitted from the axis of RF. As observed from IF the light moves
along a straight radial curve in 3-space, say along the x-axis, $x = t$, $y = 0$.
Inverting the coordinate transformation above we get

$$\tilde{x} = x\cos(\omega t) + y\sin(\omega t), \quad \tilde{y} = -x\sin(\omega t) + y\cos(\omega t).$$

Inserting $x = t$, $y = 0$ gives

$$\tilde{x} = t\cos(\omega t), \quad \tilde{y} = t\sin(\omega t).$$

This is the equation of an Archimedian spiral in parametric for, as shown in
Fig. 4.12.

Using that $r = \sqrt{\tilde{x}^2 + \tilde{y}^2}$ and the physical interpretation of the line element for
a time-like interval, $ds^2 = -d\tau^2$, we see from the line element that the proper time
interval for an observer at rest in the rotating frame, i.e. with constant values of $\tilde{x}$
and $\tilde{y}$ is related to a corresponding coordinate time interval by

$$d\tau = \sqrt{1 - r^2\omega^2}\,dt.$$

The frequency this observer measures may be written $\nu = dN/d\tau$, where dN
is the number of wave fronts passing the observer during the time interval $d\tau$. The
frequency of the emitted light is $\nu_0 = dN/dt$. Hence, the observer measures the
frequency

$$v = \frac{dN}{d\tau} = \frac{dN}{dt}\frac{dt}{d\tau} = \frac{v_0}{\sqrt{1 - r^2\omega^2}}.$$

This frequency is larger than the frequency at the axis. In the rotating frame the observer experiences a gravitational field away from the axis. Hence he concludes that *light moving downwards in a gravitational field gets an increased frequency.*

(d) We shall describe the synchronization process from the point of view of an observer at rest in the laboratory frame. A Lorentz transformation from the rotating frame to the laboratory frame gives a time difference for two neighbouring clocks with the same distance from the axis and angular coordinates ϕ and $\phi + \Delta\phi$ that are synchronized in the rotating frame,

$$\Delta t = \gamma\left(\Delta\tau + R\omega\Delta l_\phi\right) = \gamma^2 R^2 \omega \Delta\phi$$

where we have used that $\Delta\tau = 0$ for the synchronized clocks and that their distance in the rotating frame is $\Delta l_\phi = \gamma R\Delta\phi$. Integrated around the disk the angular difference is 2π. It follows that the clocks at $(R, 0)$ and $(R, 2\pi)$ will have a time difference $2\pi\gamma^2 R^2\omega$. But these clocks are at the same position on the disk. So it is not possible to Einstein-synchronize clocks globally on a rotating disk.

(e) In the cylindrical coordinate system co-moving with the rotating reference frame the spacetime line-element has the form

$$ds^2 = -\left(1 - r^2\omega^2\right)dt^2 + dr^2 + r^2 d\phi^2 + 2r^2\omega d\phi dt$$

The non-vanishing $d\phi dt$−term means that the coordinate basis vectors are not orthogonal. We shall find the vectors in an orthonormal basis field co-moving with RF.

Let us start by the time-like unit basis vector. It shall have the same direction as $\vec{e}$. Hence, according to Eq. (4.57),

$$\vec{e}_{\hat{t}} = \frac{\vec{e}_t}{|\vec{e}_t|} = \frac{\vec{e}_t}{\sqrt{\vec{e}_t \cdot \vec{e}_t}} = \frac{\vec{e}_t}{\sqrt{|g_{tt}|}} = \frac{\vec{e}_t}{\sqrt{1 - r^2\omega^2}},$$

$$\vec{e}_{\hat{r}} = \vec{e}_r, \quad \vec{e}_{\hat{\phi}} = \frac{\sqrt{1 - r^2\omega^2/c^2}}{r}\vec{e}_\phi + \frac{r\omega/c^2}{\sqrt{1 - r^2\omega^2/c^2}}\vec{e}_t.$$

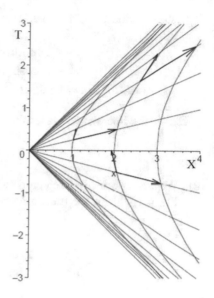

Fig. 4.13 The lines represent simultaneity planes in the accelerated frame AF. They are not simultaneous in the inertial frame IF. The curves are world lines of reference points in AF plotted with reference to IF. Three sets of basis vectors in the orthonormal basis field co-moving with AF are also shown

4.2 Uniformly accelerated system of reference

(a) It follows from the given coordinate transformation (4.91) that

$$T = \tanh(at)X, \quad X^2 - T^2 = x^2$$

For $t = $ constant the first of these equations describe a set of straight lines in the (t, X)-system that represent simultaneity planes for the reference particles of AF at different points of time. For $x = $ constant the second equation represents a set of hyperbolae. They are the world lines of fixed reference points in the accelerated reference frame as referred to the instantaneous Inertial rest frame IF of the particles at the point of time $T = 0$. These sets of lines and curves are shown in the Minkowski diagram 4.12.

(b) In IF there is Minkowski metric

$$\eta_{\hat{\mu}\hat{\nu}} = \begin{pmatrix} -1 & 1 \\ 1 & 1 \end{pmatrix}.$$

We can find the line element ds^2 by applying the transformation matrix

$$M_{\mu}^{\hat{\mu}} = \begin{pmatrix} \frac{\partial T}{\partial t} & \frac{\partial T}{\partial x} \\ \frac{\partial X}{\partial t} & \frac{\partial X}{\partial x} \end{pmatrix} = \begin{pmatrix} ax\cosh(at) & \sinh(at) \\ ax\sinh(at) & \cosh(at) \end{pmatrix}$$

to the components of the metric tensor. This gives

$$
\begin{aligned}
g_{tt} &= M_t^{\hat{\mu}} M_t^{\hat{\nu}} \eta_{\hat{\mu}\hat{\nu}} = -[ax \cosh(at)]^2 + [ax \sinh(at)]^2 = -a^2 x^2, \\
g_{tx} &= M_t^{\hat{\mu}} M_x^{\hat{\nu}} \eta_{\hat{\mu}\hat{\nu}} = -ax \cosh(at) \sinh(at) + ax \sinh(at) \cosh(at) = 0, \\
g_{xx} &= M_x^{\hat{\mu}} M_x^{\hat{\nu}} \eta_{\hat{\mu}\hat{\nu}} = [\sinh(at)]^2 + [\cosh(at)]^2 = 1.
\end{aligned}
$$

Hence, the line-element takes the form

$$
ds^2 = -a^2 x^2 dt^2 + dx^2.
$$

(c) We now consider a fixed particle in AF moving along one of the hyperbolae in Fig. 4.13. Using the transformation 4.91 we find that Its velocity in IF is

$$
v = \frac{dX}{dT} = \frac{dX/dt}{dT/dt} = \frac{xa \sinh(at)}{xa \cosh(at)} = \tanh(at).
$$

and its acceleration

$$
\frac{dv}{dT} = \frac{dt}{dT} \frac{dv}{dt} = \frac{1}{x \cosh^3(at)}.
$$

Lorentz transforming the acceleration to the instantaneous rest frame of the particle gives

$$
g = \left(1 - v^2\right)^{-3/2} \frac{dv}{dT} = \left(\frac{1}{1 - \tanh^2(at)}\right)^{3/2} \frac{1}{x \cosh^3(at)} = \frac{1}{x}.
$$

Since the particle has a constant value of x in AF it follows that the proper acceleration of the particle is constant.

(d) The time axis of an orthonormal basis co-moving with an instantaneous inertial rest frame of AF is equal to the four-velocity of the reference particles in AF,

$$
\vec{e}_{\hat{t}} = \frac{dt}{d\tau} \vec{e}_t = \frac{1}{ax} \vec{e}_t.
$$

Since the metric is diagonal the coordinate basis in AF is orthogonal, and since the values of the spatial components of the metric tensor are equal to 1, the spatial basis vector in the orthonormal basis is $\vec{e}_{\hat{x}} = \vec{e}_x$. The basis vectors $\vec{e}_{\hat{t}}$ and $\vec{e}_x$ measure unit distance in the time- and spatial directions, respectively. Hence $\vec{e}_x$ has unit length in the spatial direction, but $\vec{e}_t = ax \vec{e}_{\hat{t}}$ does not measure unit length in the time direction. Along the world line of a reference particle in AF x is constant. The time-like coordinate basis vector has unit length along the world line of a reference particle with $x = 1/a$.

We see from Figure 4.13 that AF does not cover the region $|t| > |x|$. From the line element it is seen that there is a coordinate singularity at $x = 0$. The relationship between the proper time in AF and the coordinate time is $d\tau = ax dt$, where the rate of coordinate time is position independent. Hence at $x = 0$ time does not proceed in AF.

(e) The accelerated reference frame AR is required to move in a Born rigid way. This means that all the points on the x-axis move in such a way that the rest length of all elements of the x-axis remains unchanged. As observed in IF the elements then get an increasing Lorentz contraction due to the increasing velocity.

We now think of a part of the x-axis in AF as a rod with rest length L. All the points of the rod move with constant proper acceleration. In AF the front end of the rod is at $x = x_0$, and the rear end of the rod has a position $x = x_0 - L$. The position in IF of a point with coordinate x in AF is given above by the by an equation of a hyperbola which may be written $X = \sqrt{T^2 + x^2}$. Differentiation gives the coordinate velocity as measured in IF

$$V = \frac{dX}{dT} = \frac{T}{\sqrt{T^2 + x^2}}.$$

Hence, the velocity of the front end of the rod and its rear end is

$$V_F = \frac{T}{\sqrt{T^2 + x_0^2}} \qquad V_R - \frac{T}{\sqrt{T^2 + (x_0 - L)^2}}.$$

We see that the rear end of the rod moves faster than the front end. This is due to the increasing Lorentz contraction. As observed in IF the rod becomes shorter due to the increasing velocity.

All points of the rod have to move slower than the velocity of light. The longer the rod is the faster its rear end moves. We have the limit $\lim\limits_{L \to x_0} = 1$. Hence if the rest length of the rod approaches x_0 the velocity of its rear end approaches that of light. In this limit the rear end of the rod falls together with the origin of the x-axis which also represents the limit in the backwards x-direction of the accelerated reference frame.

The motion of this limiting point can be inferred from Fig. 4.13. For negative time the reference frame moves in the negative x-direction with decreasing velocity. Then the rod has a decreasing Lorenz contraction. It gets longer, and the left end of the rod moves faster than the right one. As observed in IF the origin of the x-axis approaches the origin in IF with the velocity of light. At the point of time $T = 0$ the motion is reversed, and the limiting point moves in the positive X-direction with the velocity of light.

Fig. 4.14 World lines of
Earth, light signals from the
Earth and the spaceship

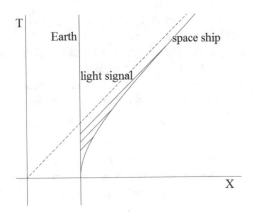

4.3 *Uniformly accelerated spaceship*

(a) A spaceship moves with constant proper acceleration $g = 10\text{m/s}^2$. The position
of the spaceship as a function of its proper time is given in Eq. (4.62),

$$X = \frac{c^2}{g} \cosh\left(\frac{g\tau}{c}\right) + k,$$

where k is a constant of integration. The spaceship starts at $X = \tau = 0$ giving
$k = -c^2/g$. Hence,

$$X = \frac{c^2}{g}\left[\cosh\left(\frac{g\tau}{c}\right) - 1\right].$$

We shall find the distance which the spaceship has moved during a proper time
$\tau = 10$years. There are 3.15×10^7s in a year, and a light year is $9.46 \times 10^{15}\text{m}$.
Hence the acceleration of gravity at the surface of the Earth may be expressed as
$g = 9.8$ m/s^2 $= 1.03$l.y./y^2. This gives $c^2/g = 0.97$l.y. and for $\tau = 10y$. we get
$g\tau/c = 10.3$. Inserting this in the formula above gives $X = 1.44 \times 10^4$l.y..

(b) In Fig. 4.14 we have shown the world lines of the spaceship and of light signals
emitted from the earth.

This figure shows that light signals emitted from points on the X-axis with negative
values of X will never arrive at the spaceship even if the spaceship always travel slower
than light. If the Earth is at a position X_E and the spaceship departs at a point of time
$T = 0$, a light signal emitted from the Earth later than $T = X_E/c = c/g$ will never
arrive at the spaceship. Hence there is a *horizon* far down in the gravitational field
experienced in the spaceship, at a position $\Delta x = X_E = c^2/g$ lower than the initial
position of the spaceship.

A light signal emitted from the Earth at a point of time T_E is received at the spaceship at a point of time τ_E. We shall find the emitter time T_E corresponding to a receiver time $\tau_E = 10$ years. The position in IF at a point of time T of a light signal emitted at a point of time T_E from the Earth at $X_E = 0$ is

$$X = c(T - T_E).$$

The position of the spaceship at this point of time is

$$X = \frac{c^2}{g}\left(\sqrt{1 + \left(\frac{gT}{c}\right)^2} - 1\right).$$

Putting the two expressions for X equal to each other and solving the resulting equation with respect to T_E gives

$$T_E = \frac{c}{g}\left(1 + \frac{gT}{c} - \sqrt{1 + \left(\frac{gT}{c}\right)^2}\right).$$

The transformation to the proper time τ of the spaceship is

$$T = \frac{c}{g}\sinh\frac{g\tau}{c}.$$

Using this in the expression for T_E gives

$$T_E = \frac{c}{g}\left(1 + \sinh\frac{g\tau_E}{c} - \cosh\frac{g\tau_E}{c}\right).$$

Inserting the acceleration of gravity at the surface of the earth, $g = 9.8m/s^2 = 1.03$l.y/y^2 gives $c^2/g = 0.97$l.y. and for $\tau = 10y$ we get $g\tau_E/c = 10.3$, which gives $T_E = 0.97$years.

The expression for the point of time of emission of the signal can be written

$$T_E = \frac{c}{g}\left(1 - e^{-\frac{g\tau_E}{c}}\right).$$

It follows that $\lim_{\tau_E \to \infty} T_E = c/g$. Hence, signals emitted later than at $T_E = c/g$ are not able to reach the spaceship.

(c) Due to the Doppler effect the frequency of the signals is less than the frequency observed at the emitter. The frequency ν measured in the spaceship is given by

$$\nu = \sqrt{\frac{1 - v/c}{1 + v/c}}\,\nu_0 = \sqrt{\frac{1 - \tanh(g\tau_E/c)}{1 + \tanh(g\tau_E/c)}}\,\nu_0,$$

where ν_0 is emitted frequency measured on the Earth, and τ_E is the point of time measured on the spaceship when the signal is received. The formula in point b) for the emission point of time may be written

$$\sqrt{\frac{1 - \tanh(g\tau_E/c)}{1 + \tanh(g\tau_E/c)}} = 1 - \frac{gT_E}{c}.$$

Hence the frequency of the received signal measured on the spaceship is

$$\nu = \left(1 - \frac{gT_E}{c}\right)\nu_0.$$

The velocity of the spaceship increases, and the observed frequency decreases due to an increasing Doppler effect. Since signals emitted later than at $T_E = c/g$ will not arrive at the spaceship, the expression for the frequency has physical meaning only for $T_E \leq c/g$. The observed frequency at the spaceship decreases towards zero at this limiting point of time.

4.4 Light emitted from a point source in a gravitational field

Consider a point like light source at the point $x = x_1, y = 0$ in a uniformly accelerated reference frame AF. Let a photon be emitted from the source at a point of time $t = 0$. It is emitted in the (x, y)-plane in a direction making an angle θ_0 with the x-axis. In the inertial laboratory frame IF the coordinates of the emission event is $T = 0$, $X = x_1$. The photon follows a null-geodesic curve which in this frame is a straight line given by

$$X = x_1 + cT \cos\theta_0, \quad Y = cT \sin\theta_0.$$

Using the transformation

$$T = x \sinh(gt/c), \quad X = x \cosh(gt/c) \quad Y = y$$

between the coordinate systems co-moving with IF and AF gives the equations for the world lines of the photon in AF

$$x = \frac{x_1}{\cosh(gt/c) - \sinh(gt/c)\cos\theta_0}, \quad y = \frac{x_1 \sinh(gt/c)\sin\theta_0}{\cosh(gt/c) - \sinh(gt/c)\cos\theta_0}.$$

Dividing the expressions for x and y by each other, we get

$$\sinh(gt/c) = \frac{y}{x \sin\theta_0}.$$

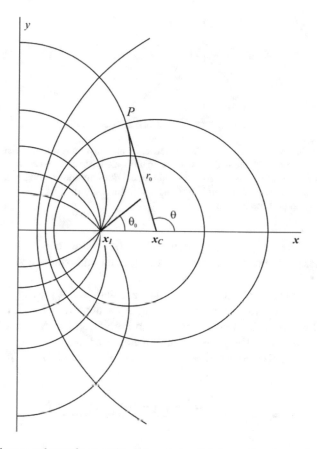

Fig. 4.15 Light rays and wave fronts emitted by a point particle in a uniformly accelerated reference frame

Inserting this into the expressions for x and y gives an equation which may be written as

$$x^2 + (y - x_1 \cot \theta_0)^2 = \left(\frac{x_1}{\sin \theta_0}\right)^2.$$

This is the equation of a circle with radius $r = x_1 / \sin \theta_0$. A set of these circular trajectories and the corresponding wave front is shown in Fig. 4.15.

4.5 Geometrical optics in a gravitational field

(a) *Asphere photographed from above.* The situation described in the text is illustrated in Fig. 4.16.

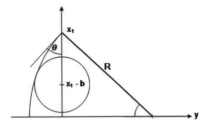

Fig. 4.16 There is a camera at $x = x_1$ on the vertical x-axis in a uniformly accelerated reference frame AF. The centre of a sphere with radius r is on the x-axis a distance b below the camera. Due to the gravitational field experienced in AF a light signal does not move along a straight line, but along a circle with radius R. Light emitted tangentially from the sphere to the camera makes an angle θ with the x-axis at the camera.

The radius of the circular light path is $c^2/g = 0.97$ light years in the gravitational field of the Earth, so the optical effects of the gravitational field are very small at the surface of the Earth.

The camera P is at a position $x = x_1$. The acceleration of gravity at the position of the camera is $g = c^2/x_1$. The centre of a sphere with radius r is a distance b below P. A ray of light is emitted tangentially from a point on the surface of the sphere such that it arrives at P. The angle θ is half the apex angle.

We see from the figure that $\sin \theta_a = x_1/R$ and $R^2 - x_1^2 = (R - r)^2 - (x_1 - b)^2$ which gives

$$R = \frac{x_1}{\sin \theta_a} = \frac{x_1^2 + r^2 - (x_1 - b)^2}{2r} = \frac{r^2 + 2x_1 b - b^2}{2r}.$$

Hence

$$\sin \theta_a = \frac{2r x_1}{r^2 + 2x_1 b - b^2}.$$

The half of the apex angle without the gravitational field is $\sin \theta_0 = r/b$. Inserting this into the above expression and using that $x_1 = c^2/g$ gives

$$\sin \theta_a = \frac{\sin \theta_0}{1 - (gb/2c^2) \cos^2 \theta_0} = \frac{r/b}{1 - (gb/2c^2)(1 - r^2/b^2)}.$$

This shows that as photographed from above the sphere appears enlarged.

(b) *A sphere photographed from below.* This situation is illustrated in Fig. 4.17.

In the present case the figure shows that $R^2 - x_1^2 = (R + r)^2 - (x_1 + b)^2$. Then a similar deduction to that in point a) gives

Fig. 4.17 Similar to the situation in a, but with the camera below the sphere

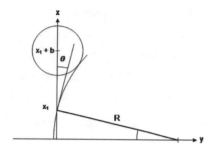

$$\sin \theta_b = \frac{\sin \theta_0}{1 + \left(gb/2c^2\right)\cos^2 \theta_0} = \frac{r/b}{1 + \left(gb/2c^2\right)\left(1 - r^2/b^2\right)}.$$

This shows that as photographed from below the sphere appears diminished.

(c) The difference in θ as photographed from above and below at the same distance is given by

$$\sin \theta_a - \sin \theta_b = \frac{\left(gb/c^2\right)\cos^2 \theta_0 \sin \theta_0}{1 - \left(gb/c^2\right)^2 \cos^4 \theta_0}.$$

With $b = 1$m we get $gb/c^2 = 1m/0.97l.y. = 1.06 \times 10^{-16}$. Hence the difference, $\Delta\theta$, between the angels θ_a and θ_b is very small. We can therefore use the approximations

$$\sin \theta_a - \sin \theta_b \approx \sin(\theta + \Delta\theta) - \sin \theta = \sin \theta \cos \Delta\theta + \cos \theta \sin \Delta\theta - \sin \theta$$
$$\approx \Delta\theta \cos \theta,$$

and

$$\Delta\theta \approx \frac{gb}{c^2} \sin \theta_0 \cos \theta_0 = \frac{gr}{c^2}\sqrt{1 - \frac{r^2}{b^2}}.$$

With $r = 0.1$m and $b = 1$m we can further use the approximation

$$\Delta\theta \approx \frac{gr}{c^2}.$$

Letting $g = 9.8$m/s^2 and $r = 0.1$m we get $\Delta\theta \approx 1,0 \times 10^{-17}$. This corresponds to $\Delta\theta \approx 2.2 \times 10^{-12}$arc seconds, which is beyond what we are able to observe.

Solutions for Chapter 5

5.1 *Dual forms*

Let$\{\vec{e}_i\}$ be a Cartesian basis in the 3-dimensional Euclidean space. Using a vector $\vec{A} = A^i \vec{e}_i$ there are two ways of constructing a form:

(i)   By constructing a 1-form from its covariant components $A_j = g_{ji} A^i$:

$$\underline{A} = A_i \underline{dx}^i.$$

(ii) By constructing a 2-form from its dual components, defined by

$$\underline{a} = \frac{1}{2} a_{ij} \underline{dx}^i \wedge \underline{dx}^j = \frac{1}{2} \varepsilon_{ijk} A^k \underline{dx}^i \wedge \underline{dx}^j$$

We write this form as $\underline{a} = \star\underline{A}$, where $\star$ means to take the *dual form*.

(a)   Given the vectors $\vec{a} = \vec{e}_x + 2\vec{e}_y - \vec{e}_z$ and $\vec{b} = 2\vec{e}_x - 3\vec{e}_y + \vec{e}_z$.

We shall find the corresponding 1-forms $\underline{A}$ and $\underline{B}$, and the dual 2-forms $\underline{a} = \star\underline{A}$ and $\underline{b} = \star\underline{B}$, and also the dual form θ to the 1-form $\underline{\sigma} = \underline{dx} - 2\underline{dy}$.

Since the space has Euclidean geometry with an orthonormal basis $g_{ij} = \delta_{ij}$, the component of the 1-forms $\underline{A}$ and $\underline{B}$ are equal to the components of the corresponding vectors $A_{\hat{i}} = \delta_{ij} A^{\hat{j}}$ and $B_{\hat{i}} = \delta_{ij} B^{\hat{j}}$.

Hence,

$$\underline{A} = \underline{dx} + 2\underline{dy} - \underline{dz} \quad \underline{B} = 2\underline{dx} - 3\underline{dy} + \underline{dz}.$$

The dual 1-forms $\underline{\alpha} = \ast\underline{A}$ and $\underline{\beta} = \ast\underline{B}$ have components $\alpha_{\hat{i}\hat{j}} = \varepsilon_{\hat{i}\hat{j}k} A^{\hat{k}}$ and $\beta_{\hat{i}\hat{j}} = \varepsilon_{\hat{i}\hat{j}k} B^{\hat{k}}$. This gives

$$\ast\underline{A} = -\underline{dx} \wedge \underline{dy} - 2\underline{dx} \wedge \underline{dz} + \underline{dy} \wedge \underline{dz}$$
$$\ast\underline{B} = \underline{dx} \wedge \underline{dy} + 3\underline{dx} \wedge \underline{dz} + 2\underline{dy} \wedge \underline{dz} \;.$$

Given a 1-form $\underline{\sigma} = \sigma_i \underline{dx}^i = \underline{dx} - 2\underline{dy}$. The 2-form $\underline{\theta} = \ast\underline{\sigma}$ has components $\alpha_{\hat{i}\hat{j}} = \varepsilon_{\hat{i}\hat{j}k} \sigma^{\hat{k}} = \alpha_{\hat{i}\hat{j}} = \varepsilon_{\hat{i}\hat{j}k} g^{\hat{k}\hat{l}} \sigma_{\hat{l}}$ which gives

$$\underline{\theta} = 2\underline{dx} \wedge \underline{dz} + \underline{dy} \wedge \underline{dz}.$$

(b)   Inserting the expressions for $\underline{A}$ and $\underline{B}$ found in point a) in $\underline{\theta} = \underline{A} \wedge \underline{B}$ we find

$$\underline{\theta} = \left(\underline{dx} + 2\underline{dy} - \underline{dz} \right) \wedge \left(2\underline{dx} - 3\underline{dy} + \underline{dz} \right) = -7\underline{dx} \wedge \underline{dy} + 3\underline{dx} \wedge \underline{dz}$$

$$- \underline{dy} \wedge \underline{dz}.$$

Calculating the vector product of $\vec{A}$ and $\vec{B}$ in the usual way we find

$$\vec{C} = \vec{A} \times \vec{B} = -\vec{e}_x - 3\vec{e}_y - 7\vec{e}_z.$$

We see that $\vec{C}$ and $\underline{\theta}$ have the same components, where C^z corresponds to θ_{xy} and so forth.

Let us show this in general. The components of the vector product are $C^{\hat{k}} = \varepsilon_{\hat{m}\hat{n}\hat{k}} A^{\hat{m}} B^{\hat{n}}$. Hence,

$$\varepsilon_{\hat{i}\hat{j}\hat{k}} C^{\hat{k}} = \varepsilon_{\hat{i}\hat{j}\hat{k}} \varepsilon_{\hat{m}\hat{n}\hat{k}} A^{\hat{m}} B^{\hat{n}} = \left(\delta_{\hat{i}\hat{m}} \delta_{\hat{j}\hat{n}} - \delta_{\hat{i}\hat{n}} \delta_{\hat{j}\hat{m}} \right) A^{\hat{m}} B^{\hat{n}} = A^{\hat{i}} B^{\hat{j}} - A^{\hat{j}} B^{\hat{i}}.$$

Since the exterior product of two 1-forms gives

$$\theta_{\hat{i}\hat{j}} = \left(\underline{A} \wedge \underline{B} \right)_{\hat{i}\hat{j}} = A_{\hat{i}} B_{\hat{j}} - A_{\hat{j}} B_{\hat{i}},$$

we see that with a metric $g_{\hat{i}\hat{j}} = \delta_{\hat{i}\hat{j}}$ there is agreement between the two expressions.

Note that if we replace $\delta_{\hat{i}\hat{m}} \delta_{\hat{j}\hat{n}} - \delta_{\hat{i}\hat{n}} \delta_{\hat{j}\hat{m}}$ by $g_{im} g_{jn} - g_{in} g_{jm}$ the deduction will bevalid in an
arbitrary basis.

(c) Direct calculation gives

$$\underline{A} \wedge *\underline{B} = \left(A_{\hat{i}} \underline{\omega}^{\hat{i}} \right) \wedge \left(\tfrac{1}{2} \varepsilon_{\hat{i}\hat{j}\hat{k}} B^{\hat{k}} \underline{\omega}^{\hat{i}} \wedge \underline{\omega}^{\hat{j}} \right)$$
$$= \left(A_x \underline{\omega}^x + A_y \underline{\omega}^y + A_z \underline{\omega}^z \right) \wedge \left(B^z \underline{\omega}^x \wedge \underline{\omega}^y - B^y \underline{\omega}^x \wedge \underline{\omega}^z + B^x \underline{\omega}^y \wedge \underline{\omega}^z \right).$$
$$= \left(A_x B^x + A_y B^y + A_z B^Z \right) \underline{\omega}^x \wedge \underline{\omega}^y \wedge \underline{\omega}^z = \left(\vec{A} \cdot \vec{B} \right) \underline{\omega}^x \wedge \underline{\omega}^y \wedge \underline{\omega}^z$$

(d) The exterior derivative of $\underline{A}$ is

$$\underline{dA} = A_{i,j} \underline{\omega}^j \wedge \underline{\omega}^i = \left(A_{y,x} - A_{x,y} \right) \underline{\omega}^x \wedge \underline{\omega}^y$$
$$+ \left(A_{z,x} - A_{x,z} \right) \underline{\omega}^x \wedge \underline{\omega}^z \left(A_{z,y} - A_{y,z} \right) \underline{\omega}^y \wedge \underline{\omega}^z.$$

The curl of a vector is

$$\nabla \times \vec{A} = \left(A^z,_y - A^y,_z \right) \vec{e}_x - \left(A^z,_x - A^x,_z \right) \vec{e}_y + \left(A^y,_x - A^x,_y \right) \vec{e}_z.$$

Hence

$$\left(\underline{dA} \right)_{ij} = \varepsilon_{ijk} \left(\nabla \times \vec{A} \right)^k$$

(e) The exterior derivative of the dual $*\underline{A}$ is

$$\underline{d} * \underline{A} = \frac{1}{2} (*\underline{A})_{ij}{}_{,k} \underline{\omega}^k \wedge \underline{\omega}^i \wedge \underline{\omega}^j = \frac{1}{2} (\varepsilon_{ijl} A^l)_{,k} \underline{\omega}^k \wedge \underline{\omega}^i \wedge \underline{\omega}^j$$

$$= A^k{}_{,k} \underline{\omega}^x \wedge \underline{\omega}^y \wedge \underline{\omega}^z.$$

The exterior derivative of a scalar field f is

$$\underline{d}f = f_{,i} \underline{\omega}^i.$$

Which has the same components as ∇f. Inserting $\underline{A} = \underline{d}f$ in the expression for $\underline{d} * \underline{A}$ we get

$$\underline{d} * \underline{d}f = \nabla^2 f \underline{\omega}^x \wedge \underline{\omega}^y \wedge \underline{\omega}^z$$

5.2 Differential operators in spherical coordinates

(a) We find the metric tensor by transforming from the metric in the Cartesian coordinate system,

$$g_{ij} = \frac{\partial x^{\hat{i}}}{\partial x^i} \frac{\partial x^{\hat{j}}}{\partial x^j} \delta_{\hat{i}\hat{j}}.$$

The coordinate transformation is

$$x = r \sin\theta \cos\phi, \quad y = r \sin\theta \sin\phi, \quad z = r \cos\theta$$

Hence, the transformation matrix has the elements

$$\begin{pmatrix} \frac{\partial x}{\partial r} & \frac{\partial x}{\partial \theta} & \frac{\partial x}{\partial \phi} \\ \frac{\partial y}{\partial r} & \frac{\partial y}{\partial \theta} & \frac{\partial y}{\partial \phi} \\ \frac{\partial z}{\partial r} & \frac{\partial z}{\partial \theta} & \frac{\partial z}{\partial \phi} \end{pmatrix} = \begin{pmatrix} \sin\theta \cos\phi & r \cos\theta \cos\phi & -r \sin\theta \sin\phi \\ \sin\theta \sin\phi & r \cos\theta \sin\phi & r \sin\theta \cos\phi \\ \cos\theta & -r \sin\theta & 0 \end{pmatrix}.$$

Inserting this into the transformation formula for the components of the metric tensor, we find the following non-vanishing components

$$g_{rr} = 1, \quad g_{\theta\theta} = r^2, \quad g_{\phi\phi} = r^2 \sin^2\theta.$$

This shows that the line-element has the form

$$dl^2 = dr^2 + r^2 d\theta^2 + r^2 \sin^2\theta d\phi^2$$

In the spherical coordinate system.

(b) The gradient of the scalar field f in the spherical coordinate system is gives by

$$\nabla f = \frac{\partial f}{\partial r}\vec{e}_{\hat{r}} + \frac{\partial f}{\partial \theta}\vec{e}_{\hat{\theta}} + \frac{\partial f}{\partial \phi}\vec{e}_{\hat{\phi}}$$

The orthonormal basis vectors are given in terms of the coordinate basis vectors as,

$$\vec{e}_{\hat{i}} = \vec{e}_i/|\vec{e}_i| = \vec{e}_i/\sqrt{\vec{e}_i \cdot \vec{e}_i} = \vec{e}_i/\sqrt{g_{ii}}.$$

This gives

$$\vec{e}_{\hat{r}} = \vec{e}_r, \quad \vec{e}_{\hat{\theta}} = \frac{1}{r}\vec{e}_\theta, \quad \vec{e}_{\hat{\phi}} = \frac{1}{r \sin \theta}\vec{e}_\phi.$$

Inserting these in the expression for the gradient gives

$$\nabla f = \frac{\partial f}{\partial r}\vec{e}_r + \frac{1}{r}\frac{\partial f}{\partial \theta}\vec{e}_\theta + \frac{1}{r \sin \theta}\frac{\partial f}{\partial \phi}\vec{e}_\phi.$$

(c) The curl of a vector is given as

$$\nabla \times \vec{A} = \frac{1}{\sqrt{g_{22}g_{33}}}\left(\frac{\partial A^3}{\partial x^2} - \frac{\partial A^2}{\partial x^3}\right)\vec{e}_1 + \frac{1}{\sqrt{g_{11}g_{33}}}\left(\frac{\partial A^1}{\partial x^3} - \frac{\partial A^3}{\partial x^1}\right)\vec{e}_2$$

$$+ \frac{1}{\sqrt{g_{11}g_{22}}}\left(\frac{\partial A^2}{\partial x^1} - \frac{\partial A^1}{\partial x^2}\right)\vec{e}_3.$$

The components of the vector field in orthonormal basis are given in terms of its coordinate component by

$$A^{\hat{r}} = \hat{A}, \quad A^{\hat{\theta}} = rA^\theta, \quad A^{\hat{\phi}} = r \sin \theta A^\phi.$$

Using this together with the components of the metric tensor from point a) in the above expression for the curl gives in spherical coordinates

$$\nabla \times \vec{A} = \frac{1}{r \sin \theta}\left(\frac{\partial(\sin \theta A^\phi)}{\partial \theta} - \frac{\partial A^\theta}{\partial \phi}\right)\vec{e}_r + \frac{1}{r \sin \theta}\left(\frac{\partial A^r}{\partial \phi} - \sin \theta \frac{\partial(rA^\phi)}{\partial r}\right)\vec{e}_\theta$$

$$+ \frac{1}{r}\left(\frac{\partial(rA^\theta)}{\partial r} - \frac{\partial A^r}{\partial \theta}\right)\vec{e}_\phi$$

(d) The divergence of a vector field is given by

$$\left(\nabla \cdot \vec{A}\right)\underline{\varepsilon} = \underline{d} * \underline{A},$$

where

$$\underline{\varepsilon} = \sqrt{|g|}\underline{\omega}^1 \wedge \underline{\omega}^2 \wedge \underline{\omega}^3,$$

is the volume form and $|g|$ the determinant of the matrix formed by the components of the metric tensor.

From Exercise 5.1 we have that the dual 2-form of a vector is given by

$$\left(*\underline{A}\right)_{ij} = \sqrt{|g|}\varepsilon_{ijk}A^k,$$

where ε_{ijk} is the Levi-Civita symbol with $\varepsilon_{r\theta\phi} = +1$. Hence, we get

$$\nabla \cdot \vec{A} = \frac{\left(\underline{d} * \underline{A}\right)_{r\theta\phi}}{\sqrt{|g|}}.$$

In spherical coordinates we have

$$\sqrt{|g|} = r^2 \sin\theta,$$

giving

$$\left(*\underline{A}\right)_{r\theta} = r^2 \sin\theta A^\phi, \quad \left(*\underline{A}\right)_{r\phi} = -r^2 \sin\theta A^\theta, \quad \left(*\underline{A}\right)_{\theta\phi} = r^2 \sin\theta A^r.$$

The exterior derivative has only one component

$$\left(\underline{d} * \underline{A}\right)_{r\theta\phi} = \left(*\underline{A}\right)_{\theta\phi,r} - \left(*\underline{A}\right)_{r\phi,\theta} + \left(*\underline{A}\right)_{r\theta,\phi}$$
$$= \sin\theta\left(r^2 A^r\right)_{,r} + r^2\left(\sin\theta\, A^\theta\right)_{,\theta} + r^2 \sin\theta\, A^\phi_{,\phi}$$

where we have used Einstein's comma-notation for partial derivatives. This gives

$$\nabla \cdot \vec{A} = \frac{\left(\underline{d} * \underline{A}\right)_{r\theta\phi}}{r^2 \sin\theta} = \frac{1}{r^2}\left(r^2 A^r\right)_{,r} + \frac{1}{\sin\theta}\left(\sin\theta A^\theta\right)_{,\theta} + A^\phi_{,\phi}.$$

We can also express the divergence of a vector in term of it components in an orthonormal basis. Using that

$$A^r = A^{\hat{r}}, \quad A^\theta = \frac{1}{r}A^{\hat{\theta}}, \quad A^\phi = \frac{1}{r \sin\theta}A^{\hat{\phi}},$$

we get

Fig 5.9 Geodesic curves on a non-rotating (*dashed line*) and rotating (*solid line*) disc

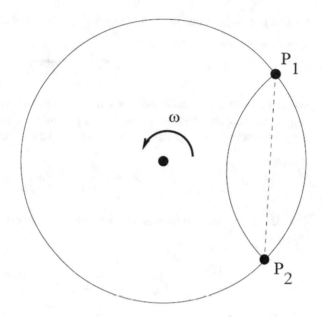

$$\nabla \cdot \vec{A} = \frac{1}{r^2}\left(r^2 A^{\hat{r}}\right)_{,r} + \frac{1}{r \sin\theta}\left(\sin\theta A^{\hat{\theta}}\right)_{,\theta} + \frac{1}{r \sin\theta} A^{\phi}_{,\phi}.$$

Inserting $\vec{A} = \nabla f$, where ∇f is given in point b), we get the expression for the Laplacian in spherical coordinates

$$\nabla^2 f = \frac{1}{r^2}\left(r^2 f_{,r}\right)_{,r} + \frac{1}{r^2 \sin\theta}\left(\sin\theta f_{,\theta}\right)_{,\theta} + \frac{1}{r^2 \sin^2\theta} f_{,\phi\phi}.$$

5.3 Spatial geodesics in a rotating frame of reference

In Fig. 5.9, we see a rotating disc. We can see two geodesic curves between P_1 and P_2. The dashed line is the geodesic for the non-rotating disc. The other curve is a geodesic for the 3-space of a rotating reference frame. We can see that the geodesic is curved inward when the disc is rotating. The curve has to curve inward since the standard measuring rods along the curve are longer there (because of less Lorentz contraction close to the axis where the velocity in the length direction of the measuring rods is less). Thus, the minimum distance, i.e. the minimum number of standard measuring rods along the curve between P_1 and P_2, will be achieved by an inwardly bent curve. We will show this mathematically, using the Lagrangian equations.

(a) We first deduce the form of the tangent vector identity for the present case. The line element for the space $d\hat{t} = dz = 0$ of the rotating reference frame is

$$dl^2 = dr^2 + \frac{r^2 d\theta^2}{1 - \frac{r^2\omega^2}{c^2}}.$$

The Lagrangian function for the spatial geodesics in the rotating frame is

$$L = \frac{1}{2}\dot{r}^2 + \frac{1}{2}\frac{r^2\dot{\theta}^2}{1 - \frac{r^2\omega^2}{c^2}}.$$

where the dot means differentiation with respect to an invariant parameter representing the arc length along the curve. In the present case the 3-vector identity for the tangent vectors of the spatial geodesic curves take the form

$$\dot{r}^2 + \frac{r^2\dot{\theta}^2}{1 - \frac{r^2\omega^2}{c^2}} = c^2.$$

(b) It is seen from the Lagrangian function that θ is cyclic ($\partial L/\partial\theta = 0$), implying that

$$p_\theta = \frac{\partial L}{\partial\dot{\theta}} = \frac{r^2\dot{\theta}}{1 - \frac{r^2\omega^2}{c^2}} = \text{constant.}$$

(c) This gives

$$\dot{\theta} = \left(1 - \frac{r^2\omega^2}{c^2}\right)\frac{p_\theta}{r^2} = \frac{p_\theta}{r^2} - \frac{\omega^2 p_\theta}{c^2}.$$

Inserting into tangent vector identity we get

$$\dot{r}^2 = 1 + \frac{\omega^2 p_\theta^2}{c^2} - \frac{p_\theta^2}{r^2}.$$

This gives the equation of the geodesic curve between P_1 and P_2 .

$$\frac{\dot{r}}{\dot{\theta}} = \pm\frac{dr}{d\theta} = \frac{r^2\sqrt{1 + \frac{\omega^2 p_\theta^2}{c^2} - \frac{p_\theta^2}{r^2}}}{p_\theta\left(1 - \frac{r^2\omega^2}{c^2}\right)}.$$

(d) The boundary conditions are $\dot{r} = 0, r = r_0$, for $\theta = 0$, where r_0 is the distance from the rotational axis to the point of the curve with minimal distance to the axis (Fig. 5.10). Inserting this into the expression for p_θ gives

$$\frac{p_\theta}{r_0} = \sqrt{1 + \frac{p_\theta^2\omega^2}{c^2}}.$$

(e) Solving this equation with respect to p_θ gives

Fig. 5.10 Geodesic curves
on a rotating disc,
coordinates

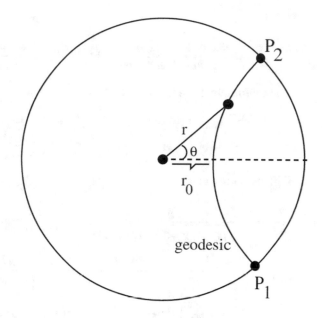

$$p_\theta = \frac{r_0}{\sqrt{1 - r_0^2\omega^2/c^2}}.$$

Inserting this into the equation of the geodesic curve, it takes the form

$$\frac{dr}{d\theta} = \frac{r\sqrt{r^2 - r_0^2}}{r_0\left(1 - r^2\omega^2/c^2\right)},$$

which may be written

$$\frac{dr}{r\sqrt{r^2 - r_0^2}} - \frac{\omega^2}{c^2}\frac{r\,dr}{\sqrt{r^2 - r_0^2}} = \frac{d\theta}{r_0}.$$

Integrating this yields

$$\theta = \pm\frac{r_0\omega^2}{c^2}\sqrt{r^2 - r_0^2} \mp \arccos\frac{r_0}{r}.$$

The curve represented by this equation is shown in Fig. 5.10.

5.4 *Christoffel symbols in a uniformly accelerated reference frame*

(a) The only non-vanishing Christoffel symbols in the coordinate system co-moving
with a uniformly accelerated reference frame are

$$\Gamma^x_{tt} = \frac{\partial x}{\partial T}\frac{\partial^2 T}{\partial t^2} + \frac{\partial x}{\partial X}\frac{\partial^2 X}{\partial t^2}, \quad \Gamma^t_{xt} = \Gamma^t_{tx} = \frac{\partial t}{\partial T}\frac{\partial^2 T}{\partial t\partial x} + \frac{\partial t}{\partial X}\frac{\partial^2 X}{\partial t\partial x}.$$

The coordinate transformation formulae are

$$\frac{gT}{c} = \left(1 + \frac{gx}{c^2}\right)\sinh\frac{gt}{c}, \quad 1 + \frac{gX}{c^2} = \left(1 + \frac{gx}{c^2}\right)\cosh\frac{gt}{c},$$

$$\frac{gT}{c} = \left(1 + \frac{gX}{c^2}\right)\tanh\frac{gt}{c}, \quad \left(1 + \frac{gX}{c^2}\right)^2 - \left(\frac{gT}{c}\right)^2 = \left(1 + \frac{gx}{c^2}\right)^2.$$

Differentiation gives

$$\frac{\partial x}{\partial T} = -\frac{gT}{\sqrt{\left(1 + gX/c^2\right)^2 - (gT/c)^2}} = -\frac{gT}{1 + gx/c^2}, \quad \frac{\partial x}{\partial X} = \frac{1 + gX/c^2}{1 + gx/c^2},$$

$$\frac{\partial t}{\partial T} = \frac{1 + gX/c^2}{\left(1 + gx/c^2\right)^2} = \frac{\cosh(gt/c)}{1 + gx/c^2}, \quad \frac{\partial ct}{\partial X} = \frac{gT/c}{\left(1 + gx/c^2\right)^2} = \frac{\sinh(gt/c)}{1 + gx/c^2},$$

$$\frac{\partial^2 T}{\partial t^2} = \frac{g}{c}\left(1 + \frac{gx}{c^2}\right)\sinh\frac{gt}{c}, \quad \frac{\partial^2 X}{\partial t^2} = g\left(1 + \frac{gx}{c^2}\right)\cosh\frac{gt}{c},$$

$$\frac{\partial^2 T}{\partial t\partial x} = \frac{g}{c^2}\cosh\frac{gt}{c}, \quad \frac{\partial^2 X}{\partial t\partial x} = \frac{g}{c}\sinh\frac{gt}{c}.$$

Inserting these expressions into the formulae of the Christoffel symbols leads to

$$\Gamma^x_{tt} = g\left(1 + gx/c^2\right), \quad \Gamma^t_{xt} = \Gamma^t_{tx} = \frac{g/c^2}{1 + gx/c^2}.$$

(b) We shall use the formula for the Christoffel symbols in terms of the derivatives of the components of the metric tensor,

$$\Gamma^\alpha_{\mu\nu} \equiv \frac{1}{2}g^{\alpha\beta}\left(g_{\beta\mu,\nu} + g_{\beta\nu,\mu} - g_{\mu\nu,\beta}\right)$$

To calculate the same Chritoffel symbols as above.

The components of the metric tensor in the coordinate system co-moving with a uniformly accelerated reference frame are

$$g_{tt} = -\left(1 + \frac{gx}{c^2}\right)^2 c^2, \quad g_{xx} = g_{yy} = g_{zz} = 1$$

only the term $\frac{\partial g_{tt}}{\partial x}$ contributes to $\Gamma^\alpha_{\mu\nu}$. Thus the only non-vanishing Christoffel symbols are

$$\Gamma_{xt}^t = \Gamma_{tx}^t = \frac{1}{2}g^{tt}\left(\frac{\partial g_{tt}}{\partial x}\right) = \frac{1}{2g_{tt}}\frac{\partial g_{tt}}{\partial x} = \frac{2(1+\frac{gx}{c^2})g}{2(1+\frac{gx}{c^2})^2 c^2} = \frac{1}{(1+\frac{gx}{c^2})}\frac{g}{c^2},$$

$$\Gamma_{tt}^x = -\frac{1}{2}g^{xx}\left(\frac{\partial g_{tt}}{\partial x}\right) = -\frac{1}{2}\left\{-2\left(1+\frac{gx}{c^2}\right)\frac{g}{c^2}c^2\right\} = \left(1+\frac{gx}{c^2}\right)g.$$

5.5 Relativistic vertical projectile motion

A particle is thrown vertically upwards with velocity v from the origin of the coordinate system in the gravitational field of a uniformly accelerated reference frame. We shall calculate the maximal height of the particle.

The line element, including only the vertical x-direction, has the form

$$ds^2 = -\left(1+\frac{gx}{c^2}\right)^2 c^2 dt^2 + dx^2$$

the Lagrange function is

$$L = \frac{1}{2}g_{\mu\nu}\dot{x}^\mu\dot{x}^\nu = -\frac{1}{2}\left(1+\frac{gx}{c^2}\right)^2 c^2\dot{t}^2 + \frac{1}{2}\dot{x}^2,$$

where the dots imply differentiation with respect to the particle's proper time τ. The initial conditions are $x(0) = 0$, $\dot{x}(0) = \gamma(c, v, 0, 0)$, where $\gamma = (1 - v^2/c^2)^{-1/2}$.

Newtonian description: $\frac{1}{2}mv^2 = mgh \Rightarrow h = \frac{v^2}{2g}$.

Relativistic description: t is a cyclic coordinate $\Rightarrow x^0 = ct$ is cyclic and $p_0 = $ constant.

$$p_0 = \frac{\partial L}{\partial \dot{x}^0} = \frac{1}{c}\frac{\partial L}{\partial \dot{t}} = -c\left(1+\frac{gx}{c^2}\right)^2\dot{t}.$$

In the present case the 4-velocity identity has the form

$$-\frac{1}{2}\left(1+\frac{gx}{c^2}\right)^2 c^2\dot{t}^2 + \frac{1}{2}\dot{x}^2 = -\frac{1}{2}c^2.$$

Since the maximum height h is reached when $\dot{x} = 0$ we have

$$\left(1+\frac{gh}{c^2}\right)\dot{t}_{x=h} = 1.$$

Since p_0 is a constant of the motion we can determine ist value from the initial condition, $p_0 = -c\dot{t}(0) = -\gamma c$, and at $x = h$:

$$p_0 = -\gamma c = -c\left(1+\frac{gh}{c^2}\right)^2\dot{t}_{x=h}$$

It follows from the two last equations that the maximal height of the projectile is

$$h = \frac{c^2}{g}(\gamma - 1).$$

In the Newtonian limit this becomes

$$h = \frac{c^2}{g}\left(\frac{1}{(1 - v^2/c^2)^{1/2}} - 1\right) \approx \frac{c^2}{g}\left(1 + \frac{1}{2}\frac{v^2}{c^2} - 1\right) \Rightarrow h \approx \frac{v^2}{2g}.$$

5.6 The geodesic equation and constants of motion

(a) Thegeodesic equation can be written

$$\frac{du_\mu}{ds} - \Gamma^\alpha_{\mu\beta} u_\alpha u^\beta = 0.$$

Inserting the expression for the Christoffel symbols give

$$\Gamma^\alpha_{\mu\beta} u_\alpha u^\beta = \Gamma_{\alpha\mu\beta} u^\alpha u^\beta = \frac{1}{2}\left(g_{\alpha\mu,\beta} + g_{\alpha\beta,\mu} - g_{\mu\beta,\alpha}\right)u^\alpha u^\beta.$$

Since the metric tensor is symmetrical this can be written

$$\Gamma^\alpha_{\mu\beta} u_\alpha u^\beta = \frac{1}{2}\left(g_{\alpha\mu,\beta} + g_{\alpha\beta,\mu} - g_{\beta\mu,\alpha}\right)u^\alpha u^\beta.$$

Since $g_{\alpha\mu,\beta} - g_{\beta\mu,\alpha}$ is antisymmetric in α and β, and $u^\alpha u^\beta$ is symmetric in α and β we have that $\left(g_{\alpha\mu,\beta} - g_{\beta\mu,\alpha}\right)u^\alpha u^\beta = 0$. Hence

$$\Gamma^\alpha_{\mu\beta} u_\alpha u^\beta = \frac{1}{2}g_{\alpha\beta,\mu} u^\alpha u^\beta.$$

Hence the geodesic equation in the form above reduces to

$$\frac{du_\mu}{ds} - \frac{1}{2}g_{\alpha\beta,\mu} u^\alpha u^\beta = 0.$$

(b) From the form of the geodesic equation in b) it follows that since the metric is independent

of t, θ and z in the static and cylindrically symmetric case, so that $g_{\mu\nu,t} = g_{\mu\nu,\theta} = g_{\mu\nu,z} = 0$, then u_t, u_θ and u_z are constants of motion in this case.

Solutions for Chapter 6

6.1 *Parallel transport*

Parallel transport of a vector $\vec{A}$ along a curve means that the covariant directional derivative of the vector along the curve vanishes,

$$\nabla_{\vec{u}}\vec{A} = 0. \tag{S6.1.1}$$

where $\vec{u}$ is the tangent vector field of the curve. Let the curve parameter be λ. Then the component equation of parallel transport along the curve takes the form

$$A^{\mu}_{;\nu}u^{\nu} = A^{\mu}_{;\nu}\frac{dx^{\nu}}{d\lambda} = 0. \tag{S6.1.2}$$

Inserting the expression $A^{\mu}_{;\nu} = A^{\mu}_{,\nu} + \Gamma^{\mu}_{\lambda\nu}A^{\lambda}$ of the covariant derivative, we get

$$A^{\mu}_{,\nu}\frac{dx^{\nu}}{d\lambda} = \frac{dA^{\mu}}{d\lambda} = -\Gamma^{\mu}_{\lambda\nu}A^{\lambda}\frac{dx^{\nu}}{d\lambda}. \tag{S6.1.3}$$

For an infinitesimal displacement along the curve the change of the vector component A^{μ} thereby becomes

$$dA^{\mu} = -\Gamma^{\mu}_{\lambda\nu}A^{\lambda}dx^{\nu}. \tag{S6.1.4}$$

(b) The result from point a) may be written as a Taylor expansion of A^{μ} to first order at the point x,

$$A^{\mu}(x + dx) = A^{\mu}(x) - \Gamma^{\mu}_{\lambda\nu}A^{\lambda}dx^{\nu}. \tag{S6.1.5}$$

The Taylor expansion is generally given as

$$A^{\mu}(x + dx) = A^{\mu}(x) + \frac{\partial A^{\mu}(x)}{\partial x^{\nu}}dx^{\nu} + \frac{1}{2}\frac{\partial^{2}A^{\mu}(x)}{\partial x^{\nu}\partial x^{\gamma}}dx^{\nu}dx^{\gamma} + \cdots \tag{S6.1.6}$$

We see that

$$\frac{\partial A^{\mu}}{\partial x^{\nu}} = -\Gamma^{\mu}_{\lambda\nu}A^{\lambda}. \tag{S6.1.7}$$

We also need the next term in the Taylor expansion,

$$\frac{\partial^{2}A^{\mu}}{\partial x^{\nu}\partial x^{\gamma}} = \frac{\partial}{\partial x^{\gamma}}\frac{\partial A^{\mu}}{\partial x^{\nu}} = \frac{\partial\left(-\Gamma^{\mu}_{\lambda\nu}A^{\lambda}\right)}{\partial x^{\gamma}} = -\Gamma^{\mu}_{\lambda\nu,\gamma}A^{\lambda} - \Gamma^{\mu}_{\lambda\nu}A^{\lambda}_{,\gamma}$$

Fig. 6.8 Infinitesimalt
parallelogram

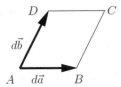

$$= -\Gamma^{\mu}_{\lambda\nu,\gamma} A^{\lambda} + \Gamma^{\mu}_{\lambda\nu} \Gamma^{\lambda}_{\alpha\gamma} A^{\alpha}. \tag{S6.1.8}$$

Hence, we get to 2.order that by parallel from x to $x + dx$ the component A^{μ} of the vector is

$$A^{\mu}(x + dx) = A^{\mu}(x) - \Gamma^{\mu}_{\lambda\nu} A^{\lambda} dx^{\nu} - \Gamma^{\mu}_{\lambda\nu,\gamma} A^{\lambda} dx^{\nu} dx^{\gamma} + \Gamma^{\mu}_{\lambda\nu} \Gamma^{\lambda}_{\alpha\gamma} A^{\alpha} dx^{\nu} dx^{\gamma}. \tag{S6.1.9}$$

Let $\vec{A}_{AB}$ be the vector $\vec{A}$ parallel transported from point A to B in the parallelogram shown in Fig. 6.8.

Using Eq. (S6.1.9) we get

$$A^{\mu}_{AB} = A^{\mu} - \Gamma^{\mu}_{\lambda\nu}(A)A^{\lambda} da^{\nu} - \Gamma^{\mu}_{\lambda\nu,\gamma}(A)A^{\lambda} da^{\nu} da^{\gamma} + \Gamma^{\mu}_{\lambda\nu}(A)\Gamma^{\lambda}_{\alpha\gamma}(A)A^{\alpha} da^{\nu} da^{\gamma}, \tag{S6.1.10}$$

where (A) means that the quantity is evaluated at the point A. Further parallel transport to C gives

$$A^{\mu}_{ABC} = A^{\mu}_{AB} - \Gamma^{\mu}_{\lambda\nu}(B)A^{\lambda} db^{\nu} - \Gamma^{\mu}_{\lambda\nu,\gamma}(B)A^{\lambda} db^{\nu} db^{\gamma} + \Gamma^{\mu}_{\lambda\nu}(B)\Gamma^{\lambda}_{\alpha\gamma}(B)A^{\alpha} db^{\nu} db^{\gamma}. \tag{S6.1.11}$$

Note that since the last parallel transport starts at the point B, we must evaluate the connection coefficients at point B. These connection coefficients deviate infinitesimally from their values in A. We want to express all the functions by their values in A. This is performed by means of a Taylor expansion. Since the connection coefficients in (S6.2.11) are all multiplied by differentials, it is sufficient to make a Taylor expansion to first order in the differentials. This gives

$$\Gamma^{\mu}_{\lambda\nu}(B) = \Gamma^{\mu}_{\lambda\nu}(A + d\vec{a}) = \Gamma^{\mu}_{\lambda\nu}(A) + \Gamma^{\mu}_{\lambda\nu,\gamma}(A)da^{\gamma} + \cdots \tag{S6.1.12}$$

Inserting Eqs. (S6.1.10) and (S6.1.12) into Eq. (S6.1.11) we get

$$\begin{aligned} A^{\mu}_{ABC} = A^{\mu}_{AB} &- \Gamma^{\mu}_{\lambda\nu} A^{\lambda} da^{\nu} - \Gamma^{\mu}_{\lambda\nu} A^{\lambda} db^{\nu} \\ &- \Gamma^{\mu}_{\lambda\nu,\gamma} A^{\lambda}\left(db^{\nu} da^{\gamma} + \frac{1}{2} da^{\nu} da^{\gamma} + \frac{1}{2} db^{\nu} db^{\gamma} \right) \\ &+ \Gamma^{\mu}_{\lambda\nu} \Gamma^{\lambda}_{\alpha\gamma} A^{\alpha}\left(db^{\nu} da^{\gamma} + \frac{1}{2} da^{\nu} da^{\gamma} + \frac{1}{2} db^{\nu} db^{\gamma} \right) \end{aligned} \tag{S6.1.13}$$

Parallel transport to C via D gives a similar expression, the only difference being that da and db are exchanged,

$$A^\mu_{ABC} = A^\mu_{AB} - \Gamma^\mu_{\lambda\nu} A^\lambda db^\nu - \Gamma^\mu_{\lambda\nu} A^\lambda da^\nu - \Gamma^\mu_{\lambda\nu,\gamma} A^\lambda \left(da^\nu db^\gamma + \tfrac{1}{2} db^\nu db^\gamma + \tfrac{1}{2} da^\nu da^\gamma \right)$$
$$+ \Gamma^\mu_{\lambda\nu} \Gamma^\lambda_{\alpha\gamma} A^\alpha \left(da^\nu db^\gamma + \tfrac{1}{2} db^\nu db^\gamma + \tfrac{1}{2} da^\nu da^\gamma \right) \,.$$

$$(S6.1.14)$$

The difference is

$$A^\mu_{ABC} - A^\mu_{ADC} = \left(\Gamma^\mu_{\lambda\nu,\gamma} - \Gamma^\mu_{\lambda\gamma,\nu} + \Gamma^\mu_{\alpha\gamma} \Gamma^\alpha_{\lambda\nu} - \Gamma^\mu_{\alpha\nu} \Gamma^\alpha_{\lambda\gamma} \right) A^\lambda da^\nu db^\gamma, \quad (S6.1.15)$$

where we have changed some of the summation indices in order to have the same indices on A, da and db in all the terms.

The difference $A^\mu_{ABC} - A^\mu_{ADC}$ corresponds to adding parallel transport of A from a to C via B and then adding the negative of the transport from A via D to C. The latter is the same as adding parallel transport from C via D to A. Hence the difference corresponds to parallel transporting A around the parallelogram. So the right-hand side is the change of the vector $\vec{A}$ by this round trip. Comparing with Eq. (6.17) we see that the quantities inside the parenthesis in equation (S6.1.15) are the components of the Riemann curvature tensor in coordinate basis. Hence,

$$\delta A^\mu = -R^\mu_{\lambda\nu\gamma} A^\lambda da^\nu db^\gamma. \quad (S6.1.16)$$

The curvature of a space makes a vector change when it is parallel transported around a closed curve in the space.

6.2 Curvature of the spatial simultaneity space of a rotating reference frame

Consider the 2. dimensional simultaneity space, $\hat{t} = \text{constant}$, $z = \text{constant}$, with line element (4.20),

$$dl^2 = dr^2 + \frac{r^2 d\theta^2}{1 - r^2\omega^2/c^2}. \quad (S6.2.1)$$

We shall calculate the Riemann curvature scalar of the 2-space described by this line-element. The 1-foms of an orthonormal basis are

$$\underline{\omega}^{\hat{r}} = \underline{d}r, \quad \underline{\omega}^{\hat{\phi}} = \frac{r}{\sqrt{1 - r^2\omega^2/c^2}} \underline{d}\phi. \quad (S6.2.2)$$

Taking the exterior derivative gives

$$\underline{d\omega}^{\hat{r}} = 0$$
$$\underline{d\omega}^{\hat{\phi}} = \left(\underline{d} \frac{r}{\sqrt{1-r^2\omega^2/c^2}} \right) \wedge \underline{d}\phi = \frac{1}{(1-r^2\omega^2/c^2)^{3/2}} \underline{d}r \wedge \underline{d}\phi = \frac{1}{r(1-r^2\omega^2/c^2)^{3/2}} \underline{\omega}^{\hat{r}} \wedge \underline{\omega}^{\hat{\phi}} \,.$$

$$(S6.2.3)$$

From Cartan's 1. structure equation

$$\underline{d\omega}^{\hat{\mu}} = -\underline{\Omega}^{\hat{\mu}}_{\hat{\nu}} \wedge \underline{\omega}^{\hat{\nu}}. \tag{S6.2.4}$$

We then have

$$\underline{\Omega}^{\hat{r}}_{\hat{\phi}} = f(r)\underline{\omega}^{\hat{\phi}} \quad \underline{\Omega}^{\hat{\phi}}_{\hat{r}} = \frac{1}{r(1 - r^2\omega^2/c^2)^{3/2}}\underline{\omega}^{\hat{\phi}} + g(r)\underline{\omega}^{\hat{r}}. \tag{S6.2.5}$$

The antisymmetry $\underline{\Omega}_{\nu\mu} = -\underline{\Omega}_{\mu\nu}$ now gives

$$f(r) = \frac{1}{r(1 - r^2\omega^2/c^2)^{3/2}}, \quad g(r) = 0. \tag{S6.2.6}$$

Hence the only non-vanishing connection forms are

$$\underline{\Omega}^{\hat{\phi}}_{\hat{r}} = \underline{\Omega}^{\hat{r}}_{\hat{\phi}} = \frac{1}{r(1 - r^2\omega^2/c^2)}\underline{\omega}^{\hat{\phi}}. \tag{S6.2.7}$$

We then calculate the curvature forms from Cartan's 2. structure equation

$$\underline{R}^{\hat{\mu}}_{\hat{\nu}} = \underline{d\Omega}^{\hat{\mu}}_{\hat{\nu}} + \underline{\Omega}^{\hat{\mu}}_{\hat{\lambda}} \wedge \underline{\Omega}^{\hat{\lambda}}_{\hat{\nu}}. \tag{S6.2.8}$$

This gives only one non-vanishing curvature form

$$\underline{R}^{\hat{\phi}}_{\hat{r}} = -\frac{3\omega^2/c^2}{(1 - r^2\omega^2/c^2)^2}\underline{\omega}^{\hat{\phi}} \wedge \underline{\omega}^{\hat{r}}. \tag{S6.2.9}$$

From

$$\underline{R}^{\hat{\mu}}_{\hat{\nu}} = R^{\hat{\mu}}_{\hat{\nu}\hat{\alpha}\hat{\beta}}\underline{\omega}^{\hat{\alpha}} \wedge \underline{\omega}^{\hat{\beta}} \tag{S6.2.10}$$

we then get the non-vanishing components of the Riemann curvature tensor

$$R^{\hat{r}}_{\hat{\phi}\hat{r}\hat{\phi}} = R^{\hat{\phi}}_{\hat{r}\hat{\phi}\hat{r}} = -\frac{3\omega^2/c^2}{(1 - r^2\omega^2/c^2)^2}. \tag{S6.2.11}$$

Contraction gives the components of the Ricci curvature tensor

$$R_{\hat{r}\hat{r}} = R^{\hat{\phi}}_{\hat{r}\hat{\phi}\hat{r}} = -\frac{3\omega^2/c^2}{(1 - r^2\omega^2/c^2)^2}, \quad R_{\hat{\phi}\hat{\phi}} = R^{\hat{r}}_{\hat{\phi}\hat{r}\hat{\phi}} = -\frac{3\omega^2/c^2}{(1 - r^2\omega^2/c^2)^2}. \tag{S6.2.12}$$

Finally the curvature scalar is the trace of the Ricci tensor

$$R = R_{\hat{\mu}}^{\hat{\mu}} = R_{\hat{r}}^{\hat{r}} + R_{\hat{\phi}}^{\hat{\phi}} = -\frac{6\omega^2/c^2}{\left(1 - r^2\omega^2/c^2\right)^2}. \tag{S6.2.13}$$

According to Eq. (6.55) this is equal to the Gaussian curvature of the surface, which is again equal to the product of the curvatures of two geodesic curves orthogonal to each other on the surface. When the Gaussian curvature is negative, these so-called principal curvatures have opposite signs, meaning that the geodesics curve in opposite ways, like they do on a saddle surface. Hence, the surface has negative curvature.

6.3 *The tidal force pendulum and the curvature of space*

(a) The particles of the pendulum are not moving geodesically. However, the forces on the particles from the rod connecting them, have no component in the direction normal to the rods, and are thus of no consequence for the period of the pendulum. We may therefore calculate the period of the tidal force pendulum using the equation of geodesic deviation,

$$\frac{D^2 \ell^\mu}{d\tau^2} + R^\mu_{\alpha\nu\beta} u^\alpha \ell^\nu u^\beta = 0. \tag{S6.3.1}$$

where τ is the proper time of the pendulum.

(b) We introduce co-moving geodesic normal coordinates with $\vec{u} = (1, 0, 0, 0)$, vanishing Christoffel symbols and origin at the mass centre of the pendulum. The coordinate basis of this coordinate system is orthonormal. The pendulum is permanently at rest at the surface of the Earth. Then it moves with non-relativistic velocity in a weak gravitational field. Hence, the proper time of the pendulum can be approximated by the coordinate time t of the Earth, and Eq. (S6.3.1) reduces to

$$\frac{d^2 \ell^{\hat{i}}}{dt^2} + R^{\hat{i}}_{\hat{0}\hat{j}\hat{0}} \ell^{\hat{j}} = 0. \tag{S6.3.2}$$

(c) From Fig. 6.6 is seen that the θ-component of the equation is

$$\frac{d^2 \ell^{\hat{\theta}}}{dt^2} + R^{\hat{\phi}}_{\hat{t}\hat{\phi}\hat{t}} \ell^{\hat{\phi}} \cos\theta - R^{\hat{r}}_{\hat{t}\hat{r}\hat{t}} \ell^{\hat{r}} \sin\theta = 0, \tag{S6.3.3}$$

where

$$\ell^{\hat{\theta}} = \ell\theta, \quad \ell^{\hat{\phi}} = \ell\sin\theta, \quad \ell^{\hat{r}} = \ell\cos\theta. \tag{S6.3.4}$$

Hence, for small angular amplitudes the equation of motion of the pendulum can be approximated by

$$\frac{\mathrm{d}^2\theta}{\mathrm{d}t^2} + \left(R_{\hat{\phi}\hat{t}\hat{\phi}\hat{t}} - R_{\hat{r}\hat{t}\hat{r}\hat{t}} \right)\theta = 0. \tag{S6.3.5}$$

This equation describes harmonic oscillations with period

$$T = \frac{2\pi}{\sqrt{R_{\hat{\phi}\hat{t}\hat{\phi}\hat{t}} - R_{\hat{r}\hat{t}\hat{r}\hat{t}}}}, \tag{S6.3.6}$$

showing that the period of the tidal force pendulum depends only upon the curvature of spacetime at its position.

At a distance R from the centre of a spherical body of mass M [6.1],

$$R_{\hat{\phi}\hat{t}\hat{\phi}\hat{t}} = GM/R^2, \quad R_{\hat{r}\hat{t}\hat{r}\hat{t}} = -2GM/R^2, \tag{S6.3.7}$$

giving

$$T = 2\pi\sqrt{\frac{R^3}{3GM}}, \tag{S6.3.8}$$

in agreement with the Newtonian calculation in Exercise 1.1 of the period of the tidal force pendulum.

[1] Misner, C.W., Thorne, K.S., Wheeler, J.A.: Gravitation, p. 821. Freeman, San Franscisco (1973)

Solutions for Chapter 7

7.1 Newtonian approximation of perfect fluid

(a) Thecovariant formulation of energy–momentum is that the covariant divergence of the energy–momentum tensor vanishes,

$$T^{\mu\nu}_{;\nu} = 0.$$

Writing out the covariant derivative we have

$$T^{\mu\nu}_{,\nu} + \Gamma^{\nu}_{\alpha\nu}T^{\mu\alpha} + \Gamma^{\mu}_{\alpha\nu}T^{\alpha\nu} = 0$$

In the Newtonian approximation we neglect the curvature of spacetime and introduce a Cartesian coordinate system with Minkowski metric and where all the components of the metric tensor are constant. Hence all the Christoffel symbols vanish, and the conservation laws reduce to

$$T^{\mu\nu}_{,\nu} = 0.$$

(b) In the Newtonian limit the components of the energy–momentum tensor are

$$T^{00} = \rho c^2, \ T^{0i} = \rho c v^i, \ T^{ij} = p\delta^{ij} + \rho v^i v^j.$$

Hence the time component of the conservation laws is

$$T^{0\nu}_{,\nu} = T^{00}_{,0} + T^{0i}_{,i} = \frac{\partial \rho}{\partial t} + \frac{\partial(\rho v^i)}{\partial x^i} = 0 \ \text{ or } \ \frac{\partial \rho}{\partial t} + \nabla \cdot (\rho \vec{v}) = 0.$$

This is the equation of continuity of the fluid.

(c) The spatial components of the conservation laws are

$$T^{i\nu}_{,\nu} = T^{i0}_{,0} + T^{ij}_{,j} = \frac{\partial(\rho v^i)}{\partial t} + \frac{\partial(p\delta^{ij} + \rho v^i v^j)}{\partial x^j}$$

$$= \frac{\partial(\rho v^i)}{\partial t} + \frac{\partial(\rho v^i v^j)}{\partial x^i} + \frac{\partial p}{\partial x_i} = 0$$

This equation can be simplified by using the equation of continuity in the form $\frac{\partial(\rho v^j)}{\partial x^j} = -\frac{\partial \rho}{\partial t}$. Hence we need to rewrite the second last term in the conservation equation as follows

$$\frac{\partial(\rho v^i v^j)}{\partial x^j} - \rho v^j \frac{\partial v^i}{\partial x^j} + v^i \frac{\partial(\rho v^j)}{\partial x^j} - \rho v^j \frac{\partial v^i}{\partial x^j} - v^i \frac{\partial \rho}{\partial t}.$$

Inserting this into the conservation equation, and using the product rule in the first term gives

$$v^i \frac{\partial \rho}{\partial t} + \rho \frac{\partial v^i}{\partial t} + \rho v^j \frac{\partial v^i}{\partial x^j} - v^i \frac{\partial \rho}{\partial t} + \frac{\partial p}{\partial x_i} = 0,$$

or

$$\rho \frac{\partial v^i}{\partial t} + \rho v^j \frac{\partial v^i}{\partial x^j} + \frac{\partial p}{\partial x_i} = 0.$$

This is usually written as

$$\rho \left(\frac{\partial v^i}{\partial t} + v^j \frac{\partial v^i}{\partial x^j} \right) = -\frac{\partial p}{\partial x_i} \ \text{ or } \ \rho \left(\frac{\partial \vec{v}}{\partial t} + (\vec{v} \cdot \nabla)\vec{v} \right) = -\nabla p.$$

which is the Euler equation of motion applied to a fluid in a region without a gravitational field, i.e. the Minkowski spacetime.

7.2   The energy–momentum tensor of LIVE

(a)   Weshall deduce the form of an energy–momentum tensor with Lorentz invariant
components as decomposed in orthonormal basis fields, i.e. with

$$T_{\hat{\mu}\hat{\nu}} = T_{\hat{\mu}'\hat{\nu}'} = \Lambda^{\hat{\alpha}}_{\hat{\mu}'}\Lambda^{\hat{\beta}}_{\hat{\nu}'} T_{\hat{\alpha}\hat{\beta}}$$

where $\Lambda^{\hat{\mu}}_{\hat{\mu}'}$ are the elements of a Lorentz transformation matrix. Let us first consider
a Lorentz transformation in the $x^{\hat{1}}$- direction,

$$\Lambda^{\hat{\mu}}_{\hat{\mu}'} = \begin{bmatrix} \gamma & \gamma v & 0 & 0 \\ \gamma v & \gamma & 0 & 0 \\ 0 & 0 & 1 & 0 \\ 0 & 0 & 0 & 1 \end{bmatrix}.$$

where $\gamma = \left(1 - v^2\right)^{-1/2}$, and we have used units so that $c = 1$. Inserting this in the
above equation leads to

$$v\left(T_{\hat{0}\hat{0}} + T_{\hat{1}\hat{1}}\right) + T_{\hat{0}\hat{1}} + T_{\hat{1}\hat{0}} = 0.$$

In a similar way transformation of $T_{\hat{0}\hat{1}}$ and $T_{\hat{1}\hat{0}}$ leads to

$$v\left(T_{\hat{0}\hat{1}} + T_{\hat{1}\hat{0}}\right) + T_{\hat{0}\hat{0}} + T_{\hat{1}\hat{1}} = 0.$$

It follows from these equations that

$$T_{\hat{0}\hat{0}} = -T_{\hat{1}\hat{1}}, \quad T_{\hat{0}\hat{1}} = -T_{\hat{1}\hat{0}}.$$

Transformation of $T_{\hat{0}\hat{2}}$ and $T_{\hat{1}\hat{2}}$ give, respectively

$$T_{\hat{0}\hat{2}} = \gamma\left(T_{\hat{0}\hat{2}} + vT_{\hat{1}\hat{2}}\right), \quad T_{\hat{1}\hat{2}} = \gamma\left(T_{\hat{1}\hat{2}} + vT_{\hat{0}\hat{2}}\right),$$

which demands that

$$T_{\hat{0}\hat{2}} = T_{\hat{1}\hat{2}} = 0.$$

In the same way one finds

$$T_{\hat{2}\hat{0}} = T_{\hat{2}\hat{1}} = T_{\hat{0}\hat{3}} = T_{\hat{1}\hat{3}} = T_{\hat{3}\hat{0}} = T_{\hat{3}\hat{1}} = 0.$$

Thus, as a result of Lorentz invariance of the components $T_{\hat{\mu}\hat{\nu}}$ under a Lorentz
transformation in the $x^{\hat{1}}$-direction the energy–momentum tensor must have the form

$$
T_{\hat{\mu}\hat{\nu}} =
\begin{bmatrix}
T_{\hat{0}\hat{0}} & T_{\hat{0}\hat{1}} & 0 & 0 \\
-T_{\hat{0}\hat{1}} & -T_{\hat{0}\hat{0}} & 0 & 0 \\
0 & 0 & T_{\hat{2}\hat{2}} & T_{\hat{2}\hat{3}} \\
0 & 0 & T_{\hat{3}\hat{2}} & T_{\hat{3}\hat{3}}
\end{bmatrix}.
$$

Demanding Lorentz invariance under a Lorentz transformation in the $x^{\hat{2}}$-direction gives the additional conditions

$$
T_{\hat{0}\hat{1}} = T_{\hat{1}\hat{0}} = T_{\hat{2}\hat{3}} = T_{\hat{3}\hat{2}} = 0, \quad T_{\hat{2}\hat{2}} = T_{\hat{0}\hat{0}}.
$$

Lastly Lorentz invariance under a Lorentz transformation in the $x^{\hat{3}}$-direction gives the additional condition

$$
T_{\hat{3}\hat{3}} = T_{\hat{0}\hat{0}}.
$$

It follows that the energy–momentum tensor for LIVE has the form

$$
T_{\hat{\mu}\hat{\nu}} = T_{\hat{0}\hat{0}}\,\mathrm{diag}(-1, 1, 1, 1) = T_{\hat{0}\hat{0}}\eta_{\hat{\mu}\hat{\nu}},
$$

where $\eta_{\hat{\mu}\hat{\nu}}$ are the components of the Minkowski metric. This is valid in orthonormal basis. Transforming to an arbitrary basis with metric $g_{\mu\nu}$, the energy–momentum tensor takes the form

$$
T_{\mu\nu} = T_{\hat{0}\hat{0}}g_{\mu\nu}.
$$

From the physical interpretation of the components of the energy–momentum tensor (including here the velocity of light) we have that $T_{\hat{0}\hat{0}} = -\rho_{\text{LIVE}}c^2$, where ρ_{LIVE} is the mass density of LIVE. Thus

$$
(T_{\text{LIVE}})_{\mu\nu} = -\rho_{\text{LIVE}}g_{\mu\nu}.
$$

(c) We now assume that LIVE can be described as a perfect fluid with energy–momentum tensor of the form (7.14),

$$
(T_{\text{LIVE}})_{\mu\nu} = \left(\rho_{\text{LIVE}} + \frac{p_{\text{LIVE}}}{c^2}\right)u_\mu u_\nu + p_{\text{LIVE}}g_{\mu\nu}.
$$

Since the energy–momentum tensor of LIVE is proportional to the metric tensor, the first term must vanish. Hence *the equation of state of* LIVE is

$$
p_{\text{LIVE}} = -\rho_{\text{LIVE}}c^2.
$$

(d) Einstein' field equations with a cosmological constant for empty space, (7.41), can be written

$$R_{\mu\nu} - \frac{1}{2}g_{\mu\nu}R = -\Lambda g_{\mu\nu}.$$

We comparing this with Einstein's field equations without a cosmological constant, (7.29),

$$R_{\mu\nu} - \frac{1}{2}g_{\mu\nu}R = \kappa T_{\mu\nu}.$$

Hence, a spacetime with LIVE, and nothing else, is mathematically equivalent to an Einstein space if LIVE is represented by an energy–momentum tensor

$$(T_{LIVE})_{\mu\nu} = -(\Lambda/\kappa)g_{\mu\nu}.$$

We see that the cosmological constant represents the density of LIVE

$$\Lambda = \kappa\rho_{LIVE}.$$

Solutions for Chapter 8

8.1 *Non-relativistic Kepler motion*

(a) According to Newtonian Lagrange dynamics the Lagrange function of a planet with mass m (considered as a particle) moving in the gravitational field of the Sun is it kinetic energy minus its potential energy. In spherical coordinates, (r, θ, ϕ), the kinetic energy is

$$T = \frac{1}{2}mv^2 = \frac{m}{2}\left(\dot{r}^2 + r^2\dot{\theta}^2 + r^2\sin^2\theta\dot{\phi}^2\right), \tag{S8.1}$$

where $\dot{r} = dr/dt$ and so forth. The potential energy is

$$V = -\frac{GMm}{r}, \tag{S8.2}$$

where M is the mass of the Sun. Hence, the Lagrange function is

$$L = \frac{m}{2}\dot{r}^2 + \frac{m}{2}r^2\dot{\theta}^2 + \frac{m}{2}r^2\sin^2\theta\dot{\phi}^2 L + \frac{GMm}{r}. \tag{S8.3}$$

Since L is does not depend upon ϕ, the momentum

$$p_\phi = \frac{\partial L}{\partial \dot\phi} = mr^2 \sin^2\theta \dot\phi \qquad (S8.4)$$

is a constant of motion. It is the angular momentum of the planet.

The Lagrange equation for θ,

$$\frac{\partial L}{\partial \theta} - \frac{d}{dt}\left(\frac{\partial L}{\partial \dot\theta}\right) = 0 \qquad (S8.5)$$

takes the form

$$mr^2 \sin\theta \cos\theta \dot\phi^2 - 2mr\dot r\dot\theta - mr^2\ddot\theta = 0 \qquad (S8.6)$$

Substituting for $\dot\phi$ from Eq. (S8.4) we get

$$\frac{2p_\phi^2 \cos\theta \dot\theta}{mr^2 \sin^3\theta} - 4mr^3\dot r\dot\theta^2 - 2mr^4\dot\theta\ddot\theta = 0 \qquad (S8.7)$$

which may be written

$$\frac{d}{dt}\left(mr^4\dot\theta^2 + \frac{p_\phi^2}{m\sin^2\theta}\right) = 0. \qquad (S8.8)$$

Integrating gives

$$mr^4\dot\theta^2 + \frac{p_\phi^2}{m\sin^2\theta} = K, \qquad (S8.9)$$

where K is a constant determined from the initial condition $\theta(0) = \pi/2$, $\dot\theta(0) = 0$. This gives $K = p_\phi^2/m$. Hence Eq. (S8.9) reduces to

$$mr^4\dot\theta^2 + \frac{p_\phi^2}{m\sin^2\theta} = \frac{p_\phi^2}{m}. \qquad (S8.10)$$

This may be written

$$m^2r^4 \sin^2\theta\dot\theta^2 = p_\phi^2(\sin^2\theta - 1) = -p_\phi^2 \cos^2\theta. \qquad (S8.11)$$

In general the left-hand side of this equation is positive, and the right-hand side is negative. Hence, the only possibility of having this equation fulfilled is that both sides vanish, which happens for $\theta = \pi/2$, $\dot\theta = 0$. This shows that the planet moves in a plane.

(b) We shall now consider motion in this plane. The Lagrange equation for r,

$$\frac{\partial L}{\partial r} - \frac{d}{dt}\left(\frac{\partial L}{\partial \dot{r}}\right) = 0 \tag{S8.12}$$

takes the form

$$\ddot{r} - r\dot{\theta}^2 - r\sin^2\theta\,\dot{\phi}^2 = -\frac{GM}{r^2}. \tag{S8.13}$$

which is the same as the r-component of Newton's 2. law applied to the planet. With $\theta = \pi/2$ the Lagrange Eq. (S8.4) for ϕ becomes

$$\dot{\phi} = \frac{p_\phi}{mr^2}. \tag{S8.14}$$

and the Lagrange Eq. (S8.13) for r reduces to

$$\ddot{r} + r\dot{\phi}^2 = -\frac{GM}{r^2}. \tag{S8.15}$$

In order to find the orbit equation we express the derivatives with respect to t in terms of derivatives with the respect to ϕ,

$$\dot{r} = \frac{dr}{dt} = \frac{dr}{d\phi}\frac{d\phi}{dt} = \dot{\phi}r' = \frac{p_\phi}{mr^2}r', \tag{S8.16}$$

$$\ddot{r} = \frac{d}{dt}\left(\frac{dr}{dt}\right) = \frac{p_\phi}{mr^2}\frac{d}{d\phi}\left(\frac{p_\phi}{mr^2}\frac{dr}{d\phi}\right) = \frac{p_\phi^2}{m^2 r^4}r'' - \frac{2p_\phi^2}{m^2 r^5}(r')^2. \tag{S8.17}$$

where $r' = dr/d\phi$. Introducing a new radial coordinate, $u = 1/r$, we get

$$r' = -\frac{u'}{u^2}, \quad r'' = -\frac{u''}{u^2} + \frac{2(u')^2}{u^3}. \tag{S8.18}$$

Inserting this into the expressions (S8.15) and (S8.16) for the radial velocity and acceleration, we get

$$\dot{r} = -\frac{p_\phi}{m}u', \quad \ddot{r} = -\frac{p_\phi^2}{m^2}u^2 u''. \tag{S8.19}$$

Substituting this into Eq. (S8.15) gives the orbit equation

$$u'' + u = \frac{GMm^2}{p_\phi^2}. \tag{S8.20}$$

This is an inhomogeneous equation for harmonic oscillations with general solution

$$u = \frac{GMm^2}{p_\phi^2}[1 + \varepsilon \sin(\phi - \phi_0)], \tag{S8.21}$$

where ε and ϕ_0 are integration constants. With the boundary condition $u(0) = GMm^2/p_\phi^2$ we have $\phi_0 = 0$.

The equation of an ellipse with major half axis a and eccentricity ε in polar coordinates with origin at one of the focal points has the form

$$u = \frac{1}{a(1 - \varepsilon^2)}(1 + \varepsilon \sin \phi). \tag{S8.22}$$

Hence with non-relativistic Kepler motion the planet moves along an elliptical path.

Equation (S8.20) has a particular solution representing circular motion with radius

$$R = \frac{p_\phi^2}{GMm^2}. \tag{S8.23}$$

Using Eq. (S8.14) the period of the motion is

$$T_0 = \frac{2\pi}{\dot{\phi}} = 2\pi \frac{mR^2}{p_\phi}. \tag{S8.24}$$

From Eq. (S8.23) we have

$$p_\phi^2 = GMm^2 R. \tag{S8.25}$$

Inserting this into Eq. (S8.24) gives

$$T_0 = 2\pi \frac{R^{3/2}}{\sqrt{GM}}. \tag{S8.26}$$

(c) We shall now consider the effect upon the motion of the planets of a flattening of the Sun. The gravitational potential of the Sun is then

$$V(r) = -\frac{GM}{r} - \frac{S}{r^3}. \tag{S8.27}$$

In this case the orbit Eq. (S8.20) is modified to

$$u'' + u = \frac{GMm^2}{p_\phi^2} + \frac{3Sm^2}{p_\phi^2}u^2. \tag{S8.28}$$

Again there is a particular solution representing circular motion. This time the radius $R = 1/u_0$ is given by

$$u_0^2 - \frac{p_\phi^2}{3Sm^2}u_0 + \frac{GM}{3S} = 0. \tag{S8.29}$$

Hence

$$p_\phi^2 = GMm^2R + \frac{3Sm^2}{R}. \tag{S8.30}$$

The period is still given by Eq. (S8.24). Inserting Eq. (S8.30) leads to

$$T = \frac{T_0}{\sqrt{1 + 3S/GMR^2}}. \tag{S8.31}$$

The flattening of the Sun makes the orbital period of the planets a little longer than in the spherical case.

(d)   We shall now assume that the planet moves along an elliptical path which deviates slightly from a circle, and calculate the precession of the perihelion due to the flattening of the Earth. Hence, $u = u_0 + u_1$ where $u_0 = 1/R$ and $|u_1| \ll |u_0|$. Inserting this into Eq. (S8.28) we get

$$u_1'' + u_0 + u_1 = \frac{m^2}{p_\phi^2}(GM + 3Su_0^2 + 6Su_0u_1 + 3Su_1^2). \tag{S8.32}$$

Calculating to 1. order in u_1 and utilizing that u_0 fulfils Eq. (S8.29) we get

$$u_1'' + u_1 = \frac{6Sm^2}{p_\phi^2 R}u_1, \tag{S8.33}$$

or

$$u_1'' + \omega^2 u_1 = 0, \quad \omega^2 = 1 - \frac{6Sm^2}{p_\phi^2 R}. \tag{S8.34}$$

This is an equation for harmonic oscillations for $6Sm^2 < p_\phi^2 R$. The solution with $u_1(0) = 0$ is

$$u_1 = \varepsilon u_0 \sin(\omega\phi), \tag{S8.35}$$

representing an ellipse with eccentricity ε. The period of the orbit is $2\pi/\omega$. For $\omega = 1$ the path is a closed ellipse with fixed orientation. For $\omega \neq 1$ the ellipse will rotate. There will be a precession of the ellipse, where the ellipse rotates an angle

$$\Delta\phi = 2\pi \left(\frac{1}{\omega} - 1 \right) \tag{S8.36}$$

for each travel around the sun. Using the expression (S8.34) for ω and inserting the expression (S8.30) for p_ϕ gives

$$\Delta\phi = 2\pi \left(\frac{\sqrt{GMR^2 + 3S}}{\sqrt{GMR^2 - 3S}} - 1 \right) \approx \frac{6\pi S}{GMR^2}. \tag{S8.37}$$

The quantity S is usually written

$$S = (1/2)J_2 GMR_{Sun}^2, \tag{S8.39}$$

where J_2 is a numerical factor determined by measurements, $J_2 = 3 \times 10^{-5}$ and $R_{Sun} = 7.0 \times 10^8 m$. For the planet Mercury the distance from the Sun is $R = 5.8 \times 10^{10} m$. This gives $\Delta\phi = 4.1 \times 10^{-8}$ radians per round trip, which corresponds to 3.5" per century, too small to explain the disagreement of 43" per century between the Newtonian prediction and measurements.

8.2 The Schwarzschild solution in isotropic coordinates

(a) In curvature coordinates the line-element of the Schwarzschild space time has the form

$$ds^2 = -\left(1 - \frac{R_S}{r} \right)c^2 dt^2 + \frac{dr^2}{1 - \frac{R_S}{r}} + r^2 d\Omega^2. \tag{S8.40}$$

We introduce a so-called radial isotropic coordinate, $\rho(r)$, where the line-element of the Schwarzschild spacetime takes the form

$$ds^2 = -\left(1 - \frac{R_S}{r(\rho)} \right)dt^2 + f^2(\rho)(d\rho^2 + \rho^2 d\Omega^2), \tag{S8.41}$$

This requires that

$$\frac{dr^2}{1 - \frac{R_S}{r}} = f^2(\rho)d\rho^2, \tag{S8.42}$$

and

$$r^2 = f^2(\rho)\rho^2. \tag{S8.43}$$

Dividing these equations by each other and taking the positive square root gives

$$\frac{d\rho}{\rho} = \frac{dr}{\sqrt{r^2 - R_S r}}.$$ (S8.44)

Integrating gives

$$\frac{\rho}{C} = r - \frac{R_S}{2} + \sqrt{r^2 - R_S r}.$$ (S8.45)

It is usual with isotropic coordinates, to choose $C = 1/2$. This gives

$$\rho = \frac{r}{2} - \frac{R_S}{4} + \frac{1}{2}\sqrt{r^2 - R_S r}.$$ (S8.46)

Solving this equation with respect to r gives

$$r = \frac{(\rho + R_S/4)^2}{4\rho}.$$ (S8.47)

From (S8.43) and (S8.47) we have

$$f(\rho) = \frac{r}{\rho} = \left(1 + \frac{R_S}{4\rho}\right)^2.$$ (S8.48)

Inserting the expressions (S8.47) and (S8.48) into Eq. (S8.41) gives the line-element of the Schwarzschild spacetime in isotropic coordinates

$$ds^2 = -\left(\frac{1 - R_S/4\rho}{1 + R_S/4\rho}\right)^2 dt^2 + \left(1 + \frac{R_S}{4\rho}\right)^4 (d\rho^2 + \rho^2 d\Omega^2).$$ (S8.49)

A series expansion of this line element to 1. order in R_S/ρ gives

$$ds^2 = -(1 - R_S/\rho)dt^2 + (1 + R_S/\rho)(d\rho^2 + \rho^2 d\Omega^2).$$ (S8.50)

(b) This coordinate system does only exist outside the horizon of a black hole. From Eq. (S8.46) we find that the value of the isotropic radius at the horizon is

$$\rho(R_S) = R_S/4.$$ (S8.51)

8.3 *Proper radial distance in the external Schwarzschild space*

The proper radial distance from the horizon to a point in the Schwarzschild spacetime with radial curvature coordinate r is

$$\hat{r} = \hat{r} = \int_{R_S}^{r} \frac{dr}{\sqrt{1 - r/R_S}} = r\sqrt{1 - \frac{R_S}{r}} + R_S \ln\left[\sqrt{\frac{r}{R_S}}\left(1 - \sqrt{1 - \frac{R_S}{r}}\right)\right]. \quad (S8.52)$$

8.4 The Schwarzschild–de Sitter metric

(a) We shall here generalize the Schwarzschild solution (8.35) to a space-time with a non-vanishing cosmological constant. We consider a static space-time with a spherically symmetric 3-space outside a spherical body (the Sun) with Schwarzschild radius R_S as described in orthonormal basis associated with curvature coordinates. Then the line-element has the form (8.1) and the non-vanishing components of the Einstein tensor are given in Eq. 8.15). Einstein's field equations are

$$E_{\hat{t}\hat{t}} = \Lambda, \quad E_{\hat{r}\hat{r}} = -\Lambda. \quad (S8.53)$$

Adding the equations and inserting the expressions (8.15) for the components of the Einstein tensor, leads to $\beta = -\alpha$ and

$$\frac{2}{r}e^{-2\beta}\beta' + \frac{1}{r^2}(1 - e^{-2\beta}) = \Lambda, \quad (S8.54)$$

which may be written as

$$\frac{d}{dr}\left[r\left(1 - e^{-2\beta}\right)\right] = \Lambda r^2. \quad (S8.55)$$

The general solution of this equation is

$$r\left(1 - e^{-2\beta}\right) = \frac{\Lambda}{3}r^3 + K, \quad (S8.56)$$

where K is an integration constant. Going to the Newtonian limit we find in the same manner as in a spacetime without a cosmological constant $K = R_S$. Thus, we get

$$e^{2\alpha} = e^{-2\beta} = 1 - \frac{R_S}{r} - \frac{\Lambda}{3}r^2, \quad (S8.57)$$

and the line-element takes the form

$$ds^2 = -\left(1 - \frac{R_S}{r} - \frac{\Lambda}{3}r^2\right)c^2dt^2 + \frac{dr^2}{1 - \frac{R_S}{r} - \frac{\Lambda}{3}r^2} + r^2d\Omega^2. \quad (S8.58)$$

Introducing a De Sitter radius $R_\Lambda = \sqrt{3/\Lambda}$, the line element takes the form

$$ds^2 = -\left(1 - \frac{R_S}{r} - \frac{r^2}{R_\Lambda^2}\right)c^2dt^2 + \frac{dr^2}{1 - \frac{R_S}{r} - \frac{r^2}{R_\Lambda^2}} + r^2d\Omega^2. \tag{S8.59}$$

(b) In globally empty space, but with LIVE, $R_\Lambda = 0$, and the line element reduces to

$$ds^2 = -\left(1 - \frac{r^2}{R_\Lambda^2}\right)c^2dt^2 + \frac{dr^2}{1 - \frac{r^2}{R_\Lambda^2}} + r^2d\Omega^2. \tag{S8.60}$$

The spacetime described by this line element is called *the DeSitter spacetime* after the Duch astronomer Willem De Sitter who found it as a solution of Einstein's field equations with a cosmological constant for empty space already in 1917.

Standard clocks at rest in this coordinate system show a time τ related to the coordinate time t by

$$d\tau = \sqrt{1 - r^2/R_\Lambda^2}\, dt. \tag{S8.61}$$

The coordinate clocks of the line-element (S8.60) are adjusted to go at the same rate as a standard clock at the origin independent of their position. Hence the standard clocks go at a slower rate the farther they are from the origin. This means that they are in a gravitational field pointing outwards from the origin. The time does not proceed at $r = R_\Lambda$.

Hence, the line-element has a singularity at $r = R_\Lambda$. In order to find its physical significance, we consider light moving radially. Along the world line of light, $ds^2 = 0$, and hence, for light emitted in the negative r-direction, i.e. towards an observer at the origin, we have

$$\frac{dr}{dt} = -\left(1 - \frac{r^2}{R_\Lambda^2}\right)c. \tag{S8.62}$$

Since the coordinate clocks go at the same rate as a standard clock at the origin, Eq. (S8.61) represents the velocity of light as measured by an observer at the origin. As measured by this observer light emitted towards him from $r = R_\Lambda$ does not proceed at all. This means the observer cannot receive information from the region outside the spherical surface with radius $r = R_\Lambda$. Hence this surface is a *horizon* for observers inside it.

In our universe observations indicate that the cosmological constant has a value which corresponds approximately to a density of LIVE equal to the critical density of the universe (see Chap. 12), $\Lambda \approx 10^{-52}$ m^{-2}, giving $R_\Lambda \approx 10^{26}$ m. This is approximately equal to ten billion light years.

8.5 The perihelion precession of Mercury and the cosmological constant

(a) Weshallnow find the orbit equation of a body moving in the gravitational field of the central mass distribution. It follows from Eq. (S8.45) that the Lagrange function of the body is

$$L = -\frac{1}{2}\left(1 - \frac{R_s}{r} - \frac{\Lambda}{3}r^2\right)\dot{t}^2 + \frac{\frac{1}{2}\dot{r}^2}{1 - \frac{R_s}{r} - \frac{\Lambda}{3}r^2} + \frac{1}{2}r^2\dot{\theta}^2 + \frac{1}{2}r^2\sin^2\theta\dot{\phi}^2 \quad (S8.63)$$

with constants of motion

$$p_t = \frac{\partial L}{\partial \dot{t}} = -\left(1 - \frac{R_s}{r} - \frac{\Lambda}{3}r^2\right)\dot{t}, \quad p_\phi = r^2\sin^2\theta\dot{\phi}. \quad (S8.64)$$

In the same way as in the case $\Lambda = 0$ the motion is planar, so we can put $\theta = \pi/2, \dot{\theta} = 0$. Then the 4-velocity identity takes the form

$$-\frac{p_t^2}{1 - \frac{R_s}{r} - \frac{\Lambda}{3}r^2} + \frac{\dot{r}^2}{1 - \frac{R_s}{r} - \frac{\Lambda}{3}r^2} + \frac{p_\phi^2}{r^2} = -1, \quad (S8.65)$$

giving

$$\dot{r}^2 = p_t^2 - \left(1 - \frac{R_s}{r} - \frac{\Lambda}{3}r^2\right)\left(1 + \frac{p_\phi^2}{r^2}\right). \quad (S8.66)$$

The orbit equation is found in the same way as with $\Lambda = 0$, and it turns out that Eq. (8.117) is generalized to

$$\frac{d^2u}{d\phi^2} + u = \frac{R_s}{2p_\phi^2} + \frac{3}{2}R_su^2 - \frac{\Lambda}{3p_\phi^2 u^3}. \quad (S8.67)$$

(c) As with $\Lambda = 0$ the solution of this equation represents a slowly rotating ellipse. We shall calculate the perihelion precession of the ellipse. A circular motion has a constant radius with $u = u_0$, where u_0 fulfils

$$u_0 = \frac{R_s}{2p_\phi^2} + \frac{3}{2}R_su_0^2 - \frac{\Lambda}{3p_\phi^2 u_0^3}, \quad (S8.68)$$

giving

$$p_\phi^2 = \frac{1}{3u_0^4}\frac{(3/2)R_su_0^3 - \Lambda}{1 - (3/2)R_su_0} \approx r_0^4\frac{3R_s/r_0^3 - \Lambda}{1 - 3R_s/2r_0}. \quad (S8.69)$$

The path of a planet is a perturbation of the circle, $u = u_0 + u_1$, where $|u_1| << |u_0|$. Inserting this into Eq. (S8.66) and calculating to 1. Order in u_1 gives

$$\frac{d^2 u_1}{d\phi^2} + u_1 = 2k u_0 u_1, \quad k = \frac{3}{2} R_S + \frac{\Lambda r_0^5}{2 p_\phi^2}. \tag{S8.70}$$

This is an equation of harmonic oscillations with general solution

$$u_1 = \varepsilon u_0 \cos[f(\phi - \phi_0)], \quad f = \sqrt{1 - 2k u_0} \approx 1 - k u_0. \tag{S8.71}$$

The period of the oscillations is $2\pi/f$. Equation (S8.71) describes a rotating ellipse. The precession angle per round trip is

$$\Delta\phi = 2\pi \left(\frac{1}{f} - 1 \right) \approx 2\pi k u_0 = \pi \left(\frac{3 R_S}{r_0} + \frac{\Lambda r_0^4}{p_\phi^2} \right). \tag{S8.72}$$

(c) Inserting the approximate value of p_ϕ^2 from Eq. (S8.69) gives

$$\Delta\phi \approx \pi \frac{3 R_S^2 + \left(r_0 - \frac{R_S}{2} \right) \Lambda r_0^3}{r_0 \left(R_S - \frac{\Lambda}{3} r_0^3 \right)}. \tag{S8.73}$$

Since $R_S \ll r_0$ the last term in the parenthesis in the numerator can be neglected. With the unit we use here the cosmological constant has dimension m^{-2}. Hence we can introduce a length characterizing the value of the cosmological constant, $R_\Lambda = 1/\sqrt{\Lambda}$. Inserting this into Eq. (S8.73) we get

$$\Delta\phi \approx \frac{3\pi}{r_0} \frac{3 R_S^2 R_\Lambda^2 + r_0^4}{3 R_S R_\Lambda^2 - r_0^3}. \tag{S8.74}$$

The cosmological constant has a value which corresponds approximately to a density of LIVE equal to the critical density of the universe (see Chap. 12), $\Lambda \approx 10^{-52}$ m^{-2}, giving $R_\Lambda \approx 10^{26}$ m. The Schwarzschild radius of the Sun is $R_S \approx 3 \times 10^3$ m, and Mercury's distance from the Sun $r_0 = 5.8 \times 10^{10}$ m. This gives $r_0^3 \ll R_S R_\Lambda^2$. Hence the expression for the perihelion precession of Mercury can be approximated by

$$\Delta\phi \approx 3\pi \frac{R_S}{r_0} + \pi \frac{r_0^3}{R_S R_\Lambda^2}. \tag{S8.75}$$

Here the first term is the general relativistic precession given in equation (8.145), and the last term is the contribution from the cosmological constant. Inserting numerical values gives $\Delta\phi \approx 5.1 \times 10^{-7} + 2 \times 10^{-23}$. Hence, the contribution from the cosmological constant is negligible.

8.6 Relativistic time effects and GPS

According to the general physical interpretation of a time-like spacetime interval the proper time interval measured on a standard clock moving in spacetime with a metric $g_{\mu\nu}$ is

$$d\tau = \left(-\frac{g_{\mu\nu}dx^\mu dx^\nu}{c^2}\right)^{1/2}. \tag{S8.76}$$

In order to find the magnitude of the relativistic effects it is sufficient to consider a satellite moving along a circular path with radius r_s in the equatorial plane. For a clock moving with velocity $v^\phi = r\frac{d\phi}{dt}$ in the Schwarzschild metric this gives

$$d\tau = \left(1 - \frac{R_S}{r_s} - \left(\frac{v^\phi}{c}\right)^2\right)^{1/2} dt, \tag{S8.77}$$

where t is the coordinate time as shown by coordinate clocks adjusted to go with a position independent rate of time. Here $R_S = 0.01$ m is the Schwarzschild radius of the Earth. The radius of the Earth is approximately $r_E = 6.4 \times 10^6$ m. Hence at the surface of the Earth $R_S/r_E = 1.6 \times 10^{-9}$.

A typical height and velocity of the GPS satellite are $r_s = 2 \times 10^4$ km and $v^\phi = 4 \times 10^3$ m/s. Hence the magnitudes of the terms inside the parenthesis are

$$\frac{R_S}{r_E} = 5 \times 10^{-10}, \quad \left(\frac{v^\phi}{c}\right)^2 = 1.7 \times 10^{-10}.$$

These are the relative magnitudes of the relativistic effects. Hence we can with good accuracy make a Taylor expansion to first order in R_S/r and $(v^\phi/c)^2$. To this order the difference between the proper time interval as measured on the satellite clocks and coordinate time interval corresponding to a coordinate interval Δt is

$$\Delta\tau = \Delta\tau_G + \Delta\tau_V = \frac{1}{2}\frac{R_S h}{r_E(r_E + h)}\Delta t - \frac{1}{2}\left(\frac{v^\phi}{c}\right)^2 \Delta t, \tag{S8.78}$$

where $\Delta\tau_G$ is the gravitational effect and $\Delta\tau_V$ the kinematical effect. Here h is the height of the satellite above the surface of the Earth. Inserting numerical values gives for a coordinate time interval corresponding to a day, $\Delta t = 24h = 8.6 \times 10^4 s$ that $\Delta\tau_G = 5.2 \times 10^{-5}s$ and $\Delta\tau_V = -7.0 \times 10^{-6}s$. The minus sign means that the kinematical effect makes the satellite clocks go slower due to the velocity-dependent time dilation. The gravitational effect acts in the opposite way, making the satellite clocks go faster that the clocks on the Earth. It should be noted that the gravitational effect is larger than the kinematical effect. The gravitational effect makes the satellite clock go faster that the Earth clock by 52 microseconds in 24 h, while the kinematical effect slows down the satellite clock with 7 microseconds.

To compute the position of an object by means of the GPS-system with a precision of 1m the GPS satellite clocks must measure time with a precision of one part in 10^{13}. Our calculation shows that to obtain such a precision both the special relativistic kinematical time dilation and the general relativistic gravitational time effect must be taken into account.

8.7 The photon sphere

We consider a photon travelling at constant radius. Due to the spherical symmetry there is no loss of generality if we choose to consider a photon which travels in the equatorial plane. We the put $dr = d\theta = 0$ and have $\theta = \pi/2$ in the Schwarzschild line-element. Furthermore since photons follow null-geodesic curves we have $ds = 0$. Hence the equation of motion of the considered photon is

$$\left(1 - \frac{R_S}{r}\right)c^2 dt^2 = r^2 d\phi^2, \tag{S8.79}$$

Giving

$$\left(\frac{d\phi}{dt}\right)^2 = \frac{c^2}{r^2}\left(1 - \frac{R_S}{r}\right). \tag{S8.80}$$

We now use the radial geodesic equation

$$\frac{d^2 r}{d\tau^2} + \Gamma^r_{\mu\nu} u^\mu u^\nu = 0. \tag{S8.81}$$

The only non-vanishing Christoffel symbols with an upper index r are

$$\Gamma^r_{tt} = \frac{c^2}{2}\left(1 - \frac{R_S}{r}\right)\frac{R_S}{r^2}, \quad \Gamma^r_{rr} = -\frac{R_S}{2\left(1 - \frac{R_S}{r}\right)r^2}, \quad \Gamma^r_{\theta\theta} = \Gamma^r_{\phi\phi} = -r\left(1 - \frac{R_S}{r}\right). \tag{S8.82}$$

Furthermore $\frac{dr}{d\tau} = \frac{d\theta}{d\tau} = \frac{d^2 r}{d\tau^2} = 0$. Hence the geodesic equation reduces to

$$\Gamma^r_{\phi\phi}\left(\frac{d\phi}{d\tau}\right)^2 = -\Gamma^r_{tt}\left(\frac{dt}{d\tau}\right)^2, \tag{S8.83}$$

or

$$\left(\frac{d\phi}{dt}\right)^2 = -\frac{\Gamma^r_{tt}}{\Gamma^r_{\phi\phi}}. \tag{S8.84}$$

Inserting the expressions (S8.82) for the Christoffel symbols leads to

$$\left(\frac{d\phi}{dt}\right)^2 = \frac{c^2 R_S}{2r^3}. \tag{S8.85}$$

Equations (S8.80) and (S8.85) give

$$\frac{c^2}{r^2}\left(1 - \frac{R_S}{r}\right) = \frac{c^2 R_S}{2r^3}. \tag{S8.86}$$

Solving this equation with respect to r gives the radius of the photon sphere in the Schwarzschild spacetime

$$r = \frac{3}{2} R_S. \tag{S8.87}$$

Solutions for Chapter 10

10.1 A spaceship falling into a black hole

(a) We shall consider a spaceship falling radially into a Schwarzschild black hole with the mass of the Sun, $M_\odot = 2,0 \times 10^{30}$ kg. The Schwarzschild radius of the black hole is

$$R_S = \frac{2GM_\odot}{c^2} - \frac{2 \cdot 6.67 \times 10^{-11} \text{ m}^3/\text{kgs}^2 \cdot 2,0 \times 10^{30} \text{ kg}}{\left(3.0 \times 10^8 \text{ m/s}\right)^2} = 3.0 \text{ km.} \tag{S10.1}$$

The line-element has the form

$$ds^2 = -\left(1 - \frac{R_S}{r}\right)c^2 dt^2 + \frac{dr^2}{1 - \frac{R_S}{r}} + r^2 d\Omega^2. \tag{S10.2}$$

Hence, the Lagrange function of the spaceship is

$$L = -\frac{1}{2}\left(1 - \frac{R_S}{r}\right)c^2\dot{t}^2 + \frac{1}{2}\frac{\dot{r}^2}{1 - \frac{R_S}{r}}. \tag{S10.3}$$

Since t is a cyclic coordinate,

$$p_t = \left(1 - \frac{R_S}{r}\right)c^2\dot{t} \tag{S10.4}$$

is a constant of motion. The 4 velocity identity takes the form

$$-\left(1 - \frac{R_S}{r}\right)c^2\dot{t}^2 + \frac{\dot{r}^2}{1 - \frac{R_S}{r}} = -c^2. \tag{S10.5}$$

Substituting for $\dot{t}$ from the previous equation we get

$$\dot{r}^2 = \frac{p_t^2}{c^2} - \left(1 - \frac{R_S}{r}\right)c^2. \tag{S10.6}$$

We consider a spaceship which falls from $r = r_0$ at $\tau = 0$, i.e. with $\dot{r}(r_0) = 0$. This gives

$$p_t^2 = \left(1 - \frac{R_S}{r_0}\right)c^4. \tag{S10.7}$$

Hence,

$$\frac{dr}{d\tau} = -c\sqrt{R_S}\sqrt{\frac{1}{r} - \frac{1}{r_0}}. \tag{S10.8}$$

Integration gives

$$\tau(r) = \frac{r_0}{c}\left(\sqrt{\left(1 - \frac{r}{r_0}\right)\frac{r}{R_S}} + \sqrt{\frac{r_0}{R_S}}\arctan\sqrt{\frac{r_0}{r} - 1}\right). \tag{S10.9}$$

The proper time taken to fall from the Earth to the Sun is

$$\tau(R_S) = \frac{r_0}{c}\left(\sqrt{1 - \frac{R_S}{r_0}} + \sqrt{\frac{r_0}{R_S}}\arctan\sqrt{\frac{r_0}{R_S} - 1}\right). \tag{S10.10}$$

Since $R_S \ll r_0$ a good approximation is

$$\tau(R_S) \approx \frac{\pi}{2}\frac{r_0}{c}\sqrt{\frac{r_0}{R_S}}, \tag{S10.11}$$

where we have used that $\lim_{x \to \infty} \arctan x = \pi/2$ Here r_0 is the distance from the Earth to the Sun, i.e. $r_0 = 1.5 \times 10^8$ km. Inserting numerical values gives $\tau(R_S) \approx 64$

days. This is the same as the Newtonian result (1.11) for the case considered here, where we neglect the gravitational field of the Earth.

(b) Let A be an observer on the spaceship and B a stationary observer at a distance r_B from the Sun. Both A and B emit signals with angular frequency ω. A receives the signals from B with an angular frequency ω_A, and B receives the signals from A with angular frequency ω_B.

We shall find ω_A and ω_B by utilizing an inertial observer C instantaneously at rest in the coordinate system at r_A. This observer is at rest relatively to B. The observer C receives signals from A with an angular frequency ω_{CA} and from B with ω_{CB}.

Let us first consider the signals from A to B via C. Since A moves away from C the signal from A to C will be redshifted due to the kinematical Doppler effect, but there is no gravitational frequency shift since A and C is at the same position during their exchange of signals. Hence

$$\omega_{CA} = \sqrt{\frac{1 - |v|/c}{1 + |v|/c}}\omega. \tag{S10.12}$$

The velocity as measured by standard clocks in a local inertial frame is given in Eq. (S10.8).

We then consider the signal from C to B. Then there is no kinematical Doppler effect since B and C are at rest relatively to each other. But there is a gravitational frequency shift,

$$\omega_B = \sqrt{\frac{(g_{tt})_C}{(g_{tt})_B}}\omega_{CA} = \sqrt{\frac{1 - R_S/r_A}{1 - R_S/r_B}}\omega_{CA}. \tag{S10.13}$$

Inserting the expression (S10.8) for the velocity into Eq. (S10.12) and combining with Eq. (S10.13) gives

$$\omega_B = \sqrt{\frac{1 - R_S/r_A}{1 - R_S/r_B}}\sqrt{\frac{1 - \sqrt{R_S}\sqrt{\frac{1}{r_A} - \frac{1}{r_0}}}{1 + \sqrt{R_S}\sqrt{\frac{1}{r_A} - \frac{1}{r_0}}}}\omega. \tag{S10.14}$$

We then consider the signals from B to the spaceship A in a similar manner. The signal is first sent from B to C. Then there is no kinematical Doppler effect, only a gravitational blue shift, since these signals move downwards in the gravitational field of the sun,

$$\omega_{CB} = \sqrt{\frac{(g_{tt})_B}{(g_{tt})_A}}\omega = \sqrt{\frac{1 - R_S/r_B}{1 - R_S/r_A}}\omega. \tag{S10.15}$$

Then the signals are emitted further from C to A. Since A moves away from C there is now a redshift,

$$\omega_A = \sqrt{\frac{1 - \sqrt{R_S}\sqrt{\frac{1}{r_A} - \frac{1}{r_0}}}{1 + \sqrt{R_S}\sqrt{\frac{1}{r_A} - \frac{1}{r_0}}}} \omega_{CB}. \tag{S10.16}$$

Combining the last two expressions we get

$$\omega_A = \sqrt{\frac{1 - \sqrt{R_S}\sqrt{\frac{1}{r_A} - \frac{1}{r_0}}}{1 + \sqrt{R_S}\sqrt{\frac{1}{r_A} - \frac{1}{r_0}}}} \sqrt{\frac{1 - R_S/r_B}{1 - R_S/r_A}} \omega. \tag{S10.17}$$

In the limit that the spaceship falls from an infinitely far away position we get

$$\lim_{r_0 \to \infty} \omega_A = \frac{\sqrt{1 - \frac{R_S}{r_B}}}{1 + \sqrt{\frac{R_S}{r_A}}} \omega. \tag{S10.18}$$

In this case there is a redshift, which means that the kinematical redshift then dominates over the gravitational blue shift.

10.2 Kinematics in the Kerr spacetime

(a) We shall consider light moving in negative and positive direction of ϕ in the Kerr spacetime in the equatorial plane $\theta = \pi/2$, as described in Boyer–Lindquist coordinates. Using that $ds^2 = 0$ along the world line of light, we then get for the angular velocity of light, $\omega_L = d\phi/dt$, from the line element (10.26),

$$\omega_L^2 - 2\omega\omega_L + \omega^2 - e^{2\nu - 2\psi} = 0. \tag{S10.19}$$

where

$$\omega = \frac{R_S a}{r^3 + a^2(r - 2R_S)}, \quad e^{\nu - \psi} = \frac{r\sqrt{r(r - R_S) + a^2}}{r^3 + a^2(r - 2R_S)}. \tag{S10.20}$$

The solution of Eq. (S10.19) is

$$\omega_L^\pm = \omega \pm e^{\nu - \psi} = \frac{R_S a \pm r\sqrt{r(r - R_S) + a^2}}{r^3 + a^2(r - 2R_S)}. \tag{S10.21}$$

The velocity of light in the equatorial plane of the Kerr spacetime is $c_{K\phi} = r\omega_L$ which approaches 1, or c, when $r \to \infty$. In the special case without rotation, $a = 0$, which represents the Schwarzschild spacetime, the coordinate velocity of light reduces to

$$c_{S\phi} = \sqrt{1 - \frac{R_S}{r}}. \tag{S10.22}$$

which is isotropic. But in the Kerr spacetime the velocity is different in the positive and the negative ϕ-direction. The time difference for light sent around the equator in the positive and negative ϕ-direction is

$$\Delta t = 2\pi \left(\frac{1}{\omega_L^-} - \frac{1}{\omega_L^+} \right) = \frac{4\pi R_S a}{r - R_S} \frac{r^3 + (r - 2R_S)a^2}{r^3 + (r + R_S)a^2}. \tag{S10.23}$$

This time difference vanishes for the non-rotating case, $a = 0$, and in general when $r \to \infty$.

(b) We are still describing particles and observers in the equatorial plane of the Kerr space. In the following a *ZAMO* in the Kerr spacetime will be describing particles with fixed r- and θ-coordinates. We introduce an orthonormal basis $(\vec{e}_{\hat{t}}, \vec{e}_{\hat{r}}, \vec{e}_{\hat{\theta}}, \vec{e}_{\hat{\phi}})$, where $\vec{e}_{\hat{t}}$ is the 4-velocity of the *ZAMO*. It follows from the form of the line element

$$ds^2 = -e^{2\nu}dt^2 + e^{2\mu}dr^2 + e^{2\lambda}d\theta^2 + e^{2\psi}(d\phi - \omega dt)^2, \tag{S10.24}$$

that the dual basis 1-forms are

$$\begin{aligned} &\underline{\omega}^{\hat{t}} = e^\nu \underline{\omega}^t, \quad \underline{\omega}^{\hat{r}} = e^\mu \underline{\omega}^r, \\ &\underline{\omega}^{\hat{\theta}} = e^\lambda \underline{\omega}^\theta, \quad \underline{\omega}^{\hat{\phi}} = e^\psi(\underline{\omega}^\phi - \omega\underline{\omega}^t). \end{aligned} \tag{S10.25}$$

Using the fundamental contraction

$$\underline{\omega}^\mu(\vec{e}_\nu) = \delta^\mu_\nu \tag{S10.26}$$

It follows that the basis-vectors of an orthonormal basis associated with the ZAMO are

$$\begin{aligned} &\vec{e}_{\hat{t}} = e^{-\nu}(\vec{e}_t + \omega\vec{e}_\phi), \quad \vec{e}_{\hat{r}} = e^{-\mu}\vec{e}_r, \\ &\vec{e}_{\hat{\theta}} = e^{-\lambda}\vec{e}_\theta, \quad \vec{e}_{\hat{\phi}} = e^{-\psi}\vec{e}_\phi. \end{aligned} \tag{S10.27}$$

We shall now find the physical velocity as measured by a ZAMO of a particle with 4-velocity components

$$U^\mu = (\dot{t}, \dot{\phi}) = \dot{t}(1, \Omega), \quad \Omega = \frac{d\phi}{dt} \tag{S10.28}$$

in the Boyer–Lindquist coordinate system. Here the dot denotes differentiation with respect to the proper time of the particle. We shall find the components of its velocity with reference to the orthonormal basis above, carried by the ZAMO. Using the tensor transformation of the vector components we get

$$U^{\hat{t}'} = e^\nu U^t = e^\nu \dot{t}, \quad U^{\hat{\phi}'} = e^\psi (U^\phi - \omega U^t) = e^\psi \dot{t}(\Omega - \omega). \tag{S10.29}$$

The physical velocity of the particle measured by the ZAMO is

$$v^{\hat{\phi}'} = \frac{U^{\hat{\phi}'}}{U^{\hat{t}'}} = e^{\psi - \nu}(\Omega - \omega). \tag{S10.30}$$

A particle at rest in the coordinate system has $\Omega = 0$ and hence the velocity

$$v_0^{\hat{\phi}'} = -e^{\psi - \nu}\omega = -\frac{R_S a}{r\sqrt{r(r - R_S) + a^2}}. \tag{S10.31}$$

We shall later need the factor

$$\hat{\gamma}' = \frac{1}{\sqrt{1 - \left(v_0^{\hat{\phi}'}\right)^2}} = \sqrt{\frac{1 + \left(1 - \frac{R_S}{r}\right)^{-1}\frac{a^2}{r^2}}{1 + \left(1 + \frac{R_S}{r}\right)\frac{a^2}{r^2}}}. \tag{S10.32}$$

In the non-rotating case, with $a = 0$, the particle is at rest relative to the ZAMO; i.e. observers at rest in the coordinate system are ZAMOs in the Schwarzschild spacetime. But in the Kerr spacetime observers at rest in the coordinate system are not ZAMOs. At $r = R_S$ the velocity of a particle at rest in the coordinate system is $v_0^{\hat{\phi}} = -1$, i. the particle moves with the velocity of light relative to the ZAMO.

(c)   We shall introduce an orthonormal basis co-moving with observers at rest in the Boyer–Lindquist coordinate system. The vectors of this orthonormal basis are

$$\vec{e}_{\hat{t}} = (-g_{tt})^{-1/2}\vec{e}_t, \quad \vec{e}_{\hat{i}} = (\gamma_{ii})^{-1/2}[\vec{e}_i - (g_{it}/g_{tt})\vec{e}_t], \tag{S10.33}$$

where

$$\gamma_{ii} = g_{ii} - g_{it}^2/g_{tt}. \tag{S10.34}$$

are the components of the spatial metric tensor. Using that

$$g_{tt} = e^{2\psi}\omega^2 - e^{2\nu} = -e^{2\nu}\left(1 - \left(v_0^{\hat{\phi}'}\right)^2\right) = -e^{2\nu}/\hat{\gamma}'^2,$$

$$g_{rr} = e^{2\mu}, \quad g_{\phi\phi} = e^{2\psi}, \quad g_{\theta\theta} = e^{2\lambda}, \quad g_{t\phi} = -e^{2\psi}\omega \qquad \text{(S10.35)}$$

we get

$$\gamma_{rr} = g_{rr} = e^{2\mu} \quad \gamma_{\theta\theta} = g_{\theta\theta} = e^{2\lambda},$$

$$\gamma_{\phi\phi} = g_{\phi\phi} - \frac{g_{\phi t}^2}{g_{tt}} = e^{2\psi} + \frac{\omega^2 e^{2\psi}}{e^{2\nu}\hat{\gamma}^{-2}} = e^{2\psi}\left(1 + \hat{\gamma}^2\left(v_0^{\hat{\phi}'}\right)^2\right) = \hat{\gamma}^2 e^{2\psi}. \quad \text{(S10.36)}$$

Inserting the expressions (S10.35) and (S10.36) into Eq. (S10.33) gives

$$\vec{e}_{\hat{t}} = \hat{\gamma}e^{-\nu}\vec{e}_t, \quad \vec{e}_{\hat{r}} = e^{-\mu}\vec{e}_r,$$

$$\vec{e}_{\hat{\theta}} = e^{-\lambda}\vec{e}_r, \quad \vec{e}_{\hat{\phi}} = \hat{\gamma}^{-1}e^{-\psi}\vec{e}_\phi + \hat{\gamma}e^{-\nu}v_0^{\hat{\phi}'}\vec{e}_t. \qquad \text{(S10.37)}$$

The vector $\vec{e}_{\hat{t}}$ is the 4-velocity of an observer at rest in the BL-coordinate system, i.e. of a *static* observer. Using the fundamental contraction (S10.26) we get the corresponding basis 1-forms,

$$\underline{\omega}^{\hat{t}} = \hat{\gamma}^{-1}e^\nu\underline{\omega}^t - \hat{\gamma}e^\psi v_0^{\hat{\phi}'}\underline{\omega}^\phi, \underline{\omega}^{\hat{r}} = e^\mu\underline{\omega}^r, \underline{\omega}^{\hat{\theta}} = e^\lambda\underline{\omega}^\theta, \underline{\omega}^{\hat{\phi}} = \hat{\gamma}e^\psi\underline{\omega}^\phi. \quad \text{(S10.38)}$$

(d) The transformation matrix leading from the coordinate basis vectors in the Boyer–Lindquist system to the ZAMO can be read off from equation (S10.37),

$$\left(M^\mu_{\hat{\mu}}\right) = \begin{pmatrix} \hat{\gamma}e^{-\nu} & 0 & 0 & \hat{\gamma}e^{-\nu}v_0^{\hat{\phi}'} \\ 0 & e^{-\mu} & 0 & 0 \\ 0 & 0 & e^{-\lambda} & 0 \\ 0 & 0 & 0 & \hat{\gamma}e^\psi \end{pmatrix}. \qquad \text{(S10.39)}$$

The inverse transformation matrix is

$$\left(M^{\hat{\mu}}_{\mu}\right) = \begin{pmatrix} \hat{\gamma}^{-1}e^\nu & 0 & 0 & -\hat{\gamma}e^\psi v_0^{\hat{\phi}'} \\ 0 & e^\mu & 0 & 0 \\ 0 & 0 & e^{-\lambda} & 0 \\ 0 & 0 & 0 & \hat{\gamma}e^\psi \end{pmatrix}. \qquad \text{(S10.40)}$$

The components in the orthonormal basis (S10.37) of the Boyer–Linquist system of the 4-velocity of aa particle with the 4-velocity (S10.28) are found by applying the transformation (S10.40), $U^{\hat{\mu}} = M^{\hat{\mu}}_{\mu}U^\mu$, which gives,

$$U^{\hat{t}} = \hat{\gamma}^{-1}e^{\nu}U^t - \hat{\gamma}e^{\psi}v_0^{\hat{\phi}'}U^{\phi} = \hat{\gamma}^{-1}e^{\nu}\dot{t} - \hat{\gamma}e^{\psi}v_0^{\hat{\phi}'}\Omega\dot{t}, \quad U^{\hat{\phi}} = \hat{\gamma}e^{\psi}U^{\phi} = \hat{\gamma}e^{\psi}\Omega\dot{t}.$$
$$\text{(S10.41)}$$

The physical velocity of the particle as observed by an observer at rest in the Boyer–Lindquist coordinate system is

$$v^{\hat{\phi}} = \frac{U^{\hat{\phi}}}{U^{\hat{t}}} = \frac{e^{\psi-\nu}\Omega}{\hat{\gamma}^{-2} + e^{2\psi-2\nu}\omega\Omega}. \tag{S10.42}$$

Using Eqs. (S10.30) and (S10.31) this can be written as

$$v^{\hat{\phi}} = \frac{v^{\hat{\phi}'} - v_0^{\hat{\phi}'}}{1 - v^{\hat{\phi}'}v_0^{\hat{\phi}'}}, \tag{S10.43}$$

which is the relativistic formula for velocity addition.

10.3 A gravitomagnetic clock effect

Following [10.1] and [10.2] we consider two clocks moving freely in opposite directions in the equatorial plane of the Kerr spacetime outside a rotating body. The clocks move along a path with $r = $ constant and $\theta = \pi/2$.

(a) In this case the radial geodesic equation

$$\frac{d^2r}{d\tau^2} + \Gamma^r_{\alpha\beta}\frac{dx^{\alpha}}{d\tau}\frac{dx^{\beta}}{d\tau} = 0 \tag{S10.44}$$

reduces to

$$\Gamma^r_{tt}dt^2 + 2\Gamma^r_{\phi t}d\phi dt + \Gamma^r_{\phi\phi}d\phi^2 = 0.. \tag{S10.45}$$

Inserting the expression (5.65) for the Christoffel symbols we obtain

$$g_{tt,r}\left(\frac{dt}{d\phi}\right)^2 + 2g_{\phi t,r}\frac{dt}{d\phi} + g_{\phi\phi,r} = 0. \tag{S10.46}$$

From the line element (10.28) we see that in the equatorial plane of the Kerr spacetime the non-vanishing components of the metric are

$$g_{tt} = -\left(1 - \frac{R_S}{r}\right), \quad g_{\phi t} = \frac{R_S a}{r}, \quad g_{\phi\phi} = r^2 + a^2 + \frac{R_S a^2}{r}. \tag{S10.47}$$

Performing the differentiations and inserting the result into equation (S10.46) gives

$$\left(\frac{dt}{d\phi}\right)^2 - 2a\frac{dt}{d\phi} + a^2 - \frac{2r^3}{R_S} = 0. \tag{S10.48}$$

The solution of this equation is

$$\frac{dt}{d\phi} = a \pm \frac{1}{\omega_K}, \tag{S10.49}$$

where ω_K is the Keplerian angular frequency,

$$\omega_K^2 = \frac{2R_S}{r^3} = \frac{GM}{r^3}, \tag{S10.50}$$

and equation (S10.50) is an expression of Kepler's 3. law.

(b) To find the gravitomagnetic clock effect one integrates equation (S10.49) over 2π for co-rotating and counter-rotating clocks. This gives

$$t_\pm = T_K \pm 2\pi a, \tag{S10.51}$$

where

$$T_K = \frac{2\pi}{\omega_K} \tag{S10.52}$$

is the Keplerian period. The time difference in the travel time around the central rotating body is

$$\Delta t = 4\pi a = 4\pi J/M, \tag{S10.53}$$

where J is the angular momentum of the central body. Inserting the velocity of light we have,

$$\Delta t = 4\pi J/Mc^2. \tag{S10.54}$$

The mass of the Earth is $m = 6 \times 10^{26}$ kg and its angular momentum $J = 10^{34}$ kg m^2/s. Hence outside the Earth the travel time difference for clocks travelling freely around the Equatior in opposite directions is 2×10^{-7} s.

References

10.1 Cohen, J.M., Mashoon, B.: Standard clocks, interferometry, and gravitomagnetism. Phys. Lett. **A181**, 353–358 (1993).

10.2. Faruque, S.B.: Gravitomagnetic clock effect in the orbit of a spinning particle orbiting the Kerr black hole. Phy. Lett. **A327**, 95–97 (2004).

Solutions for Chapter 11

11.1 *The Schwarzschild-de Sitter metric*

In the static spherically symmetric case the line element can be written as in Eq. (8.1). Then there are only 2 independent field equations, which can be taken to be the $\hat{t}\hat{t}$−and $\hat{r}\hat{r}$−equations. The components of the Einstein curvature tensor are given in equation (8.15). Then the $\hat{t}\hat{t}$−and $\hat{r}\hat{r}$−field equations (7.43) with a cosmological constant for empty space take the form

$$\frac{2}{r}e^{-2\beta}\beta' + \frac{1}{r^2}(1 - e^{-2\beta}) = \Lambda, \tag{S11.1}$$

$$\frac{2}{r}e^{-2\beta}\alpha' - \frac{1}{r^2}(1 - e^{-2\beta}) = -\Lambda. \tag{S11.2}$$

Adding the equations we get with a suitable coordinate condition $\beta = -\alpha$. Equation (S11.1) can be written as

$$\frac{d}{dr}\left[r\left(1 - e^{-2\beta}\right)\right] = \Lambda. \tag{S11.3}$$

The general solution of this equation is

$$e^{-2\beta} = 1 - \frac{K}{r} - \frac{\Lambda}{3}r^2. \tag{S11.4}$$

Demanding that the solution reduces to the exterior Schwarzschild solution with vanishing cosmological constant, we get $K = R_S$. Introducing the length $R_H = \sqrt{3/\Lambda}$ the line element can then be written as

$$ds^2 = -\left(1 - \frac{R_S}{r} + \frac{r^2}{R_H^2}\right)c^2 dt^2 + \frac{dr^2}{1 - \frac{R_S}{r} + \frac{r^2}{R_H^2}} + r^2 d\Omega^2, \tag{S11.5}$$

which describes the Schwarzschild-De Sitter spacetime.

11.2 *A spherical domain wall in empty space described by the Israel formalism*

Theenergy–momentum tensor of a domain wall is

$$S_{ij} = -\sigma^{(3)}g_{ij}. \tag{S11.6}$$

Hence, the only non-vanishing mixed components of the energy–momentum tensor are

$$S_\theta^\theta = S_\phi^\phi = -\sigma. \tag{S11.7}$$

In empty space the equation of continuity of the domain wall reduces to (11.58) which then gives

$$\frac{d\sigma}{d\tau} = 0 \quad \sigma = \text{constant} \tag{S11.8}$$

Hence, the mass density of the domain wall remains constant during radial motion even if the area of the domain wall changes.

Inserting Eq. (S11.6) into Eq. (11.55) gives

$$[K_{\theta\theta}] - \kappa\left(-\sigma^{(3)}g_{\theta\theta} + \frac{3}{2}\sigma^{(3)}g_{\theta\theta}\right) - \frac{\kappa}{2}\sigma R^2. \tag{S11.9}$$

This equation has the same form as Eq. (11.74) for dust, but there σR^2 =constant, while here σ =constant. The equation of motion (11.89) for dust is still valid, but with M_S replaced by $4\pi\sigma R^2$,

$$M = \left(\sqrt{1 + \dot{R}^2} - 2\pi\sigma R\right)4\pi\sigma R^2. \tag{S11.10}$$

A static spherical domain wall has

$$M = (1 - 2\pi\sigma R)4\pi\sigma R^2, \tag{S11.11}$$

which may be written

$$R_S = R_{SS}(1 - R_{SS}/4R). \tag{S11.12}$$

Here R_S is the Schwarzschild radius of the domain wall as measured by an observer far from it, R_{SS} is Schwarzschild radius as measured by an observer at the wall, and R is its radius.

Solutions for Chapter 12

12.1 *Gravitational collapse*

(a) We shall here study particles falling freely from rest at a position infinitely far from a black hole with Schwarzschild radius R_S. In Exercise 10.1 we considered a spaceship falling from a position r_0 outside a black hole and into the hole. Hence we shall here consider the limit $r_0 \to \infty$.

We found the following expression for the radial component of the 4-velocity

$$\dot{r}^2 = \frac{p_t^2}{c^2} - \left(1 - \frac{R_S}{r}\right)c^2. \tag{S12.1}$$

The boundary condition $\dot{r}(\infty) = 0$ gives $p_t = c^2$. Hence

$$\dot{r} = -c\sqrt{\frac{R_S}{r}}. \tag{S12.2}$$

New integration with $r(0) = \rho$ gives

$$\tau = \frac{2}{3c\sqrt{R_S}}\left(\rho^{3/2} - r^{3/2}\right). \tag{S12.3}$$

Inserting this expression into Eq. (S12.2) gives

$$dt = -c\sqrt{R_S}\,\frac{\sqrt{r}}{r - R_S}\,dr. \tag{S12.4}$$

Integration with the boundary condition $t(0) = \tau$ gives

$$ct = c\tau + R_S \ln\frac{\sqrt{r/R_S} + 1}{\left|\sqrt{r/R_S} - 1\right|} - 2\sqrt{R_S r}. \tag{S12.5}$$

Solving the expression (S12.3) with respect to r gives

$$r = \left(\rho^{3/2} - \frac{3c\sqrt{R_S}}{2}\tau\right)^{2/3} = \rho\left(1 - \frac{3c\sqrt{R_S}}{2}\tau\rho^{-3/2}\right). \tag{S12.6}$$

Inserting this into Eq. (S12.5) gives the relationship between the proper time of a freely falling particle and the coordinate time, which is the same as the proper time of an observer at rest far away from the black hole. Note that ρ has a constant value for a freely falling particle.

(b) We shall now assume that there is Schwarzschild spacetime outside the black hole, and find the form of the spacetime line-element, using τ as time coordinate and ρ as radial coordinate. The radial coordinate ρ is commoving with the freely falling particles. Solving Eq. (S12.6) with respect to ρ gives

$$\rho^{3/2} = r^{3/2} + \frac{3}{2}c\sqrt{R_S}\tau. \tag{S12.7}$$

We see that ρ increases with τ. As seen from the freely falling particles points with fixed Schwarzschild coordinates move outwards.

Taking the differential of t, using Eqs. (S12.5) and (S12.4) we get

$$dt = \frac{\partial t}{\partial \tau}d\tau + \frac{\partial t}{\partial r}dr = d\tau - c\sqrt{R_S}\,\frac{\sqrt{r}}{r - R_S}\,dr. \tag{S12.8}$$

Inserting this into the Schwarzschild line-element (8.35) gives

$$ds^2 = -\left(1 - \frac{R_S}{r}\right)c^2 d\tau^2 + 2\sqrt{\frac{R_S}{r}}cd\tau dr + dr^2 + r^2 d\Omega^2. \tag{S12.9}$$

Taking the differential of Eq. (S12.6) gives

$$dr = \frac{\partial r}{\partial \rho}d\rho + \frac{\partial r}{\partial \tau}d\tau = \sqrt{\frac{\rho}{r}}d\rho - \sqrt{\frac{R_S}{r}}d\tau. \tag{S12.10}$$

Inserting this into Eq. (S12.8) and using Eq. (S12.6) gives

$$ds^2 = -c^2 d\tau^2 + \left[1 - \frac{3}{2}\sqrt{R_S}c\tau\rho^{-\frac{3}{2}}\right]^{-\frac{2}{3}}d\rho^2 + \left[1 - \frac{3}{2}\sqrt{R_S}c\tau\rho^{-\frac{3}{2}}\right]^{\frac{4}{3}}\rho^2 d\Omega^2. \tag{S12.11}$$

This metric is regular at the Schwarzschild radius $r = R_S$ which has an increasing co-moving radial coordinate,

$$\rho_S = R_S\left(1 + \frac{3c}{2R_S}\tau\right)^{2/3}, \tag{S12.12}$$

which increases with time. But the metric is singular at

$$\rho = \left(\frac{3\sqrt{R_S}}{2}c\tau\right)^{2/3}, \tag{S12.13}$$

which corresponds to the centre of the black hole, $r = 0$.

(c) We shall now consider a collapsing star which has a position-dependent energy density $\rho(\tau)$, assuming that the pressure is zero. Assume further that the interior spacetime can be described by a Friedmann metric with Euclidean spatial geometry ($k = 0$). We shall find the metric inside the star.

We assume that the star is homogeneous so that the density of the star is constant in space, but increases with time since the star collapses. The total mass of the star is assumed to be constant. The surface of the star has a radius given by equation (S12.6). Let the radius of the star at an arbitrary point of time be R and at $\tau = 0$ be R_0. Hence the density of the star is

$$\rho(\tau) = \frac{M}{(4/3)\pi R^3} = \frac{6M}{4\pi R_0^3\left(2 - 3\sqrt{R_S}c\tau R_0^{-3/2}\right)^2}. \tag{S12.14}$$

Inserting this into the Friedmann equation (12.46) for a flat universe with vanishing cosmological constant gives,

$$\frac{\dot{a}^2}{a^2} = \frac{A^2}{\left(1 - \frac{3}{2}Ac\tau\right)^2}, \quad A^2 = \frac{R_S}{R_0^3}. \tag{S12.15}$$

Since the star collapses we take the negative root, getting

$$\frac{da}{a} = -\frac{Acd\tau}{1 - \frac{3}{2}Ac\tau}. \tag{S12.16}$$

Integration with the initial condition $a(0) = 1$ gives

$$a(\tau) = \left(1 - \frac{3}{2}Ac\tau\right)^{2/3}. \tag{S12.17}$$

Hence the metric in the collapsing star is

$$ds^2 = -c^2 d\tau^2 + \left(1 - \frac{3}{2}\frac{\sqrt{R_S}}{R_0^{3/2}}c\tau\right)^{4/3}\left(d\rho^2 + \rho^2 d\Omega^2\right). \tag{S12.18}$$

12.2 *The volume of a closed Lemaître-Friedmann-Robertson-Walker (LFRW) universe*

The volume of the region contained inside a radius $r = aR_0\chi$ in a closed LFRW universe is

$$V = 4\pi a^3 R_0^3 \int_0^\chi \sin^2\chi \, d\chi = 2\pi a^3 R_0^3\left(\chi - \frac{1}{2}\sin 2\chi\right). \tag{S12.19}$$

12.3 *Conformal time*

The conformal time η of the Isotropic and homogeneous universe models is defined by

$$dt = a(\eta)d\eta. \tag{S12.20}$$

Inserting this into the line-element (12.1) gives

$$ds^2 = a(\eta)^2\left(-c^2 d\eta^2 + R_0^2 d\chi^2 + r(\chi)^2 d\Omega^2\right). \tag{S12.21}$$

Light moving radially have $ds^2 = d\Omega^2 = 0$, and hence a coordinate velocity

$$\frac{d\chi}{d\eta} = \frac{c}{R_0}.$$ (S12.22)

The physical velocity of the light is

$$\frac{R_0 d\chi}{d\eta} = c.$$ (S12.23)

12.4 Lookback time and the age of the universe

(a) Accordingto Eq. (12.111) the Hubble parameter at the emission point of time is given as a function of the redshift of the source by

$$H(z) = H_0 \left[\Omega_{rad0}(1+z)^4 + \Omega_{m0}(1+z)^3 + \Omega_k(1+z)^2 + \Omega_{L0} \right]^{1/2}.$$ (S12.24)

Using Eq. (12.113) the lookback time of a source with redshift z is

$$t_{LB}(z) = t_0 - t_E(z) = \int_0^\infty \frac{dz}{(1+z)H(z)} - \int_z^\infty \frac{dz}{(1+z)H(z)} = \int_0^z \frac{dz}{(1+z)H(z)},$$ (S12.25)

giving

$$t_{LB}(z) = t_H \int_0^z \frac{dz}{(1+z)\sqrt{\Omega_{rad0}(1+z)^4 + \Omega_{m0}(1+z)^3 + \Omega_k(1+z)^2 + \Omega_{L0}}}$$ (S12.26)

where $t_H = 1/H_0$ is the Hubble age of the universe, and Ω_{L0} is the present value of the density parameter of LIVE.

The age of the universe is

$$t_0 = t_{LB}(\infty) = t_H \int_0^\infty \frac{dz}{(1+z)\sqrt{\Omega_{rad0}(1+z)^4 + \Omega_{m0}(1+z)^3 + \Omega_k(1+z)^2 + \Omega_{L0}}}.$$ (S12.27)

(b) An empty universe (the Milne universe) has $\Omega_{rad0} = \Omega_{m0} = \Omega_{L0} = 0$, $\Omega_k = 1$ giving

$$t_{LB}(z) = t_H \int_0^z \frac{dz}{(1+z)^2} = t_H \frac{z}{1+z}. \tag{S12.28}$$

Hence the age of the Milne universe is equal to its Hubble age, $t_0 = t_{LB}(\infty) = t_H$.

(c) The Einstein–de Sitter universe is a flat, mass-dominated universe with $\Omega_{rad0} = \Omega_{L0} = \Omega_k = 0$, $\Omega_{m0} = 1$. Then the integral (S12.26) reduces to

$$t_{LB}(z) = t_H \int_0^z \frac{dz}{(1+z)^{5/2}} = \frac{2}{3} t_H \left[1 - \frac{1}{(1+z)^{3/2}} \right]. \tag{S12.29}$$

The age of this universe is

$$t_0 = t_{LB}(\infty) = \frac{2}{3} t_H. \tag{S12.30}$$

(d) A matter-dominated universe with curved 3-space has

$$t_{LB}(z) = t_H \int_0^z \frac{dz}{(1+z)^2 \sqrt{\Omega_{m0} z + 1}}. \tag{S12.31}$$

For positive spatial curvature, $\Omega_{m0} > 1$, this gives

$$t_{LB}(z) = \frac{t_H}{\Omega_{m0} - 1} \left[\frac{\Omega_{m0}}{\sqrt{\Omega_{m0} - 1}} \arcsin \frac{\sqrt{\Omega_{m0} - 1}(\sqrt{1 + \Omega_{m0} z} - 1)}{\Omega_{m0} \sqrt{1 + z}} + \frac{\sqrt{1 + \Omega_{m0} z}}{1 + z} - 1 \right], \tag{S12.32}$$

The age of the universe is

$$t_0 = t_{LB}(\infty) = \frac{t_H}{\Omega_{m0} - 1} \left(\frac{\Omega_{m0}}{\sqrt{\Omega_{m0} - 1}} \arccos \sqrt{\frac{1}{\Omega_{m0}} - 1} \right). \tag{S12.33}$$

For negative curvature, $\Omega_{m0} < 1$, the integral (S12.30) gives

$$t_{LB}(z) = \frac{t_H}{1 - \Omega_{m0}} \left[1 - \frac{\sqrt{1 + \Omega_{m0} z}}{1 + z} - \frac{\Omega_{m0}}{\sqrt{1 - \Omega_{m0}}} \arcsinh \frac{\sqrt{1 - \Omega_{m0}}(\sqrt{1 + \Omega_{m0} z} - 1)}{\Omega_{m0} \sqrt{1 + z}} \right]. \tag{S12.34}$$

The present age of this universe is

$$t_0 = t_{LB}(\infty) = \frac{t_H}{1 - \Omega_{m0}}\left(1 - \frac{\Omega_{m0}}{\sqrt{1 - \Omega_{m0}}}\text{arccosh}\sqrt{\frac{1}{\Omega_{m0}}}\right). \qquad (S12.35)$$

(e) We shall now determine the age of a matter-dominated universe from the parametric solution (S12.140)–(S12.144). We shall first consider the universe model with negatively curved 3-space. The present value of the conformal time is fund from the requirement $a(\eta_0) = 1$ which gives

$$\frac{c\eta_0}{R_0} = \text{arccosh}\left(\frac{2 - \Omega_{m0}}{\Omega_{m0}}\right). \qquad (S12.36)$$

Inserting this into Eq. (S12.141) gives

$$t_0 = t(\eta_0) = \frac{t_H}{1 - \Omega_{m0}}\left[1 - \frac{1}{2}\frac{\Omega_{m0}}{\sqrt{1 - \Omega_{m0}}}\text{arccosh}\left(\frac{2 - \Omega_{m0}}{\Omega_{m0}}\right)\right] \qquad (S12.37)$$

This seems to be in conflict with equation (S12.35). However applying the identity

$$\text{arccosh}x = \frac{1}{2}\text{arccosh}(2x^2 - 1) \qquad (S12.38)$$

with $x = 1/\sqrt{\Omega_{m0}}$, shows that the expressions are identical.

Using Eqs. (S12.143) and (S12.144) we find the age of a universe with a positively curved 3-space

$$t_0 = t(\eta_0) = \frac{t_H}{\Omega_{m0} - 1}\left[\frac{1}{2}\frac{\Omega_{m0}}{\sqrt{\Omega_{m0} - 1}}\text{arccos}\left(\frac{2 - \Omega_{m0}}{\Omega_{m0}}\right) - 1\right]. \qquad (S12.39)$$

(f) For a flat universe with cold matter and LIVE$\Omega_{rad0} = \Omega_k = 0$, and the integral (S12.26) reduces to

$$t_{LB}(z) = t_H \int_0^z \frac{dz}{(1 + z)\sqrt{\Omega_{m0}(1 + z)^3 + \Omega_{L0}}}, \qquad (S12.40)$$

which gives

$$t_{LB} = \frac{2t_H}{3\sqrt{\Omega_{L0}}}\text{arctanh}\frac{\sqrt{\Omega_{L0}}\left(\sqrt{\Omega_{L0} + \Omega_{m0}(1 + z)^3} - 1\right)}{\sqrt{\Omega_{L0} + \Omega_{m0}(1 + z)^3} - \Omega_{L0}}. \qquad (S12.41)$$

The age of this universe is equal to the look back time of an object with infinitely great redshift,

$$t_0 = t_{LB}(\infty) = \frac{2}{3} t_H \frac{\text{arctanh} \sqrt{\Omega_{L0}}}{\sqrt{\Omega_{L0}}} \tag{S12.42}$$

in agreement with Eqs. (12.164) and (12.168). Since $\lim\limits_{x \to 0} \frac{\text{arctan} x}{x} = 1$ we get

$$t_{LB} = \frac{2}{3} t_H \tag{S12.43}$$

for a flat matter-dominated universe with $\Omega_{L0} = 0$, in agreement with Eq. (S12.30).

(g) In the case of a flat LIVE-dominated universe with $\Omega_{m0} = 0, \Omega_{L0} = 1$ the integral (S12.140) gives

$$t_{LB} = t_H \ln(1 + z). \tag{S12.44}$$

12.5 The LFRW universe models with a perfect fluid

Inthis problem we will investigate FRW models with a perfect fluid. We will assume that the perfect fluid obeys the equation of state

$$p = w\rho c^2, \tag{S12.45}$$

where $-1 \le w \le 1$.

(a) The Friedmann equations (12.46) and (12.57) with vanishing cosmological constant are

$$3 \frac{\dot{a}^2 + kc^2 / R_0^2}{a^2} = \kappa \rho \tag{S12.46}$$

and

$$\ddot{a} = -\frac{\kappa}{6} a(\rho + 3\frac{p}{c^2}). \tag{S12.47}$$

Inserting equation (S12.45) the last equation takes the form

$$\ddot{a} = -\frac{\kappa}{6} a(1 + 3w)\rho. \tag{S12.48}$$

The integral (12.106) of the energy–momentum conservation equation with the normalization $a(t_0) = 1$ takes the form

$$\rho = \rho_0 a^{-3(1+w)}. \tag{S12.49}$$

(b) For $w > -1/3$ Eq. (S12.48) gives $\ddot{a} < 0$, i.e. decelerated expansion. Hence at a point of time the expansion will stop, $\dot{a} = 0$ and the universe will start to collapse. It follows from equation (S12.46) that this is only possible for $k > 0$, i.e. for a closed universe.

(c) We now assume that the spatial curvature is Euclidean, $k = 0$. Inserting (S12.49) into (S12.46) then gives

$$a^{1+3w}\dot{a}^2 = \frac{\kappa}{3}\rho_0. \tag{S12.50}$$

Integration for $w \neq -1$ with the conditions $a(0) = 0$, $a(t_0) = 1$ gives

$$a(t) = \left(\frac{t}{t_0}\right)^{\frac{2}{3(1+w)}}. \tag{S12.51}$$

The Hubble parameter is

$$H = \frac{\dot{a}}{a} = \frac{2}{3(1+w)}\frac{1}{t}, \tag{S12.52}$$

and the deceleration parameter is

$$q = -\frac{a\ddot{a}}{\dot{a}^2} = \frac{1}{2}(1+3w). \tag{S12.53}$$

It follows from Eqs. (S12.49) and (S12.51) that the time evolution of the mass density is

$$\rho(t) = \rho_0 \left(\frac{t_0}{t}\right)^2. \tag{S12.54}$$

(d) For the present universe model the time evolution of the radius of the particle horizon is given in Eq. (12.194),

$$l_{\text{PH}} = \frac{3(1+w)}{1+3w}t. \tag{S12.55}$$

This shall be expressed in terms of the redshift and the Hubble age of the universe. Using Eq. (S12.51) the relation between the redshift and the cosmic time is

$$1 + z = \frac{1}{a} = \left(\frac{t_0}{t}\right)^{\frac{2}{3(1+w)}}. \tag{S12.56}$$

Hence

$$t = t_0(1 + z)^{-\frac{3}{2}(1+w)}. \tag{S12.57}$$

Inserting this into Eq. (S12.55) gives

$$l_{PH} = \frac{3(1 + w)}{1 + 3w} t_0(1 + z)^{-\frac{3}{2}(1+w)}. \tag{S12.58}$$

The relationship between the age of the universe and its Hubble age is given in equation (12.193),

$$t_0 = \frac{2}{3(1 + w)} t_H. \tag{S12.59}$$

Inserting this into equation (S12.54) gives

$$l_{PH} = \frac{2}{1 + 3w} t_H(1 + z)^{-\frac{3}{2}(1+w)}. \tag{S12.60}$$

(e) For a dust-dominated universe, $w = 0$, the above results give $\rho a^3 = \text{constant}$, which means that the mass of dust in a co-moving volume is constant. The time evolution of the scale factor, the Hubble parameter and the deceleration parameter is $a(t) = (t/t_0)^{2/3}$, $H = \frac{2}{3}\frac{1}{t}$ and $q = 1/2$, the first two in agreement with Eqs. (12.142) and (12.145). For this universe model the radius of the particle horizon evolve as $l_{PH} = 3t = 2t_H(1 + z)^{-(3/2)}$.

For a radiation-dominated universe model, $w = 1/3$, we get $\rho = \rho_0 a^{-4}$. Hence the energy of radiation in a co-moving volume decreases with time. This is due to the work performed by the radiation upon the region outside a co-moving volume. The time evolution of the scale factor, the Hubble parameter and the deceleration parameter is $a(t) = (t/t_0)^{1/2}$, $H = \frac{1}{2}\frac{1}{t}$ and $q = 1$. The time evolution of the radius of the particle horizon is $l_{PH} = 2t = 2t_H(1 + z)^{-(3/2)} = t_H(1 + z)^{-2}$.

For a LIVE-dominated universe, $w = -1$, we get $\rho = \rho_0$; i.e. the density of the vacuum energy is constant in the expanding universe. Hence the vacuum energy in a co-moving volume increases. This is due to the negative work performed at the boundary of the co-moving volume due to the negative pressure of the vacuum energy. Hence vacuum energy is extracted from an infinitely remote region into the regions at a finite distance from the origin.

Integration of equation (S12.46) for $w = -1$ gives $a(t) = e^{H_0 t}$. The Hubble parameter is constant, $H = H_0$, and the deceleration parameter is $q = -1$. This universe model has no particle horizon.

(f) We shall find a general formula for the deceleration parameter of a LFRW universe model,

$$q = -\frac{\ddot{a}}{aH^2}. \tag{S12.61}$$

From $\dot{a} = aH$ we have $\ddot{a} = \dot{a}H + a\dot{H}$. Hence

$$q = -1 - \frac{\dot{H}}{H^2}. \tag{S12.62}$$

Differentiating Eq. (12.129) we get

$$-\frac{\dot{H}}{H_0^2} = \frac{3}{2}\frac{\Omega_{m0}}{a^3} + \frac{2\Omega_{rad0}}{a^4} + \frac{\Omega_{k0}}{a^2}. \tag{S12.63}$$

Using that $a(t_0) = 1$ and that $\Omega_{k0} = 1 - \Omega_{m0} - \Omega_{rad0} - \Omega_{L0}$, we find that the present value of the deceleration parameter is

$$q_0 = \frac{\Omega_{m0}}{2} + \Omega_{rad0} - \Omega_{L0}. \tag{S12.64}$$

12.6 Age—density relation for a radiation-dominated universe

The age of a radiation-dominated universe is given by equation (S12.27) with $\Omega_{m0} = \Omega_{L0} = 0$, and $\Omega_{rad0} + \Omega_k = 0$, i.e.

$$t_0 = t_H \int_0^\infty \frac{dz}{(1+z)^2\sqrt{\Omega_{rad0}(1+z)^2 + 1 - \Omega_{rad0}}}. \tag{S12.65}$$

This gives

$$t_0 = \frac{t_H}{1 + \sqrt{\Omega_{rad0}}}. \tag{S12.66}$$

12.7 Redshift—luminosity relation for matter-dominated universe: Mattig's fomula

It follows from $1 + z = \frac{1}{a}$ that $dz = -\frac{\dot{a}}{a^2}dt = -\frac{H}{a}dt$. The equation of motion for light moving radially towards the origin is $R_0 a d\chi = -dt$. Hence we obtain a relation between the difference in redshift of to objects and the difference of their radial coordinate,

$$R_0 d\chi = \frac{dz}{H(z)}. \tag{S12.67}$$

Inserting the expression (12.111) of the Hubble parameter gives the redshift-distance relation

$$R_0 H_0 d\chi = \left[\Omega_{\text{rad}0}\left(1 + z^4\right) + \Omega_{m0}(1 + z)^3 + \Omega_{k0}(1 + z)^2 + \Omega_{L0}\right]^{-1/2} dz.$$
(S12.68)

This formula is valid in curved as well as flat universe models. In a matter-dominated universe it gives

$$R_0\chi = \frac{1}{H_0} \int\limits_0^z \frac{dz}{(1 + z)\sqrt{1 + \Omega_{m0}z}}.$$
(S12.69)

The so-called *distance function* is

$$D(\chi) = R_0 S_k(\chi).$$
(S12.70)

Performing the integral (S12.69) and inserting the result into equation (S12.70) gives the distance function of a matter-dominated universe is

$$D(z) = \frac{2}{H_0} \frac{\Omega_{m0}z - (2 - \Omega_{m0})\left(\sqrt{1 + \Omega_{m0}z} - 1\right)}{\Omega_{m0}^2(1 + z)}.$$
(S12.71)

This is called Mattig's formula.

The luminosity distance to an object with redshift z is

$$d_L(z) = (1 + z)D(z) = \frac{2}{H_0} \frac{\Omega_{m0}z - (2 - \Omega_{m0})\left(\sqrt{1 + \Omega_{m0}z} - 1\right)}{\Omega_{m0}^2}.$$
(S12.72)

This is the redshift–luminosity relation for matter-dominated universe. For the Einstein–de Sitter universe, with $\Omega_{m0} = 1$, this relation reduces to

$$d_L = \frac{2c}{H_0}(1 + z - \sqrt{1 + z}).$$
(S12.73)

12.8 *Newtonian approximation with vacuum energy*

(a) The time-time component of Einstein's field equations is

$$G_{tt} + \Lambda g_{tt} = 8\pi G\rho.$$
(S12.74)

In orthonormal basis $g_{\hat{t}\hat{t}} = -1$. Hence the field equations without a cosmological constant corresponds to those with a cosmological constant if the cosmological constant is intepreted to represent the constant density ρ_L of LIVE according to

$$\Lambda = -8\pi G \rho_L. \tag{S12.75}$$

When equations (9.15) and (9.16) of the linear field approximation are generalized to include a cosmological constant we obtain

$$\nabla^2 \phi + \Lambda = 4\pi G \rho, \tag{S12.76}$$

which may be written

$$\nabla^2 \phi = 4\pi G (\rho + 2\rho_L). \tag{S12.77}$$

(b) In the spherically symmetric case this equation takes the form

$$\frac{1}{r^2} \frac{d}{dr} \left(r^2 \frac{d\phi}{dr} \right) = 4\pi G (\rho + 2\rho_L). \tag{S12.78}$$

Assume there is a particle with mass m at the origin. In the space outside the particle Eq. (S12.78) then reduces to

$$\frac{1}{r^2} \frac{d}{dr} \left(r^2 \frac{d\phi}{dr} \right) = 8\pi G \rho_L. \tag{S12.79}$$

The solution of this equation is

$$\phi = -\frac{Gm}{r} - \frac{4}{3} \pi G \rho_L r^2. \tag{S12.80}$$

The acceleration of gravity is

$$\mathbf{g} = -\frac{Gm}{r^2} \mathbf{e}_r + \frac{8}{3} \pi G \rho_L r \mathbf{e}_r. \tag{S12.81}$$

Hence, the acceleration of gravity vanishes at a radius

$$r_0 = \left(\frac{3m}{8\pi \rho_L} \right)^{1/3}. \tag{S12.82}$$

The mass of LIVE inside this surface is

$$m_L = \frac{4\pi}{3} \rho_L r_0^3. \tag{S12.83}$$

It follows that the mass of the LIVE inside the surface with vanishing acceleration of gravity is only half as large as the mass at the centre, $m_L(r_0) = m/2$.

As an illustration we calculate the transition radius for the Solar system in the present Universe. Then the density of the LIVE is $\rho_L = \Omega_L \rho_{cr}$, where the mass parameter of LIVE is $\Omega_L = 0.7$ and the critical density is $\rho_{cr} \approx 10^{-26}$ kg/m^3. The mass of the Sun is $m = 2 \times 10^{30}$ kg. Inserting these quantities into Eq. (13) gives a transition radius of approximately 300 l.y. Since the distances in the solar system are much smaller than this, the cosmological constant is of negligible significance for solar system gravitational effects.

12.9 Universe models with constant deceleration parameter

The deceleration parameter is

$$q = -\frac{a\ddot{a}}{\dot{a}^2}. \tag{S12.84}$$

Integration with constant q and the boundary conditions $a(0) = 0$, $a(t_0) = 1$ gives

$$a = \left(\frac{t}{t_0}\right)^{1/(1+q)}. \tag{S12.85}$$

In a flat universe filled by a perfect fluid with equation of state $p = w\rho$ energy–momentum conservation implied Eq. (12.106),

$$\rho = \rho_0 a^{-3(1+w)}. \tag{S12.86}$$

For a flat universe with vanishing cosmological constant equation (12.47) then takes the form'

$$3\frac{\dot{a}^2}{a^2} = \kappa\rho = \kappa\rho_0 a^{-3(1+w)}. \tag{S12.87}$$

Integration with $a(0) = 0$ gives

$$a(t) = \left(\frac{t}{t_0}\right)^{\frac{2}{3(1+w)}}. \tag{S12.88}$$

Comparison with Eq. (S12.85) shows that the solution with constant deceleration parameter is identical to the time evolution of a plat universe filled by a perfect fluid with equation of state

$$p = \frac{1}{3}(2q - 1)\rho c^2. \tag{S12.89}$$

12.10 Density parameters as functions of the redshift

The density parameter in a universe with radiation matter in the form of dust, and LIVE is

$$\Omega = \frac{\rho}{\rho_{cr}} = \frac{\kappa}{3} \frac{\rho}{H^2} = \Omega_{rad} + \Omega_m + \Omega_L. \tag{S12.90}$$

It follows from Eq. (12.106) that

$$\rho_{rad} = \rho_{rad0}(1 + z)^4, \; \rho_m = \rho_{m0}(1 + z)^3, \; \rho_L = \rho_{L0}. \tag{S12.91}$$

Combining this with Eqs. (12.111) and (S12.90) can be written

$$\Omega(z) = \frac{\Omega_{rad0}(1 + z)^4 + \Omega_{m0}(1 + z)^3 + \Omega_{L0}}{\Omega_{rad0}(1 + z)^4 + \Omega_{m0}(1 + z)^3 + \Omega_k(1 + z)^2 + \Omega_{L0}}. \tag{S12.92}$$

where

$$\Omega_{rad0} + \Omega_{m0} + \Omega_k + \Omega_{L0} = 1. \tag{S12.93}$$

The density parameters of radiation, matter and LIVE as functions of the redshift are

$$\Omega(z) = \frac{\Omega_{rad0}(1 + z)^4}{\Omega_{rad0}(1 + z)^4 + \Omega_{m0}(1 + z)^3 + \Omega_k(1 + z)^2 + \Omega_{L0}}, \tag{S12.94}$$

$$\Omega(z) = \frac{\Omega_{m0}(1 + z)^3}{\Omega_{rad0}(1 + z)^4 + \Omega_{m0}(1 + z)^3 + \Omega_k(1 + z)^2 + \Omega_{L0}}, \tag{S12.95}$$

$$\Omega(z) = \frac{\Omega_{L0}}{\Omega_{rad0}(1 + z)^4 + \Omega_{m0}(1 + z)^3 + \Omega_k(1 + z)^2 + \Omega_{L0}}. \tag{S12.96}$$

12.11 LFRW universe with radiation and matter

We shall deduce the solutions of Einstein's field equations for LFRW-universes with dust and radiation, following [1]. We introduce conformal time in the same way as in Eq. (12.126). Then Friedmann's 1. equation takes the form

$$\left(\frac{da}{d\eta}\right)^2 = \frac{\kappa}{3}(\rho_{m0}a + \rho_{r0}) - ka^2. \tag{S12.97}$$

This equation can be written as

$$\left(\frac{da}{d\eta}\right)^2 = 2\alpha a + \beta^2 - ka^2, \tag{S12.98}$$

where

$$\alpha = \frac{\kappa}{6}\rho_{m0} = \frac{1}{2}\Omega_{m0}H_0^2, \quad \beta = \sqrt{\frac{\kappa}{3}\rho_{rad0}} = H_0\sqrt{\Omega_{rad0}}. \tag{S12.99}$$

The solutions of Eq. (S12.98) with the initial condition $a(0) = 0$ are
For $k = 1$:

$$a = \alpha(1 - \cos\eta) + \beta\sin\eta, \quad t = \alpha(\eta - \sin\eta) + \beta(1 - \cos\eta). \tag{S12.100}$$

For $k = 0$:

$$a = \frac{\alpha}{2}\eta^2 + \beta\eta, \quad t = \frac{\alpha}{6}\eta^3 + \frac{\beta}{2}\eta^2. \tag{S12.101}$$

For $k = -1$:

$$a = \alpha(\cosh\eta - 1) + \beta\sinh\eta, \quad t = \alpha(\sinh\eta - \eta) + \beta(\cosh\eta - 1). \tag{S12.102}$$

12.12 Event horizons in de Sitter universe models

The *event horizon* represents the barrier between the future events that can be observed, and those that cannot. It is a spherical surface around an observer. The event horizon sets up a limit in the future observable universe, since in the future the observer will be able to obtain information only from events which happen inside their event horizon. According to its definition the coordinate radius, r_{EH}, of the event horizon is given by

$$\int_0^{r_{EH}} \frac{dr}{\sqrt{1 - k(r/R_0)^2}} = \int_t^\infty \frac{cdt}{a(t)}. \tag{S12.103}$$

The proper distance to the horizon is

$$l_{EH} = a(t)\int_0^{r_{EH}} \frac{dr}{\sqrt{1 - k(r/R_0)^2}} = a(t)\int_t^\infty \frac{cdt}{a(t)}. \tag{S12.104}$$

The scale factors of the de Sitter universe models are given in Eq. (12.160). We have 3 cases
$k = 1$: In this case the scale factor is

$$a(t) = \frac{R_0}{c}\sqrt{\frac{3}{\Lambda}}\cosh\left(\sqrt{\frac{\Lambda}{3}}t\right). \tag{S12.105}$$

Inserting this into Eq. (S12.103) gives

$$\int_0^{r_{EH}} \frac{dr}{\sqrt{1-(r/R_0)^2}} = \frac{c^2}{R_0}\sqrt{\frac{3}{\Lambda}} \int_t^\infty \frac{dt}{\cosh(\sqrt{\Lambda/3}t)}, \tag{S12.106}$$

which leads to

$$\arcsin\left(\frac{r_{EH}}{R_0}\right) = \frac{6c^2}{\Lambda R_0^2} \arctan\left(e^{-\sqrt{\Lambda/3}t}\right). \tag{S12.107}$$

Hence the coordinate radius of the event horizon is

$$r_{EH} = R_0 \sin\left[\frac{6c^2}{\Lambda R_0^2} \arctan\left(e^{-\sqrt{\Lambda/3}t}\right)\right]. \tag{S12.108}$$

$k = 0$: The flat de Sitter universe has scale factor

$$a(t) = e^{(\Lambda/3)t}. \tag{S12.109}$$

In this case Eq. (S12.104) takes the form

$$\int_0^{r_{EH}} dr = \int_t^\infty \frac{cdt}{e^{\sqrt{\Lambda/3}t}}, \tag{S12.110}$$

giving

$$r_{EH} = c\sqrt{\Lambda/3}e^{-\sqrt{\Lambda/3}t}. \tag{S12.111}$$

$k = -1$: For this model the scale factor is

$$a(t) = \frac{R_0}{c}\sqrt{\frac{3}{\Lambda}} \sinh\left(\sqrt{\frac{\Lambda}{3}}t\right). \tag{S12.112}$$

Equation (S12.104) then takes the form

$$\int_0^{r_{EH}} \frac{dr}{\sqrt{1+(r/R_0)^2}} = \frac{c^2}{R_0}\sqrt{\frac{3}{\Lambda}} \int_t^\infty \frac{dt}{\sinh(\sqrt{\Lambda/3}t)}, \tag{S12.113}$$

which leads to

$$\operatorname{arcsinh}\left(\frac{r_{EH}}{R_0}\right) = -\frac{3c^2}{\Lambda R_0^2} \ln\left[\tanh\left(\frac{1}{2}\sqrt{\frac{\Lambda}{3}}t\right)\right]. \tag{S12.114}$$

Hence

$$r_{EH} = R_0 \sinh\left\{-\frac{3c^2}{\Lambda R_0^2} \ln\left[\tanh\left(\frac{1}{2}\sqrt{\frac{\Lambda}{3}}t\right)\right]\right\}.$$ (S12.115)

12.13 *Flat universe model with radiation and LIVE*

(a) We shall find the scale factor as function of cosmic time following the procedure in [1]. It follows from Eq. (12.106) that in a universe containing non-interacting radiation and LIVE the density of LIVE is constant, and the density of the radiation decreases as the fourth power of the expansion factor. Also in a flat universe the sum of the densities of radiation and LIVE is equal to the critical density. Hence the total density can be written

$$\rho = \rho_{L0} + \rho_{rad0}a^{-4} = 1 - \rho_{rad0} + \rho_{rad0}a^4$$ (S12.116)

with the usual normalization of the scale factor, $a(t_0) = 1$. Then the first Friedmann equation takes the form

$$a^2\dot{a}^2 = \frac{\kappa}{3}\left((\rho_{cr0} - \rho_{rad0})a^4 + \rho_{rad0}\right).$$ (S12.117)

which can be written as

$$a^2\dot{a}^2 = (\omega/2)^2\left(a^4 + A^4\right),$$ (S12.118)

where

$$\omega = 2\sqrt{\frac{\kappa}{3}(\rho_{cr0} - \rho_{rad0})} = 2H_0\sqrt{1 - \Omega_{rad0}},$$

$$A = \left(\frac{\rho_{rad0}}{\rho_{cr0} - \rho_{rad0}}\right)^{1/4} = \left(\frac{\Omega_{rad0}}{1 - \Omega_{rad0}}\right)^{1/4}.$$ (S12.119)

Introducing the function $y = (a/A)^2$ Eq. (S12.118) takes the form

$$\dot{y}^2 = \omega^2\left(y^2 + 1\right).$$ (S12.120)

The solution with the initial condition $y(0) = 0$ is

$$y = \sinh(\omega t).$$ (S12.121)

Hence the scale factor is

$$a(t) = A\sqrt{\sinh(\omega t)}.$$ (S12.122)

(b) The Hubble parameter and the critical density are

$$H = \frac{\omega}{2} \coth(\dot{\omega}t) \tag{S12.123}$$

and

$$\rho_{cr} = (3/\kappa)H^2 = \rho_{L0} \coth^2(\omega t). \tag{S12.124}$$

Hence the density parameter of LIVE and radiation is

$$\Omega_L = \tanh^2(\omega t), \quad \Omega_{rad0} = \frac{1}{\cosh(\omega t)}. \tag{S12.125}$$

A graph essentially of the Hubble parameter as a function of time is shown in Fig. 12.22.

The graph shows that the Hubble parameter approaches the constant value $\omega/2$ for $\omega t > 2$. Also the scale factor then approaches an exponential function. Hence this model approaches the flat de Sitter universe for large times. This is a consequence of the fact that the density of LIVE is constant while the density of radiation decreases during the expansion, so that the model ends up by being LIVE dominated.

(c) The age of the universe is determined from the condition $a(t_0) = 1$. Equation (S12.122) then gives

$$t_0 = (1/\omega)\mathrm{arcsinh}(1/A^2) = \frac{t_H}{2\sqrt{1-\Omega_{rad0}}}\mathrm{arcsinh}\sqrt{\frac{1-\Omega_{rad0}}{\Omega_{rad0}}}. \tag{S12.126}$$

An alternative expression follows from Eq. (S12.125),

$$t_0 = \frac{t_H}{2}\frac{\mathrm{arctanh}\sqrt{1-\Omega_{rad0}}}{\sqrt{1-\Omega_{rad0}}}. \tag{S12.127}$$

The acceleration of the scale factor is

$$\ddot{a} = \frac{A\omega^2}{2} \sinh^{1/2}(\omega t)\left[1 - \frac{1}{2}\coth^2(\omega t)\right]. \tag{S12.128}$$

There is a transition from decelerated expansion due to the attractive gravity of the radiation to accelerated expansion due to the repulsive gravity of LIVE at a point of time t_1 given by $\ddot{a}(t_1) = 0$. This leads to

$$\coth(\omega t_1) = \sqrt{2}. \tag{S12.129}$$

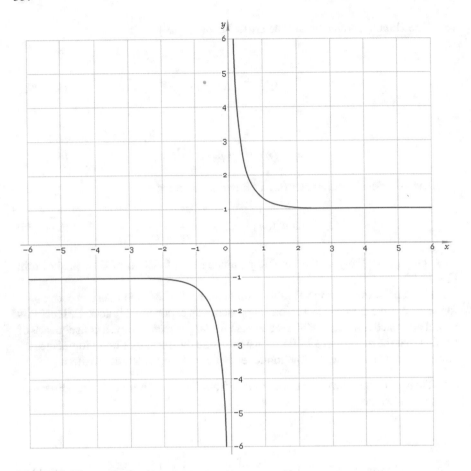

Fig. 12.22 The graph of $\coth x$

Combining this with Eq. (S12.127) gives

$$t_1 = t_0 \frac{\operatorname{artanh}\left(1/\sqrt{2}\right)}{\operatorname{artanh}\sqrt{1 - \Omega_{rad0}}}. \tag{S12.130}$$

The present density parameter of the cosmic background radiation is $\Omega_{rad0} \approx 10^{-4}$. Inserting this in the above equation gives $t_1 \approx 0.17 t_0 = 2.3$ billion years. Hence neglecting dust, a flat universe with LIVE and radiation equal to the CMB of our universe, will experience a transition from decelerated expansion to accelerated expansion 2.3 billion years after the Big bang.

It follows from Eqs. (S12.121) and (S12.129) that the observed redshift of an object emitting the observed radiation at the point of time t_1 is

$$z_1 = \frac{1}{a(t_1)} - 1 = \frac{1}{A\sqrt{\sinh(\omega t_1)}} - 1 = \frac{1}{\sqrt{2}A} - 1 = \frac{1}{\sqrt{2}}\sqrt{\frac{1 - \Omega_{rad0}}{\Omega_{rad0}}} - 1.$$

$$\text{(S12.131)}$$

Inserting $\Omega_{rad0} \approx 10^{-4}$ gives $z_1 = 6.1$.

12.14 The De Sitter Universe

(a) The formula for the cosmic redshift of light emitted at a point of time t_e and received at a point of time t_0 is: $z_1 = \frac{1}{a(t_e)} - 1$. The scale factor of the "flat" de Sitter universe is $a(t) = e^{Ht}$. This gives $z_1 - e^{H(t_0 - t_e)} - 1$. Light moves along a null geodetic curve, $ds^2 = 0$.

For light moving radially this gives $dr = -ce^{-Ht}dt$. Integration leads to

$$r = \frac{c}{H}\left(e^{-Ht_e} - e^{-Ht_0}\right) = \frac{c}{H}e^{-Ht_0}\left(e^{H(t_e - t_0)} - 1\right), \qquad \text{(S12.132)}$$

Giving

$$z_1 = (H/c)e^{Ht_0}r_e. \qquad \text{(S12.133)}$$

(b) The 4-acceleration of an arbitrary particle is

$$\vec{a} = \frac{d\vec{u}}{d\tau} = \left(\frac{du^\mu}{d\tau} + \Gamma^\mu_{\alpha\beta}u^\alpha u^\beta\right)\vec{e}_\mu. \qquad \text{(S12.134)}$$

With the present line-element the 4 velocity of a *reference* particle is $u^\mu = \frac{dx^\mu}{d\tau} = (c, 0, 0, 0)$ all the time. Hence $\frac{du^\mu}{d\tau} = 0$. Thus the expression for the 4-acceleration reduces to $\vec{a} = \Gamma^\mu_{tt}c^2\vec{e}_\mu$. These Christoffel symbols are

$$\Gamma^\mu_{tt} = \frac{1}{2}g^{\mu\nu}\left(\frac{\partial g_{vt}}{\partial t} + \frac{\partial g_{tv}}{\partial t} - \frac{\partial g_{tt}}{\partial x^\nu}\right). \qquad \text{(S12.135)}$$

Since the metric is diagonal the first two terms vanish, and since $g_{tt} = -c^2$ the last term vanish. Hence the four-acceleration of a reference particle in this reference frame is $\vec{a} = 0$. This means that the reference particles are freely falling. Hence an observer with constant radial coordinate r does not experience any acceleration of gravity.

(c) For a static metric the frequency change of light measured locally at the position of the emitter, e, and the receiver, r, comes from the position dependence of the rate of time in a gravitational field. Let $\Delta\tau_e$ be the period of light emitted at R as measured locally by a standard clock at rest, and Δt_e the as measured by a coordinate clock. The corresponding quantities as measured at the receives are $\Delta\tau_r$ and Δt_r. Then

$$\Delta\tau_e = \sqrt{-(g_{tt})_e}\Delta t_e, \quad \Delta\tau_r = \sqrt{-(g_{tt})_r}\Delta t_r. \qquad (S12.136)$$

A static metric means that the coordinate clocks tick with a position independent rate, $\Delta t_e = \Delta t_r$. Hence the proper periods of the light at the emitter and the receiver are related by

$$\Delta\tau_r = \sqrt{\frac{(g_{tt})_r}{(g_{tt})_e}}\Delta\tau_e. \qquad (S12.137)$$

Locally the wavelength and the period are related by $\lambda = c\Delta\tau$. Hence the gravitational redshift (for light moving upwards, blueshift downwards) is

$$z = \frac{\lambda_r}{\lambda_e} - 1 = \sqrt{\frac{(g_{tt})_r}{(g_{tt})_e}} - 1. \qquad (S12.138)$$

With the given static form of the de Sitter metric we get

$$z_2 = \frac{1}{\sqrt{1 - H^2 R_e^2}} - 1. \qquad (S12.139)$$

The reason that the redshifts in (a) and (c) are different, is that the experiments are not identical, since the emitter in (a) moves relative to the receiver, but the emitter in (c) is at rest relative to the receiver.

(d) From point (b) we have that the 4-acceleration of a reference particle permanently at rest in the coordinate system is

$$\vec{a} = \Gamma^\mu_{TT}c^2\vec{e}_\mu = -\frac{1}{2}g^{\mu\nu}\frac{\partial g_{TT}}{\partial x^\nu}\vec{e}_\mu = -\frac{1}{2}g^{RR}\frac{\partial g_{TT}}{\partial R}\vec{e}_R = -\frac{1}{2g_{RR}}\frac{\partial g_{TT}}{\partial R}\vec{e}_R. \qquad (S12.140)$$

Inserting $g_{TT} = -(c^2 - H^2 R^2)$, $g_{RR} = (1 - H^2 R^2/c^2)^{-1}$ gives

$$\vec{a} = -(1 - H^2 R^2/c^2)H^2 R\vec{e}_R. \qquad (S12.141)$$

The particle accelerates in the negative R-direction. Hence the observers at rest in this reference frame experiences an acceleration of gravity in the positive R-direction.

(e) It follows from the given coordinate transformation that a particle with $r = r_0$ will have a motion with

$$R = r_0 e^{HT} \sqrt{1 - H^2 R^2/c^2} \quad \text{or} \quad R = \frac{r_0 e^{HT}}{\sqrt{1 + \frac{H^2 r_0^2}{c^2} e^{2HT}}}. \tag{S12.142}$$

Relatively to the (T, R)-system the reference particles of the (t, r)-system have an accelerated outwards motion.

(f) In the (t, r)-system the redshift is explained as an expansion effect, but in the (T, R)-system the redshift is explained as a gravitational effect. The light which moves inwards towards the observer at the origin, moves upwards in the outwards directed gravitational field that is experienced in this reference frame, because the reference particles are not falling freely.

12.15 The Milne Universe

(a) This line element describes an expanding, isotropic and homogeneous universe model, a FLRW-universe. It has negative spatial curvature, and thus infinitely large spatial extension, and the scale factor is $a(t) = t/t_0$.

(b) The Hubble parameter is $H(t) = 1/t$ with a present value $H(t_0) = 1/t_0$. Hence the age of this universe is equal to the inverse of the present value of its Hubble parameter.

(c) The deceleration parameter is $q = a\ddot{a}/\dot{a}^2 - 0$. Hence this universe model expands with a constant velocity.

(d) Given the coordinate transformation $T = t\sqrt{1 + \left(\frac{r}{ct_0}\right)^2}$, $R = \frac{rt}{t_0}$. It follows that

$$c^2 T^2 - R^2 = c^2 t^2 + \left(\frac{rt}{t_0}\right)^2 - \left(\frac{rt}{t_0}\right)^2 = c^2 t^2. \tag{S12.143}$$

Hence

$$t = \sqrt{T^2 - R^2/c^2}. \tag{S12.144}$$

Furthermore it follows from the given coordinate transformation that

$$\frac{R}{cT} = \frac{r}{\sqrt{c^2 t_0^2 + r^2}}. \tag{S12.145}$$

Solving this equation with respect to r gives

$$r = \frac{ct_0 R}{\sqrt{c^2 T^2 - R^2}}.$$ (S12.146)

Equations (d2) and (d4) are the inverse transformation.

(e) The world lines of the reference frame E in which the coordinates t, r are co-moving are given by $r = r_0$=constant. It follows from Eq. (ds) that these world lines are given by the curves described by $cT = K_1 R$, $K_1 = \sqrt{1 + (ct_0/r_0)}$ in a Minkowski diagram referring to the T, R-system. These curves are straight lines.

According to Eq. (d1) the simultaneity curves of the t, r-system, $t = a$=constant, are hyperbolae.

The world lines and simultaneity curves of the t, r-system relative to the T, R-system are shown in the Minkowski diagram below.

12.16 Natural Inflation

(a) We here consider a natural inflation model with potential

$$V(\phi) = V_0 \left(1 + \cos \tilde{\phi}\right),$$ (S12.147)

where $\tilde{\phi} = \phi/M$, and M is the spontaneous symmetry breaking scale. Writing the Einstein gravitational constant as $\kappa = 1/M_P^2$, where M_P is the Planck mass, and inserting the potential (S12.147) into Eq. (12.304) we find the slow-roll parameters

$$\varepsilon = \frac{b}{2}\frac{1 - \cos \tilde{\phi}}{1 + \cos \tilde{\phi}}, \quad \eta = -b\frac{\cos \tilde{\phi}}{1 + \cos \tilde{\phi}}, \quad b = \left(\frac{M_P}{M}\right)^2.$$ (S12.148)

Here ϕ is the initial value of the field giving rise to N e-folds. The parameter b represents the symmetry breaking scale and $b < 1$ for $M > M_P$.

The spectral parameters are

$$\delta_{ns} = b\frac{3 - \cos \tilde{\phi}}{1 + \cos \tilde{\phi}}, \quad n_T = -b\frac{1 - \cos \tilde{\phi}}{1 + \cos \tilde{\phi}}, \quad r = 8b\frac{1 - \cos \tilde{\phi}}{1 + \cos \tilde{\phi}}.$$ (S12.149)

(b) It follows from these equations that

$$r = 4(\delta_{ns} - b), n_T = -\frac{r}{8},$$ (S12.150)

The symmetry breaking parameter b can be determined from observations from the first of the relationships (S12.150),

$$b = \delta_{ns} - \frac{r}{4}.$$ (S12.151)

The observed value $\delta_{ns} = 0.032$ and $r \geq 0$ gives $b < 0.02$ and hence $M \geq 5.6 M_P$. This shows that the symmetry breaking energy is higher than the Planck energy which is a weakness of the original natural inflation model.

(c) Inserting the potential (S12.147) into equation (12.322) and performing the integration gives the number of e-folds

$$N = \frac{1}{b} \ln \frac{1 - \cos\left(\tilde{\phi}_f\right)}{1 - \cos\left(\tilde{\phi}\right)} \tag{S12.152}$$

It is usual to specify that the inflationary era ends when $\varepsilon = 1$. Inserting $\phi = \phi_f$ in the first of the Eq. (S12.148) with $\varepsilon = 1$ we get

$$\cos\left(\tilde{\phi}_f\right) = -\frac{2 - b}{2 + b}. \tag{S12.153}$$

Inserting this into Eq. (S12.152) gives

$$\cos\left(\tilde{\phi}\right) = 1 - \frac{4}{2 + b} e^{-bN}. \tag{S12.154}$$

It follows from Eqs. (6.5.22) and (6.5.29) that

$$\delta_{ns} = b \frac{(2 + b)e^{bN} + 2}{(2 + b)e^{bN} - 2}, n_T = -\frac{2b}{(2 + b)e^{bN} - 2}, r = \frac{16b}{(2 + b)e^{bN} - 2}. \tag{S12.155}$$

Inserting $\delta_{ns} = 0.032$ and $N = 50$ into equation the first of the expressions (S12.155) we get $b = 0.02$ and $bN = 1$, and the last of the equations gives $r \approx 0.09$ which is a little larger than the observational requirement $r < 0.04$.

Index

© Springer Nature Switzerland AG 2020
Ø. Grøn, *Introduction to Einstein's Theory of Relativity*,
Undergraduate Texts in Physics, https://doi.org/10.1007/978-3-030-43862-3

Printed in the United States
By Bookmasters